T0297631

Multimodal Behavior Analysis in the Wild

Computer Vision and Pattern Recognition Series

Series Editors

Horst Bischof Institute for Computer Graphics and Vision, Graz University of Technology, Austria
Kyoung Mu Lee Department of Electrical and Computer Engineering, Seoul National University, Republic of Korea
Sudeep Sarkar Department of Computer Science and Engineering, University of South Florida, Tampa, United States

Also in the Series:

Lin and Zhang, Low-Rank Models in Visual Analysis: Theories, Algorithms and Applications, 2017, ISBN: 9780128127315
Zheng et al., Statistical Shape and Deformation Analysis: Methods, Implementation and Applications, 2017, ISBN: 9780128104934
De Marsico et al., Human Recognition in Unconstrained Environments: Using Computer Vision, Pattern Recognition and Machine Learning Methods for Biometrics, 2017, ISBN: 9780081007051
Saha et al., Skeletonization: Theory, Methods and Applications, 2017, ISBN: 9780081012918

Multimodal Behavior Analysis in the Wild

Advances and Challenges

Edited by

Xavier Alameda-Pineda

Elisa Ricci

Nicu Sebe

ACADEMIC PRESS

An imprint of Elsevier

Academic Press is an imprint of Elsevier
125 London Wall, London EC2Y 5AS, United Kingdom
525 B Street, Suite 1650, San Diego, CA 92101, United States
50 Hampshire Street, 5th Floor, Cambridge, MA 02139, United States
The Boulevard, Langford Lane, Kidlington, Oxford OX5 1GB, United Kingdom

Copyright © 2019 Elsevier Ltd. All rights reserved.

No part of this publication may be reproduced or transmitted in any form or by any means, electronic or mechanical, including photocopying, recording, or any information storage and retrieval system, without permission in writing from the publisher. Details on how to seek permission, further information about the Publisher's permissions policies and our arrangements with organizations such as the Copyright Clearance Center and the Copyright Licensing Agency, can be found at our website: www.elsevier.com/permissions.

This book and the individual contributions contained in it are protected under copyright by the Publisher (other than as may be noted herein).

Notices

Knowledge and best practice in this field are constantly changing. As new research and experience broaden our understanding, changes in research methods, professional practices, or medical treatment may become necessary.

Practitioners and researchers must always rely on their own experience and knowledge in evaluating and using any information, methods, compounds, or experiments described herein. In using such information or methods they should be mindful of their own safety and the safety of others, including parties for whom they have a professional responsibility.

To the fullest extent of the law, neither the Publisher nor the authors, contributors, or editors, assume any liability for any injury and/or damage to persons or property as a matter of products liability, negligence or otherwise, or from any use or operation of any methods, products, instructions, or ideas contained in the material herein.

Library of Congress Cataloging-in-Publication Data
A catalog record for this book is available from the Library of Congress

British Library Cataloguing-in-Publication Data
A catalogue record for this book is available from the British Library

ISBN: 978-0-12-814601-9

For information on all Academic Press publications
visit our website at https://www.elsevier.com/books-and-journals

ELSEVIER
Book Aid
International
Working together
to grow libraries in
developing countries
www.elsevier.com • www.bookaid.org

Publisher: Mara Conner
Acquisition Editor: Tim Pitts
Editorial Project Manager: Ali Afzal-Khan
Production Project Manager: Sruthi Satheesh
Designer: Matthew Limbert

Typeset by VTeX

Contents

List of Contributors

Xavier Alameda-Pineda
Inria Grenoble Rhone-Alpes, Perception Team, France

Stefano Alletto
University of Modena and Reggio Emilia, Department of Engineering "Enzo Ferrari", Modena, Italy

Vasileios Argyriou
Kingston University, Faculty of Science, Engineering and Computing, Surrey, UK

Claudio Baecchi
University of Florence, Firenze, Italy

Lorenzo Baraldi
University of Modena and Reggio Emilia, Department of Engineering "Enzo Ferrari", Modena, Italy

Marco Bertini
University of Florence, Firenze, Italy

Marc Bolaños
University of Barcelona, Department of Mathematics and Computer Science, Barcelona, Spain
Computer Vision Center, Barcelona, Spain

Pierre Bour
Kingston University, Faculty of Science, Engineering and Computing, Surrey, UK

Luca Brayda
Istituto Italiano di Tecnologia, RBCS, Genova, Italy

Alessio Brutti
Center for Information Technology, Fondazione Bruno Kessler, Trento, Italy

Laura Cabrera Quiros
Delft University of Technology, Intelligent Systems, the Netherlands
Instituto Tecnológico de Costa Rica, Electronic Engineering Department, Costa Rica

Victor Campos
Barcelona Supercomputing Center, Barcelona, Catalonia, Spain

Alejandro Cartas
University of Barcelona, Department of Mathematics and Computer Science, Barcelona, Spain

Andrea Cavallaro
Centre for Intelligent Sensing, Queen Mary University London, London, UK

Shih-Fu Chang
Columbia University, New York City, NY, USA

Wen-Sheng Chu
Carnegie Mellon University, Pittsburgh, PA, USA

Jeffrey F. Cohn
Department of Psychology, University of Pittsburgh, Pittsburgh, PA, USA

Marcella Cornia
University of Modena and Reggio Emilia, Department of Engineering "Enzo Ferrari", Modena, Italy

Emile Cribelier
Kingston University, Faculty of Science, Engineering and Computing, Surrey, UK

Rita Cucchiara
University of Modena and Reggio Emilia, Department of Engineering "Enzo Ferrari", Modena, Italy

Alberto Del Bimbo
University of Florence, Firenze, Italy

Antoine Deleforge
Inria Nancy - Grand Est, Villers-lès-Nancy, France

Eyal Dim
The University of Haifa, Israel

Mariella Dimiccoli
University of Barcelona, Department of Mathematics and Computer Science, Barcelona, Spain

Itir Onal Ertugrul
Robotics Institute, Carnegie Mellon University, Pittsburgh, PA, USA

Anna Esposito
Università degli Studi della Campania L. Vanvitelli, Dipartimento di Psicologia, Caserta, Italy

Andrea Ferracani
University of Florence, Firenze, Italy

Sharon Gannot
Bar-Ilan University, Faculty of Engineering, Ramat-Gan, Israel

Maite Garolera
Consorci Sanitari de Terrassa, Barcelona, Spain

Ekin Gedik
Delft University of Technology, Intelligent Systems, the Netherlands

Olga Gelonch
Consorci Sanitari de Terrassa, Barcelona, Spain

Jeffrey M. Girard
Language Technology Institute, Carnegie Mellon University, Pittsburgh, PA, USA

Laurent Girin
Univ. Grenoble Alpes, CNRS, Grenoble-INP, GIPSA-Lab, Saint Martin d'Hères, France
INRIA Grenoble Rhône-Alpes, Perception Group, Montbonnot-Saint-Martin, France

Xavier Giro-i-Nieto
Barcelona Supercomputing Center, Barcelona, Catalonia, Spain
Universitat Politecnica de Catalunya, Barcelona, Catalonia, Spain

Luca Giuliani
Università degli Studi di Genova, DIBRIS, Genova, Italy
Istituto Italiano di Tecnologia, RBCS, Genova, Italy

Furkan Gürpınar
Boğaziçi University, Department of Computer Engineering, Bebek, Istanbul, Turkey

Zakia Hammal
Robotics Institute, Carnegie Mellon University, Pittsburgh, PA, USA

Hayley Hung
Delft University of Technology, Intelligent Systems, the Netherlands
CWI, Distributed and Interactive Systems, the Netherlands

László A. Jeni
Robotics Institute, Carnegie Mellon University, Pittsburgh, PA, USA

Kristiina Jokinen
AIST/AIRC, Tokyo, Japan

Brendan Jou
Columbia University, New York City, NY, USA

Heysem Kaya
Namık Kemal University, Department of Computer Engineering, Çorlu, Tekirdağ, Turkey

Walter Kellermann
Multimedia Communications and Signal Processing, Telecommunications Laboratory, University Erlangen-Nuremberg, Erlangen, Germany

Tsvi Kuflik
The University of Haifa, Israel

Xiaofei Li
INRIA Grenoble Rhône-Alpes, Perception Group, Montbonnot-Saint-Martin, France

Nicoletta Noceti
Università degli Studi di Genova, DIBRIS, Genova, Italy

Francesca Odone
Università degli Studi di Genova, DIBRIS, Genova, Italy

Gabriel Oliveira-Barra
University of Barcelona, Department of Mathematics and Computer Science, Barcelona, Spain

Marcello Pelillo
European Centre for Living Technology, Venice, Italy
DAIS, Venice, Italy

Francesco Perrone
University of Glasgow, School of Computing Science, Glasgow, UK

Emily Mower Provost
University of Michigan, Computer Science and Engineering, Ann Arbor, MI, USA

Petia Radeva
University of Barcelona, Department of Mathematics and Computer Science, Barcelona, Spain

Elisa Ricci
University of Trento, Department of Information Engineering and Computer Science, Italy
Fondazione Bruno Kessler, Technology of Vision, Italy

Giorgio Roffo
University of Glasgow, School of Computing Science, Glasgow, UK

Albert Ali Salah
Boğaziçi University, Department of Computer Engineering, Bebek, Istanbul, Turkey
Nagoya University, Future Value Creation Research Center (FV-CRC), Nagoya, Japan

Alexander Schmidt
Multimedia Communications and Signal Processing, Telecommunications Laboratory, University Erlangen-Nuremberg, Erlangen, Germany

Björn Schuller
Imperial College London, Department of Computing, London, UK
University of Augsburg, Chair of Embedded Intelligence for Health Care and Wellbeing, Augsburg, Germany

Filomena Scibelli
Università degli Studi della Campania L. Vanvitelli, Dipartimento di Psicologia, Caserta, Italy

Nicu Sebe
University of Trento, Department of Information Engineering and Computer Science, Italy

Lorenzo Seidenari
University of Florence, Firenze, Italy

Giuseppe Serra
University of Udine, Department of Mathematics, Computer Science and Physics, Udine, Italy

Joan Sosa-Garciá
Università degli Studi di Genova, DIBRIS, Genova, Italy

Estefania Talavera
University of Barcelona, Department of Mathematics and Computer Science, Barcelona, Spain
Computer Vision Center, Barcelona, Spain
University of Groningen, Intelligent Systems Group, AG Groningen, the Netherlands

Mohammad Tayarani
University of Glasgow, School of Computing Science, Glasgow, UK

Qi Tian
The University of Texas at San Antonio, Department of Computer Science, San Antonio, TX, USA

Jordi Torres
Barcelona Supercomputing Center, Barcelona, Catalonia, Spain

Andrea Trucco
Università degli Studi di Genova, DITEN, Genova, Italy

Panagiotis Tzirakis
Imperial College London, Department of Computing, London, UK

Tiberio Uricchio
University of Florence, Firenze, Italy

Sebastiano Vascon
European Centre for Living Technology, Venice, Italy
DAIS, Venice, Italy

Alessandro Vinciarelli
University of Glasgow, School of Computing Science, Glasgow, UK

Dong-Bach Vo
University of Glasgow, School of Computing Science, Glasgow, UK

Graham Wilcock
CDM Interact, Helsinki, Finland

Lingxi Xie
The Johns Hopkins University, Department of Computer Science, Baltimore, MD, USA

Stefanos Zafeiriou
Imperial College London, Department of Computing, London, UK
University of Oulu, Center for Machine Vision and Signal Analysis, Oulu, Finland

Biqiao Zhang
University of Michigan, Computer Science and Engineering, Ann Arbor, MI, USA

About the Editors

Xavier Alameda-Pineda received his PhD from INRIA and University of Grenoble in 2013. He was a post-doctoral researcher at CNRS/GIPSA-Lab and at the University of Trento in the deep and structural learning group. He is a research scientist at INRIA working on signal processing and machine learning for scene and behavior understanding using multimodal data. He is the winner of the best paper award of ACM MM 2015, the best student paper award at IEEE WASPAA 2015, the best scientific paper award on image, speech, signal and video processing at IEEE ICPR 2016, and the ACM SIGMM rising star award in 2018. He is member of IEEE and of ACM SIGMM.

Elisa Ricci is an associate professor at University of Trento and a researcher at Fondazione Bruno Kessler. She received her PhD from the University of Perugia in 2008. She has since been a postdoctoral researcher at Idiap and FBK, Trento and a visiting researcher at University of Bristol. Her research interests are directed along developing machine learning algorithms for video scene analysis, human behavior understanding and multimedia content analysis. She is a program chair of ACM Multimedia 2020 and was an area chair of ACM MM 2016 and 2018, ECCV 2016, and ICCV 2017. She received the IBM Best Student Paper Award at ICPR 2014.

Nicu Sebe is a full professor at the University of Trento, Italy, where he is leading research in the areas of multimedia information retrieval and human behavior understanding. He was a general co-chair of FG 2008, ACM MM 2013, and ACM ICMR 2017, a program chair of CIVR 2007 and 2010, ACM MM 2007 and 2011, and of ECCV 2016, ICCV 2017 and ICPR 2020. He is a senior member of IEEE and ACM and a fellow of IAPR.

Multimodal behavior analysis in the wild: An introduction

0.1 ANALYZING HUMAN BEHAVIOR IN THE WILD FROM MULTIMODAL DATA

Due to its importance in many applications, the automatic analysis of human behavior has been a popular research topic in the last decades. Understanding human behavior is relevant in many fields, such as assistive robotics, human–computer interaction, surveillance and security, to cite only a few.

The automatic extraction of behavioral cues is an extremely challenging task involving several disciplines ranging from machine learning, signal processing, computer vision, social psychology, etc. Thanks to the recent progress in the area of Artificial Intelligence and deep learning, significant advances have been made in the last few years in the development of systems for human behavior analysis. For instance, technologies for speech recognition and machine translation have significantly improved and they are now able to work in a wide range of real-world settings. Similarly, several advances have been made in the robotics field, witnessed by the advent on the market of robots which accurately recognize and mimic human emotions. More surprisingly, in the last years technologies have appeared which are able to interpret people behaviors even more precisely than human observers. For instance, computer vision researchers have developed systems which can estimate physiological signals (e.g. heart and respiration rate) by analyzing subtle skin color variations from face videos [18] or which can track the position of a moving person behind a wall from the shadow arising on the ground at the base of the wall's edge [6]. Despite this progress, still many current technologies for human behavior analysis have limited applicability and are not robust enough to operate in arbitrary conditions and in real-world settings. In other words, the path towards automatically understanding human behaviors 'in the wild' is still to be discovered.

It is a well-known fact that the automatic analysis of human behavior can benefit from harnessing multiple modalities. While earlier work on behavior understanding focused on an unimodal setting, typically considering only visual or audio data, more recent approaches leverage multimodal information. Investigating methods to process multimodal data is of utmost importance as multiple modalities provide a more complete representation of human behavior. Moreover, data gathered with different sensors can be incomplete or corrupted by noise. In synthesis, unimodal approaches fail to provide a robust and accurate representation of behavioral patterns and smart multimodal fusion strategies are required. However, from the machine learning point of view, multimodal human behavior analysis is a challenging task since learning the complex relationship across modalities is non-trivial.

The analysis of human behavior from multimodal data has been encouraged in the last few years by the emergence on the market of novel devices, such as wearable watches or smartphones. These devices typically include several sensors (e.g. camera, microphone, accelerometers, bluetooth, etc.), i.e., they are inherently multimodal. Additionally, the diffusion at consumer level of systems such as drones, low-cost infrared or wearable cameras has opened the possibility of studying human behavior considering other types of data, such as images in an ego-vision setting, 2.5D data, or birds-eye view videos. These technologies provide complementary information to traditional sensing modalities, such as distributed camera and microphone networks. For instance, when analyzing social scenes, wearable sensing devices can be exploited in association with data from traditional cameras to localize people and estimate their interactions [3]. Similarly, egocentric videos can be used together with images from surveillance cameras for the purpose of automatic activity recognition or analyzed jointly with audio signals for robust people re-identification [7].

Besides the widespread diffusion of novel devices, in the last decade the study of human behavior from multimodal data has been also encouraged by the emergence of new methodologies. In particular, research work from social signal processing [20,8] has allowed for significantly advances of the field. Studies have clearly stressed the importance of non-verbal cues such as gestures, gaze, and emotional patterns in human communication and the need of designing methodologies for inferring these cues by processing multimodal data. Similar studies demonstrated the interest of fusing auditory and visual data for lower-level tasks [4,5]. Furthermore, social signal processing has enabled one to improve technologies for the automatic analysis of human behavior thanks to the integrations of concepts from social psychology into machine learning algorithms (e.g. the concepts of proxemics and kinesics are now used in many approaches for automatically detecting groups in social scenes [11,19]).

Significant advances in the field of understanding human behavior have also been achieved thanks to the (re-)discovery of deep neural networks. Deep learning has significantly improved the accuracy of many systems for extracting behavioral cues under real-world conditions. For instance, in computer vision deep models have been applied with success to the tasks of activity recognition [22], gaze estimation [16], group analysis, etc. Some of the technologies described in this book adopting deep learning architectures have been deployed in real-world settings (e.g. the audio-visual systems described in Chapter 8 have been used by museum visitors). In addition, several research studies have proposed deep learning-based strategies for fusing multimodal data, outperforming previous approaches based on traditional machine learning models.

The fast and broad progress in Artificial Intelligence has not only enabled great advances in the analysis of human behavior but has also opened new possibilities for generating realistic human-like behavioral data [23,14,17]. Notable examples relate to the synthesis of realistic-looking images of people, to the generation of human-sounding speech as well as to the design of robots that emulate human emotions. Furthermore, the successes achieved with deep learning have also encouraged the research community to address new challenges. For instance, several recent studies tackled the problem of activity forecasting and behavior prediction [1,21]. Other work focused on the rethinking the action–perception loop and devising end-to-end trainable architecture to directly predict actions from raw data [24,15]. In the area of human behavior analysis these studies can be extremely relevant and would ultimately lead for instance to creating unified models for jointly analyzing human social behaviors and controlling intelligent vehicles and social robots.

While the progress in the study of human behaviors has been considerable, recent work has also pointed out many limitations of current methodologies and systems. For instance, the adoption of deep learning in several applications has highlighted the need for large-scale datasets. Indeed, datasets can be a very limiting issue depending on the application at hand for different reasons, such as labeling cost, privacy, synchronization problems, etc. The research community is pushing towards handling this problem and several datasets have been made available in the last few years for studying human behaviors in the wild. Notable attempts are for instance the efforts made by researchers involved in the Chalearn initiative [13] or in other dataset collection campaigns [10,12,9,2]. Besides the issues with data, several open challenges involve the design of algorithms for inferring behavioral cues. In particular, understanding human behavior requires approaches which operate at different levels of granularity and which are able to infer both low-level cues (e.g. detecting people position or pose) and high-level information (e.g. group dynamics and social interactions, emotional patterns). However, devising methods which deal with tasks at different levels of granularity in isolation is largely suboptimal. Future research efforts should be made devoted to addressing the problem of human behavior analysis in a more holistic manner.

0.2 SCOPE OF THE BOOK

The main objective of this book is to present an overview of recent advances in the area of behavioral signal processing. A special focus is given to describing the strengths and weaknesses of current methods and technologies which (i) analyze human behaviors by exploiting different data modalities and (ii) are deployed in real-world scenarios. In other words, the two prominent characteristics of the book are the *multimodality* and the *in the wild* perspective. Regarding multimodality, the book presents state-of-the-art human behavior understanding methods which exploit information coming from different sensors. Audio and video being the most popular modalities used for analyzing human behavior and activities, they have a privileged role in the manuscript. However, the book also covers methodologies and applications where emerging modalities such as accelerometer or proximity data are exploited for behavior understanding. Regarding the *in the wild* aspect, the book aims to describe the current usage, limitations and challenges of systems combining multimodal data, signal processing and machine learning for the understanding of behavioral cues in real-world scenarios. Importantly, the book covers tasks at different levels of complexity, from low level (speaker detection, sensorimotor links, source separation), through middle level (conversational group detection, activity recognition) to high level (affect and emotion recognition).

This book is intended to be a resource for experts and practitioners interested on the state of the art and future research challenges of multimodal behavioral analysis in the wild. It is suitable

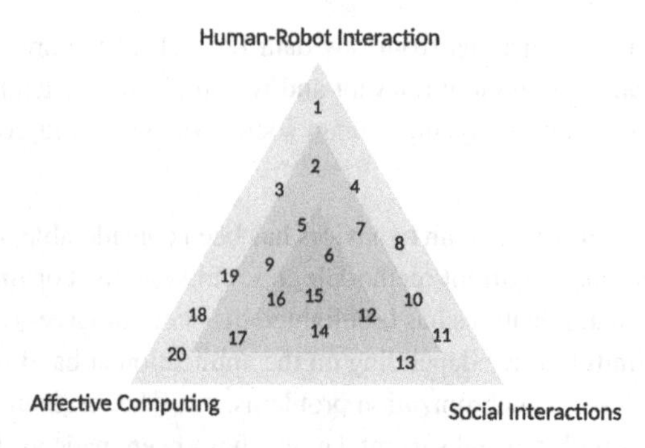

■ **FIGURE 0.1** Overview of the structure of the book chapters.

for researchers and graduate students in the fields of computer vision, audio processing, pattern recognition, multimedia analysis, machine learning, robotics, and social signal processing. The chapters of the book are organized according to three main directions, corresponding to three different application domains as illustrated in Fig. 0.1.

The first series of chapters mostly deal with the problem of behavior understanding and multimodal fusion in the context of robotics and Human–Robot Interaction (HRI). In particular, Chapter 1 focuses on the development of dialog systems for robotic platforms and addresses two important challenges: how to move from closed-domain to open-domain dialogues, and how to create multimodal (audio-visual) dialog systems. The authors describe an approach to jointly tackle these two problems by proposing a Constructive Dialog Model and show how they handle the topic shifts considering Wikipedia as external resource. Chapter 2 describes a robust methodology for audio-motor integration applied to robot hearing. In robotics, audio signal processing in the wild amounts to deal with sounds recorded by a system that moves and whose actuators produce noise. This creates additional challenges in sound source localization, signal enhancement and recognition. But the specificity of such platforms also brings interesting opportunities: can information about the robot actuators' states be meaningfully integrated in the audio processing pipeline to improve performance and efficiency? While robot audition grew to become an established field, methods that explicitly use motor-state information as a complementary modality to audio are scarcer. This chapter proposes a unified view of this endeavor, referred to as audio-motor integration. A literature review and two learning-based methods for audio-motor integration in robot audition are presented, with application to single-microphone sound source localization and ego-noise reduction on real data. Chapter 3 reviews the literature related to multichannel audio source separation in real-life environments. The authors explore some of the major achievements in the field and discuss some of the remaining challenges. Several important issues, e.g. moving sources and/or microphones, varying numbers of sources and sensors, high reverberation levels, spatially diffuse sources, synchronization, etc., are extensively discussed. Many applicative scenarios, such as smart assistants, cellular phones, hearing aids and robots, are presented together with the most prominent associated methodologies. The chapter concludes with open challenges and interesting future guidelines on the topic.

A second series of chapters describe methodologies for fusing multimodal data collected with wearable technologies. In particular, Chapter 4 describes the development of novel wearable glasses aimed to assist users with limited technology skills and disabilities. The glasses process audio-visual data and are equipped with technologies for visual object recognition to support users with low vision as well as with algorithms for enhancing speech signals for people with hearing loss impairment. The chapter further illustrates the results of a user study conducted with people with disabilities in real-world settings. Chapter 5 also focuses on analyzing audio-visual data from wearable sensors and describes an approach for person re-identification where information from audio signals is exploited to complement image streams in the case of challenging conditions (e.g., rapid changes in camera pose, self-occlusions, motion blur, etc.). Similarly, Chapter 6 considers video streams collected with wearable cameras. The authors address the problem of recognizing activities from visual lifelogs and, after outlining the main challenges of this task, they perform a detailed review of state-of-the-art methods and show the results of an extensive experimental comparison. An interesting application of visual lifelogs is described in Chapter 7. The authors present an approach to automatically analyzing images collected from wearable cameras in order to extract a nonredundant set of frames useful for the purpose of memory stimulation in patients with neurodegenerative diseases (Alzheimer, Mild Cognitive Impairment, etc.). Chapters 8 and 9 describe how wearable technologies, alone or in combination with more traditional static and distributed sensors, can be used to analyze visitor behavior in museums. In particular, these chapters address the challenges of interpreting raw multimodal data for the purpose of visitor tracking and improving tourist experience.

Chapters 10–15 mostly describe recent methodologies and open problems in the analysis of social scenes. Specifically, Chapters 10 and 11 present approaches which exploit data from wearable sensors for the purpose of understanding social interactions. In particular, Chapter 10 addresses the problem of discovering conversational groups (more precisely, F-formations) in egocentric videos depicting social gatherings and presents an algorithm based on Structural SVM. Chapter 11 also focuses on the challenges of analyzing conversational scenes and, in particular, illustrates the limitations of current datasets publicly available for the automated analysis of human social behavior. The authors also describe the conceptual and practical issues inherent to the data collection process, with a specific focus on the multimodality and the 'in the wild' perspective. The problem of analyzing social interactions and detecting conversational groups is also addressed in Chapter 12. In particular, a methodology for recognizing F-formations derived from game theory is described but, differently from Chapter 10, the approach is tested on data from static surveillance cameras. Chapter 13 also addresses the problem of analyzing social interactions. The authors point out that understanding nonverbal behavioral cues (e.g., facial expressions, gaze, gestures, etc.) is important both from a human science perspective, as it helps to understand how people work, and from a technological point of view, because it allows one to design systems can make sense of social and psychological phenomena. Chapter 14 focuses on crowd analysis, different from the work presented in the previous chapters, which deals with social scenes with a small number of people. The chapter describes the challenges of understanding crowd behaviors in realistic settings and provides an overview of state of the art approaches for analyzing visual data and detecting motion patterns, tracking people, recognizing activities and spotting anomalous behaviors. A methodological contribution is presented in Chapter 15, where the problem of learning robust and invariant representations in visual recognition systems is considered. This issue is of utmost importance when deploying systems operating in the wild.

The last series of chapters describe methodologies for detecting affective and emotional patterns in real-world settings. In particular, Chapter 16 focuses on the problem of visual affect recognition. The chapter addresses the challenges of bridging the 'affective gap' between visual features and semantic concepts. Following the Adjective Noun Pair (ANP) paradigm, i.e. considering mid-level representations of pairs of noun-adjectives such as 'scary face', 'beautiful women', etc., the authors present an approach for sentiment and emotion prediction which operates by embedding ANP constructs in a latent space.

Chapter 17 addresses the problem of video-based emotion recognition 'in the wild' and describes an approach for fusing audio-visual data. The method uses summarizing functionals of complementary visual descriptors in conjunction with audio features. The audio and visual data are fused within a least squares classifier framework. The authors report state-of-the-art results on the EmotiW Challenge. Chapter 18 also considers the problem of emotion recognition from audiovisual signals in real-world environments. This chapter highlights the differences between affect recognition in real-world and laboratory settings, provides an overview of state of the art methodologies, and illustrates an audio-visual continuous emotion recognition system based on deep learning. Similarly, Chapter 19 focuses on affective facial computing with a special emphasis on addressing the 'in the wild' aspect. Specifically, the authors consider the generalizability of facial computing technologies across various domains and propose a review of several previous studies on the topic. The outcome of their study is that the ability of current systems to generalize across domains is limited and that, in this context, transfer learning and domain adaptation methodologies are a precious resource. Finally, Chapter 20 discusses the problem of emotion recognition from behavioral data, with special emphasis on distinguishing between self-reported and perceived emotions. The authors further analyze how this aspect influences the design of systems for emotion recognition and outline recent advances and challenges in this research topic.

0.3 SUMMARY OF IMPORTANT POINTS

This book aims to describe recent works in the area of human behavior analysis, with special emphasis on studies considering *multimodal data*. Besides providing an overview on state-of-the-art research in the field, the book highlights the main challenges associated with the automatic analysis of human behaviors *in real-world scenarios*, discussing the limitations of existing technologies. The chapters of the book describe a large variety of methodologies to extract behavioral cues from multimodal data and consider different applications. This clearly demonstrates that the topic addressed by the book can be of interest for a large set of researchers and graduate students working in different fields.

REFERENCES

[1] Alexandre Alahi, Vignesh Ramanathan, Kratarth Goel, Alexandre Robicquet, Amir A. Sadeghian, Li Fei-Fei, Silvio Savarese, Learning to predict human behavior in crowded scenes, in: Group and Crowd Behavior for Computer Vision, 2017, pp. 183–207.

[2] Xavier Alameda-Pineda, Jacopo Staiano, Ramanathan Subramanian, Ligia Batrinca, Elisa Ricci, Bruno Lepri, Oswald Lanz, Nicu Sebe, Salsa: a novel dataset for multimodal group behavior analysis, IEEE Trans. Pattern Anal. Mach. Intell. 38 (8) (2016) 1707–1720.

[3] Xavier Alameda-Pineda, Yan Yan, Elisa Ricci, Oswald Lanz, Nicu Sebe, Analyzing free-standing conversational groups: a multimodal approach, in: ACM International Conference on Multimedia, 2015, pp. 5–14.

[4] Yutong Ban, Laurent Girin, Xavier Alameda-Pineda, Radu Horaud, Exploiting the complementarity of audio and visual data in multi-speaker tracking, in: IEEE/CVF ICCV Workshop on Computer Vision for Audio-Visual Media, 2017.

[5] Yutong Ban, Xiaofei Li, Xavier Alameda-Pineda, Laurent Girin, Radu Horaud, Accounting for room acoustics in audio-visual multi-speaker tracking, in: IEEE International Conference on Acoustic, Speech and Signal Processing, 2018.

[6] Katherine L. Bouman, Vickie Ye, Adam B. Yedidia, Frédo Durand, Gregory W. Wornell, Antonio Torralba, William T. Freeman, Turning corners into cameras: principles and methods, in: International Conference on Computer Vision, 2017.

[7] Alessio Brutti, Andrea Cavallaro, Online cross-modal adaptation for audio-visual person identification with wearable cameras, IEEE Trans. Human-Mach. Syst. 47 (1) (2017) 40–51.

[8] Judee K. Burgoon, Nadia Magnenat-Thalmann, Maja Pantic, Alessandro Vinciarelli, Social Signal Processing, Cambridge University Press, 2017.

[9] Laura Cabrera-Quiros, Andrew Demetriou, Ekin Gedik, Leander van der Meij, Hayley Hung, The MatchNMingle dataset: a novel multi-sensor resource for the analysis of social interactions and group dynamics in-the-wild during free-standing conversations and speed dates, IEEE Trans. Affect. Comput. (2018), https://doi.org/10.1109/TAFFC.2018.2848914.

[10] Tatjana Chavdarova, Pierre Baqué, Stéphane Bouquet, Andrii Maksai, Cijo Jose, Louis Lettry, Pascal Fua, Luc Van Gool, François Fleuret, The wildtrack multi-camera person dataset, arXiv preprint, arXiv:1707.09299, 2017.

[11] Marco Cristani, R. Raghavendra, Alessio Del Bue, Vittorio Murino, Human behavior analysis in video surveillance: a social signal processing perspective, Neurocomputing 100 (2013) 86–97.

[12] Dima Damen, Hazel Doughty, Giovanni Maria Farinella, Sanja Fidler, Antonino Furnari, Evangelos Kazakos, Davide Moltisanti, Jonathan Munro, Toby Perrett, Will Price, et al., Scaling egocentric vision: the epic-kitchens dataset, arXiv preprint, arXiv:1804.02748, 2018.

[13] Sergio Escalera, Xavier Baró, Hugo Jair Escalante, Isabelle Guyon, Chalearn looking at people: a review of events and resources, in: International Joint Conference on Neural Networks, 2017, pp. 1594–1601.

[14] Agrim Gupta, Justin Johnson, Li Fei-Fei, Silvio Savarese, Alexandre Alahi, Social GAN: socially acceptable trajectories with generative adversarial networks, in: IEEE/CVF Conference on Computer Vision and Pattern Recognition, 2018.

[15] Guan-Horng Liu, Avinash Siravuru, Sai Prabhakar, Manuela Veloso, George Kantor, Learning end-to-end multimodal sensor policies for autonomous navigation, arXiv preprint, arXiv:1705.10422, 2017.

[16] Adria Recasens, Aditya Khosla, Carl Vondrick, Antonio Torralba, Where are they looking?, in: Advances in Neural Information Processing Systems, 2015, pp. 199–207.

[17] Aliaksandr Siarohin, Enver Sangineto, Stéphane Lathuilière, Nicu Sebe, Deformable GANs for pose-based human image generation, in: IEEE/CVF Conference on Computer Vision and Pattern Recognition, 2018.

[18] Sergey Tulyakov, Xavier Alameda-Pineda, Elisa Ricci, Lijun Yin, Jeffrey F. Cohn, Nicu Sebe, Self-adaptive matrix completion for heart rate estimation from face videos under realistic conditions, in: IEEE/CVF Conference on Computer Vision and Pattern Recognition, 2016, pp. 2396–2404.

[19] Jagannadan Varadarajan, Ramanathan Subramanian, Samuel Rota Bulò, Narendra Ahuja, Oswald Lanz, Elisa Ricci, Joint estimation of human pose and conversational groups from social scenes, Int. J. Comput. Vis. 126 (2–4) (2018) 410–429.

[20] Alessandro Vinciarelli, Maja Pantic, Hervé Bourlard, Social signal processing: survey of an emerging domain, Image Vis. Comput. 27 (12) (2009) 1743–1759.
[21] Carl Vondrick, Deniz Oktay, Hamed Pirsiavash, Antonio Torralba, Predicting motivations of actions by leveraging text, in: IEEE/CVF Conference on Computer Vision and Pattern Recognition, 2016, pp. 2997–3005.
[22] Jindong Wang, Yiqiang Chen, Shuji Hao, Xiaohui Peng, Lisha Hu, Deep learning for sensor-based activity recognition: a survey, Pattern Recognit. Lett. (2018), https://doi.org/10.1109/CVPR.2016.327.
[23] Wei Wang, Xavier Alameda-Pineda, Dan Xu, Pascal Fua, Elisa Ricci, Nicu Sebe, Every smile is unique: landmark-guided diverse smile generation, in: IEEE/CVF Conference on Computer Vision and Pattern Recognition, 2018, pp. 7083–7092.
[24] Huazhe Xu, Yang Gao, Fisher Yu, Trevor Darrell, End-to-end learning of driving models from large-scale video datasets, in: IEEE/CVF Computer Vision and Pattern Recognition, 2017.

Xavier Alameda-Pineda*, Elisa Ricci†,‡, Nicu Sebe†
*Inria Grenoble Rhone-Alpes, Perception Team, France †University of Trento, Department of Information Engineering and Computer Science, Italy ‡Fondazione Bruno Kessler, Technology of Vision, Italy

Chapter 1

Multimodal open-domain conversations with robotic platforms

Kristiina Jokinen*, Graham Wilcock†
*AIST/AIRC, Tokyo, Japan †CDM Interact, Helsinki, Finland

CONTENTS

1.1 INTRODUCTION

From a historical point of view, the development of natural language conversational systems has accelerated in recent years due to advances in computational facilities and multimodal dialog modeling, availability of big data, statistical modeling and deep learning, and increased demand in commercial applications. In a somewhat simplified manner, we can say that the capability of conversational systems has improved in roughly 20-year time-spans if seen from the viewpoint of technological advancements: from ELIZA's imitation of human-like properties in the early 1960s, via systems that understand spoken natural language and various multimodal acts in limited domains developed in the 1980s, to interactive systems that are part of everyday environments in the 21st century. Embodied conversational agents,

Multimodal Behavior Analysis in the Wild. https://doi.org/10.1016/B978-0-12-814601-9.00025-0
Copyright © 2019 Elsevier Ltd. All rights reserved.

chatbots, Siri, Amazon Alexa, Google Home, etc. are examples of the multitude of interactive systems that aim to provide natural language capabilities for question answering and to search for useful information in the cloud.

The rapid development of robot technology has had a huge impact on interaction research and in particular, on developing social robotics, i.e. human–robot applications, where the robot can provide natural language communication with a user, and be able to observe and understand the user's needs and emotions. This enables a novel type of interaction where the robot is not just a tool to do things, but an agent to communicate with: social robots can interact with human users in natural language, and support companionship and peer-type assistance which feature information-providing as well as "chatting" and sensitivity to social aspects of interaction. Co-located acting and free observations of the partner are both beneficial and challenging for interaction modeling. Interaction becomes richer and more natural, but also more complicated: learning the various social signals and constructing a shared context for the interaction (cf. [15]).

Social robotics emphasizes the robot's communication skills besides its autonomous decision-making and moving around in the environment. Social robots show more human-like interaction and try to act in a proactive manner so as to support human interest and activity. Consequently, social robotics has had a huge impact on interaction technology. Communication is simultaneously visual, verbal, and vocal, i.e. humans not only utter words, but use various vocalizations (laughs, coughs), head, gaze, hands, and the whole body to convey messages. In order to understand human behavior and communicative needs, the robot should observe the user's multimodal signals and be able to generate reasonable behavior patterns in interactive situations. The main hypothesis is that the more engaging the interaction is in terms of communicative competence, the better results the interaction produces, whether or not the task that the user is involved in with the robot concerns friendly chatting or some more structured task.

The chapter is structured as follows. The next section describes the Constructive Dialog Model which forms the foundation for our work. Section 1.2 discusses issues in moving from closed-domain dialogs to open-domain dialogs, including how to manage topic shifts and how to use Wikipedia as a knowledge source. Section 1.3 addresses multimodal interaction with the Nao robot by speech, gesturing and face-tracking, and multimodal aspects of topic modeling. Section 1.4 briefly presents two future research directions, the use of domain ontologies in dialog systems and the need to integrate robots with the Internet of Things. Section 1.5 presents conclusions.

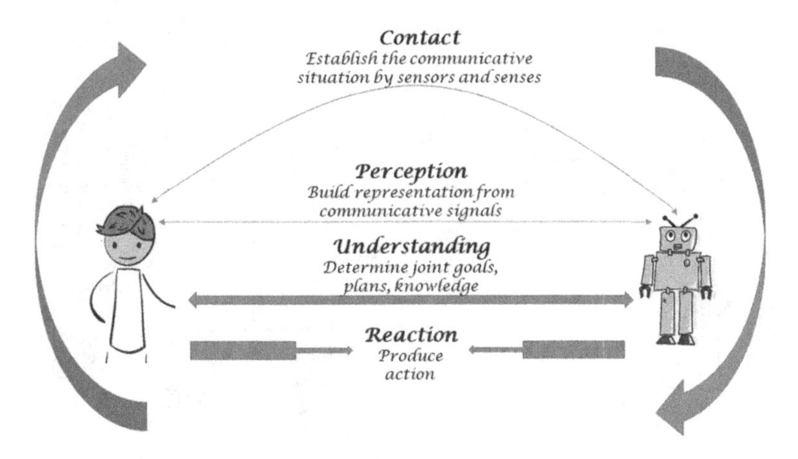

■ **FIGURE 1.1** The communication cycle in the Constructive Dialog Model.

1.1.1 Constructive Dialog Model

Conversational interactions are cooperative activities through which the interlocutors build common ground (Clark and Schaefer [8]). Cooperation indicates the interlocutors' basic willingness to engage in the conversation, and manifests itself in smooth turn-taking and coherent replies. The agents react to the situation according to their beliefs, intentions and interpretation of the situation, and they use multimodal signals to indicate how the basic enablements of communication are fulfilled.

In Fig. 1.1, interaction in the Constructive Dialog Model (CDM) [15] is seen as a cycle which starts with the participants being in contact, observing the partner's intent to communicate, interpreting the partner's communicative content, and producing their own reaction to the message in an appropriate manner. Fig. 1.1 shows the communication cycle with the basic enablements of communication, which concern Contact, Perception, Understanding, and Reaction (Allwood [1], Jokinen [15]).

Contact refers to the participants' mutual awareness of their intention to communicate, i.e. being close enough to be able to communicate or having a means to communicate such as a phone or skype if not in a face-to-face situation. Perception relates to the participants' perception of communicative signals as a message with an intent. Understanding concerns the participants' cognitive processes to interpret the message in the given context. Reaction is the speakers' observable behavior which manifests their reaction to the new changed situation in which the agents find themselves.

In the CDM system architecture in Fig. 1.2, signal detection and signal analysis modules implement Contact and Perception, respectively, for

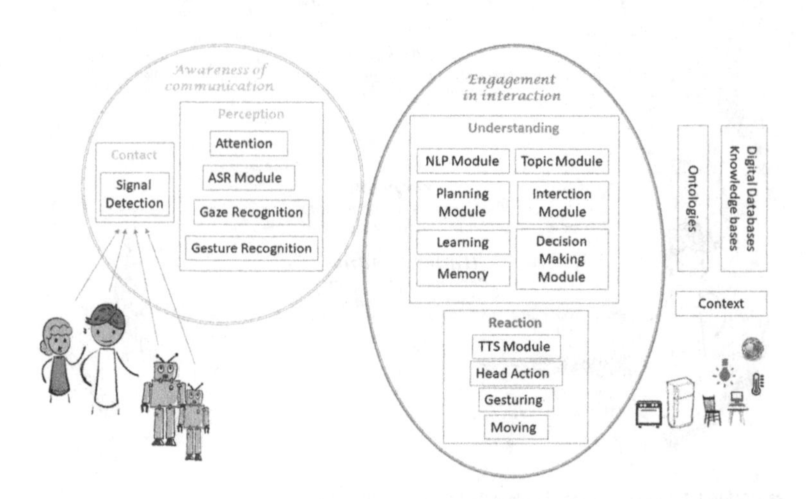

■ **FIGURE 1.2** An implementation of the CDM model in a system architecture.

speech, gesture, and gaze recognition, and these components are responsible for interpreting the user awareness. Understanding is implemented by the decision-making and related modules, while Reaction corresponds to the production and coordination of utterances and motoric actions, including internal updates. Together these two are responsible for the system's engagement with the user. The dialog management is based on dialog states (also called mental states) which are representations of the situation the robot is in and the situation it believes the user is in.

Many neurocognitive studies show how activation in the brain is triggered by the mere appearance of a human in the vicinity of a person, while attention is directed to a human face (Levitski et al. [22]). The robot agent obtains information from the environment via its sensors and the dialog component integrates them into the system knowledge base through its recognition and decision-making processes. The perception of the partner concerns recognition of the signals as having some communicative meaning: the face belongs to a particular person, the sounds belong to a particular language, and gesturing has communicative content.

Interpretation of the signals concerns their further processing to form a meaningful semantic representation in the given context. The new information entered into the system will trigger a reaction, i.e. cognitive processes which evaluate the new information with respect to the agent's own goals and the decision-making process which results in carrying out an action that in turn triggers a similar analysis and generation process in the partner. If the speaker is repeatedly exposed to a certain kind of communicative situation and if the speaker's communicative action results in a successful goal

achievement, the same action will be used again, to maximize benefits in the future.

To construct common ground, the interlocutors thus pay attention to signals that indicate the partner's reaction to the conveyed message, their emotional state, and the new information in the partner's speech. Non-verbal signals such as pauses, intonation, nods, smiles, frowns, eye-gaze, gesturing etc. are effectively used to signal the speaker's understanding and emotions (Feldman and Rim [12]). Studies on embodied conversational agents have widely focused on various aspects of interaction, multimodality, and culturally competent communication (see e.g. Andre and Pelachaud [4], Jokinen et al. [16,17]). For instance, in human–human interactions, gestures and head movements complement language expressions and enable the interlocutors to coordinate the dialog in a tacit manner. Gesturing is synchronized with speech, and besides the pointing gestures and iconic gestures that refer to objects and describe events, gesturing also provides rhythmic accompaniment to speech (co-gesturing) which contributes to the fluency of expression and construction of shared understanding.

A related question is the robot's understanding of the relation between language and the physical environment, i.e. the embodiment of linguistic knowledge via action and interaction. The connection between linguistic expressions and objects in the environment is called grounding. In dialog modeling, the term grounding is usually used to refer to the interlocutors' actions that enable them to build mutual understanding of the referential elements. Grounding in interactions can be studied with respect to the notion of affordance (Jokinen [15]): the interlocutors' actions (speech and gesturing) should readily support a natural and smooth way of communication. Interactions between humans and robots as well as between humans and intelligent environments should enable easy recognition of various communicatively important objects, and the different objects must be distinguished from each other so as to be correctly referred to.

The CDM framework is applied in the WikiTalk open-domain robot dialog system (Jokinen and Wilcock [21]) where both the human and the robot can initiate topics and ask questions on a wide variety of topics. It is also applied in human–robot interaction such as newspaper-reading or story-telling in elder-care and educational activities. It aims at acquiring a good level of knowledge about the user and his/her context and can thus enable an open-domain conversation with the user, presenting useful and interesting information. As the robot can observe user behavior, it can infer a user's emotion and interest levels, and tailor its presentation accordingly.

1.2 OPEN-DOMAIN DIALOGS

In traditional closed-domain dialog systems, such as flight reservation systems, the system asks questions in order to achieve a specified dialog goal. Finite state machines can be used for this kind of closed-domain form-filling dialog. The system asks questions, which are predefined for the specific domain, in order to achieve the dialog goal by filling in the required fields in the form. It is easy to change the information within the domain database, for example about flights and destinations, but it is very difficult to change to a different domain because the questions are specific to the domain.

In order to advance from closed-domain to open-domain dialogs, WikiTalk [26] uses Wikipedia as its source of world knowledge. By exploiting ready-made paragraphs and sentences from Wikipedia, the system enables a robot to talk about thousands of different topics (there are 5 million articles in English Wikipedia). WikiTalk is open-domain in the sense that the currently talked-about topic can be any topic that Wikipedia has an article about, and the user can switch topics at any time and as often as desired. In an open-domain system it is extremely important to keep track of the current topic and to have smooth mechanisms for changing to new topics.

1.2.1 Topic shifts and topic trees

An important feature that enables an interactive system to manage dialogs in a natural manner is its ability to handle smooth topic shifts, i.e. to be able to provide a relevant continuation in the current dialog state. The underlying problem is that knowing what is relevant depends on the overall organization of knowledge.

The organization of knowledge into related topics has often been done with the help of topic trees. Originally "focus trees" were proposed by McCoy and Cheng [23] to trace foci in natural language generation systems. The branches of the tree describe what sort of topic shifts are cognitively easy to process and can be expected to occur in dialogs: random jumps from one branch to another are not very likely to occur, and if they do, they should be appropriately marked. McCoy and Cheng [23] dealt with different types of focusing phenomena by referring to a model of the conceptual structure of the domain of discourse. They introduced the notion of focus tree and argued that the tree structure is more flexible in managing focus shifts than a stack: instead of pushing and popping foci in a particular order into and from the stack, the tree allows traversal of the branches in a different order, and the coherence of the text can be determined on the basis of the distance of the focus nodes in the tree.

The focus tree is a subgraph of the world knowledge, built in the course of the discourse on the basis of the utterances that have occurred so far. The tree both constrains and enables prediction of what is likely to be talked about next, and thus provides a top-down approach to dialog coherence. The topic (focus) is a means to describe thematically coherent discourse structure, and its use has been mainly supported by arguments regarding anaphora resolution and processing effort. Focus shifting rules are expressed in terms of the type of relationships which occur in the domain. In language generation, they provide information about whether or not a topic shift is easy to process (and, similarly, whether or not the hearer will expect some kind of marker), and in language analysis they help to decide on what sort of topic shifts are likely to occur. Jokinen, Tanaka and Yokoo [19] applied the idea of focus tree in spoken dialog processing. They made the distinctions between topical vs. non-topical informational units, i.e. what the utterance is about vs. what is in the background, and new vs. old information in the dialog context.

Grosz, Weinstein and Joshi [14] distinguished between global and local coherence, as well as between global focus and centering, respectively. The former refers to the ways in which larger segments of discourse relate to each other, and accordingly, global focus refers to a set of entities that are relevant to the overall discourse. The latter deals with individual sentences and their combinations into larger discourse segments, and accordingly, centering refers to a more local focusing process which identifies a single entity as the most central one in an individual sentence. Each sentence can thus be associated with a single backward-looking center which encodes the notion of global focusing and a set of forward-looking centers which encodes the notion of centering.

The organization of knowledge has always been one of the big questions, but we can now look for help with this question from the internet. In fact we can assume that world knowledge is somehow stored in the internet and we wish to take advantage of this. Previously, topic trees were hand-coded which is time-consuming and subjective. Automatic clustering programs were also used but were not entirely satisfactory. Our approach to topic trees exploits the organization of domain knowledge in terms of topic types found in the web, and more specifically in Wikipedia.

We use topic information in predicting the likely content of the next utterance, and thus we are more interested in the topic types that describe the information conveyed by utterances than the actual topic entity. Consequently, instead of tracing salient entities in the dialog and providing heuristics for different shifts of attention, we seek a formalization of the information structure of utterances in terms of the new information that is

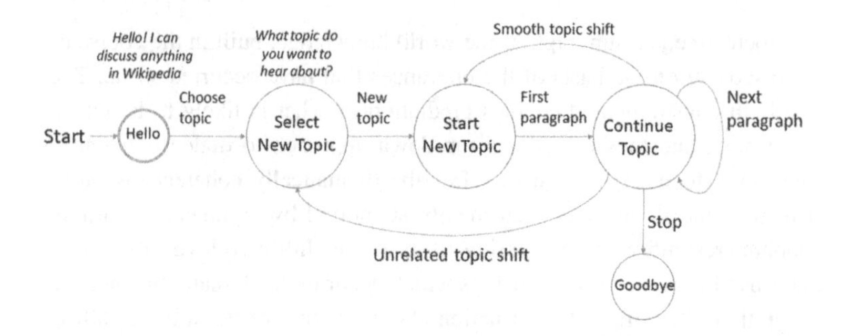

■ **FIGURE 1.3** WikiTalk: simplified FSM including smooth topic shifts.

exchanged. Wikipedia provides an extensive, freely available, open-domain and constantly growing knowledge source. We therefore use Wikipedia to produce robot contributions in open-domain dialogs.

1.2.2 Dialogs using Wikipedia

In WikiTalk, topic-tracking and topic-shifting are managed with the help of the concepts of Topic and NewInfo, in accordance with the Constructive Dialog Model [15]. The distinctive contribution of WikiTalk is that NewInfos, the pieces of new information to be conveyed to the partner, are associated with hyperlinks in the Wikipedia texts. The hyperlinks are extracted and the system predicts that the user will often want to shift the topic to one of the NewInfos. Using these hyperlinks the system can change topics smoothly according to the human's changing interests.

In order to support open-domain dialogs, dialog control in WikiTalk is different from closed-domain systems. The open-domain dialog control is described by Wilcock [26], and it is summarized here. A finite-state machine (FSM) is used, but the set of states are not domain-specific such as "get departure day" or "get destination city" in a closed-domain system such as flight reservations. In WikiTalk the states manage the flow of dialog interaction with the user by managing topic-tracking and topic-shifting. This makes it possible for a finite number of states to manage dialogs with an infinite number of topics. A simplified diagram of these states is shown in Fig. 1.3.

Like closed-domain dialog systems, an open-domain system needs a way to get started and a way to finish. These are handled by a Hello state and a Goodbye state (Fig. 1.3). As well as saying "Hello" and "Goodbye", these states can include other behaviors, for example a mobile robot can stop moving around for the duration of the conversation, or a humanoid robot can stand up or sit down to talk, as desired.

After the Hello state, the system transitions to the selectNewTopic state. The topic can be selected from a list of favorites or from a history list of recently talked-about topics, or a phonetic spelling mechanism can be used to find a totally new topic. WikiTalk allows the favorites to be easily modified, and the robot reminds the user about favorites or recent topics. As well as talking about topics requested by the user, the robot suggests new topics using the daily "Did you know?" items in Wikipedia.

When a new topic has been selected, the startNewTopic state connects to Wikipedia online and gets the newest version of the article for the topic. As Wikipedia articles are designed to be viewed in web browsers, with images, info tables, footnotes, and special symbols, the articles are cleaned up and reformatted to make the text suitable for a speech synthesizer. The text is also split into chunks of a suitable size for spoken dialog interaction. Instead of deep processing of the Wikipedia texts using information extraction, question answering or summarization techniques, we adopt a shallow processing approach in which selected chunks of the texts are read out aloud. The focus of the system is on identifying the new information and on managing the topic chains and topic shifts.

After a new topic is started, the continueTopic state presents chunks of the topic text as desired by the user. If the user explicitly asks for more, or implicitly shows interest in the topic, the system continues with the next chunk. After hearing a chunk about a given topic, the user can ask for it to be repeated, or go back to the previous chunk, or can interrupt the current chunk and skip to the next chunk on the same topic, or can switch to a different topic.

If the user wishes to switch to a totally new topic (an "unrelated topic-shift"), the system transitions back to the selectNewTopic state, but if the user wishes to make a "smooth topic-shift" to a related topic, the system does this by using hyperlinks extracted from the Wikipedia articles to transition directly to the startNewTopic state (see Fig. 1.3) with the desired topic. This smooth topic-shift capability enables the robot to follow the interests of the individual user, who can navigate freely around Wikipedia and have the robot talk about any topic. As there are millions of articles in Wikipedia, this can reasonably be called an open-domain dialog system.

A video (Fig. 1.4) showing a robot talking fluently about Shakespeare and Julius Caesar among other topics can be seen at https://www.youtube.com/watch?v=NkMkImATfYQ. With this multilingual version of WikiTalk the robot also switches between English and Japanese. The dialog in the video is described in detail by [27].

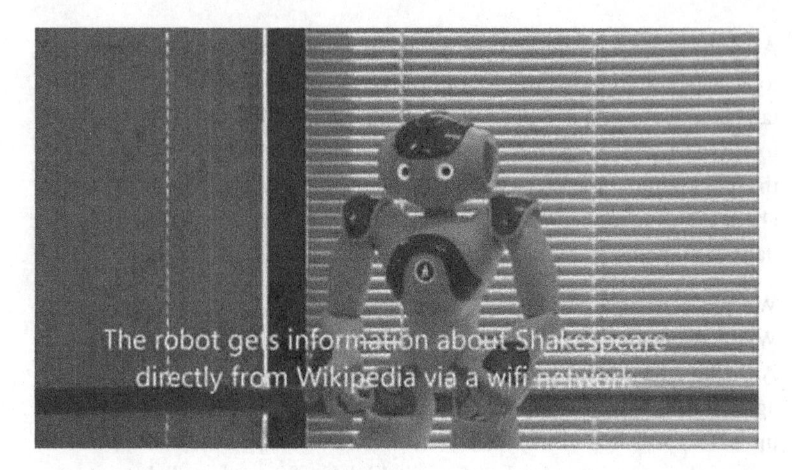

The robot gets information about Shakespeare
directly from Wikipedia via a wifi network

■ **FIGURE 1.4** The robot talks about Shakespeare and Julius Caesar.

1.3 MULTIMODAL DIALOGS

Interaction technology has to address the engagement of the user in the interaction. The system has to manage the interaction so that there is a natural conversation rather than a monolog on a particular topic. For instance, in teaching and learning situations such conversational capability is important. This requires dynamic tracking not only of dialog topics but also of the user's focus of attention and of the user's level of interest in the topic. Techniques for attention-tracking and interest-tracking in interactive situations are important parts of the system.

Engagement is central for successful learning. The robot will not get tired and it can continue as long as the student wishes to. The robot can also repeat the same information, and will be at its best in this kind of repetitive tasks. On the other hand, the robot also enables short interactions: it can ask what the student has been doing lately, and it can also tell a short story about some interesting topic so as to provide light entertainment to the math teaching task. Such activity with a robot also focuses the child's attention to interaction, and as the robot is not just a tool but an interactive agent, this can be assumed to develop the student's self-motivation, initiative taking skills, and social interaction. Short interactions can produce a "togetherness" feeling, but the robot's less expressive and emotional appearance also supports a feeling of easiness in interaction as has been shown to be the case e.g. when the robot interacts with children; cf. Aliz-e project (http://www.aliz-e.org/) and experiments with the Nao robot teaching interaction skills to autistic children (http://www.telegraph.co.uk/technology/news/10632937/The-robot-teacher-connecting-with-autistic-children.html or http://spectrum.

ieee.org/automaton/robotics/humanoids/aldebaran-robotics-nao-robot-autism-solution-for-kids).

An important property of a humanoid robot is situatedness: rather than having an animated agent on the computer screen, the robot brings the interaction into the present situation, and acts in the same space with the learner. The co-location of the human user and the robot agent can improve the interactive services in two ways: first it can help the human user in understanding the content by directing the student's attention and monitoring their multimodal signals, and second, it can engage the user in the interaction and learning experience by mere presence as the human user is socially involved in the interactive situation.

In the context of human–robot interaction, Fujie et al. [13] argued that communication efficiency can be improved by combining linguistic and paralinguistic information. Bohus and Horvitz [6] have demonstrated the use of multimodal signaling in multi-party conversations where the participants enter and leave the interactive situation freely (the interaction is with an animated agent, not with a robot agent).

1.3.1 Multimodal WikiTalk for robots

After prototyping WikiTalk on a robot simulator as illustrated in [20], the system was implemented on the Aldebaran NAO robot as described by [9] and [21]. Using a humanoid robot enabled us to include multimodal communication features, especially face-tracking and gesturing, which were integrated with the spoken conversation system. This required an event-driven system to manage speech events and action events, as shown in Fig. 1.5.

The robot needs to estimate the user's level of interest in the topic, and the human's proximity and gaze both are important for this. The robot also integrates nodding, gesturing and its body posture with its speech during the conversation. Issues in synchronizing gestures with speech are discussed by [24].

As an example of multimodal gesturing, beat gestures are small vertical hand movements that have a pragmatic function of emphasizing and giving rhythm to the speech. These beat gestures are usually synchronized with NewInfos, and serve a similar role as intonation in distinguishing new, not expected information from the old, expected topic information. With this multimodal communication management, the visual gestures emphasize the least known elements so that the partner will more easily notice the new information and understand it. Interaction management is closely related to information presentation: planning and generating appropriate responses, giving feedback, and managing topic shifts.

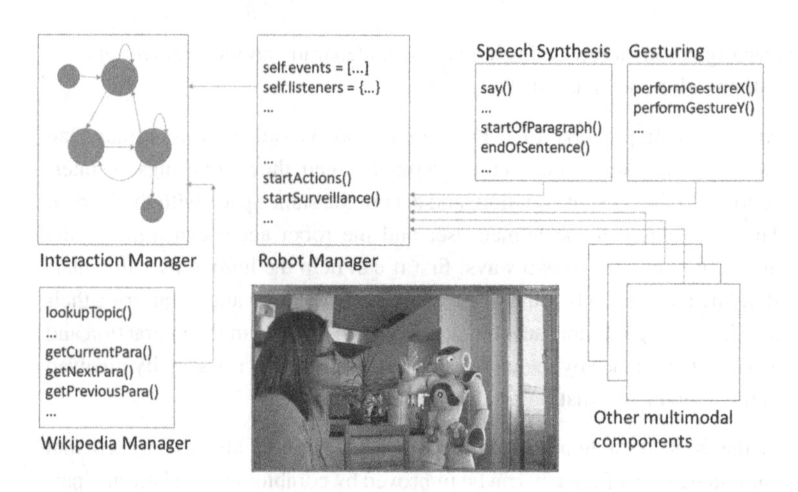

■ **FIGURE 1.5** Multimodal event-driven architecture on the robot.

Assessing the level of interest of the user has two sides: how to detect whether the human partner is interested in the topic or not, and what the system should do as a result. Detecting the level of interest is part of the system's external interface, and deciding what to do about it is part of the system's internal management strategy. In order to assess the interest level correctly, the external interface should not be limited to verbal feedback, but should include intonation, eye-gaze, gestures, body language and other factors. The internal strategy for reacting appropriately must decide what to do not only if the user is clearly interested or clearly not interested, but also how to continue when the interest level is unclear, which may be a more difficult decision.

Information about the evaluation of the Nao robot system based on the recorded user testing sessions at the 8th International Summer Workshop on Multimodal Interfaces, Metz, 2012 is given in [9] and [3].

1.3.2 Multimodal topic modeling

Conventional topic modeling and text classification have been mainly focused on static documents (i.e. documents collected from archived journals, logging, on-line chat and so on) in single modality environments (Alvarez-Melis and Saveski [2], Blei [5]). The differences between static and conversational dynamic documents can be summarized by the variations between writing and speaking. Written text is often formalized and restricted to express a clear topic, the written document is longer with enhanced coherence, detail, and clarity. On the other hand, talking is spontaneous and speaker-dependent, hence it encapsulates a high level of diversity and ambiguity in unstructured documents. The possibility of topic modeling in such complicated situations has been explored by Nguyen et al. [25] and Yeh et al. [28], but these studies have not taken into account the multimodal perspective of interactive dialog since the actual speakers have to rely on both visual and auditory information from their opponents to infer the context of the discourse.

Another study by Brethes et al. [7] proposed a method for extracting information from visual modality that supports richer human–robot interaction. Moreover, an important feature that enables automatic interactive systems to manage dialogs in a natural manner is their ability to handle smooth topic shifts in different situations and in different cultural contexts, i.e. to provide relevant continuation in the current dialog state. This ability requires the agent to acquire deeper understanding of the dialog context and to track the conversational states from multiple perspectives. Many relevant mechanisms have been proposed in [19] and [21], but an exploration of topic-correlated multimodal features remains an open research area with great potential.

1.4 FUTURE DIRECTIONS

1.4.1 Dialogs using domain ontologies

As intelligent agents and social robots are now envisaged as companions and assistants at workplaces and homes, one of the main challenges that such agents must address is the acquisition of knowledge in order to act and also to understand the human partner's actions. In particular, to improve efficiency and quality of the services that social robots provide in such contexts, it is important that they are equipped with knowledge that enables them to act and communicate naturally.

According to socio-cognitive models of human activity (Engström et al. [10,11]), for the agents to coordinate their actions, they need two types of knowledge: knowledge about the activity and tasks (facts about the task and how it is performed in a given context), as well as knowledge about their roles and relation with the other actors (self-understanding in the given operational context). The role does not only mean having a "stance" or a particular "character", but it also means being aware of the tacit rules for reasoning and ethical consideration that guide the participants' social relations and decision making. In multimodal interactions, the main issue is the knowledge acquisition problem, which requires that the system has the capability to learn through interaction: it is able to recognize new relevant information in the communicative acts, and update its knowledge bases appropriately. Moreover, a social robot must convince the user that its responses are reliable and communicatively adequate, i.e. the robot has the role of a trustworthy partner who provides truthful information.

These issues pose considerable challenges for dialog management. The current technology allows us to automate dialogs which are robust and based on big-data processing and straightforward question-answering interactions with the users. However, also models for the application and the general world knowledge are necessary and thus there is a need for extending the robot's knowledge base with domain ontologies which enable reasoning on actions and entities. The knowledge can be achieved by learning, and also by providing the robot with real-size knowledge-bases. If the system is allowed some initial knowledge of the domain, we can also combine the advantages of having initial bootstrap information (learning will not start from scratch), and of controlled flexibility (knowledge updates are related to an initial knowledge state which the user can understand and feel in control of).

The latter can be initiated by studying human activities and structuring the tacit knowledge possessed by the participants with the help of a crowdsourcing methodology that engages the participants in interaction with each other, where they explain their knowledge through interaction. The type and sequence of typical processes and activities involved in the tasks together with risks involved, and the roles and responsibilities of the actors who can also provide evaluation of the activities can thus be visualized, and the awareness of the issues increased. The beginning of such work has been described by Jokinen et al. [18], where the goal-oriented knowledge of the activities is connected to operational procedures that aim to support efficiency and quality of services, especially in various community activities in educational and cultural settings, e.g. nursing care services and health promotion through music and dance therapy. The goal is to enable non-tangible things such as the knowledge and awareness of the actions to connect with companionable agents.

1.4.2 IoT and an integrated robot architecture

The interaction environment consists not only of the immediate social context of the conversations but also of the physical environment in which the human and the robot agent operate. Social interaction takes place via natural language communication usually in face-to-face situations, and traditionally this environment has been a separate context for dialog modeling updated according to the agents' dialog acts. Due to advances in IoT technology, the context for human–robot interaction has widened to include the physical and digital environment in which human–human and human–robot interactions occur.

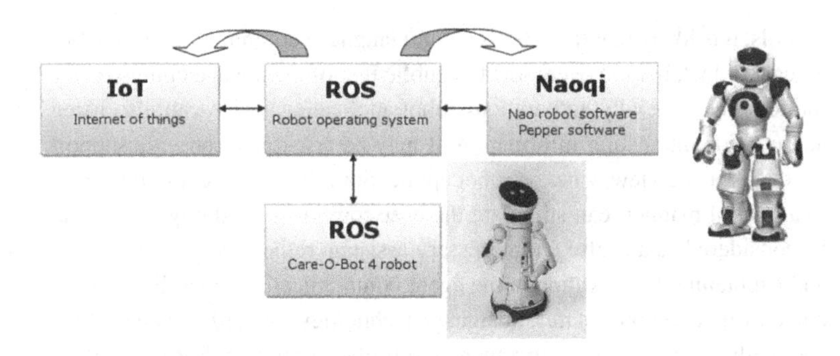

■ **FIGURE 1.6** Integrated robot architecture with IoT, ROS and Naoqi.

The use of multiple robots from different manufacturers, using different operating software, requires a common framework such as ROS (Robot Operating System). Some robots (such as Nao) have their own operating software (Naoqi on Nao) but can be connected to ROS via a ROS-Naoqi bridge module. Some robots (such as Care-O-Bot 4) use ROS directly. ROS can also be connected to IoT via a ROS-IoT bridge module. An integrated robot architecture for these connections is shown in Fig. 1.6.

Smart environments can themselves take an active part in dialog management, and become another "partner" in the conversation between the situated agents: smart home devices can alert the user of harmful events, remind of regular activities, or intervene when something unexpected is about to happen. Also, situated agents can change and operate on the smart environment, and influence the operation of the environment not only by manipulating the objects but by communicating with it. Interaction thus requires interaction capabilities from both partners. In the rapidly changing IoT world, such interaction will become more normal, and will affect human users in two ways: making life look more complex and difficult, but also making it easier to cope with everyday activities, being of assistance to users by extending human cognitive, physical and interaction capabilities and possibilities.

1.5 CONCLUSION

The work contributes to the area of social robotics by extending a social robot's dialog capability from the Wikipedia articles to any digital domain that is available in the web, for the purpose of using the information in interactive agent dialogs. Such natural language interfaces have many application possibilities, and the current popular areas deal with various elder-care and robot teacher applications. To address the Eurobarometer challenges of using robots as companions and care-takers in elderly people care or at schools, it is likely that the robot's natural language communication capability can, in fact, help to overcome the public fear of the novel technology. By introducing a friendly autonomous robot application which can also listen and provide interesting information, it may be possible to provide support for the opposite view, where a robot application, capable of communicating in a natural manner, can stimulate the user cognitively and physically, and be considered as a useful and necessary assistant rather than a complicated and frightening tool. Although the agent communicative capability can become livelier and push natural language technology forward, the algorithms and methods still need improvement and further designing. For instance, in the case of our robot application, the answer selection could have a separate priority within the same class, and the dialog design can provide more question and answer types. Furthermore, the agent could have multiple agendas with larger decision space.

Future work will also include more extensive evaluation with the robot agent. The current system has only been evaluated with respect to its operation and first impressions by the users, but a more systematic user study is planned. Also non-verbal communication can be studied and included in the agent's communicative capability so as to provide visual cues of the new and important information. For instance, the system should recognize the user's nodding and gesturing, and be able to adjust its dialogs on the basis of this kind of feedback. What is important in our case is to capture the user's attentional state on the basis of possible topics that the user may direct their attention to and which will then be realized as a topic continuation or topic shift.

REFERENCES

[1] Jens Allwood, Linguistic Communication as Action and Cooperation: A Study in Pragmatics, Gothenbg. Monogr. Linguist., vol. 2, University of Gothenburg, 1976.

[2] David Alvarez-Melis, Martin Saveski, Topic modeling in twitter: aggregating tweets by conversations, in: ICWSM, 2016.

[3] Dimitra Anastasiou, Kristiina Jokinen, Graham Wilcock, Evaluation of WikiTalk – user studies of human–robot interaction, in: Proceedings of 15th International Conference on Human–Computer Interaction (HCII 2013), Las Vegas, 2013.

[4] Elisabeth André, Catherine Pelachaud, Interacting with embodied conversational agents, in: F. Cheng, K. Jokinen (Eds.), Speech Technology: Theory and Applications, Springer, 2010, pp. 123–150.

[5] David M. Blei, Probabilistic topic models, Commun. ACM 55 (4) (April 2012) 77–84.

[6] Dan Bohus, Eric Horvitz, Model for multiparty engagement in open-world dialogue, in: Proceedings of SIGDIAL 2009, 2009, pp. 225–234.

[7] Ludovic Brèthes, Paulo Menezes, Frédéric Lerasle, Jean-Bernard Hayet, Face tracking and hand gesture recognition for human–robot interaction, in: Proceedings of IEEE International Conference on Robotics and Automation, 2004, pp. 1901–1906.

[8] Herbert H. Clark, Edward F. Schaefer, Contributing to discourse, Cogn. Sci. 13 (1989).

[9] Adam Csapo, Emer Gilmartin, Jonathan Grizou, JingGuang Han, Raveesh Meena, Dimitra Anastasiou, Kristiina Jokinen, Graham Wilcock, Multimodal conversational interaction with a humanoid robot, in: Proceedings of 3rd IEEE International Conference on Cognitive Infocommunications (CogInfoCom 2012), Kosice, 2012, pp. 667–672.

[10] Yrjö Engeström, Learning by Expanding, Cambridge University Press, 2014.

[11] Yrjö Engeström, Reijo Miettinen, Raija-Leena Punamäki, Perspectives on Activity Theory, Cambridge University Press, 1999.

[12] Robert S. Feldman, Bernard Rimé, Fundamentals of Nonverbal Behavior, Cambridge University Press, Cambridge, 1991.

[13] Shinya Fujie, Kenta Fukushima, Tetsunori Kobayashi, A conversation robot with back-channel feedback function based on linguistic and non-linguistic information, in: Proceedings of 2nd International Conference on Autonomous Robots and Agents (ICARA-2004), 2004, pp. 379–384.

[14] Barbara J. Grosz, Scott Weinstein, Aravind K. Joshi, Centering: a framework for modeling the local coherence of discourse, Comput. Linguist. 21 (2) (June 1995) 203–225.

[15] Kristiina Jokinen, Constructive Dialogue Modelling: Speech Interaction and Rational Agents, John Wiley & Sons, 2009.

[16] Kristiina Jokinen, Kazuaki Harada, Masafumi Nishida, Seiichi Yamamoto, Turn-alignment using eye-gaze and speech in conversational interaction, in: Proceedings of 11th International Conference on Spoken Language Processing (Interspeech 2010), Makuhari, Japan, 2010.

[17] Kristiina Jokinen, Masafumi Nishida, Seiichi Yamamoto, Collecting and annotating conversational eye-gaze data, in: Proceedings of Multimodal Corpora: Advances in Capturing, Coding and Analyzing Multimodality (MMC 2010), Valletta, Malta, 2010.

[18] Kristiina Jokinen, Satoshi Nishimura, Ken Fukuda, Takuichi Nishimura, Dialogues with IoT Companions: enabling human interaction with intelligent service items, in: Proceedings of the International Conference on Companion Technologies, Ulm, Germany, 2017.

[19] Kristiina Jokinen, Hideki Tanaka, Akio Yokoo, Context management with topics for spoken dialogue systems, in: Proceedings of Joint International Conference of Computational Linguistics and the Association for Computational Linguistics (COLING-ACL'98), Montreal, Canada, 1998, pp. 631–637.

[20] Kristiina Jokinen, Graham Wilcock, Constructive interaction for talking about interesting topics, in: Proceedings of Eighth International Conference on Language Resources and Evaluation (LREC 2012), Istanbul, 2012.

[21] Kristiina Jokinen, Graham Wilcock, Multimodal open-domain conversations with the Nao robot, in: Joseph Mariani, Sophie Rosset, Martine Garnier-Rizet, Laurence Devillers (Eds.), Natural Interaction with Robots, Knowbots and Smartphones: Putting Spoken Dialogue Systems into Practice, Springer, 2014, pp. 213–224.

[22] Andres Levitski, Jenni Radun, Kristiina Jokinen, Visual interaction and conversational activity, in: Proceedings of the 4th Workshop on Eye Gaze in Intelligent Human Machine Interaction: Eye Gaze and Multimodality, Santa Monica, USA, 2012.

[23] Kathleen F. McCoy, Jeannette Cheng, Focus of attention: constraining what can be said next, in: C.L. Paris, W.R. Swartout, W.C. Mann (Eds.), Natural Language Generation in Artificial Intelligence and Computational Linguistics, Kluwer Academic Publishers, 1991, pp. 103–124.

[24] Raveesh Meena, Kristiina Jokinen, Graham Wilcock, Integration of gestures and speech in human–robot interaction, in: Proceedings of 3rd IEEE International Conference on Cognitive Infocommunications (CogInfoCom 2012), Kosice, 2012, pp. 673–678.

[25] Viet-An Nguyen, Jordan Boyd-Graber, Philip Resnik, Deborah Cai, Jennifer Midberry, Yuanxin Wang, Modeling topic control to detect influence in conversations using nonparametric topic models, Mach. Learn. 95 (2014) 06.

[26] Graham Wilcock, WikiTalk: a spoken Wikipedia-based open-domain knowledge access system, in: Proceedings of the COLING 2012 Workshop on Question Answering for Complex Domains, Mumbai, 2012, pp. 57–69.

[27] Graham Wilcock, Kristiina Jokinen, Multilingual WikiTalk: Wikipedia-based talking robots that switch languages, in: Proceedings of the 16th Annual SIGdial Meeting on Discourse and Dialogue, Prague, 2015.

[28] Jui-Feng Yeh, Chen-Hsien Lee, Yi-Shiuan Tan, Liang-Chih Yu, Topic model allocation of conversational dialogue records by latent Dirichlet allocation, in: Proceedings of Signal and Information Processing Association Annual Summit and Conference (APSIPA), Istanbul, 2014, p. 1.

Chapter 2

Audio-motor integration for robot audition

Antoine Deleforge*, Alexander Schmidt†, Walter Kellermann†

*Inria Nancy - Grand Est, Villers-lès-Nancy, France †Multimedia Communications and Signal Processing, Telecommunications Laboratory, University Erlangen-Nuremberg, Erlangen, Germany

CONTENTS

2.1 INTRODUCTION

The most natural way for humans to communicate is by speech. For this reason, building robots which can interact with humans via speech is an important goal in robotics, which has received growing research interest over the past 20 years [2,28,36,37,43,44,46,56,57]. Examples of desired high-level features for such communicating robots are the ability to look towards the person they interact with [58], speech recognition in noisy, multi-source and reverberant environments [69] or *speech diarization, i.e.*, the identification of who talks to whom and when [51]. These high-level abilities can be associated to lower-level audio signal processing tasks such as sound-source localization and tracking [2,57], sound-source separation and speech enhancement [21] or dereverberation [47]. None of these tasks

Multimodal Behavior Analysis in the Wild. https://doi.org/10.1016/B978-0-12-814601-9.00012-2
Copyright © 2019 Elsevier Ltd. All rights reserved.

is specific to robot audition and they have been extensively studied over the past decades in contexts as varied as hearing aids, voice-controlled assistants in cars, smartphones or smart homes, audio signal restoration, or live music recording. But what makes robot audition fundamentally different from these fields?

A distinctive feature is that, by the definition of a robot, the microphones are mounted on a system equipped with *actuators*, *i.e.*, motors. These actuators may impact received auditory signals in two different ways:

1. They may **change the microphone positions** if the latter are mounted on a mobile part such as a humanoid robot's head.
2. They may **create acoustic noise** at the microphones, referred to as *ego-noise*.

Sensor mobility may be used as an asset by actively placing sensors in order to improve sound-source localization [9,43,48,55] or speech enhancement [4,45,67]. This is referred to as *active audition*. On the other hand, moving sensors also prevent from the use of classical audio signal processing tools that assume a static sound propagation model from sources to microphones, such as beamforming [21]. Besides, ego-noise may significantly impair auditory scene analysis, especially when microphones are placed near the actuators [36,75]. But contrary to other audio signal processing applications involving noise and movement, robots can benefit from *proprioceptors*. These sensors provide information on the current *motor state*, *i.e.*, the translational and rotational position, speed and acceleration of the robot's actuators. Can this additional modality be used to the benefit of robot audition in an unconstrained environment? Despite its potential usefulness, methods explicitly integrating it into the audio processing pipeline are scarce [20,33,60,61]. We refer to this endeavor as *audio-motor integration*.

This chapter proposes a unified view on audio-motor integration methodologies for robot audition. Section 2.2 starts by a literature overview of related work in the fields of psychophysics and robotics. Then two learning-based audio-motor integration models and their applications to real-world tasks are presented. Section 2.3 introduces a single-channel sound-source localization method based on head movements. Section 2.4 presents a general ego-noise reduction framework exploiting proprioceptors and dictionary learning. Finally, a conclusion and future perspectives on audio-motor integration for robot audition is presented in Section 2.5. An illustration of the different components of a typical auditory-motor system, their associated literature, and the parts addressed in this chapter is showed in Fig. 2.1.

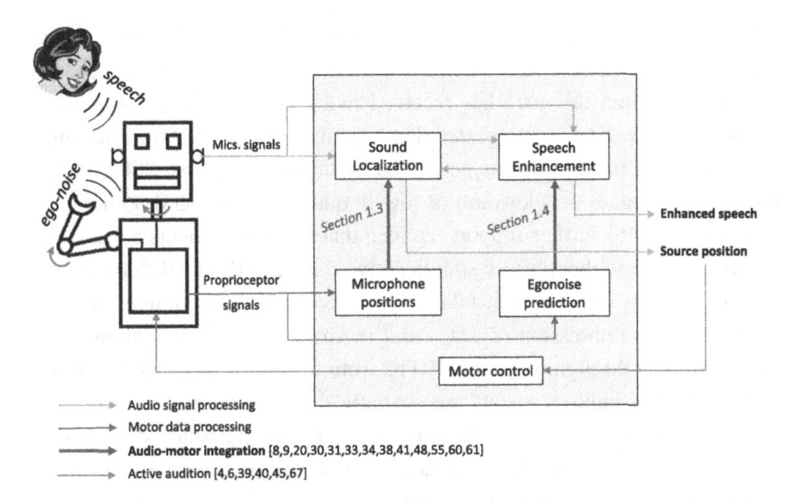

FIGURE 2.1 Illustration of an auditory-motor system whose goal is to enhance a speech source and to turn the head towards it in the presence of ego-noise.

2.2 AUDIO-MOTOR INTEGRATION IN PSYCHOPHYSICS AND ROBOTICS

The majority of studies on audio-motor integration, both in psychophysics and robotics, focus on sound-source localization. From a psychophysical point of view, it is known since the experiments of Lord Rayleigh in 1907 that humans use *binaural* cues in order to estimate the direction of a sound source [64]. Two types of binaural cues seem to play an essential role, namely *interaural level differences* (ILD) and *interaural time differences* (ITD), the frequency domain counterpart of the latter being referred to as *interaural phase differences* (IPDs). Both ILDs and IPDs are known to be subject-dependent and frequency-dependent cues. This is captured by the so-called head related transfer functions (HRTFs), determined by the shape of the head, pinna and torso. HRTFs filter the sound propagating from a source to eardrums and depend on the source direction. It is known that spatial information provided by interaural-difference cues within a restricted band of frequency is spatially ambiguous, particularly along vertical and front–back axis [42]. This suggests that humans make use of full spectral information for 2D sound-source localization [25]. This is confirmed by biological models of the auditory system hypothesizing the existence of neurons dedicated to the computation of interaural cues in specific frequency bands [74]. A large number of computational models were developed for robust sound localization and tracking based on ITD, ILD, IPD and HRTFs in the context of robot audition; see for example the recent reviews [2,57].

While the features mentioned above are generally used assuming static sensors, sound localization features could also be extracted using sensor motions, although this idea has received much less attention. Early psychophysical experiments suggested that head motions are useful for disambiguating potential confusions generated by the human pinna's filter [73], notably to estimate the elevation of low frequency sounds [53]. Other experiments [66,76] further support the idea that head movements are useful for localization, although less significantly so than ITD and ILD [42]. Interestingly, [32] suggests that taking into account head motions is useful to improve the impression of *3D sound* in virtual reality. They show that when listening through different HRTFs from his/her own, a listener often complains that auditory events are spatially diffuse, and makes incorrect judgments about the source locations. The experimental study of [32] on human subjects demonstrates that head motion can overcome HRTF mismatches, increasing perceived location accuracy.

Despite these psychophysical evidences, very few computational sound localization studies incorporate head motion. In [33], several ITD values obtained from different motor states with a two-microphone device with two degrees of freedom (translation, rotation) are used for 2D sound-source direction and distance estimation. The authors of [38,41] propose to map binaural cues to the azimuths of multiple sound sources using Gaussian mixture models [41] or deep neural networks [38] and two static head positions are used to resolve front–back ambiguities. In [9,48,55] mobile robots are used to actively collect several viewpoints of one or several emitting sound sources in order to accurately estimate their azimuths and distances. More broadly, studies tackling the problem of audio-based simultaneous localization and mapping (SLAM) have recently emerged [19,34]. Interestingly, [34] does it with a single microphone at multiple viewpoints thanks to acoustic echoes. Note that the above-mentioned studies rely on static localization techniques applied from different viewpoints. In contrast, [8] computes the average binaural cross-correlation of a continuously rotating head to estimate a source's azimuth in an anechoic scenario. To the best of the authors' knowledge, exploiting continuous sensor movement for 2D sound–sound direction estimation with a single microphone, as showed in Section 2.3 of this chapter, has not been proposed before.

Complementarily to sound-source localization, [45] introduced the idea of *active robot audition* (see Fig. 2.1) where the head of a robot is directed towards estimated source locations in order to enhance the signal of interest via, *e.g.* beamforming. This idea was recently further exploited in [4, 67] using a so-called *robomorphic array*, where microphones are placed on a robot's limbs. In [39], it is showed that turning the head of a binaural robotic system towards an estimated source direction enhances sound localization performance. From a psychophysical viewpoint, it is well known that humans use active audition to improve sound perception notably to address the so-called *cocktail party problem* [24]. Other approaches focus on the goal of orienting a system towards a sound source, without necessarily exploiting this for enhancement. This ability is referred to as phonotaxis, and it was implemented on a rat-inspired robot equipped with mobile ears in [6]. More recently, sound has been used in feedback-control loops, *i.e.*, *audio-servoing*. In [40] motor commands are used to align perceived ITDs to a target value, resulting in the robot placing itself on a line with constant ITD value.

Another category of methods consists in using the motor capabilities of a robot to help it *learn* sound-source localization techniques [5,7,12,15,27, 65], much like humans and mammals do during early stages of development. Several psychophysical studies suggest that the link between auditory features and source locations would not be hard-coded in the brain but rather learned [3,77] or re-learned [26] from experience. One example of such learning processes is the sensori-motor theory of perception, originally laid by Poincaré [54] and more recently investigated in [49]. This theory suggests that experiencing the sensory consequences of voluntary motor actions is necessary for an organism to learn the perception of space. In [3] a psychophysical sensori-motor model of sound-source localization using HRTF datasets of bats and humans was proposed. Similar ideas were implemented on robots using reinforcement learning [5], linear regression [27], locally-linear regression [12], look-up tables [15] or manifold learning [7, 65]. Interestingly, [65] uses multiple view points and diffusion kernels to learn the estimation of the azimuth angle of a white-noise source using a single microphone.

Finally, a few robotic studies use audio-motor integration for *ego-noise* reduction. The idea is to map the current motor-state of a robot to some estimates of the corresponding acoustic noise produced by the motors. For instance, the ego-noise power-spectral density can be predicted this way using an artificial neural network [31] or a nearest neighbor search [30] based on a training dataset. In [20] the same idea is applied to predict multichannel ego-noise covariance matrices using Gaussian process regression. These statistics can be used to cancel the noise via Wiener filtering, under a local stationarity assumption. In [60,61], a motor data-guided dictionary-learning framework is presented, allowing one to model noise which is non-stationary both spatially and spectrally. This framework and exemplary results on real data are presented in Section 2.4 of this chapter.

2.3 SINGLE-MICROPHONE SOUND LOCALIZATION USING HEAD MOVEMENTS

In this section, we present a method that uses motor movements to extract auditory features enabling 2D sound-source localization with a single microphone placed on an acoustic dummy head[1]. The mapping from features to direction is learned from a training dataset, building on the audio-motor, learning-based sound-source localization framework of [12].

2.3.1 HRTF model and dynamic cues

Let $s(f,t) \in \mathbb{C}$ denote the signal emitted by a static sound source and $y(f,t) \in \mathbb{C}$ denote the signal recorded by a microphone placed on an acoustic dummy head in the short-time Fourier domain, with f and t denoting frequency and time indices, respectively. We assume that the emitted signal is stationary over the considered time interval. The acoustic head acts as a rigid body causing reflections and shadowing of the source signal. The corresponding linear filter, referred to as the HRTF, is denoted $a(f,\phi,\psi)$ and depends on the frequency f and on the relative azimuth ϕ and elevation ψ of the source in the microphones' frame[2]. Note that in the case of microphone movements, ϕ and ψ will depend on time as well. We have

$$y(f,t) = a\big(f, \phi(t), \psi(t)\big) s(f,t). \tag{2.1}$$

We define the *received log-power* spectrogram by

$$p(f,t) = \log\big|y(f,t)\big|^2 = \log\big|a\big(f,\phi(t),\psi(t)\big)\big|^2 + \log\big|s(f,t)\big|^2. \tag{2.2}$$

Since the emitted signal $s(f,t)$ is assumed to be stationary, its power spectral density $\mathbb{E}_t\{|s(f,t)|^2\}$ does not depend on time, and hence its instantaneous estimate $|s(f,t)|^2$ should be independent of t on average.

Let us now consider that the microphone is mounted on a robotic head with 2 degrees of freedom: pan (azimuth rotation) and tilt (elevation rotation), such as the one showed in Fig. 2.2. Any motor command ξ will change the relative direction (ϕ,ψ) of the source in the microphone's frame. A command ξ can be identified with a function associating a time t to a motor state $\xi(t)$. This motor state can be accessed through the proprioceptors of the robot. We define a *dynamic cue* $\boldsymbol{\tau}(\xi) = \{\tau(f,\xi)\}_{f=1}^{F} \in \mathbb{R}^F$ by the *expected*

---

[1] This study extends the unpublished technical report [14].

[2] Assuming that the sound source is placed in the far field, the dependency on source distance can be neglected. As showed in [50], this typically occurs for distances > 1.8 m on binaural systems.

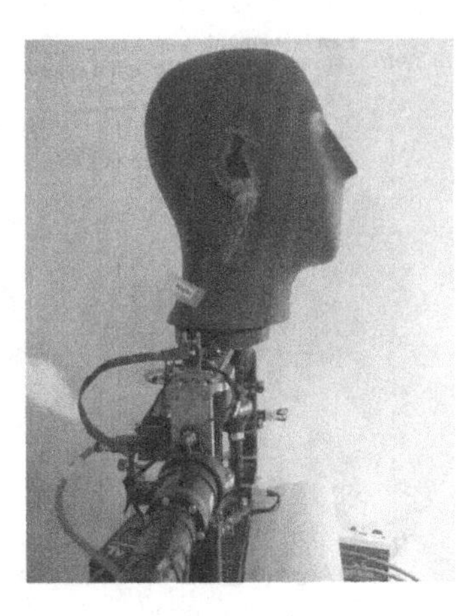

■ **FIGURE 2.2** The POPEYE setup used in the CAMIL dataset [14] and in our experiments. A binaural dummy head is mounted on a motor system with two degrees of freedom: pan (left–right) and tilt (up–down).

temporal derivatives of $p(f,t)$ at all frequencies while the robot performs command ξ:

$$\tau(f,\xi) = \mathbb{E}_t\left\{\frac{\partial p(f,t)}{\partial t}\right\}, \quad f = 1\ldots F, \tag{2.3}$$

where F is the number of frequency bands considered. As mentioned, the stationarity assumption on $s(f,t)$ implies that $\log|s(f,t)|^2$ is time-independent on average and hence its expected derivative is zero. On the other hand, the motor command ξ significantly changes the relative source directions ϕ and ψ, implying that $\log|a(f,\phi(t),\psi(t))|^2$ has a non-negligible expected derivative. Hence, $\tau(f,\xi)$ is a spatial cue that does not depend on the emitted signal and only depends on the source direction (ϕ,ψ).

In practice, $\tau(\xi)$ can be approximated at each frequency by the slope of the least-square linear regression between discrete time indexes and the received log-energies $\{p(f,t)\}_{t=1}^{T}$, where T is the number of time frames considered. The validity of this approximation relies on the fact that typical HRTFs are approximately locally-linear with respect to source directions, as demonstrated in [12]. Hence, using motor commands with constant angular velocities, $p(f,t)$ should vary approximately linearly in time, justifying the use of linear regression. The slope of the least-square linear regression is then a least mean square estimator of the expected temporal derivative, assuming zero-mean perturbations. This is illustrated in Fig. 2.3 using the CAMIL dataset version 0.1[3] [14]. This dataset was recorded with the setup

[3] Data available at https://team.inria.fr/perception/the-camil-dataset/.

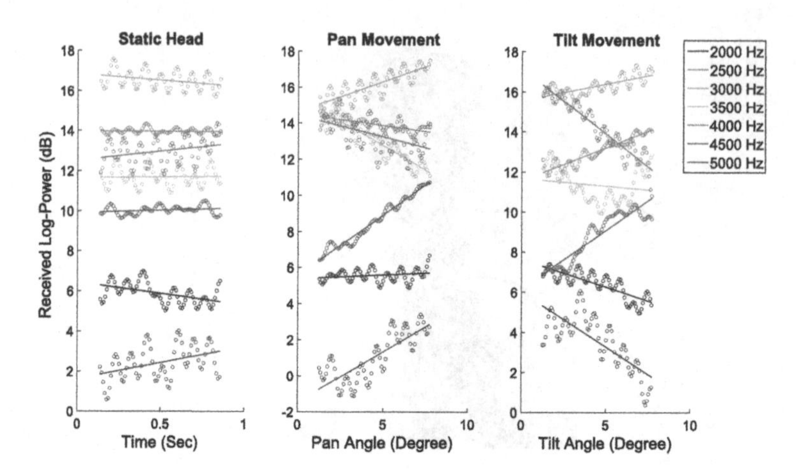

FIGURE 2.3 These figures represent the received log-power (2.2) by a single acoustic dummy-head microphone as a function of time (or head-angle) at different frequencies. The least-square linear regression of each curve is represented by a solid line. The emitted signal is an 800 ms random mixture of 600 sine waves with frequencies ranging from 50 Hz to 6000 Hz. Left: the head is static. Middle: the head performs a pan movement at constant 9°/s velocity. Right: the head performs a tilt movement at constant 9°/s velocity.

of Fig. 2.2. As can be seen, the stationarity of the emitted signal implies that the received log-power is roughly constant over time when the head is static (left). On the other hand, head movements induce near linear variations of the log-power. The expected derivative of these variations can be approximated by the slope of their least-square linear regression (solid lines). The sampling frequency of the signal was 48 kHz and the sliding short-time Fourier Hamming window was set to 200 ms with 95% overlap. This resulted in 8193 discrete positive frequencies between 0 and 24 kHz and $T = 101$ time frames per second of signal. In practice, only cues corresponding to frequencies between 1500 and 6000 Hz were kept, as they showed to be the most useful for localization. This resulted in $F = 1537$ frequency bands used in the proposed dynamic cues.

2.3.2 Learning-based sound localization

Once dynamic cues $\boldsymbol{\tau}(\boldsymbol{\xi})$ are computed, they need to be mapped to a corresponding source direction. While computationally demanding physical models exist to generate HRTF filters based on source directions and accurate 3D head models [78], we are interested here in the other way around mapping, *i.e.*, from dynamic cues to source directions. This cannot be easily obtained in practice, in particular when head movements are considered. Due to the non-feasibility of fully modeling the physics of sound propagation in realistic settings, an alternative approach has recently emerged and is referred to as *supervised* or *learning-based* sound-source localization [12,

16,65]. These methods bypass the use of an explicit, approximate physical model by directly learning a mapping from audio features to spatial properties using an appropriate training dataset.

The CAMIL dataset version 0.1 [14] consists of recordings made with a binaural dummy head (Fig. 2.2) in the presence of a sound source (loudspeaker) placed at 16,200 annotated relative directions in the microphones' frame. The source is placed 2.7 m from the receiver in all recordings, and the reverberation time of the room is around 400 ms. For each direction, three recordings of 1 s each are available: (i) no head movement, (ii) a 9° pan movement rightwards at constant speed, and (iii) a 9° tilt movement downwards at constant speed. The emitted source signals are designed to be approximately stationary and correspond to random linear combinations of 600 sine waves with frequencies ranging from 50 Hz to 6000 Hz and random phase offsets. During motor commands, the annotated relative source position corresponds to the one half-way through the movement. For the single microphone experiments, only the left microphone channel is used.

The training dataset is composed of N pairs $\{\boldsymbol{\tau}_n(\boldsymbol{\xi}), z_n\}_{n=1}^N \subseteq \mathbb{R}^F \times \mathbb{R}^2$ where $z_n = (\phi_n, \psi_n)$ is the nth source direction and $\boldsymbol{\xi}$ is a fixed command throughout the dataset. Three motor commands are considered: *pan*, *tilt* or the successive combination of both, each at constant angular velocity (9°/s). Only source directions corresponding to azimuth angles between −90° and +90° and elevation angles between −45° and +45° are kept, resulting in 3812 directions out of which $N = 2859$ were kept for training and the 953 others for testing.

This training set must be used so that given a new test observation $\tilde{\boldsymbol{\tau}}(\boldsymbol{\xi})$, an associated source direction $\tilde{z}$ can be estimated. To achieve this, we use the high- to low-dimensional regression method Gaussian locally-linear mapping (GLLiM[4] [17]) proposed in [13]. GLLiM is a probabilistic method that estimates Q local affine transformations from a low-dimensional space (here, the space of source directions) to a high-dimensional space (here, the space of dynamic cues) using a Gaussian mixture model. This mapping is then reversed through Bayesian inversion, yielding an efficient estimator of $\tilde{z}$ given $\tilde{\boldsymbol{\tau}}(\boldsymbol{\xi})$. GLLiM was notably successfully applied to supervised binaural sound-source localization using either real [12,16] or simulated [22] training sets. Here, a fixed value $Q = 50$, diagonal and equal noise covariance matrices and equal mixture weights are used in all experiments (see [13] for details of the GLLiM method).

[4]The code for this method is available at https://team.inria.fr/perception/gllim_toolbox/.

Table 2.1 Azimuth and elevation estimation errors on the testing set using a single microphone and different motor commands. Results are presented in the form Avg±Std (Out %) where Avg±Std denote the mean and standard deviations of *inlying* absolute angular errors and Out denotes the percentage of *outliers*. Outlying errors are defined as those larger than 30°

Command	Pan	Tilt	Pan + Tilt
Azimuth (°)	3.00 ± 3.3 (0.5%)	4.02 ± 3.5 (0.1%)	3.26 ± 3.2 (0.0%)
Elevation (°)	2.50 ± 2.8 (0.3%)	1.62 ± 1.3 (0.2%)	1.71 ± 1.6 (0.0%)

Table 2.2 Azimuth and elevation estimation errors on the testing set using one static microphone and dynamic cues or two static microphones and binaural (ILPD) cues. Results are presented in the same form as Table 2.1

Method	Dynamic cues (1 mic.)	ILD + IPD [12] (2 mic.)
Azimuth (°)	14.0 ± 8.7 (65%)	2.04 ± 1.6 (0.0%)
Elevation (°)	13.5 ± 8.5 (40%)	1.26 ± 1.0 (0.0%)

2.3.3 **Results**

The proposed method was trained on a random subset of $N = 2859$ cue-to-direction pairs and tested on the remaining 953 dynamic cues. This was done so that the emitted training sounds are all distinct from the emitted test sounds and the training source directions are distinct from the test source directions. Results are shown in Table 2.1. As can be seen, audio-motor integration and small head movements enable the localization of a sound source in 2D with high precision ($< 4°$ with combined commands) using a *single microphone*. This is impossible with any of the existing sound-source localization methods in the literature, to the best of the authors' knowledge. For comparison, results obtained with a static head and dynamic cues (one microphone) or binaural cues (ILD and IPD, two microphones, as in [12]) are showed in Table 2.2. Unsurprisingly, using dynamic cues with a static head yields localization results similar to randomness, with over 50% of outliers. This is because the absence of movement removes spatial information from $\boldsymbol{\tau}(\boldsymbol{\xi})$, bringing (2.3) close to 0. On the other hand, using traditional binaural cues with two static microphones, as done in, *e.g.*, [12], yields comparable results to the ones obtained with the proposed dynamic cues and a single moving microphone. This validates the feasibility of single-microphone sound-source localization by audio-motor integration.

An important limit of this approach is that it requires a carefully annotated training dataset, which is likely to be room- and system-dependent. Besides, the method strongly relies on the assumption that emitted signals are approximately stationary during the emission period (≈ 0.8 s in our experiments). Robust extensions of this method that discard time–frequency points with low energy would be an interesting route to investigate. Besides engineering applicability, these results are the first ones to corroborate psychophysical evidence suggesting that small continuous head movements may help localizing sounds [53,66,73,76], using an artificial system.

2.4 EGO-NOISE REDUCTION USING PROPRIOCEPTORS

This section presents a general audio-motor integration framework for ego-noise reduction, initially presented in [60] and extended in [61]. This framework enables the modeling of noise signals that are non-stationary both spatially and spectrally by building on a phase-optimized dictionary-learning method[5] [17], which is also outlined here for completeness.

2.4.1 Ego-noise: challenges and opportunities

When a robot is moving, its rotating joints as well as the moving parts of its body cause significant noise which is referred to as ego-noise. This corrupts recordings and therefore degrades performance of, *e.g.*, a speech recognizer. To relieve this problem, a suitable noise reduction mechanism is required. This task is particularly challenging because the noise involved is often louder than the signals of interest. Moreover, it is highly non-stationary as the robot performs different movements with varying speeds and accelerations. Furthermore, ego-noise cannot be modeled as a single static point interferer as the joints are located all over the body of the robot.

All this discourages the use of traditional statistical noise reduction techniques such as (multichannel) Wiener filtering or beam forming [21,36]. On the bright side, however, two robot-specific opportunities may be exploited. First, ego-noise will usually be strongly structured both spatially and spectrally (*e.g.*, Fig. 2.4, bottom-right) because it is produced by an automated system restricted to a limited number of degrees of freedom. Second, important extra information may be exploited in addition to the audio signals, namely, the instantaneous motor state of the robot, *e.g.*, the joints' angles and angular velocities collected by proprioceptors (*e.g.*, Fig. 2.4, top-right).

2.4.2 Proprioceptor-guided dictionary learning

The existence of a strong spectral structure in ego-noise motivates the use of dictionary learning methods. The idea is to express structure in terms of

[5] The code for this method is available at https://robot-ears.eu/po_ksvd/.

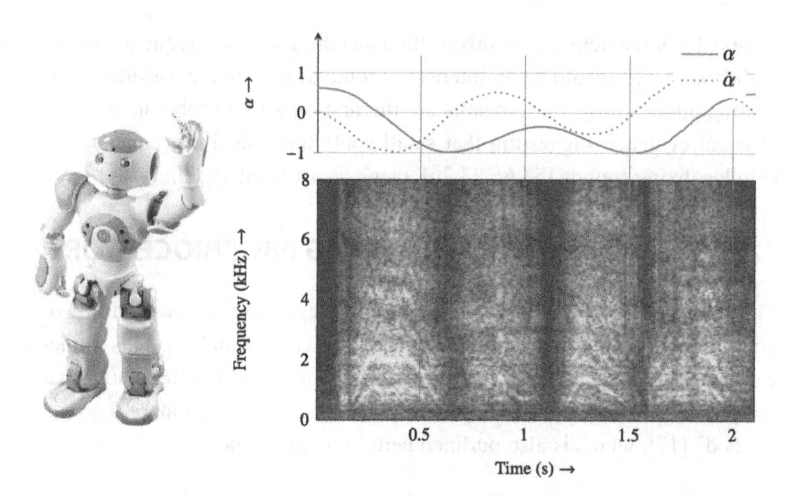

■ **FIGURE 2.4** Left: the robot NAO. Bottom-right: ego-noise spectrogram of NAO waving the left arm, showing distinctive spectral structures. Note that stationary fan noise components have been removed by a multichannel Wiener filter [36]. Top-right: corresponding motor data, *i.e.*, the angle α and first derivative $\dot{\alpha}$ of the left shoulder pitch joint involved in the movement (both normalized).

sparsity in a particular basis. More precisely, if $\mathbf{Y} = [\boldsymbol{y}_1, \ldots, \boldsymbol{y}_T] \in \mathbb{C}^{P \times T}$ represents T examples of P-dimensional signals, there must exist a set of K atoms or a *dictionary* $\mathbf{D} = [\boldsymbol{d}_1, \ldots, \boldsymbol{d}_K] \in \mathbb{C}^{P \times K}$ such that each signal is a linear combination of only a few atoms, *i.e.*, $\mathbf{Y} \approx \mathbf{DX}$ where $\mathbf{X}$ has sparse columns. Estimating $\mathbf{D}$ and $\mathbf{X}$ from $\mathbf{Y}$ is a sparse instance of matrix factorization. In audio signal processing, it is natural to seek such a factorization in the non-negative power spectral density (PSD) domain, since the magnitude spectra of natural sounds such as speech often feature redundancy and sparsity. This approach gave rise to a large number of methods for audio signal representations and extractions within the framework of *non-negative matrix factorization* (NMF) [63,72]. Extensions of NMF to complex-valued multichannel spectrograms have later been proposed [52,59]. Most of these extensions assume a simple spatial structure: the modeled signal consists in a mixture of a few fixed, point sources, *i.e.*, with constant steering vectors. Such models are not appropriate for ego-noise which features complex spatial structures. In [17], a model including both complex spectral and spatial structures was proposed, via *phase-optimized* dictionary learning. This model is summarized in Section 2.4.3 of this chapter.

The general concept of dictionary-based noise reduction comprises the following steps. First, a noise dictionary $\mathbf{D}_{\text{noise}}$ is learned from a set of noise-only examples $\mathbf{Y}_{\text{noise}}$ via the factorization $\mathbf{Y}_{\text{noise}} \approx \mathbf{D}_{\text{noise}}\mathbf{X}_{\text{noise}}$ with $\mathbf{X}_{\text{noise}}$ being maximally sparse. Then, given a test observation $\tilde{\boldsymbol{y}} \in \mathbb{C}^P$ containing both noise and target signals, its noisy part is estimated by finding $\tilde{x} \in \mathbb{C}^K$ such that $\tilde{y} \approx \mathbf{D}_{\text{noise}}\tilde{x}$. The enhanced target signal is then given by $\tilde{y} - \mathbf{D}_{\text{noise}}\tilde{x}$.

Besides methods purely based on the structure of audio signals, another approach stipulates exploiting available proprioceptor, *i.e.*, motor data, and map them to a noise model. In [31], the time-varying noise power spectral density (PSD) is estimated by a deep neural network (DNN). The DNN is fed by motor data, which incorporates not only current, but also past sensor values. In [30], PSD noise templates are used for spectral subtraction. In the learning step, only ego-noise is present and each point in the motor data space is associated with a certain spectral noise template. In the testing (or working) phase, the approach uses a nearest-neighbor criterion to find the best matching template, which is then subtracted from the magnitude spectrum of the recording. In [20], multichannel noise covariance matrices are predicted from motor states via Gaussian process regression.

In [60], we introduced an alternative and generic audio-motor integration framework to fuse the information brought by an advanced learned structured audio model on the one hand, and instantaneous motor data on the other hand. The key idea is to replace the computationally costly search in the dictionary, which is intractable for large dictionaries (NP-hard [11]), by a classification procedure guided by current proprioceptor data $\boldsymbol{\alpha}$. These data are fed into support vector machines (SVMs) [62] to efficiently find suitable entries in the ego-noise dictionary $\mathbf{D}_{\text{noise}}$. We showed that this approach reduces computational complexity while simultaneously improving performance. This approach and some results are outlined in Section 2.4.4.

2.4.3 Phase-optimized dictionary learning

Although ego-noise is highly non-stationary, it has distinctive spectral and spatial characteristics. The basic idea of a dictionary representation is to capture such characteristics by a collection of prototype signals, called atoms, collected in a dictionary. In our case, the structured ego-noise signal should be represented by a linear combination of a few atoms at each time frame. If these atoms are specifically designed to represent signals sharing spectral and spatial characteristic of ego-noise only, subtracting these atoms should remove the noise while preserving the residual signal of interest such as speech. We briefly summarize here the recent approach [17] that automatically learns a multichannel dictionary capturing both spatial and spectral characteristics of a training signal. In the following we represent a multichannel signal in the spectral domain by concatenation of the M channels per frequency bin, giving a signal vector of dimension $P = MF$, where F represents the number of frequency bins per channel. Then, we denote the dictionary by $\mathbf{D} = [\boldsymbol{d}_1, \ldots, \boldsymbol{d}_K] \in \mathbb{C}^{P \times K}$ containing K atoms $\boldsymbol{d}_k \in \mathbb{C}^P$. Moreover, the dictionary is corrected by a time-varying phase matrix $\boldsymbol{\Phi}_t \in \mathbb{C}^{F \times T}$ at each time frame t, where each element has unit complex modulus. The phase-corrected dictionary is then given by

$$\mathbf{D}\{\boldsymbol{\Phi}_t\} := \begin{pmatrix} \boldsymbol{d}_{1,1} & \ldots & \boldsymbol{d}_{1,K} \\ \boldsymbol{d}_{2,1} & \ldots & \ldots \\ \vdots & \ddots & \vdots \\ \boldsymbol{d}_{F,1} & \ldots & \boldsymbol{d}_{F,K} \end{pmatrix} \odot \begin{pmatrix} \phi_{1,1,t} & \ldots & \phi_{1,K,t} \\ \phi_{2,1,t} & \ldots & \ldots \\ \vdots & \ddots & \vdots \\ \phi_{F,1,t} & \ldots & \phi_{F,K,t} \end{pmatrix} \tag{2.4}$$

where each element $\boldsymbol{d}_{f,k} \in \mathbb{C}^M$ captures the spectral value of atom k at frequency bin f as well as the relative phases and gains between the M channels. Here, $\odot$ denotes a modified Hadamard product, where each vector $\boldsymbol{d}_{f,k} \in \mathbb{C}^M$ in the matrix $\mathbf{D}$ is multiplied by a global phase term $\phi_{f,k,t} \in \mathbb{C}$ in $\boldsymbol{\Phi}_t$. A given multichannel spectrogram frame $\boldsymbol{y}_t$ should then be approximated by $\boldsymbol{y}_t \approx \mathbf{D}\{\boldsymbol{\Phi}_t\}\boldsymbol{x}_t$, where the vector $\boldsymbol{x}_t \in \mathbb{C}^K$ picks the atoms from the dictionary. Since only a few atoms should be used, $\boldsymbol{x}_t$ is constrained to be sparse, *i.e.*, it should contain at most $S_{\max}$ nonzero elements, where $S_{\max}$ is referred to as the *sparsity level*. The overall problem can then be written as

$$\begin{aligned} &\underset{\mathbf{D},\boldsymbol{\Phi},\mathbf{X}}{\operatorname{argmin}} \sum_{t=1}^{T} \|\boldsymbol{y}_t - \mathbf{D}\{\boldsymbol{\Phi}_t\}\boldsymbol{x}_t\|_2^2 \quad \text{subject to} \\ &\|\boldsymbol{x}_t\|_0 \le S_{\max},\ x_{kt} \ge 0 \text{ and } |\phi_{ft,k}|^2 = 1 \quad \forall\ f,k,t. \end{aligned} \tag{2.5}$$

Here, $\|\cdot\|_2$ and $\|\cdot\|_0$ denote the ℓ_2 and ℓ_0 norm, respectively. The latter counts the number of nonzero elements in $\boldsymbol{x}_t$. The minimization (2.5) is done with respect to (w.r.t.) different arguments, depending on which stage of the algorithm is considered. The training and testing stages are outlined in the following (see [17] for details).

- **Training:** (2.5) is minimized w.r.t. $\mathbf{D}$, $\boldsymbol{x}_t$ and $\boldsymbol{\Phi}_t$. In the training stage, $\mathbf{D}_{\text{noise}}$ should be learned using a set of training examples $\mathbf{Y}_{\text{noise}}$. For this, [17] proposes the phase-optimized K-SVD (PO-KSVD) algorithm. It can be viewed as a phase-optimized complex extension of the popular dictionary learning method K-SVD [1]. It alternates between a sparse coding step and a dictionary update step.
- **Testing:** (2.5) is minimized w.r.t. $\boldsymbol{x}_t$ and $\boldsymbol{\Phi}_t$ for each new test observation $\tilde{\boldsymbol{y}}_t$ while the pre-trained dictionary $\mathbf{D}_{\text{noise}}$ is fixed. The best fitting entries from dictionary $\mathbf{D}_{\text{noise}}$ are searched and subtracted from $\tilde{\boldsymbol{y}}_t$, which may contain ego-noise and speech, for example. The NP-hard problem of finding the best combination of atoms is done using an extension of orthogonal matching pursuit (OMP, after [68]) due to

its empirically good performance. The extension is called PO-OMP (for phase-optimized OMP).

2.4.4 Audio-motor integration via support vector machines

While the knowledge of the robot's instantaneous motor state should intuitively be beneficial for ego-noise reduction, a crucial question is at which stage motor data should be included. As described in the previous section, one of the main bottle-necks of the multichannel dictionary method in [17] is the testing phase, where an NP-hard sparse coding problem is approximately solved by the costly iterative PO-OMP procedure. Hence, we propose to replace the entire testing stage by a novel and more efficient motor-guided atom selection method while keeping the dictionary learning stage of [17], for which computational time is less of an issue.

The physical state of a 1-dimensional robot joint can be described by its position in terms of an angle α_t at a given time stamp τ_t. Furthermore, from successive angle stamps we can calculate a discrete-time approximation of its first angle derivatives, *i.e.*, the angle speed

$$\dot{\alpha}_t = \frac{\alpha_t - \alpha_{t-1}}{\tau_t - \tau_{t-1}}.$$

For each joint, we collect the recorded and calculated angle data in a feature vector $\boldsymbol{\alpha}_t = (\alpha_t, \dot{\alpha}_t)$, which is, a bit loosely, referred to as motor data in the following. Note that using the joints' acceleration is also an option, but this is disregarded here as it did not prove useful in our experiments. Assume a spectrum $\boldsymbol{y}_t$ is observed at a time frame τ_t in which ego-noise alone is present. Additionally, the motor data $\boldsymbol{\alpha}_t$ for this frame is available. The proposed concept stipulates the association of this motor data point with those atoms which PO-OMP would select in a pre-trained ego-noise dictionary to represent spectrum $\boldsymbol{y}_t$. Note that the order in which atoms are chosen is unimportant. For example, if at time step t PO-OMP selects atoms 8, 5 and 7 in the dictionary and at time step $t+1$ the output is 7, 8 and 5, both motor data samples $\boldsymbol{\alpha}_t$ and $\boldsymbol{\alpha}_{t+1}$ are associated with the set of atom $\{8, 7, 5\}$. The curly brackets $\{\cdot\}$ emphasize that the order of selected atoms is not considered. The number of possible atom combinations for any given atom set is then

$$N_C = \binom{K}{S_{\max}}. \tag{2.6}$$

In the following, the N_C possible atom sets are denoted as $\hat{\mathbf{D}}_q \in \mathbb{C}^{P \times S_{\max}}$, $q = 1, \ldots, N_C$, where the choice of notation emphasizes that each $\hat{\mathbf{D}}_q$ con-

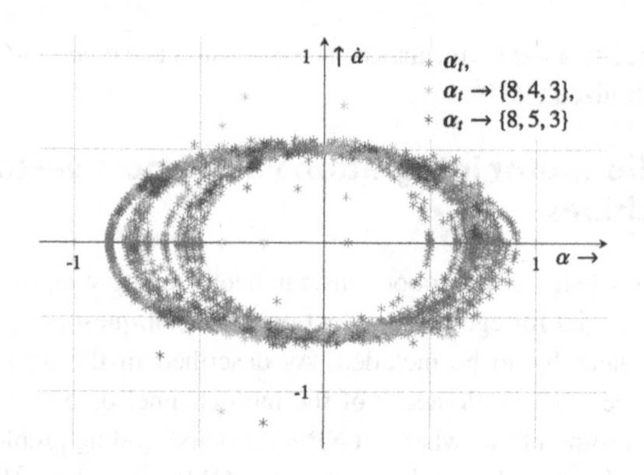

FIGURE 2.5 Motor data points $\boldsymbol{\alpha}_t = (\alpha_t, \dot{\alpha}_t)$ for a right shoulder pitch movement. Points highlighted in blue and red are those associated with a certain set of selected atoms in the dictionary. They appear to form clusters.

tains a selection of atoms from **D**. Our plan in the following is to decide on a set $\hat{\mathbf{D}}_q$ based on a motor data sample $\boldsymbol{\alpha}_t$ and use all atoms in $\hat{\mathbf{D}}_q$ for ego-noise suppression.

Fig. 2.5 shows some motor data points in the $(\alpha, \dot{\alpha})$ plane which are associated to a certain set of atoms as an example. They appear to form clusters. The non-linearity of the clusters' contours motivates the use of a kernel method. It is reasonable to classify these points using a classifier $\mathcal{C}_q(\boldsymbol{\alpha}) \in \{-1, +1\}$ deciding if a new incoming motor data point $\boldsymbol{\alpha}$ falls into the clustering area of atom set q. If yes, $\mathcal{C}_q(\boldsymbol{\alpha}) = 1$, if not $\mathcal{C}_q(\boldsymbol{\alpha}) = -1$ holds. All in all, a number of N_C such classifiers must be trained. For notational convenience, they are represented in vector form:

$$\mathcal{C}(\boldsymbol{\alpha}) = \left[\mathcal{C}_1(\boldsymbol{\alpha}), \ldots, \mathcal{C}_{N_C}(\boldsymbol{\alpha})\right]. \tag{2.7}$$

We propose to model the data points to cluster as following an unknown probability density function (pdf). By estimating its support, the above described clustering problem is solved. To do so, we use a method from the broad range of support vector machines (SVM). The 1-Class-SVM [62] is a method based on Gaussian kernels that estimates a classifier $\mathcal{C}(\cdot)$ whose decision boundaries can be shown to be the support of a pdf that generated the training data with high probability. The reader is referred to [62] for more details on this classification method.

Note that the decision regions of trained classifiers can partly overlap. Formally, an ambiguity is given if for an input motor data vector α_t more than two classifiers return $+1$. To handle this, all N_C classifiers are associated with a weighting factor $w_1, \ldots, w_{N_C}$, being identical to the number of in-

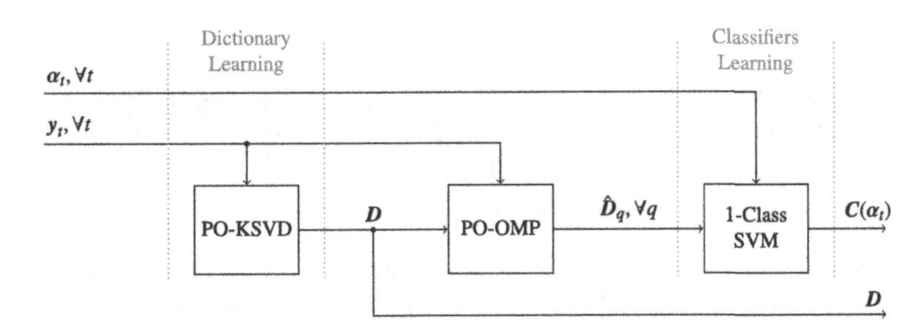

■ **FIGURE 2.6** Illustration of the training phase of the proposed audio-motor integration framework for ego-noise reduction, using training samples $\boldsymbol{y}_t$, $t = 1, \ldots, T$, and associated motor data $\boldsymbol{\alpha}_t$, $t = 1, \ldots, T$.

volved data points in each of the N_C trainings. By this, a decision region gets a larger weight when it contains more data points. Each of the K entries of the dictionary **D** gets a counter, initialized with zero. It is then iterated over the K atoms: the counter is increased by w_q if $\mathcal{C}_q(\alpha_t) = 1$ and has the currently investigated atom in its recommendation. The final decision is then given by choosing those atoms that have the $S_{\max}$ largest weights.

To summarize, the proposed proprioceptor-guided multichannel dictionary methodology uses the following steps:

- **Training:** The input consists of spectrogram frame samples $\boldsymbol{y}_t, t = 1, \ldots, T$ containing ego-noise only. Each sample is associated with a motor data vector α_t, $t = 1, \ldots, T$. After **D** is learned using PO-KSVD [17] (recall that we do not use motor data for this), PO-OMP is performed with the same samples $\boldsymbol{y}_t$ as input. The selected atoms per sample and their associated motor vector are then processed in the second training step which learns the 1-Class SVMs. This gives N_C classifiers, as defined by (2.7). Each of the classifiers is associated to one specific set of atoms from **D**. Fig. 2.6 gives a schematic overview of the training phase.
- **Testing:** The input consists of a new incoming noisy observation $\boldsymbol{y}_t$ containing both ego-noise and a target signal to denoise, and a corresponding motor data sample $\boldsymbol{\alpha}_t$. The latter is used to decide immediately on a set of atoms $\hat{\mathbf{D}}_q, q = 1, \ldots, N_C$ using the trained classifiers. The iterative search in the dictionary is unnecessary, so that the proposed algorithm can be expected to be of significantly lower complexity than PO-OMP without motor data. What remains is only the calculation of the gains for all entries in $\hat{\mathbf{D}}_q$, collected in vector form $\hat{\boldsymbol{x}}_t$ and the phase optimization, resulting in the phase matrix $\hat{\boldsymbol{\Phi}}_t$. Determining those unknowns corresponds to the very last step of PO-OMP [17], when all

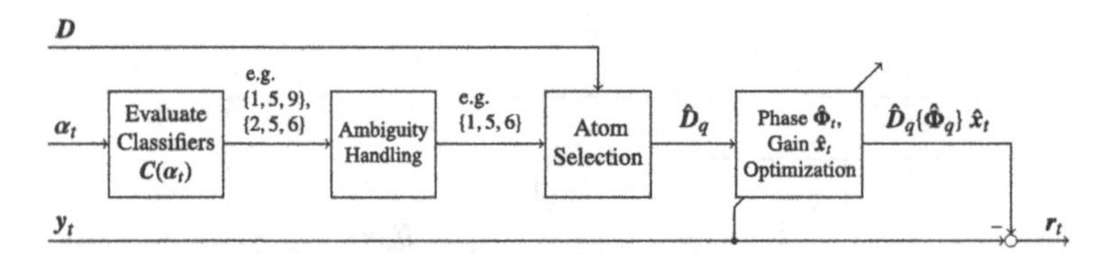

■ **FIGURE 2.7** Illustration of the test-phase of the proposed audio-motor integration framework for ego-noise reduction. Atoms $\hat{\mathbf{D}}_q$ are selected from $\mathbf{D}$ on the basis of motor data $\boldsymbol{\alpha}_t$ only. The incoming audio data sample $\mathbf{y}_t$ is a mixture of ego-noise and a target signal.

atoms have been selected. Fig. 2.7 gives a schematic overview of the testing phase.

2.4.5 **Results**

We present an experiment performed with the robot NAO [23] (see Fig. 2.4, left). NAO has four microphones which are all located in the head. Furthermore, the robot has 26 joints, 2 in the head, 12 in the arms, 12 in the legs. We perform exclusively movements of the right arm that involve 6 joints. This gives a feature vector of dimension 12, $\boldsymbol{\alpha} \in \mathbb{R}^{12}$. The sampling frequency for all recordings is $Fs = 16$ kHz, the short-time Fourier transform (STFT) domain uses a Hamming window of length 64 ms and an overlap of 50%. NAO performed its movements in a room with moderate reverberation ($T60 = 200$ ms). Each dictionary used in the following was trained with 30 s of recording. The stationary noise from a cooling fan was removed before the training started. For this, we employed a speech distortion weighted multichannel Wiener filter (MWF) [21,36]. It needs the power spectral density matrix of pure fan noise as input, which can be easily estimated for constant rotation speed of the fan when the robot is not moving. For testing, 200 utterances from the GRID corpus [10] were recorded with the fan switched off. The loudspeaker was positioned at a 1 m distance of NAO, at a height of 1.5 m. The recorded utterances were added to out-of-training movement noise. These mixtures were then used to evaluate the ego-noise suppression algorithms described above after applying the MWF to suppress the fan noise. The classifiers were trained on 2800 motor data samples in total. To find the best parameter ν and γ, we started a sweep over different settings for both variables. Each setting was cross-checked on a set of data points which was excluded from the training. The overall performance of the ego-noise suppression is measured in terms of signal-to-inference ratio (SIR in dB) and signal-to-distortion ratio (SDR, in dB), as defined in [71]. While SIR measures the overall noise cancellation, SDR also incorporates information about how much speech is distorted by the suppression algorithm. Additionally, we measure the keyword speech recognition rate (RR),

Table 2.3 Comparison of the proposed audio-motor integrated ego-noise reduction method with two baselines and unprocessed signals, using different metrics

	SIR [dB]	SDR [dB]	RR [%]
Proposed [60]	**14.71**	**2.64**	**73.0**
PO-OMP [17]	14.46	2.57	71.8
NMF [35]	2.51	0.8	45.2
Unprocessed	−5.48	−8.15	36.1

using pocketsphinx [29] on the GRID corpus [10], as defined by the CHiME challenge [70].

We tested different parameter constellations for K and $S_{\max}$. The best results were obtained for a dictionary size of $K = 20$ with sparsity level $S_{\max} = 3$ for this scenario. We parametrized the SVM with a sparsity regularizer $\nu = 2^2$ and a Gaussian kernel width $\gamma = 2^{-2}$ (see [62] for details of these parameters).

Table 2.3 compares the results obtained with the proposed audio-motor integration method compared to the results obtained with PO-OMP (no proprioceptor data is used). Both the audio-motor and PO-OMP approach clearly outperform the unprocessed recordings in all metrics used. For comparison, we also give suppression results of one-channel NMF [35]. Although NMF brings about an improvement, best results are obtained using PO-OMP and the audio-motor approach. The latter clearly reproduces results of PO-OMP and slightly even outperforms it. This can be explained by the fact that PO-OMP sometimes wrongly estimates atoms due to the presence of speech (recall that PO-OMP uses audio data only). As expected, the needed calculation time in Matlab for the classifiers approach is approximately 30% below that of PO-OMP as the search in the dictionary is unnecessary. Note that the theoretical number of possible atom sequences and therefore classifiers is given by (2.6), i.e., $\binom{20}{3} = 1140$ in our case. Interestingly, only 252 classifiers had to be trained in the given case as only 252 atom sets appeared. Therefore, (2.6) is indeed only an upper bound. Nevertheless, we noted that the computational bottleneck of this approach remains the non-convex estimation of the phase-corrections matrix $\mathbf{\Phi}_t$. This difficult and general problem in audio, referred to as *phase unmixing*, is the subject of current research [18].

2.5 CONCLUSION AND PERSPECTIVES

This chapter attempted to present a unified perspective on audio-motor integration with the key question in mind: *How can audio and motor modalities be combined to enhance robot audition?* After summarizing the challenges and opportunities specific to robot audition, we surveyed the literature on audio-motor integration, from both psychophysics and robotics viewpoints. We then presented two recent examples of learning-based audio-motor integration framework for robotics. In the first one, addressing the fundamental problem of acoustic source localization, the expected derivative of received audio signal log-power in one microphone with respect to motor-state during a motor command is used to derive spatial features. We showed how these features enable 2D sound direction estimation with a single-microphone, a task impossible to achieve without audio-motor integration. In the second example, addressing the generic problem of ego-noise reduction for acoustic signal enhancement, we presented a general audio-motor integration framework to fuse the information brought about by a dictionary-based structured audio model on the one hand, and instantaneous motor data on the other hand. We showed that using motor data reduced the computational complexity while improving performance.

Despite psychophysical evidence pointing out the usefulness of audio-motor integration with humans, only few computational and engineering studies exploit this opportunity. However, with the increasing prominence of robots in our daily lives, and with their fast-developing capability to naturally interact with humans via voice, it is likely that research in computational audio-motor integration will keep gaining momentum. A number of crucial questions will need to be answered along this development:

- How could one step out of the classical static-source, static-microphone signal processing framework and, in particular, how to model fast and complex motions?
- What is the optimum auditory representation that can be predicted from motor states? Which motor-space features are best suited for audio prediction?
- How could one *close the sensori-motor loop* by deriving long- and short-term optimal actions that ease the enhancement of recorded audio signals?
- Which information may be extracted from dynamic audio inputs for simultaneous localization and mapping, and how to extract them?

We hope that this chapter helps to direct further research interest towards these exciting challenges with great practical relevance on the horizon.

REFERENCES

[1] Michal Aharon, Michael Elad, Alfred Bruckstein, K-SVD: an algorithm for designing overcomplete dictionaries for sparse representation, IEEE Trans. Signal Process. 54 (11) (2006) 4311–4322.

[2] Sylvain Argentieri, Patrick Danes, Philippe Souères, A survey on sound source localization in robotics: from binaural to array processing methods, Comput. Speech Lang. 34 (1) (2015) 87–112.
[3] Murat Aytekin, Cynthia F. Moss, Jonathan Z. Simon, A sensorimotor approach to sound localization, Neural Comput. 20 (3) (2008) 603–635.
[4] Hendrik Barfuss, Walter Kellermann, An adaptive microphone array topology for target signal extraction with humanoid robots, in: 2014 14th International Workshop on Acoustic Signal Enhancement (IWAENC), IEEE, 2014, pp. 16–20.
[5] Erik Berglund, Joaquin Sitte, Sound source localisation through active audition, in: 2005 IEEE/RSJ International Conference on Intelligent Robots and Systems (IROS 2005), IEEE, 2005, pp. 653–658.
[6] Mathieu Bernard, Steve N'Guyen, Patrick Pirim, Bruno Gas, Jean-Arcady Meyer, Phonotaxis behavior in the artificial rat Psikharpax, in: International Symposium on Robotics and Intelligent Sensors, IRIS2010, Nagoya, Japan, 2010, pp. 118–122.
[7] Mathieu Bernard, Patrick Pirim, Alain de Cheveigné, Bruno Gas, Sensorimotor learning of sound localization from an auditory evoked behavior, in: 2012 IEEE International Conference on Robotics and Automation (ICRA), IEEE, 2012, pp. 91–96.
[8] Jonas Braasch, Samuel Clapp, Anthony Parks, Torben Pastore, Ning Xiang, A binaural model that analyses acoustic spaces and stereophonic reproduction systems by utilizing head rotations, in: The Technology of Binaural Listening, Springer, 2013, pp. 201–223.
[9] Gabriel Bustamante, Patrick Danés, Thomas Forgue, Ariel Podlubne, Towards information-based feedback control for binaural active localization, in: 2016 IEEE International Conference on Acoustics, Speech and Signal Processing (ICASSP), IEEE, 2016, pp. 6325–6329.
[10] Martin Cooke, Jon Barker, Stuart Cunningham, Xu Shao, An audio-visual corpus for speech perception and automatic speech recognition, J. Acoust. Soc. Am. 120 (5) (2006) 2421–2424.
[11] Geoff Davis, Stephane Mallat, Marco Avellaneda, Adaptive greedy approximations, Constr. Approx. 13 (1) (1997) 57–98.
[12] Antoine Deleforge, Florence Forbes, Radu Horaud, Acoustic space learning for sound-source separation and localization on binaural manifolds, Int. J. Neural Syst. 25 (01) (2015) 1440003.
[13] Antoine Deleforge, Florence Forbes, Radu Horaud, High-dimensional regression with gaussian mixtures and partially-latent response variables, Stat. Comput. 25 (5) (2015) 893–911.
[14] Antoine Deleforge, Radu Horaud, Learning the Direction of a Sound Source Using Head Motions and Spectral Features, Technical report, INRIA, 2011.
[15] Antoine Deleforge, Radu Horaud, The cocktail party robot: sound source separation and localisation with an active binaural head, in: Proceedings of the Seventh Annual ACM/IEEE International Conference on Human–Robot Interaction, ACM, 2012, pp. 431–438.
[16] Antoine Deleforge, Radu Horaud, Yoav Y. Schechner, Laurent Girin, Co-localization of audio sources in images using binaural features and locally-linear regression, IEEE/ACM Trans. Audio Speech Lang. Process. 23 (4) (2015) 718–731.
[17] Antoine Deleforge, Walter Kellermann, Phase-optimized K-SVD for signal extraction from underdetermined multichannel sparse mixtures, in: 2015 IEEE International Conference on Acoustics, Speech and Signal Processing (ICASSP), IEEE, 2015, pp. 355–359.
[18] Antoine Deleforge, Yann Traonmilin, Phase unmixing: multichannel source separation with magnitude constraints, in: 2017 IEEE International Conference on Acoustics, Speech and Signal Processing (ICASSP), IEEE, 2017, pp. 161–165.

[19] Christine Evers, Alastair H. Moore, Patrick A. Naylor, Acoustic simultaneous localization and mapping (A-SLAM) of a moving microphone array and its surrounding speakers, in: 2016 IEEE International Conference on Acoustics, Speech and Signal Processing (ICASSP), IEEE, 2016, pp. 6–10.

[20] Koutarou Furukawa, Keita Okutani, Kohei Nagira, Takuma Otsuka, Katsutoshi Itoyama, Kazuhiro Nakadai, Hiroshi G. Okuno, Noise correlation matrix estimation for improving sound source localization by multirotor UAV, in: 2013 IEEE/RSJ International Conference on Intelligent Robots and Systems (IROS), IEEE, 2013, pp. 3943–3948.

[21] Sharon Gannot, Emmanuel Vincent, Shmulik Markovich-Golan, Alexey Ozerov, A consolidated perspective on multimicrophone speech enhancement and source separation, IEEE/ACM Trans. Audio Speech Lang. Process. 25 (4) (2017) 692–730.

[22] Clément Gaultier, Saurabh Kataria, Antoine Deleforge, VAST: the virtual acoustic space traveler dataset, in: International Conference on Latent Variable Analysis and Signal Separation, Springer, 2017, pp. 68–79.

[23] David Gouaillier, Vincent Hugel, Pierre Blazevic, Chris Kilner, Jérôme Monceaux, Pascal Lafourcade, Brice Marnier, Julien Serre, Bruno Maisonnier, Mechatronic design of Nao humanoid, in: IEEE International Conference on Robotics and Automation, ICRA'09, 2009, IEEE, 2009, pp. 769–774.

[24] Simon Haykin, Zhe Chen, The cocktail party problem, Neural Comput. 17 (9) (2005) 1875–1902.

[25] P.M. Hofman, A.J. Van Opstal, Spectro-temporal factors in two-dimensional human sound localization, J. Am. Stat. Assoc. 103 (5) (1998) 2634–2648.

[26] Paul M. Hofman, Jos G.A. Van Riswick, A. John Van Opstal, Relearning sound localization with new ears, Nat. Neurosci. 1 (5) (1998) 417–421.

[27] Jonas Hornstein, Manuel Lopes, José Santos-Victor, Francisco Lacerda, Sound localization for humanoid robots-building audio-motor maps based on the HRTF, in: 2006 IEEE/RSJ International Conference on Intelligent Robots and Systems, IEEE, 2006, pp. 1170–1176.

[28] Jie Huang, Noboru Ohnishi, Noboru Sugie, Building ears for robots: sound localization and separation, Artif. Life Robot. 1 (4) (1997) 157–163.

[29] David Huggins-Daines, Mohit Kumar, Arthur Chan, Alan W. Black, Mosur Ravishankar, Alexander I. Rudnicky, Pocketsphinx: a free, real-time continuous speech recognition system for hand-held devices, in: 2006 IEEE International Conference on Acoustics, Speech and Signal Processing, ICASSP 2006, Proceedings, vol. 1, 2006, IEEE, 2006.

[30] Gökhan Ince, Kazuhiro Nakadai, Tobias Rodemann, Yuji Hasegawa, Hiroshi Tsujino, Jun-ichi Imura, Ego noise suppression of a robot using template subtraction, in: 2009 IEEE/RSJ International Conference on Intelligent Robots and Systems, IROS 2009, IEEE, 2009, pp. 199–204.

[31] Akinori Ito, Takashi Kanayama, Motoyuki Suzuki, Shozo Makino, Internal noise suppression for speech recognition by small robots, in: Ninth European Conference on Speech Communication and Technology, 2005.

[32] Masaharu Kato, Hisashi Uematsu, Makio Kashino, Tatsuya Hirahara, The effect of head motion on the accuracy of sound localization, Acoust. Sci. Technol. 24 (5) (2003) 315–317.

[33] Laurent Kneip, Claude Baumann, Binaural model for artificial spatial sound localization based on interaural time delays and movements of the interaural axis, J. Acoust. Soc. Am. 124 (5) (2008) 3108–3119.

[34] Miranda Kreković, Ivan Dokmanić, Martin Vetterli, Echoslam: simultaneous localization and mapping with acoustic echoes, in: 2016 IEEE International Conference on Acoustics, Speech and Signal Processing (ICASSP), IEEE, 2016, pp. 11–15.

[35] Yifeng Li, Alioune Ngom, Versatile sparse matrix factorization and its applications in high-dimensional biological data analysis, in: IAPR International Conference on Pattern Recognition in Bioinformatics, Springer, 2013, pp. 91–101.

[36] Heinrich W. Löllmann, Hendrik Barfuss, Antoine Deleforge, Stefan Meier, Walter Kellermann, Challenges in acoustic signal enhancement for human–robot communication, in: Proceedings of Speech Communication, 11. ITG Symposium, VDE, 2014, pp. 1–4.

[37] Heinrich W. Löllmann, Alastair H. Moore, Patrick A. Naylor, Boaz Rafaely, Radu Horaud, Alexandre Mazel, Walter Kellermann, Microphone array signal processing for robot audition, in: Hands-free Speech Communications and Microphone Arrays (HSCMA), 2017, IEEE, 2017, pp. 51–55.

[38] Ning Ma, Tobias May, Guy J. Brown, Exploiting deep neural networks and head movements for robust binaural localization of multiple sources in reverberant environments, IEEE/ACM Trans. Audio Speech Lang. Process. 25 (12) (2017) 2444–2453.

[39] Ning Ma, Tobias May, Hagen Wierstorf, Guy J. Brown, A machine-hearing system exploiting head movements for binaural sound localisation in reverberant conditions, in: 2015 IEEE International Conference on Acoustics, Speech and Signal Processing (ICASSP), IEEE, 2015, pp. 2699–2703.

[40] Aly Magassouba, Nancy Bertin, François Chaumette, Sound-based control with two microphones, in: 2015 IEEE/RSJ International Conference on Intelligent Robots and Systems (IROS), IEEE, 2015, pp. 5568–5573.

[41] Tobias May, Ning Ma, Guy J. Brown, Robust localisation of multiple speakers exploiting head movements and multi-conditional training of binaural cues, in: 2015 IEEE International Conference on Acoustics, Speech and Signal Processing (ICASSP), IEEE, 2015, pp. 2679–2683.

[42] John C. Middlebrooks, David M. Green, Sound localization by human listeners, Annu. Rev. Psychol. 42 (1) (1991) 135–159.

[43] Kazuhiro Nakadai, Tino Lourens, Hiroshi G. Okuno, Hiroaki Kitano, Active audition for humanoid, in: AAAI/IAAI, 2000, pp. 832–839.

[44] Kazuhiro Nakadai, Hiroshi G. Okuno, Hiroaki Kitano, Real-time sound source localization and separation for robot audition, in: Seventh International Conference on Spoken Language Processing, 2002.

[45] Kazuhiro Nakadai, Hiroshi G. Okuno, Hiroaki Kitano, Robot recognizes three simultaneous speech by active audition, in: IEEE International Conference on Robotics and Automation, 2003, Proceedings ICRA'03, vol. 1, IEEE, 2003, pp. 398–405.

[46] Kazuhiro Nakadai, Toru Takahashi, Hiroshi G. Okuno, Hirofumi Nakajima, Yuji Hasegawa, Hiroshi Tsujino, Design and implementation of robot audition system HARK-open source software for listening to three simultaneous speakers, Adv. Robot. 24 (5–6) (2010) 739–761.

[47] Patrick Naylor, Nikolay D. Gaubitch, Speech Dereverberation, Springer Science & Business Media, 2010.

[48] Quan V. Nguyen, Francis Colas, Emmanuel Vincent, François Charpillet, Long-term robot motion planning for active sound source localization with Monte Carlo tree search, in: Hands-free Speech Communications and Microphone Arrays (HSCMA), 2017, IEEE, 2017, pp. 61–65.

[49] J. Kevin O'Regan, Alva Noë, A sensorimotor account of vision and visual consciousness, Behav. Brain Sci. 24 (5) (2001) 939–973.

[50] Makoto Otani, Tatsuya Hirahara, Shiro Ise, Numerical study on source-distance dependency of head-related transfer functions, J. Acoust. Soc. Am. 125 (5) (2009) 3253–3261.

[51] Kazuhiro Otsuka, Shoko Araki, Kentaro Ishizuka, Masakiyo Fujimoto, Martin Heinrich, Junji Yamato, A realtime multimodal system for analyzing group meetings by combining face pose tracking and speaker diarization, in: Proceedings of the 10th International Conference on Multimodal Interfaces, ACM, 2008, pp. 257–264.
[52] Alexey Ozerov, Cédric Févotte, Multichannel nonnegative matrix factorization in convolutive mixtures for audio source separation, IEEE Trans. Audio Speech Lang. Process. 18 (3) (2010) 550–563.
[53] Stephen Perrett, William Noble, The effect of head rotations on vertical plane sound localization, J. Acoust. Soc. Am. 102 (4) (1997) 2325–2332.
[54] Henri Poincaré, The Foundations of Science: Science and Hypothesis, the Value of Science, Science and Method, Science Press, New York, 1929, G.B. Halsted, trans. of La valeur de la science, 1905.
[55] Alban Portello, Patrick Danes, Sylvain Argentieri, Acoustic models and Kalman filtering strategies for active binaural sound localization, in: 2011 IEEE/RSJ International Conference on Intelligent Robots and Systems (IROS), IEEE, 2011, pp. 137–142.
[56] Rajkishore Prasad, Hiroshi Saruwatari, Kiyohiro Shikano, Robots that can hear, understand and talk, Adv. Robot. 18 (5) (2004) 533–564.
[57] Caleb Rascon, Ivan Meza, Localization of sound sources in robotics: a review, Robot. Auton. Syst. 96 (2017) 184–210.
[58] Jordi Sanchez-Riera, Xavier Alameda-Pineda, Johannes Wienke, Antoine Deleforge, Soraya Arias, Jan Čech, Sebastian Wrede, Radu Horaud, Online multimodal speaker detection for humanoid robots, in: 2012 12th IEEE–RAS International Conference on Humanoid Robots (Humanoids), IEEE, 2012, pp. 126–133.
[59] Hiroshi Sawada, Hirokazu Kameoka, Shoko Araki, Naonori Ueda, Multichannel extensions of non-negative matrix factorization with complex-valued data, IEEE Trans. Audio Speech Lang. Process. 21 (5) (2013) 971–982.
[60] Alexander Schmidt, Antoine Deleforge, Walter Kellermann, Ego-noise reduction using a motor data-guided multichannel dictionary, in: 2016 IEEE/RSJ International Conference on Intelligent Robots and Systems (IROS), IEEE, 2016, pp. 1281–1286.
[61] Alexander Schmidt, Heinrich Loellmann, Walter Kellermann, A novel ego-noise suppression algorithm for acoustic signal enhancement in autonomous systems, in: 2018 IEEE International Conference on Acoustics, Speech and Signal Processing (ICASSP), IEEE, 2018.
[62] Bernhard Schölkopf, John C. Platt, John Shawe-Taylor, Alex J. Smola, Robert C. Williamson, Estimating the support of a high-dimensional distribution, Neural Comput. 13 (7) (2001) 1443–1471.
[63] Paris Smaragdis, Judith C. Brown, Non-negative matrix factorization for polyphonic music transcription, in: 2003 IEEE Workshop on Applications of Signal Processing to Audio and Acoustics, IEEE, 2003, pp. 177–180.
[64] J.W. Strutt (Lord Rayleigh), On the perception of the direction of sound, Proc. R. Soc. Lond., A Contain. Pap. Math. Phys. Character 83 (559) (1909) 61–64.
[65] Ronen Talmon, Israel Cohen, Sharon Gannot, Supervised source localization using diffusion kernels, in: 2011 IEEE Workshop on Applications of Signal Processing to Audio and Acoustics (WASPAA), IEEE, 2011, pp. 245–248.
[66] Willard R. Thurlow, John W. Mangels, Philip S. Runge, Head movements during sound localization, J. Acoust. Soc. Am. 42 (2) (1967) 489–493.
[67] Vladimir Tourbabin, Hendrik Barfuss, Boaz Rafaely, Walter Kellermann, Enhanced robot audition by dynamic acoustic sensing in moving humanoids, in: 2015 IEEE International Conference on Acoustics, Speech and Signal Processing (ICASSP), IEEE, 2015, pp. 5625–5629.
[68] Joel A. Tropp, Anna C. Gilbert, Signal recovery from random measurements via orthogonal matching pursuit, IEEE Trans. Inf. Theory 53 (12) (2007) 4655–4666.

[69] Jean-Marc Valin, Shun'ichi Yamamoto, Jean Rouat, François Michaud, Kazuhiro Nakadai, Hiroshi G. Okuno, Robust recognition of simultaneous speech by a mobile robot, IEEE Trans. Robot. 23 (4) (2007) 742–752.
[70] Emmanuel Vincent, Jon Barker, Shinji Watanabe, Jonathan Le Roux, Francesco Nesta, Marco Matassoni, The second 'chime' speech separation and recognition challenge: an overview of challenge systems and outcomes, in: 2013 IEEE Workshop on Automatic Speech Recognition and Understanding (ASRU), IEEE, 2013, pp. 162–167.
[71] Emmanuel Vincent, Rémi Gribonval, Cédric Févotte, Performance measurement in blind audio source separation, IEEE Trans. Audio Speech Lang. Process. 14 (4) (2006) 1462–1469.
[72] Tuomas Virtanen, Monaural sound source separation by nonnegative matrix factorization with temporal continuity and sparseness criteria, IEEE Trans. Audio Speech Lang. Process. 15 (3) (2007) 1066–1074.
[73] Hans Wallach, The role of head movements and vestibular and visual cues in sound localization, J. Exp. Psychol. 27 (4) (1940) 339.
[74] D. Wang, G.J. Brown, Computational Auditory Scene Analysis: Principles, Algorithms and Applications, IEEE Press, 2006.
[75] Lin Wang, Andrea Cavallaro, Ear in the sky: ego-noise reduction for auditory micro aerial vehicles, in: 2016 13th IEEE International Conference on Advanced Video and Signal Based Surveillance (AVSS), IEEE, 2016, pp. 152–158.
[76] Frederic L. Wightman, Doris J. Kistler, Resolution of front–back ambiguity in spatial hearing by listener and source movement, J. Acoust. Soc. Am. 105 (5) (1999) 2841–2853.
[77] B.A. Wright, Y. Zhang, A review of learning with normal and altered sound-localization cues in human adults, Int. J. Audiol. 45 (S1) (2006) 92–98.
[78] Tian Xiao, Qing Huo Liu, Finite difference computation of head-related transfer function for human hearing, J. Acoust. Soc. Am. 113 (5) (2003) 2434–2441.

Chapter 3

Audio source separation into the wild

Laurent Girin*,†, Sharon Gannot‡, Xiaofei Li†

*Univ. Grenoble Alpes, CNRS, Grenoble-INP, GIPSA-Lab, Saint Martin d'Hères, France †INRIA Grenoble Rhône-Alpes, Perception Group, Montbonnot-Saint-Martin, France ‡Bar-Ilan University, Faculty of Engineering, Ramat-Gan, Israel

CONTENTS

3.1 INTRODUCTION

Source separation is a topic of signal processing that has been of major interest for decades. It consists of processing an observed mixture of signals so as to extract the elementary signals composing this mixture. In the context of audio processing, it refers to the extraction of signals simultaneously emitted by several sound sources, from the audio recordings of the resulting mixture signal. It has major applications, going from speech enhancement as a front-end for telecommunication systems and automatic speech recognition, to demixing and remixing of music. Despite recent progress using deep learning techniques, single-channel multi-source recordings are still regarded particularly difficult to separate. In this chapter we deal with audio source separation in the wild, and we address multichannel recordings obtained using multiple microphones in a natural environment, as opposed to mixtures created by mixing software which generally do not match the acoustics of real environments, e.g. music production in studio. Typically, we discuss such problems as having to separate the speech signals emitted simultaneously by different persons sharing the same acoustic enclosure, considering also ambient noise and other interfering sources such as domestic apparatuses.

Multimodal Behavior Analysis in the Wild. https://doi.org/10.1016/B978-0-12-814601-9.00022-5
Copyright © 2019 Elsevier Ltd. All rights reserved.

Even when using multichannel recordings, source separation in general is a difficult problem that belongs to the general class of inverse problem. As such, it is often ill-posed, in particular when the number of sensors used to capture the mixture signals is lower than the number of emitting sources. Consequently, in the signal processing community in general, and in the audio processing community in particular, the source separation problem has often been addressed within quite controlled configurations, i.e. "laboratory" studies, that are carefully designed to allow a proper evaluation protocol and an in-depth inspection of the behavior of the proposed techniques. As we will detail in this chapter, this often comes in contrast to robust, in-the-wild configurations, where the source separation algorithms are confronted with the complexity of real-world data, and thus "do not work so well". In this chapter, we describe such limitations of multichannel audio source separation (MASS) and present a review of the studies that have been proposed to overcome those limitations, trying to make MASS techniques progressively go from laboratories into the wild.[1]

Research in speech enhancement and speaker separation has followed two convergent paths, starting with microphone array processing (also referred to as *beamforming*) and blind source separation, respectively. These communities are now strongly interrelated and routinely borrow ideas from each other. Hence, in this chapter we discuss the two paradigms interchangeably. We will explore several important practical scenarios, e.g. moving sources and/or microphones, varying number of sources and sensors, high reverberation levels, spatially diffuse sources, and synchronization problems. Several applications such as smart assistants, cellular phones, hearing devices and robots, which have recently gained growing research and industrial interest, will be discussed.

This chapter is organized as follows. In Section 3.2, we briefly present the fundamentals of multichannel audio source separation. In Section 3.3 we list the current major limitations of MASS that prevent a large deployment of MASS technique in the wild, and we present approaches that have been proposed in the literature to overcome these limitations.

3.2 MULTICHANNEL AUDIO SOURCE SEPARATION

In this section, we briefly present the fundamentals of multichannel audio source separation. This presentation is limited to the basic material that is

[1] Of course, this is just a general view of the academic studies in the field as a whole. Some researchers in the field have considered with great attention the practical aspects of audio source separation techniques.

necessary to understand the following discussion on MASS in the wild. Indeed, the goal of this chapter is not to extensively present the theoretical foundations and principles of source separation and beamforming, even limited to the audio context: Many publications have addressed this issue, including books [13,26,33,61,89] and overview papers [20,48,106].

Hundreds of multichannel audio signal enhancement techniques have been proposed in the literature over the last 40 years along two historical research paths. *Microphone array processing* emerged from the theory of sensor array processing for telecommunications and it focused mostly on the localization and enhancement of speech in noisy or reverberant environments [14,18,25,49,84], while MASS was later popularized by the machine learning community and it addressed "cocktail party" scenarios involving several sound sources mixed together [26,89,106,113,136,137]. These two research tracks have converged in the last decade and they are hardly distinguishable today. Source separation techniques are not necessarily blind anymore and most of them exploit the same theoretical tools, impulse response models and spatial filtering principles as speech enhancement techniques.

The formalization of the MASS problem begins with the formalization of the mixture signal. The most general expression for a linear mixture of J source signals recorded by I microphones is

$$\mathbf{x}(t) = \sum_{j=1}^{J} \mathbf{y}_j(t) + \mathbf{b}(t) \in \mathbb{R}^I, \tag{3.1}$$

where $\mathbf{y}_j(t) \in \mathbb{R}^I$ is the multichannel image of the jth source signal $s_j(t)$ [128], taking into account the effect of acoustic propagation from the position of the emitted source to the microphones (each entry $y_{ij}(t)$ of $\mathbf{y}_j(t)$ is the image of $s_j(t)$ at microphone i). $\mathbf{b}(t)$ is a sensor noise term. In most studies on MASS, the effect of acoustic propagation from source j to microphone i is modeled as a linear time-invariant filter of impulse response $a_{ij}(t)$, and we have

$$\mathbf{x}(t) = \sum_{j=1}^{J} \sum_{\tau=0}^{L_a-1} \mathbf{a}_j(\tau) s_j(t-\tau) + \mathbf{b}(t). \tag{3.2}$$

The vector $\mathbf{a}_j(\tau)$ contains all responses $a_{ij}(t)$ for $i \in [1, I]$, which are assumed to have the same length L_a for convenience. Depending on the application, the goal of MASS is to estimate either the source images $\mathbf{y}_j(t)$ or the (monochannel) source signals $s_j(t)$ from the observation of $\mathbf{x}(t)$.

State-of-the-art MASS methods generally start with a time–frequency (TF) decomposition of the temporal signals, usually by applying the short-time

Fourier transform (STFT) [29]. This is for two main reasons. First, model-based approaches can take advantage of the very particular *sparse* structure of audio signals in the TF plane [115]: A small proportion of source TF coefficients have a significant energy. Source signals are thus generally much less overlapping in the TF domain than in the time domain, naturally facilitating the separation. Second, it is common to consider that at each frequency, the time-domain convolutive mixing process (3.2) is transformed by the STFT into a simple product between the source STFT coefficients and the discrete Fourier transform (DFT) coefficients of the mixing filter, see e.g. [4,91,107,109,110,146,147] and many other studies:

$$\mathbf{x}(f,n) \approx \sum_{j=1}^{J} \mathbf{a}_j(f)s_j(f,n) + \mathbf{b}(f,n) = \mathbf{A}(f)\mathbf{s}(f,n) + \mathbf{b}(f,n), \quad (3.3)$$

where $\mathbf{x}(f,n)$, $s_j(f,n)$ and $\mathbf{b}(f,n)$ are the STFT of $\mathbf{x}(\tau)$, $\mathbf{b}(\tau)$ and $s_j(\tau)$, respectively, and $\mathbf{a}_j(f)$ gathers the DFT of the entries of $\mathbf{a}_j(\tau)$, known as the acoustic transfer functions (ATFs). The ATF vectors are concatenated in the matrix $\mathbf{A}(f)$ and the source signals $s_j(f,n)$ are stacked in the vector $\mathbf{s}(f,n)$. In many practical scenarios, $\mathbf{A}(f)$ is substituted by the respective relative transfer function (RTF) matrix, for which each column is normalized by its first entry. Under this normalization the first row of $\mathbf{A}(f)$ is all '1's and the source signals $s_j(f,n)$ are substituted by their image on the first microphone.[2]

As further discussed in Section 3.3.3, (3.3) is an approximation that is valid if the length of the mixing filters impulse responses is shorter than the length of the STFT window analysis. In the literature and in the following this approximation is referred to as the multiplicative transfer function (MTF) approximation [9] or the narrowband approximation [71].

MASS methods can then be classified into four (non-exclusive) categories [48,137]. Firstly, separation methods based on independent component analysis (ICA) consist in estimating the demixing filters that maximize the independency of separated sources [26,61]. TF-domain ICA methods have been largely investigated [103,110,127]. Unfortunately, ICA-based methods are subject to the well-known scale ambiguity and source permutation problems across frequency bins [3], which must generally be solved as a post-processing step [62,119,120]. In addition, these methods cannot be directly applied to under-determined mixtures.

---

[2]One can select other microphones as the reference microphone or use other methods of normalization. Selecting the reference microphone might have an impact on the MASS performance. This topic is out of the scope of this chapter.

Secondly, methods based on sparse component analysis (SCA) and binary masking rely on the assumption that only one source is active at each TF point [4,91,118,146] (most methods consider only one active source at each TF bin though in principle the SCA and ICA approaches can be combined by considering up to I active sources in each TF bin). These methods often rely on some sort of source clustering in the TF domain, generally based on spatial information extracted from prior mixing filter identification. Therefore, for this kind of methods, the source separation problem is often linked to the source localization problem (where are the emitting sources?).

Thirdly, more recent methods are based on probabilistic generative models in the STFT domain and associated parameter estimation and source inference algorithms [137]. The latter are mostly based on the well-known expectation–maximization (EM) methodology [30] and the like (i.e. iterative alternating optimization techniques). One popular approach is to model the source STFT coefficients with a complex-valued local Gaussian model (LGM) [37,45,82,148], often combined with a nonnegative matrix factorization (NMF) model [74] applied to the source power spectral density (PSD) matrix [6,44,107,109], which is reminiscent of pioneering works such as [12]. This allows one to drastically reduce the number of model parameters and (to some extent) to alleviate the source permutation problem. The sound sources are generally separated using Wiener filters constructed from the learned parameters. Such approach has been extended to super-Gaussian (or heavy-tailed) distributions to model the audio signal sparsity in the TF domain [98], as well as to a fully Bayesian framework by considering prior distributions for the (source and/or mixing process) model parameters [66]. Note that, as for SCA/binary masking methods, the estimation of mixing parameters and the source separation itself are two different steps, often processed alternately within the EM principled methodology.

Methods belonging to the fourth category can be broadly classified as beamforming methods, which is roughly equivalent to *linear spatial filtering*. A beamformer is a vector $\mathbf{w}(f) = \left[w_1(f), \ldots, w_I(f)\right]^T$ comprising one complex-valued weight per microphone, that is applied to $\mathbf{x}(n, f)$. The output $\mathbf{w}^H(f)\mathbf{x}(n, f)$ can be transformed back into the time-domain by an inverse STFT. Beamformers originally referred to spatial filters based on the direction of arrival (DOA) of the source signal and were only later generalized to any linear spatial filters. DOA-based beamformers are still widely used, especially when simplicity of the implementation and its robustness are of crucial importance. However, the performance of these beamformers is expected to degrade in comparison with modern beamformers that take the entire acoustic path into account [46]. The beamformer weights are set according to a specific optimization criterion. Many beamforming criteria can be found in the general literature [134]. In the speech processing community, the minimum variance distortionless response (MVDR) beamformer [46], the maximum signal-to-noise ratio (MSNR) beamformer [142], the multichannel Wiener filter (MWF) [34], specifically its speech distortion weighted variant (SDW-MWF) [35], and the linearly constrained minimum variance (LCMV) beamformer [92], are widely used.

One may state that, so far, the beamforming approach led to more effective industrial real-world applications than the "generic" MASS approach. This lies in the difference between (a) enhancing a spatially fixed dominant target speech signal from background noise (possibly composed of several sources) and (b) clearly separating several source signals, all considered as signals of interest, that are mixed with a similar power (the cocktail party problem). The first problem is generally simpler to solve than the second one. In other words, the second one can be seen as an extension of the first one. Anyway, as we will see now, problems of in-the-wild processing remain challenging in each case.

3.3 MAKING MASS GO FROM LABS INTO THE WILD

3.3.1 Moving sources and sensors

In a realistic in-the-wild scenario, sound sources are often moving. Sometimes they are moving slightly, e.g. the small movements of the head of a sat speaker. Sometimes they are moving a lot, e.g. a person speaking while walking in a room. Sensors can also move, e.g. microphones embedded within a mobile robot. In the same vein, the acoustic environment can also change over time, e.g. we close a window, an object is placed in between a source and the microphones, or the separation system has to operate in another room. All those changes imply changes in the acoustic propagation of sources to microphones, i.e. changes in the mixing filters. Yet the vast majority of MASS methods described in the signal processing literature deals with the assumption of fixed sources, fixed sensors and fixed environment, i.e. technically speaking the mixing filters are considered as *time-invariant* (at least over the duration of the processed recordings), as expressed in (3.2). Into-the-wild MASS methods should consider the more realistic case of *time-varying* mixtures corresponding to source-to-microphone channels that can change over time, which would account for possible source or microphone motions and environment changes. For example, in many human–robot interaction scenarios, there is a strong need to consider mixed speech signals emitted by moving speakers and perturbed by reverberation that can change over time and by non-stationary ambient noise.

Studies dealing with moving sources or moving sensors or changing environment actually exist, but they are quite sparse compared to the large number of studies with time-invariant filters. Early attempts addressing the separation of time-varying mixtures basically consisted in block-wise adaptations of time-invariant methods: An STFT frame sequence is split into blocks, and a time-invariant MASS algorithm is applied to each block. Hence, block-wise adaptations assume time-invariant filters within blocks. The separation parameters are updated from one block to the next and the separation result over a block can be used to initialize the separation of the next block. Frame-wise algorithms can be considered as particular cases of block-wise algorithms, with single-frame blocks, and hybrid methods may combine block-wise and frame-wise processing. Notice that, depending on the implementation, some of these methods may run online.

Interestingly, most of the block-wise approaches use ICA, either in the temporal domain [1,59,70,116] or in the Fourier domain [99,100]. In addition to being limited to over-determined mixtures, block-wise ICA methods need to account for the source permutation problem, not only across frequency bins, as usual, but across successive blocks as well. Examples of block-wise adaptation of binary-masking or LGM-based methods are more scarce. As for binary masking, a block-wise adaptation of [4] was proposed in [83]. This method performs source separation by clustering the observation vectors in the source image space. As for LGM, [126] describes an online block- and frame-wise adaptation of the general LGM framework proposed in [109]. One important problem, common to all block-wise approaches, is the difficulty to choose the block size. Indeed, the block size must assume a good trade-off between local channel stationarity (short blocks) and sufficient data to infer relevant statistics (long blocks). The latter constraint can drastically limit the dynamics of either the sources or the sensors [83]. Other parameters such as the step-size of the iterative update equations may also be difficult to set [126]. In general, systematic convergence towards a good separation solution using a limited amount of signal statistics remains an open issue.

A more principled approach consists in modeling the mixing filter as a time-varying process and considering the MASS in the angle of an adaptive process, in the spirit of early works on adaptive filtering [145]. For example, an early iterative and sequential approach for speech enhancement in reverberant environment was proposed in [144]. This method used the EM framework to jointly estimate the desired speech signal and the required (deterministic) parameters, namely the speech auto-regressive coefficients, and the speech and noise mixing filters taps. Only the case of a 2×2 mixture was addressed. A subspace tracking recursive LCMV beamforming method for extracting multiple moving sources was proposed in [94]. This method is applicable only to over-determined mixture.

As for under-determined time-varying convolutive mixtures, a method using binary masking within a probabilistic LGM framework was proposed in [57]. The mixing filters were considered as latent variables following a Gaussian distribution with mean vector depending on the DOA of the corresponding source. The DOA was modeled as a discrete latent variable taking values from a finite set of angles and following a discrete hidden Markov model (HMM). A variational expectation–maximization (VEM) algorithm was derived to perform the inference, including forward–backward equations to estimate the DOA sequence.

In [65,67], the transfer function of the mixing filters were considered as continuous latent variables ruled by a first-order linear dynamical system (LDS) with Gaussian noise [17], in the spirit of [47]. This model was used in combination with a source LGM + NMF model, still to process under-determined time-varying convolutive mixtures. This approach may be seen as a generalization of [107] to moving sources/microphones. As in [57], a VEM algorithm was developed for the joint estimation of the model parameters and inference of the latent variables. Here, a Kalman smoother was used for the inference of the time-varying mixing filters, which were combined with estimated source PSDs to build separating Wiener filters. This model can be more effective than the discrete DOA-dependent HMM model of [57] in reverberant conditions, since the relationship between the transfer function and the source DOA can be quite complex, and Wiener filters are more general than binary masks.

A VEM approach for beamforming (hence, over-determined scenario) that is specifically designed for dynamic scenarios can be found in [73,90,121]. In these methods, the speech signal is modeled as an LGM in the STFT domain and the RTFs as a first-order Markov model. The posterior distribution of the speech signal and the channel is recursively estimated in the E-step.

In comparison to the block-wise adaptation methodology described in [126], explicit time-varying mixture models have the potential to exploit the information available within the whole sequence of input mixture frames. They were generally proposed in batch mode but they can be adapted to online processing, e.g. by replacing a Kalman smoother with a Kalman filter [144].

DOA estimation plays an important role in the design of beamforming methods. DOA tracking of multiple speakers based on a discrete set of angles and HMM is given in [117]. DOA estimation procedures for single source (or alternating sources) scenarios, which are based on variants of the recursive least squares (RLS) methodology, are presented in [39]. Recursive versions of the EM procedure are utilized for multiple speakers tracking in [123]. A simple, yet effective method for localizing multiple sources in reverberant environments is the steered-response power with phase transform (SRP-PHAT) [31]. The MUltiple Signal Classification (MUSIC) algorithm [122] or more specifically, the root-MUSIC variant, allow for fast adaptation of LCMV beamformers by exploiting instantaneous DOA estimates [131,132].

Finally, we can mention that robots, as moving platforms, open new opportunities and challenges for the sound source localization and separation tasks. An open source robot audition system, named 'HARK', is described in [101]. The localization module is based on successive application of the MUSIC algorithm [122], while the separation stage uses a geometry-assisted MASS method [111]. Bayesian methods for tracking multiple sources are also gaining interest in the literature, e.g. using particle filters [40,133] and probability hypothesis density (PHD) filters [41]. Two recent European projects, the "Ears"[3] project and the "Two!Ears"[4] project, explored new algorithms for enhancing the auditory capabilities of humanoid robots [85] and link them with decision and action [19].

3.3.2 Varying number of (active) sources

In a realistic in-the-wild scenario, sound sources are often intermittent, i.e. they do not emit sounds all the time. As an example of major importance we can mention a natural conversational regime between several speakers that includes speech turns. Depending on the context and content of the conversation, the speech signals can have very low to very strong time overlap. The sound sources may not even be present in the scene all the time, e.g. a person that goes in and out of a room, occasionally speaking or turning on and off a sounding device. Yet, the vast majority of MASS methods described in the signal processing literature deals with the assumption of a fixed number of sources over time. In addition, this fixed number of sound sources is often assumed to be known and all sources are assumed to be continuously active, i.e. they emit during all the time of the processed recording sequence. The situation of the literature with this respect is similar to the previous section: A few studies with the number of active sources varying in time do exist but they are largely outnumbered by studies with a fixed number of constantly active sources.

One straightforward manner to address this problem is to proceed to the estimation of the number of sources present in the scene and/or the number of

---

[3] https://robot-ears.eu/.

[4] http://twoears.eu/project/.

active sources as a pre-processing step before going into the separation problem. A method based on speech sparsity in the STFT domain is presented in [5]. A variational EM approach for complex Watson mixture models is presented in [36]. In the beamforming context, identifying the number of speakers, as well as the number of available sensors, necessitates an update of the weights of the beamformer. An efficient implementation, based on low-rank update of correlation matrices, is presented in [95].

The detection of the number of active sources and associating an "identity" (i.e. a label) to each detected source is related to the so-called *diarization* problem. Indeed, in speech processing, speaker diarization refers to the task of detecting who speaks and when in an audio stream [2,135]. In many (dialog) applications, the speakers are assumed to take distinct speech turns, i.e. they speak one after the other, and speaker diarization thus amounts to signal segmentation and speaker recognition. In a source separation context, the automatic detection of the number and "identity" of simultaneously active sources can thus be considered as an additional multisource diarization task to be considered jointly with the separation task, or within the separation task. Indeed, both processes are complementary: Knowing the source diarization is expected to ease the separation process, for example by enabling to adapt the separation system to the actual number of active sources and to the speaker characteristics; in turn diarization is easier using separated source signals than using mixed source signals.

Processing of speech intermittency for MASS appears in [26] for the instantaneous mixing case. For convolutive mixtures, [58,108] presented a framework for joint processing of MASS and diarization, where factorial Hidden Markov models were used to model the activity of the sources. Unfortunately, due to its factorial nature, the model does not account for correlations on the activity of different sources, i.e. the activities of the different sources are assumed independent with each other, which is questionable for natural conversations for example. Very recently, joint processing of the two tasks have been proposed in [64] (for over-determined mixtures) and in [68, 69]. The models in [68] and [69] combine a diarization state model (that encodes the combination of active sources within a given set of maximum size N) with the multichannel LGM + NMF model of [107] and with the full-rank spatial covariance matrix model of [37], respectively. In contrast to [58,108], modeling the activity of all sources jointly using a diarization state enables to exploit the potential correlations on speaker activity.

Note that estimating the number of active sources present in the scene is a good example of problem common to source separation and source localization. Strategies have been developed for the automatic detection of the number of sources within a (probabilistic) source localization framework, e.g. [42,81,141]. Obviously, such strategies may be exploited in or extended to source separation, as already explored in [117,118].

3.3.3 Spatially diffuse sources and long mixing filters

The vast majority of current state-of-the-art MASS methods considers convolutive mixtures of sources, as expressed by (3.2): each source image within the mixture signal recorded at the microphones is assumed to be the result of the convolution of a small "point" source signal with the impulse response of the source-to-microphone acoustic path. This formulation implies that each sound source is assumed to be spatially concentrated at a single point of the acoustic space. This is fine to some extent for speech signals, but this is more questionable for "large" sound sources such as wind, trucks, or large musical instruments, which emit sound in a large region of space. In that later case, each source image is better considered as a spatially distributed source.

Moreover, if the source signal propagates in highly reverberant enclosures, the late tail of the room impulse response (RIR) is perceived as arriving from all directions. If the reverberation time T_{60}, defined as the elapsed time until the reverberation level has decreased by 60 dB from its initial value, is large, the reverberant sound field is said to be diffuse, homogeneous and isotropic. The normalized spatial correlation between the received signal at two different microphones i and i', and at frequency f, is given in closed-form for spherical symmetry [28,72] by

$$\begin{aligned}\Omega_{ii'}(f) &= \frac{\mathbb{E}^{\text{spat}}(r_{ij}(f) r^*_{i'j}(f))}{\sqrt{\mathbb{E}^{\text{spat}}(|r_{ij}(f)|^2)}\sqrt{\mathbb{E}^{\text{spat}}(|r_{i'j}(f)|^2)}} \\ &= \frac{\sin(2\pi f \ell_{ii'}/c)}{2\pi f \ell_{ii'}/c}\end{aligned} \tag{3.4}$$

where $\mathbb{E}^{\text{spat}}$ denotes *spatial expectation* over all possible absolute positions of the sources and of the microphone array in the room, and $\ell_{ii'}$ denotes the distance between the microphones. A closed-form result also exists for cylindrical symmetry [27]. A simulator of both sound fields can be found in [53].

To address this problem, the authors of [37] proposed to use a full-rank (FR) spatial covariance matrix (SCM) for characterizing the spatial distribution of the source images (across channels), instead of the rank-1 matrix corresponding to the MTF model [107]. This FR-SCM model is assumed to represent diffuse sources better than the MTF approximation and it is compliant with the vision of a diffuse source as a sum of subsources with identical PSD distributed in a large region of the physical space. The model parameters are estimated using an expectation–maximization (EM) algorithm. Note that this approach does not attempt to explicitly model the mixture process but rather focuses on the properties of the resulting source images. The FR-SCM model was further used and improved in [6,38].

Moreover, even for point sources, the processing of convolutive mixtures in the STFT domain is confronted to a severe limitation with respect to in-the-wild scenarios: The length of room impulse responses (RIRs) is in general (much) larger than the length of the STFT analysis window, which can severely question the validity of the approximation (3.3). Typical values for the STFT window length for speech mixtures are within 16–64 ms, in order to adapt to the global non-stationarity and local stationarity of speech signals. At the same time, the typical T_{60} reverberation time of usual home/office rooms is within 200–600 ms, and large meeting rooms or auditorium can have a T_{60} larger than 1 s. The ratio between RIR length and STFT window length can thus easily be within 10–50 instead of being lower than 1. Therefore, (3.3) can be a quite poor approximation, even for moderately reverberant environments, if the sources are positioned further away from the microphones.[5] The MTF approximation is still widely used to address convolutive mixtures problems due to its practical interest: The fact that a time-domain convolutive mixture becomes an independent instantaneous mixture at each frequency bin f facilitates the technical development of solutions to the separation problem. While this can be a reasonable choice of model, we stress that the validity of the MTF approximation should be verified prior to its application. In some cases, a mixed MTF and full-rank models should be considered [37].

Here again, compared to the impressive amount of papers on MASS methods for convolutive mixtures based on (3.3), only a few have addressed solutions to overcome the limitation of the MTF approximation. Although there are only a few, these solutions can be classified into two general approaches: Methods mixing time-domain (TD) and TF-domain processing, and methods that totally remain in the TF-domain.

As for the first approach, the method of [71] consists in modeling the sources in the TF domain while keeping a TD representation of the convolutive mixture using the inverse transform expression. The TD source signals are estimated using a Lasso optimization technique (hence the method is called W-Lasso for wideband Lasso) with a regularization term on STFT source

[5] Actually, the ratio between the coherent direct-path and the reverberation tail, the so-called direct-to-reverberant ratio (DRR) plays an important role examining the validity of MTF assumption.

coefficients to take into account source sparsity in the TF domain. In [7] an improved W-Lasso with a re-weighted scheme is presented. W-Lasso achieves quite good source separation performance in reverberant environments, at the price of a tremendous computation time. Also, only semi-blind separation with known mixing filters was addressed in [71], which is poorly satisfying with regard to going towards separation in the wild. A similar hybrid TD/TF approach was recently followed in [76,77]. The source signals were represented using either the modified discrete cosine transform (MDCT), which is real-valued and critically sampled, or the odd-frequency STFT (OFSTFT). A probabilistic LGM + NMF model was used for the source coefficients, which were inferred from TD mixture observations using a VEM algorithm. This led to very interesting results, at the price of huge computation. Here also, most experiments were conducted in a semi-blind setup with known mixing filters since the joint estimation of the filters' impulse responses remains difficult.

As for the approaches that work totally in the TF domain, let us first mention that the authors of [37] have shown that, in addition to modeling diffuse sources, their method is able to circumvent *to some extent* the discussed limitations of the MTF approximation. Other methods have investigated TF mixture models more accurate than (3.3). Fundamentally, the time-domain convolution can be exactly represented as a two-dimensional filtering in the TF domain [50]. This representation was used for linear system identification in [10] as an alternative to MTF, under the name of cross-band filters (CBFs). Using CBFs, a source image STFT coefficient is represented as a summation over frequency bins of multiple convolutions between the input source STFT coefficients and the TF-domain filter impulse response, along the frame axis. This exact representation becomes an approximation when we limit the number of bins either in the frequency-wise summation or in the frame-wise convolution. In particular, considering only the current frequency bin, i.e. a unique convolution along the STFT frame axis, is a reasonable practical approximation, referred to as the convolutive transfer function (CTF) model [130]. Using this CTF model, the mixture model (3.2) writes in the STFT domain:

$$\mathbf{x}(f,n) = \sum_{j=1}^{J} \sum_{n'=0}^{Q_a-1} \mathbf{a}_j(f,n') s_j(f,n-n') + \mathbf{b}(f,n). \tag{3.5}$$

Here, the ith entry of $\mathbf{a}_j(f,n)$, denoted $a_{ij}(f,n)$, is not the DFT of $a_{ij}(t)$ (nor is it its STFT), but it is its CTF. The CTF contains several STFT-frame-wise filter taps and its expression is a bit more complicated than the DFT, though easily computable from $a_{ij}(t)$, see [10].

The full CBF representation was considered for solving the MASS problem in [11], in combination with a high-resolution NMF (HR-NMF) model of the source signal. A variational EM (VEM) algorithm was proposed to estimate the filters and infer the source signals. Unfortunately, due to the model complexity, this method was observed to perform well only in an oracle setup where both filters and source parameters are initialized from the individual source images. Therefore, in the current state of knowledge, the CBFs seem difficult to integrate into a realistic MASS framework and one has to resort to CTF-like approximations.

An MVDR beamformer, implemented in a generalized sidelobe canceler (GSC) structure, that utilizes the CTF model was proposed in [129]. It was shown to outperform the GSC beamformer which uses the MTF approximation [46]. It can be noted that, as opposed to the full-rank model, the CTF approximation allows for coherent processing and can therefore implement an almost perfect null towards a point interfering source. The ability of the FR model to suppress the interference signal is limited by the number of microphones and their constellation and cannot exceed I^2 [112] for a fully diffuse signal.

Interestingly, the pioneering work [8] combined an STFT-domain convolutive model very similar to (3.5) with a Gaussian mixture model (GMM) of source signals. In this paper, the STFT-domain convolution was intuited from empirical observations and was referred to as "subband filters" (no theoretical justification nor references were provided). Because of the overall complexity of the model (especially the large number of GMM components that is necessary to accurately represent speech signals), the author resorted to a VEM algorithm for parameters estimation. In [56], an STFT-domain convolutive model also very similar to (3.5) was used together with an HMM on source activity. However, the optimization method used to estimate the parameters and infer the source signal is quite complex.

A Lasso-type optimization applied to the MASS problem was considered in [79] within the CTF framework. More specifically, the ℓ_2-norm model fitting term of Lasso was defined at each frequency bin with the STFT-domain convolutive mixture (3.5) instead of the TD convolutive mixture (3.2) as done in [71]. In parallel, the ℓ_1-norm regularizer of Lasso was kept so as to exploit the sparsity audio signals in the TF domain. Because the number of filter frames Q_a in (3.5) is much lower than the length L_a of the TD filter impulse response in (3.2), the computation cost in [79] is drastically reduced compared to [71]. This was obtained at the price of a quite moderate loss in separation performance, showing the good accuracy of the CTF approximation. However, as for [71], this was done only in a semi-blind setup with known filters. To address the use of CTF in a fully blind scenario, the mixture model (3.5) was plugged into a probabilistic framework with an LGM for source STFT coefficients in [80]. An exact EM algorithm was developed for joint estimation of the parameters (source parameters and CTF filter coefficients) and inference of the source STFT coefficients. The joint estimation of source STFT coefficients and CTF mixing filter coefficients was recently addressed as a pure optimization problem in [43].

Another attempt to address the problem of long filters is presented in [124]. In this work, the RTF is split into an early part which is coherently processed and a late part which is treated as an additive noise. This noise is reduced by a combination of an MVDR beamformer and a postfilter. In [125], a nested GSC scheme was proposed that treats the long RIRs jointly as a coherent phenomenon, using CTF modeling, and as a diffused sound field. Different blocks of the proposed scheme use different RIR models.

In parallel with the above attempts to model long mixing filters, as briefly mentioned in the previous sections, several more or less recent studies have considered the modeling of the mixing filters/process as latent variables, possibly in a fully Bayesian framework, either to better account for uncertainty in filter estimation or to introduce prior knowledge on these filters (for example the approximate knowledge of source DOA or the specific structure of room acoustic impulse responses). Because of room limitation, we do not describe those works and only add [21,52,75] to the already cited references [38,57,67,109].

It is clear from the above discussion that many models of the mixing filters are used in the literature. It is still in open question, which of the models is the most appropriate. Most probably, the answer to this question depends on the scenario.

3.3.4 **Ad hoc microphone arrays**

Classical microphone arrays usually consist of a condensed set of microphone mounted on a single device. Establishing wireless acoustic sensor networks (WASNs), comprising multiple cooperative devices (e.g., cellphone, tablet, hearing aid, smart watch) may increase the chances to find a subset of the microphones that is close to a relevant sound source. Consequently, WASNs may demonstrate higher separation capabilities than a single-device solution. The wide-spread availability of devices equipped with multiple microphones makes the vision closer to reality.

The distributed and ad hoc nature of WASNs arises new challenges, e.g. transmission and processing constraints, synchronization between nodes and dynamic network topology. Addressing these new challenges is a prerequisite for fully exploiting the potential of WASNs.

Several families of distributed algorithms, which only require the transmission of a fused version of the signals received by each node, were proposed in [93]. The distributed adaptive node-specific signal estimation (DANSE) family of algorithms consists of distributed version of SDW-MWF [15] and LCMV beamformers [16]. A distributed version of the GSC beamformer is presented in [96]. A randomized gossip implementation of the delay and sum beamformer is presented in [149], and a diffusion adaptation method for distributed MVDR beamformer in [105]. The problem of synchronizing the clock drifts in several nodes is addressed in e.g. [24,114,140,143].

The full potential of ad hoc microphone arrays to separate source in the wild is yet to be explored.

3.4 CONCLUSIONS AND PERSPECTIVES

In this review, we have presented several ways to make MASS and beamforming techniques go from laboratories to real-life scenarios. Laboratory studies are often based on a set of assumptions on the source signals and/or the mixture process that may not be totally realistic (e.g. static, point sources, and spatially stable microphone constellation). In this section, we will briefly explore a few families of devices that already work in real-life scenarios. We will conclude this section and the entire chapter by a perspective on the future of the research in the field.

In recent years, we have witnessed the penetration, in an accelerating pace, of smart audio devices to the consumer electronics market. These devices, designed to work in adverse conditions, include personal assistants embedded in smart phones, portable computers and most notably, smart loudspeakers, e.g. Amazon Echo ("Alexa"),[6] Microsoft Invoke (with "Cortana"),[7] Apple HomePod (with "Siri")[8] and Google Home [78].

Basically, these smart loudspeakers demonstrate that tremendous progress has already been made in middle-range devices, capable of executing automatic speech recognition (ASR) engines in noisy environments. Smart loudspeakers are equipped with several microphones (six for Apple HomePod, seven for Amazon Echo and Microsoft Invoke, and only two microphones for Google Home). Algorithmically, these devices consist of a denoising (mostly using a steered beamformer), dereverberation and echo cancelation

[6]https://www.slideshare.net/AmazonWebServices/designing-farfield-speech-processing-systems-with-intel-and-amazon-alexa-voice-service-alx305-reinvent-2017.
[7]https://news.harman.com/releases/harman-reveals-the-harman-kardon-invokeTM-intelligent-speaker-with-cortana-from-microsoft.
[8]https://www.apple.com/homepod/.

stages. The devices usually employ localization (or DOA estimation) algorithms to provide the steering direction, as an important prerequisite to the application of the beamformer. The acquired localization information is also used for indicating the direction of the detected source with respect to the device. As smart loudspeakers acquire and enhance a speech signal in a noisy and reverberant environment, they provide a living example of in-the-wild beamforming. Yet, their performance may be still limited to home scenarios with a predominant and reasonably spatially stable speaker relatively close to the device (as opposed to the above-mentioned adverse scenarios with several active and moving sources with low DRR).

Hearing aids [32] are another example of successful application of MASS/beamforming technologies, aiming at speech quality and intelligibility improvement, as well as enhancing the spatial awareness of the hearing aid wearer (in binaural setting). Hearing devices impose severe real-time constraints on the applied algorithms (latency shorter than 10 ms). Moreover, robustness and reliability are of major importance to prevent potential hearing damage to the hearing impaired person. Binaural cue preservation can be obtained by calculating a common gain to both hearing devices [63] or by applying a beamformer that incorporates binaural information into the optimization criterion, e.g. MWF [97] or LCMV [54]. Beamforming-based binaural processing is usually regarded computationally more expensive than the common gain approach. An important issue in designing a binaural enhancement algorithm is to determine the source of interest. In many cases, the beamformer is steered towards the look direction of the hearing aid wearer.

Most cellular phones are nowadays equipped with multiple microphones (3–4) and they usually work in adverse conditions demonstrating reasonable performance. A few systems already employ microphone networks, e.g. smart home and smart cities.

The quest for realistic solutions, capable of processing a large amount of sound sources in real-life environments and in dynamical scenarios of various characters, still continues. Many of the theoretical and practical questions are still open and there are performance gaps to be filled for many scenarios such as under-determined mixtures with many simultaneously active sources, multiple moving sources and moving sensors (e.g. robots, cellular phones), high-power and non-stationary noise (e.g. from heavy machinery and drilling noise in mines), and binaural hearing (for both hearing impaired people and robots imitating the human auditory system [87]).

Recent years have witnessed a revolution in MASS techniques. Nowadays, *deep learning* solutions seem to be the new El Dorado for audio processing. Still, most studies deal with single-channel denoising/enhancement/separation algorithms [22,51,55,86,102,139]. More recently, multichannel processing solutions that employ deep learning [23,104], as well as robust ASR systems [60], have been proposed. Deep learning has also influenced the hearing aid industry [138] and the development of binaural algorithms [150]. An improved localization strategy that utilizes active head movements and deep learning is proposed in [88]. Despite the impressive performance gains obtained by deep learning based speech processing approaches, the field is still in its infancy and major breakthroughs are expected in the foreseeable future.

As an outcome from this review, it is evident that significant progress is still required for obtaining robust and reliable source separation in difficult real-life scenarios, especially under severe online constraints. We anticipate that future solutions will combine ideas from both the array processing/source separation and machine learning paradigms. As always, only such combined solutions, together with practical know-how, are capable of advancing the already established solutions towards comprehensive audio source separation methods that work in the wild.

REFERENCES

[1] R. Aichner, H. Buchner, S. Araki, S. Makino, On-line time-domain blind source separation of nonstationary convolved signals, in: Int. Conf. Independent Component Analysis and Blind Source Separation (ICA), Nara, Japan, 2003.

[2] X. Anguera Miro, S. Bozonnet, N. Evans, C. Fredouille, G. Friedland, O. Vinyals, Speaker diarization: a review of recent research, IEEE Trans. Audio Speech Lang. Process. 20 (2) (2012) 356–371.

[3] S. Araki, R. Mukai, S. Makino, T. Nishikawa, H. Saruwatari, The fundamental limitation of frequency domain blind source separation for convolutive mixtures of speech, IEEE Trans. Speech Audio Process. 11 (2) (2003) 109–116.

[4] S. Araki, H. Sawada, R. Mukai, S. Makino, Underdetermined blind sparse source separation for arbitrarily arranged multiple sensors, Signal Process. 87 (8) (2007) 1833–1847.

[5] S. Arberet, R. Gribonval, F. Bimbot, A robust method to count and locate audio sources in a multichannel underdetermined mixture, IEEE Trans. Signal Process. 58 (1) (2010) 121–133.

[6] S. Arberet, A. Ozerov, N.Q.K. Duong, E. Vincent, R. Gribonval, F. Bimbot, P. Vandergheynst, Nonnegative matrix factorization and spatial covariance model for under-determined reverberant audio source separation, in: IEEE International Symposium on Signal Processing and Its Applications (ISSPA), Kuala Lumpur, Malaysia, 2010.

[7] S. Arberet, P. Vandergheynst, R. Carrillo, J.-P. Thiran, Y. Wiaux, Sparse reverberant audio source separation via reweighted analysis, IEEE Trans. Audio Speech Lang. Process. 21 (7) (2013) 1391–1402.

[8] H. Attias, New EM algorithms for source separation and deconvolution with a microphone array, in: IEEE International Conference on Acoustics, Speech and Signal Processing (ICASSP), 2003.

[9] Y. Avargel, I. Cohen, On multiplicative transfer function approximation in the short-time Fourier transform domain, IEEE Signal Process. Lett. 14 (5) (2007) 337–340.
[10] Y. Avargel, I. Cohen, System identification in the short-time Fourier transform domain with crossband filtering, IEEE Trans. Audio Speech Lang. Process. 15 (4) (2007) 1305–1319.
[11] R. Badeau, M.D. Plumbley, Multichannel high-resolution NMF for modeling convolutive mixtures of non-stationary signals in the time–frequency domain, IEEE/ACM Trans. Audio Speech Lang. Process. 22 (11) (2014) 1670–1680.
[12] L. Benaroya, L.M. Donagh, F. Bimbot, R. Gribonval, Non negative sparse representation for Wiener based source separation with a single sensor, in: IEEE International Conference on Acoustics, Speech, and Signal Processing (ICASSP), 2003.
[13] J. Benesty, J. Chen, Y. Huang, Microphone Array Signal Processing, Springer, 2008.
[14] J. Benesty, S. Makino, J. Chen (Eds.), Speech Enhancement, Springer, 2005.
[15] A. Bertrand, M. Moonen, Distributed adaptive node-specific signal estimation in fully connected sensor networks – part I: sequential node updating, IEEE Trans. Signal Process. 58 (2010) 5277–5291.
[16] A. Bertrand, M. Moonen, Distributed node-specific LCMV beamforming in wireless sensor networks, IEEE Trans. Signal Process. 60 (2012) 233–246.
[17] C. Bishop, Pattern Recognition and Machine Learning, Springer, 2006.
[18] M.S. Brandstein, D.B. Ward (Eds.), Microphone Arrays: Signal Processing Techniques and Applications, Springer, 2001.
[19] G. Bustamante, P. Danès, T. Forgue, A. Podlubne, J. Manhes, An information based feedback control for audio-motor binaural localization, Auton. Robots 42 (2) (2018) 477–490.
[20] J.F. Cardoso, Blind signal separation: statistical principles, Proc. IEEE 9 (10) (1998) 2009–2025.
[21] A.T. Cemgil, C. Févotte, S. Godsill, Variational and stochastic inference for Bayesian source separation, Digit. Signal Process. 2007 (17) (2007) 891–913.
[22] S.E. Chazan, J. Goldberger, S. Gannot, A hybrid approach for speech enhancement using MoG model and neural network phoneme classifier, IEEE/ACM Trans. Audio Speech Lang. Process. 24 (12) (2016).
[23] S.E. Chazan, J. Goldberger, S. Gannot, DNN-based concurrent speakers detector and its application to speaker extraction with LCMV beamforming, in: IEEE International Conference on Audio and Acoustic Signal Processing (ICASSP), Calgary, Alberta, Canada, 2018.
[24] D. Cherkassky, S. Gannot, Blind synchronization in wireless acoustic sensor networks, IEEE/ACM Trans. Audio Speech Lang. Process. 25 (3) (2017) 651–661.
[25] I. Cohen, J. Benesty, S. Gannot (Eds.), Speech Processing in Modern Communication: Challenges and Perspectives, Springer, 2010.
[26] P. Comon, C. Jutten (Eds.), Handbook of Blind Source Separation – Independent Component Analysis and Applications, Academic Press, 2010.
[27] R.K. Cook, R.V. Waterhouse, R.D. Berendt, S. Edelman, M.C. Thompson Jr., Measurement of correlation coefficients in reverberant sound fields, J. Acoust. Soc. Am. 27 (6) (1955) 1072–1077.
[28] H. Cox, Spatial correlation in arbitrary noise fields with application to ambient sea noise, J. Acoust. Soc. Am. 54 (5) (1973) 1289–1301.
[29] R.E. Crochiere, L.R. Rabiner, Multi-Rate Signal Processing, Prentice Hall, Englewood Cliffs, NJ, 1983.
[30] A.P. Dempster, N.M. Laird, D.B. Rubin, et al., Maximum likelihood from incomplete data via the EM algorithm, J. R. Stat. Soc. 39 (1) (1977) 1–38.
[31] J.H. DiBiase, H.F. Silverman, M.S. Brandstein, Robust localization in reverberant rooms, in: Microphone Arrays, Springer, 2001, pp. 157–180.
[32] H. Dillon, Hearing Aids, Thieme, 2012.

[33] P. Divenyi (Ed.), Speech Separation by Humans and Machines, Springer Verlag, 2004.
[34] S. Doclo, M. Moonen, GSVD-based optimal filtering for single and multimicrophone speech enhancement, IEEE Trans. Signal Process. 50 (9) (2002) 2230–2244.
[35] S. Doclo, A. Spriet, J. Wouters, M. Moonen, Speech distortion weighted multichannel Wiener filtering techniques for noise reduction, in: Speech Enhancement, Signals and Communication Technology, Springer, Berlin, 2005, pp. 199–228.
[36] L. Drude, A. Chinaev, D.H. Tran Vu, R. Haeb-Umbach, Source counting in speech mixtures using a variational EM approach for complex Watson mixture models, in: IEEE International Conference on Acoustics, Speech, and Signal Processing (ICASSP), Florence, Italy, 2014.
[37] N. Duong, E. Vincent, R. Gribonval, Under-determined reverberant audio source separation using a full-rank spatial covariance model, IEEE Trans. Audio Speech Lang. Process. 18 (7) (2010) 1830–1840.
[38] N. Duong, E. Vincent, R. Gribonval, Spatial location priors for Gaussian model based reverberant audio source separation, EURASIP J. Adv. Signal Process. 2013 (149) (2013).
[39] T.G. Dvorkind, S. Gannot, Time difference of arrival estimation of speech source in a noisy and reverberant environment, Signal Process. 85 (1) (2005) 177–204.
[40] C. Evers, Y. Dorfan, S. Gannot, P.A. Naylor, Source tracking using moving microphone arrays for robot audition, in: IEEE International Conference on Acoustics, Speech and Signal Processing (ICASSP), New Orleans, Louisiana, 2017.
[41] C. Evers, A.H. Moore, P.A. Naylor, Localization of moving microphone arrays from moving sound sources for robot audition, in: European Signal Processing Conference (EUSIPCO), Budapest, Hungary, 2016.
[42] M.F. Fallon, S.J. Godsill, Acoustic source localization and tracking of a time-varying number of speakers, IEEE Trans. Audio Speech Lang. Process. 20 (4) (2012) 1409–1415.
[43] F. Feng, Séparation aveugle de sources: de l'instantané au convolutif, Ph.D. thesis, Université Paris Sud, 2017.
[44] C. Févotte, N. Bertin, J.-L. Durrieu, Nonnegative matrix factorization with the Itakura–Saito divergence. With application to music analysis, Neural Comput. 21 (3) (2009) 793–830.
[45] C. Févotte, J.-F. Cardoso, Maximum likelihood approach for blind audio source separation using time–frequency Gaussian source models, in: IEEE Workshop Applicat. Signal Process. to Audio and Acoust. (WASPAA), New Paltz, NJ, 2005.
[46] S. Gannot, D. Burshtein, E. Weinstein, Signal enhancement using beamforming and nonstationarity with applications to speech, IEEE Trans. Signal Process. 49 (8) (2001) 1614–1626.
[47] S. Gannot, M. Moonen, On the application of the unscented Kalman filter to speech processing, in: IEEE Int. Workshop on Acoustic Echo and Noise Control (IWAENC), Kyoto, Japan, 2003.
[48] S. Gannot, E. Vincent, S. Markovich-Golan, A. Ozerov, A consolidated perspective on multimicrophone speech enhancement and source separation, IEEE/ACM Trans. Audio Speech Lang. Process. 25 (4) (2017) 692–730.
[49] S.L. Gay, J. Benesty (Eds.), Acoustic Signal Processing for Telecommunication, Kluwer, 2000.
[50] A. Gilloire, M. Vetterli, Adaptive filtering in subbands with critical sampling: analysis, experiments, and application to acoustic echo cancellation, IEEE Trans. Signal Process. 40 (8) (1992) 1862–1875.
[51] E. Girgis, G. Roma, A. Simpson, M. Plumbley, Combining mask estimates for single channel audio source separation using deep neural networks, in: Conference of the International Speech Communication Association (INTERSPEECH), 2016.

[52] L. Girin, R. Badeau, On the use of latent mixing filters in audio source separation, in: International Conference on Latent Variable Analysis and Signal Separation (LVA/ICA), Grenoble, France, 2017.
[53] E. Habets, S. Gannot, Generating sensor signals in isotropic noise fields, J. Acoust. Soc. Am. 122 (2007) 3464–3470.
[54] E. Hadad, S. Doclo, S. Gannot, The binaural LCMV beamformer and its performance analysis, IEEE/ACM Trans. Audio Speech Lang. Process. 24 (3) (2016) 543–558.
[55] J.R. Hershey, Z. Chen, J. Le Roux, S. Watanabe, Deep clustering: discriminative embeddings for segmentation and separation, in: IEEE International Conference on Acoustics, Speech, and Signal Processing (ICASSP), Shanghai, China, 2016.
[56] T. Higuchi, H. Kameoka, Joint audio source separation and dereverberation based on multichannel factorial hidden Markov model, in: IEEE International Workshop on Machine Learning for Signal Processing (MLSP), 2014.
[57] T. Higuchi, N. Takamune, N. Tomohiko, H. Kameoka, Underdetermined blind separation and tracking of moving sources based on DOA-HMM, in: IEEE International Conference on Acoustics, Speech, and Signal Processing (ICASSP), Florence, Italy, 2014.
[58] T. Higuchi, H. Takeda, N. Tomohiko, H. Kameoka, A unified approach for underdetermined blind signal separation and source activity detection by multichannel factorial hidden Markov models, in: Conference of the International Speech Communication Association (INTERSPEECH), Singapore, 2014.
[59] K.E. Hild II, D. Erdogmus, J.C. Principe, Blind source separation of time-varying, instantaneous mixtures using an on-line algorithm, in: IEEE International Conference on Acoustics, Speech, and Signal Processing (ICASSP), Orlando, Florida, 2002.
[60] T. Hori, Z. Chen, H. Erdogan, J.R. Hershey, J. Le Roux, V. Mitra, S. Watanabe, Multi-microphone speech recognition integrating beamforming, robust feature extraction, and advanced NN/RNN backend, Comput. Speech Lang. 46 (2017) 401–418.
[61] A. Hyvärinen, J. Karhunen, E. Oja (Eds.), Independent Component Analysis, Wiley and Sons, 2001.
[62] M.Z. Ikram, D.R. Morgan, A beamforming approach to permutation alignment for multichannel frequency-domain blind speech separation, in: IEEE International Conference on Acoustics, Speech and Signal Processing (ICASSP), Orlando, Florida, 2002.
[63] A.H. Kamkar-Parsi, M. Bouchard, Instantaneous binaural target PSD estimation for hearing aid noise reduction in complex acoustic environments, IEEE Trans. Instrum. Meas. 60 (4) (2011) 1141–1154.
[64] B. Kleijn, F. Lim, Robust and low-complexity blind source separation for meeting rooms, in: Int. Conf. on Hands-free Speech Communication and Microphone Arrays, San Francisco, CA, 2017.
[65] D. Kounades-Bastian, L. Girin, X. Alameda-Pineda, S. Gannot, R. Horaud, A variational EM algorithm for the separation of moving sound sources, in: IEEE Workshop Applicat. Signal Process. to Audio and Acoust. (WASPAA), New Paltz, NJ, 2015.
[66] D. Kounades-Bastian, L. Girin, X. Alameda-Pineda, S. Gannot, R. Horaud, An inverse-Gamma source variance prior with factorized parameterization for audio source separation, in: IEEE International Conference on Acoustics, Speech and Signal Processing (ICASSP), Shanghai, China, 2016.
[67] D. Kounades-Bastian, L. Girin, X. Alameda-Pineda, S. Gannot, R. Horaud, A variational EM algorithm for the separation of time-varying convolutive audio mixtures, IEEE/ACM Trans. Audio Speech Lang. Process. 24 (8) (2016) 1408–1423.

[68] D. Kounades-Bastian, L. Girin, X. Alameda-Pineda, S. Gannot, R. Horaud, An EM algorithm for joint source separation and diarization of multichannel convolutive speech mixtures, in: IEEE International Conference on Acoustics, Speech and Signal Processing (ICASSP), New Orleans, Louisiana, 2017.

[69] D. Kounades-Bastian, L. Girin, X. Alameda-Pineda, S. Gannot, R. Horaud, Exploiting the intermittency of speech for joint separation and diarization, in: IEEE Workshop on Applications of Signal Processing to Audio and Acoustics (WASPAA), New Paltz, NJ, 2017.

[70] A. Koutras, E. Dermatas, G. Kokkinakis, Blind speech separation of moving speakers in real reverberant environments, in: IEEE International Conference on Acoustics, Speech, and Signal Processing (ICASSP), Istanbul, Turkey, 2000.

[71] M. Kowalski, E. Vincent, R. Gribonval, Beyond the narrowband approximation: Wideband convex methods for under-determined reverberant audio source separation, IEEE Trans. Audio Speech Lang. Process. 18 (7) (2010) 1818–1829.

[72] H. Kuttruff, Room Acoustics, Taylor & Francis, 2000.

[73] Y. Laufer, S. Gannot, A Bayesian hierarchical model for speech enhancement, in: IEEE International Conference on Audio and Acoustic Signal Processing (ICASSP), Calgary, Alberta, Canada, 2018.

[74] D.D. Lee, H.S. Seung, Learning the parts of objects by non-negative matrix factorization, Nature 401 (1999) 788–791.

[75] S. Leglaive, R. Badeau, G. Richard, Multichannel audio source separation with probabilistic reverberation priors, IEEE Trans. Audio Speech Lang. Process. 24 (12) (2016).

[76] S. Leglaive, R. Badeau, G. Richard, Multichannel audio source separation: variational inference of time–frequency sources from time-domain observations, in: International Conference on Acoustics, Speech and Signal Processing (ICASSP), New-Orleans, Louisiana, 2017.

[77] S. Leglaive, R. Badeau, G. Richard, Separating time–frequency sources from time-domain convolutive mixtures using non-negative matrix factorization, in: IEEE Workshop on Applications of Signal Processing to Audio and Acoustics (WASPAA), New Paltz, NY, 2017.

[78] B. Li, T.N. Sainath, A. Narayanan, J. Caroselli, M. Bacchiani, A. Misra, I. Shafran, H. Sak, G. Pundak, K. Chin, K.C. Sim, R.J. Weiss, K.W. Wilson, E. Variani, C. Kim, O. Siohan, M. Weintraub, E. McDermott, R. Rose, M. Shannon, Acoustic modeling for Google Home, in: Conference of the International Speech Communication Association (INTERSPEECH), Stockholm, Sweden, 2017.

[79] X. Li, L. Girin, R. Horaud, Audio source separation based on convolutive transfer function and frequency-domain Lasso optimization, in: IEEE International Conference on Acoustics, Speech and Signal Processing (ICASSP), New Orleans, Louisiana, 2017.

[80] X. Li, L. Girin, R. Horaud, An EM algorithm for audio source separation based on the convolutive transfer function, in: IEEE Workshop on Applications of Signal Processing to Audio and Acoustics (WASPAA), New Paltz, NY, 2017.

[81] X. Li, L. Girin, R. Horaud, S. Gannot, Multiple-speaker localization based on direct-path features and likelihood maximization with spatial sparsity regularization, IEEE/ACM Trans. Audio Speech Lang. Process. 25 (10) (2017) 1007–2012.

[82] A. Liutkus, B. Badeau, G. Richard, Gaussian processes for underdetermined source separation, IEEE Trans. Signal Process. 59 (7) (2011) 3155–3167.

[83] B. Loesch, B. Yang, Online blind source separation based on time–frequency sparseness, in: IEEE International Conference on Acoustics, Speech, and Signal Processing (ICASSP), Taipei, Taiwan, 2009.

[84] P.C. Loizou, Speech Enhancement: Theory and Practice, CRC Press, 2007.

[85] H. Löllmann, A. Moore, P. Naylor, B. Rafaely, R. Horaud, A. Mazel, W. Kellermann, Microphone array signal processing for robot audition, in: IEEE Int. Conf. on Hands-free Speech Communications and Microphone Arrays (HSCMA), San Francisco, CA, 2017.

[86] Y. Luo, Z. Chen, J.R. Hershey, J. Le Roux, N. Mesgarani, Deep clustering and conventional networks for music separation: stronger together, in: IEEE International Conference on Acoustics, Speech and Signal Processing (ICASSP), New Orleans, Lousiana, 2017.

[87] R.F. Lyon, Human and Machine Hearing: Extracting Meaning from Sound, Cambridge University Press, 2017.

[88] N. Ma, T. May, G.J. Brown, Exploiting deep neural networks and head movements for robust binaural localization of multiple sources in reverberant environments, IEEE/ACM Trans. Audio Speech Lang. Process. 25 (12) (2017) 2444–2453.

[89] S. Makino, T.-W. Lee, H. Sawada (Eds.), Blind Speech Separation, Springer, 2007.

[90] S. Malik, J. Benesty, J. Chen, A Bayesian framework for blind adaptive beamforming, IEEE Trans. Signal Process. 62 (9) (2014) 2370–2384.

[91] M. Mandel, R.J. Weiss, D.P.W. Ellis, Model-based expectation–maximization source separation and localization, IEEE Trans. Audio Speech Lang. Process. 18 (2) (2010) 382–394.

[92] S. Markovich, S. Gannot, I. Cohen, Multichannel eigenspace beamforming in a reverberant noisy environment with multiple interfering speech signals, IEEE Trans. Audio Speech Lang. Process. 17 (6) (2009) 1071–1086.

[93] S. Markovich-Golan, A. Bertrand, M. Moonen, S. Gannot, Optimal distributed minimum-variance beamforming approaches for speech enhancement in wireless acoustic sensor networks, Signal Process. 107 (2015) 4–20.

[94] S. Markovich-Golan, S. Gannot, I. Cohen, Subspace tracking of multiple sources and its application to speakers extraction, in: IEEE International Conference on Acoustics, Speech, and Signal Processing (ICASSP), Dallas, TX, 2010.

[95] S. Markovich-Golan, S. Gannot, I. Cohen, Low-complexity addition or removal of sensors/constraints in LCMV beamformers, IEEE Trans. Signal Process. 60 (3) (2012) 1205–1214.

[96] S. Markovich-Golan, S. Gannot, I. Cohen, Distributed multiple constraints generalized sidelobe canceler for fully connected wireless acoustic sensor networks, IEEE Trans. Audio Speech Lang. Process. 21 (2) (2013) 343–356.

[97] D. Marquardt, E. Hadad, S. Gannot, S. Doclo, Theoretical analysis of linearly constrained multi-channel Wiener filtering algorithms for combined noise reduction and binaural cue preservation in binaural hearing aids, IEEE/ACM Trans. Audio Speech Lang. Process. 23 (12) (2015) 2384–2397.

[98] N. Mitianoudis, M.E. Davies, Audio source separation of convolutive mixtures, IEEE Trans. Speech Audio Process. 11 (5) (2003) 489–497.

[99] R. Mukai, H. Sawada, S. Araki, S. Makino, Robust real-time blind source separation for moving speakers in a room, in: IEEE International Conference on Acoustics, Speech, and Signal Processing (ICASSP), 2003.

[100] K. Nakadai, H. Nakajima, Y. Hasegawa, H. Tsujino, Sound source separation of moving speakers for robot audition, in: IEEE International Conference on Acoustics, Speech, and Signal Processing (ICASSP), Taipei, Taiwan, 2009.

[101] K. Nakadai, T. Takahashi, H. Okuno, H. Nakajima, Y. Hasegawa, H. Tsujino, Design and implementation of robot audition system 'HARK' – open source software for listening to three simultaneous speakers, Adv. Robot. 24 (5–6) (2010) 739–761.

[102] A. Narayanan, D. Wang, Ideal ratio mask estimation using deep neural networks for robust speech recognition, in: IEEE International Conference on Acoustics, Speech and Signal Processing (ICASSP), Vancouver, Canada, 2013.

[103] F. Nesta, P. Svaizer, M. Omologo, Convolutive BSS of short mixtures by ICA recursively regularized across frequencies, IEEE Trans. Audio Speech Lang. Process. 19 (3) (2011) 624–639.

[104] A. Nugraha, A. Liutkus, E. Vincent, Multichannel audio source separation with deep neural networks, IEEE/ACM Trans. Audio Speech Lang. Process. 24 (9) (2016) 1652–1664.

[105] M. O'Connor, W.B. Kleijn, Diffusion-based distributed MVDR beamformer, in: IEEE International Conference on Acoustics, Speech and Signal Processing (ICASSP), Florence, Italy, 2014.

[106] P. O'Grady, B.A. Pearlmutter, S. Rickard, Survey of sparse and non-sparse methods in source separation, Int. J. Imaging Syst. Technol. 15 (1) (2005) 18–33.

[107] A. Ozerov, C. Févotte, Multichannel nonnegative matrix factorization in convolutive mixtures for audio source separation, IEEE Trans. Audio Speech Lang. Process. 18 (3) (2010) 550–563.

[108] A. Ozerov, C. Févotte, M. Charbit, Factorial scaled hidden Markov model for polyphonic audio representation and source separation, in: IEEE Workshop on Applications of Signal Processing to Audio and Acoustics (WASPAA), 2009.

[109] A. Ozerov, E. Vincent, F. Bimbot, A general flexible framework for the handling of prior information in audio source separation, IEEE Trans. Audio Speech Lang. Process. 20 (4) (2012) 1118–1133.

[110] L. Parra, C. Spence, Convolutive blind separation of non-stationary sources, IEEE Trans. Speech Audio Process. 8 (3) (2000) 320–327.

[111] L.C. Parra, C.V. Alvino, Geometric source separation: merging convolutive source separation with geometric beamforming, IEEE Trans. Speech Audio Process. 10 (6) (2002) 352–362.

[112] A.T. Parsons, Maximum directivity proof for three-dimensional arrays, J. Acoust. Soc. Am. 82 (1) (1987) 179–182.

[113] M.S. Pedersen, J. Larsen, U. Kjems, L.C. Parra, Convolutive blind source separation methods, in: Springer Handbook of Speech Processing, Springer, 2008, pp. 1065–1094.

[114] P. Pertilä, M.S. Hämäläinen, M. Mieskolainen, Passive temporal offset estimation of multichannel recordings of an ad-hoc microphone array, IEEE Trans. Audio Speech Lang. Process. 21 (11) (2013) 2393–2402.

[115] M. Plumbley, T. Blumensath, L. Daudet, R. Gribonval, M.E. Davies, Sparse representations in audio and music: from coding to source separation, Proc. IEEE 98 (6) (2010) 995–1005.

[116] R.E. Prieto, P. Jinachitra, Blind source separation for time-variant mixing systems using piecewise linear approximations, in: IEEE International Conference on Acoustics, Speech, and Signal Processing (ICASSP), Philadelphia, PN, 2005.

[117] N. Roman, D. Wang, Binaural tracking of multiple moving sources, IEEE Trans. Audio Speech Lang. Process. 16 (4) (2008) 728–739.

[118] N. Roman, D. Wang, G.J. Brown, Speech segregation based on sound localization, J. Acoust. Soc. Am. 114 (4) (2003) 2236–2252.

[119] H. Sawada, S. Araki, R. Mukai, S. Makino, Grouping separated frequency components by estimating propagation model parameters in frequency-domain blind source separation, IEEE Trans. Audio Speech Lang. Process. 15 (5) (2007) 1592–1604.

[120] H. Sawada, R. Mukai, S. Araki, S. Makino, A robust and precise method for solving the permutation problem of frequency-domain blind source separation, IEEE Trans. Speech Audio Process. 12 (5) (2004) 530–538.

[121] D. Schmid, G. Enzner, S. Malik, D. Kolossa, R. Martin, Variational bayesian inference for multichannel dereverberation and noise reduction, IEEE/ACM Trans. Audio Speech Lang. Process. 22 (8) (2014) 1320–1335.

[122] R. Schmidt, Multiple emitter location and signal parameter estimation, IEEE Trans. Antennas Propag. 34 (3) (1986) 276–280.
[123] O. Schwartz, S. Gannot, Speaker tracking using recursive EM algorithms, IEEE/ACM Trans. Audio Speech Lang. Process. 22 (2) (2014) 392–402.
[124] O. Schwartz, S. Gannot, E. Habets, Multi-microphone speech dereverberation and noise reduction using relative early transfer functions, IEEE/ACM Trans. Audio Speech Lang. Process. 23 (2) (2015) 240–251.
[125] O. Schwartz, S. Gannot, E.P. Habets, Nested generalized sidelobe canceller for joint dereverberation and noise reduction, in: IEEE International Conference on Audio and Acoustic Signal Processing (ICASSP), Brisbane, Australia, 2015.
[126] L. Simon, E. Vincent, A general framework for online audio source separation, in: Int. Conf. on Latent Variable Analysis and Signal Separation (LVA/ICA), Tel-Aviv, Israel, 2012.
[127] P. Smaragdis, Blind separation of convolved mixtures in the frequency domain, Neurocomputing 22 (1) (1998) 21–34.
[128] N. Sturmel, A. Liutkus, J. Pinel, L. Girin, S. Marchand, G. Richard, R. Badeau, L. Daudet, Linear mixing models for active listening of music productions in realistic studio conditions, in: Proc. Convention of the Audio Engineering Society (AES), Budapest, Hungary, 2012.
[129] R. Talmon, I. Cohen, S. Gannot, Convolutive transfer function generalized sidelobe canceler, IEEE Trans. Audio Speech Lang. Process. 17 (7) (2009) 1420–1434.
[130] R. Talmon, I. Cohen, S. Gannot, Relative transfer function identification using convolutive transfer function approximation, IEEE Trans. Audio Speech Lang. Process. 17 (4) (2009) 546–555.
[131] O. Thiergart, M. Taseska, E. Habets, An informed LCMV filter based on multiple instantaneous direction-of-arrival estimates, in: IEEE International Conference on Acoustics, Speech and Signal Processing (ICASSP), Vancouver, Canada, 2013.
[132] O. Thiergart, M. Taseska, E. Habets, An informed parametric spatial filter based on instantaneous direction-of-arrival estimates, IEEE/ACM Trans. Audio Speech Lang. Process. 22 (12) (2014) 2182–2196.
[133] J.-M. Valin, F. Michaud, J. Rouat, Robust localization and tracking of simultaneous moving sound sources using beamforming and particle filtering, Robot. Auton. Syst. 55 (3) (2007) 216–228.
[134] H.L. Van Trees, Detection, Estimation, and Modulation Theory, vol. IV, Optimum Array Processing, Wiley, New York, 2002.
[135] D. Vijayasenan, F. Valente, H. Bourlard, Multistream speaker diarization of meetings recordings beyond MFCC and TDOA features, Speech Commun. 54 (1) (2012) 55–67.
[136] E. Vincent, N. Bertin, R. Gribonval, F. Bimbot, From blind to guided audio source separation: how models and side information can improve the separation of sound, IEEE Signal Process. Mag. 31 (3) (2014) 107–115.
[137] E. Vincent, M.G. Jafari, S.A. Abdallah, M.D. Plumbley, M.E. Davies, Probabilistic modeling paradigms for audio source separation, in: Machine Audition: Principles, Algorithms and Systems, 2010, pp. 162–185.
[138] D. Wang, Deep learning reinvents the hearing aid, IEEE Spectr. 54 (3) (2017) 32–37.
[139] D. Wang, J. Chen, Supervised speech separation based on deep learning: an overview, arXiv preprint arXiv:1708.07524, 2017.
[140] L. Wang, S. Doclo, Correlation maximization-based sampling rate offset estimation for distributed microphone arrays, IEEE/ACM Trans. Audio Speech Lang. Process. 24 (3) (2016) 571–582.
[141] L. Wang, T.-K. Hon, J.D. Reiss, A. Cavallaro, An iterative approach to source counting and localization using two distant microphones, IEEE/ACM Trans. Audio Speech Lang. Process. 24 (6) (2016) 1079–1093.

[142] E. Warsitz, R. Haeb-Umbach, Blind acoustic beamforming based on generalized eigenvalue decomposition, IEEE Trans. Audio Speech Lang. Process. 15 (5) (2007) 1529–1539.

[143] S. Wehr, I. Kozintsev, R. Lienhart, W. Kellermann, Synchronization of acoustic sensors for distributed ad-hoc audio networks and its use for blind source separation, in: IEEE Int. Symposium on Multimedia Software Engineering, Miami, FL, 2004.

[144] E. Weinstein, A.V. Oppenheim, M. Feder, J.R. Buck, Iterative and sequential algorithms for multisensor signal enhancement, IEEE Trans. Signal Process. 42 (4) (1994) 846–859.

[145] B. Widrow, J.R. Glover, J.M. McCool, J. Kaunitz, C.S. Williams, R.H. Hearn, J.R. Zeidler, J.R.E. Dong, R.C. Goodlin, Adaptive noise cancelling: principles and applications, Proc. IEEE 63 (12) (1975) 1692–1716.

[146] S. Winter, W. Kellermann, H. Sawada, S. Makino, MAP-based underdetermined blind source separation of convolutive mixtures by hierarchical clustering and l1-norm minimization, EURASIP J. Appl. Signal Process. 2007 (1) (2007) 81.

[147] O. Yilmaz, S. Rickard, Blind separation of speech mixtures via time–frequency masking, IEEE Trans. Signal Process. 52 (7) (2004) 1830–1847.

[148] T. Yoshioka, T. Nakatani, M. Miyoshi, H. Okuno, Blind separation and dereverberation of speech mixtures by joint optimization, IEEE Trans. Audio Speech Lang. Process. 19 (1) (2011) 69–84.

[149] Y. Zeng, R.C. Hendriks, Distributed delay and sum beamformer for speech enhancement via randomized gossip, IEEE/ACM Trans. Audio Speech Lang. Process. 22 (1) (2014) 260–273.

[150] X. Zhang, D. Wang, Deep learning based binaural speech separation in reverberant environments, IEEE/ACM Trans. Audio Speech Lang. Process. 25 (5) (2017) 1075–1084.

Chapter 4

Designing audio-visual tools to support multisensory disabilities

Nicoletta Noceti*, Luca Giuliani*,†, Joan Sosa-Garciá*, Luca Brayda†, Andrea Trucco‡, Francesca Odone*
*Università degli Studi di Genova, DIBRIS, Genova, Italy †Istituto Italiano di Tecnologia, RBCS, Genova, Italy ‡Università degli Studi di Genova, DITEN, Genova, Italy

CONTENTS

4.1 INTRODUCTION

According to the World Health Organization[1] (WHO), around 253 million people live with vision impairment worldwide—more than the 80% are aged 50 years or above—and 360 million people have disabling hearing loss—with approximately one third of people over 65 years of age.

Living with disabilities may affect communication and social inclusion, with a significant impact on everyday life, causing feelings of loneliness and isolation. For this reason, in the last few years, there have been made

[1] http://www.who.int/blindness/world_sight_day/2017/en/.

Multimodal Behavior Analysis in the Wild. https://doi.org/10.1016/B978-0-12-814601-9.00013-4
Copyright © 2019 Elsevier Ltd. All rights reserved.

major efforts in designing and developing devices and systems able to assist users with visual impairment—to address navigation and obstacle avoidance [27,29,50], wayfinding [7,9,44,52], automatic reading, accessible access to web, apps, and displays [15,32,33,38,49,51], object recognition [8,37,43,48], multiple object detection [28]—and hearing loss—with spatial sound segregation techniques provided by prosthesis [14], cochlear implants [46], and also alternative devices as necklaces [47] or glasses [4,30].

In this chapter we discuss the technologies devised in *Glassense*,[2] a regional project developed within the *SI4Life Ligurian Regional Hub—Research and Innovation—Life Sciences*. The problem of assistive technologies supporting sensory disabilities is particularly central in the Liguria Region, which has the eldest population in Italy, with an average age of almost 50 years and more than 40% of the population beyond 55 years old. The aim of *Glassense* was to develop a proof of concept prototype of a sensorized pair of glasses to assist users with limited technology skills and possibly multiple disabilities. The potential beneficiaries of this technology are primarily older users with low vision (hereafter referred to as Blind or Visually Impaired individuals, BVI) and/or hypoacusis. The idea behind this contribution is to show that persons with partial sensory loss can take benefit from interactive technologies that seamlessly give selective information, without overwhelming the available senses. We focus on visual disability because of the lack of non-visual information concerning objects of everyday use; then we focus on partial hearing loss because of the shortage of directional hearing aids in speech understanding tasks.

It is worth introducing some User Stories (USs) that may help in clarifying scenarios where the devised technologies can be of profit and support the users.

US1. *Tina is at home. She is a 60 years old lady with inherited low vision. Today, she would like to cook some pasta for dinner. She is looking in the cupboard for her husband's favorite, fusilli. She finds five boxes that are very similar to her low visual perception. She gently shakes the box but the noise is not distinctive of the type of pasta. To find the right type she has to open all the boxes.*

US2. *Tony is at the grocery store. He is a 65 years old retired school teacher with low vision caused by a retinopathy. He would like to buy a packet of cookies for his nephew who is coming to visit him in the afternoon. His nephew has very specific tastes and only enjoys one type of biscuit. He asks for directions and locates the appropriate stand. Then, he grabs a packet of*

[2] *Glassense: Wearable technologies for sensory supplementation.*

biscuits, but he is not sure they are his nephew's favorites. Tony has to wait for a sighted person who can help him making his purchase.

US3. *Lisa is inside a crowded restaurant. She is 70 years old and is discussing with her son's family at the table, while surrounded by other customers: her three nephews and their parents are telling her how beautiful the seaside is. However, she cannot understand the story since everybody is talking at the same time. She pretends to get the meaning, but she is very sad because her hearing aids do not help in such cases. And she feels bad about asking to speak one at a time. She would like her hearing aids to help her pick up the voice of one person only.*

US4. *Andrew is 45 and has worked in a data center for many years. The huge room he usually works in is full of very noisy server racks. His hearing skills are normal but he can barely understand his colleagues when talking close to the rack fans. He really would like to isolate noise from speech since misunderstandings at the workplace can be very annoying.*

US5. *George is 60 and he is progressively losing his sight. He is also wearing hearing aids. Even if he is not blind, his acuity is less than 6/18, therefore he has low vision. He would like to be supported by a technology that enhances his residual vision, since he cannot read lips, and that possibly helps his hearing aids in noisy environments. Because two senses do not work at best, he feels isolated from society.*

Inspired by these use cases and in agreement with stakeholders, we identified a list of user requirements corresponding to an equal number of constraints the system should satisfy:

- *No feedback delay*, to perform assistive tasks in real time (computation should be done locally on the glasses to avoid communication delay).
- *A natural interface*, to be accepted by a wide audience of users including older users, with I/O functionalities as easy as possible.
- *A simple training procedure* to favor personalizations (e.g. for the visual recognition task, the insertion of new objects in the "knowledge" of the user).
- *Unobtrusion.* Any assistive feature enhancing the residual sensory modalities should not be distractive or overwhelm the already limited sensory load.

Following these indications, we developed assistive technologies—which are now embedded on the glasses—as follows.

To provide solutions for visual loss assistance, we developed an object recognition engine leveraging a *collaborative principle* that allows the user to obtain hints on how to discriminate between different objects lying in the peri-personal space of the user and with a similar tactile sensation [42]. The method is based on a classical image retrieval scheme, and the recognition is triggered with an audio command by the user after having gathered information on the object category, with the use of manipulation—to understand the shape—and/or audio cues—e.g. by shaking the object to produce a sound from its content.

To assist people with partial hearing loss, the glasses aim to increase spatial awareness. We imagined to enhance the voice of the persons that a hearing impaired user is talking to, and aimed at enhancing the acoustic sound sources in front of the user. We targeted speech, i.e. the main acoustic content that serves for social interaction. Following the requirements, enhanced speech has to be sent in real time to the acoustic prostheses of hearing impaired people. We present here a complementary input source to existing acoustic prostheses, that helps to solve the so-called *cocktail party problem* [16].

The proposed system has been field-tested in the application domain of interest: we present an experimental evaluation of the functionalities with visual and/or impaired users, identified by the staff of *Istituto David Chiossone for the blind* (http://www.chiossone.it). The outcome of this analysis reveals the appropriateness of our approaches from both the standpoint of the performance, and of the acceptance and the pleasantness of the users.

The remainder of this paper is organized as follows. We start with Section 4.2, where we provide an account of existing approaches tailored for the application of interest for the project. Section 4.3 introduces the *Glassense system*, followed by Sections 4.4 and 4.5, where we provide a more detailed description of the embedded functionalities. A discussion of field tests with impaired users is reported in Section 4.6, while Section 4.7 is left to a final discussion.

4.2 RELATED WORKS

In this section we provide an account of previous approaches related to the assistive technologies of interest for the project.

Hearing aid. Modern prostheses are able to improve sound perception in most daily occasions. These devices perform noise reduction and dereverberation, but the enhancement is very limited in crowdy environments. The brain usually processes binaural cues to locate sources of interest while isolating possible disturbances [18], solving the so-called *"cocktail party problem"*. Hearing loss and the use of acoustical prostheses modify the cited cues in a way that forces the person to make a greater mental effort. Previous work like [13] shows that the use of one microphone per ear is better than having one microphone only in the most damaged ear. Moreover, it has been shown that binaural multi-microphone hearing aids can decrease the effect of competing spatial sound sources [11]. The use of arrays of microphones allows one to apply beamforming, a spatial signal processing technique that exploits constructive and destructive interferences to filter sounds coming from specific directions. Some of these solutions directly exploit prostheses [14] or cochlear implant [46] microphones as two elements arrays. Widrow in 2001 [47] proposed a microphone array within a necklace, which was shown to be effective, but it is not available on the market. Boone in 2006 [4], as a follow up of Merks's work [31], proposed a pair of glasses equipped with two arrays of microphones, one for each temple. The device is currently available on the market with the name of *earing glasses* and is produced by Varibel company. In 2011 Mens [30] analyzed the performance of the hearing glasses using several beamforming modalities. The device acts as a substitute for a hearing prosthesis and a configuration accounting for the specific subject hearing loss profile is required.

In *Glassense*, inspired by Boone's work, we developed a generic system that has the purposed of being a complement, and not a substitute, to any existing already personalized hearing prosthesis.

Object recognition for BVI individuals. Given the peculiarities of the application domain we consider, we specifically focus on visual recognition methods for BVI. In this sense, we start observing that we are witness to a rapid growth in the number of apps and dedicated systems. Many of them were launched as market products, and the technical details are missing most of the times, impeding the reproducibility.

The main ingredients of the majority of existing object recognition methods are image descriptions based on local features (SURF or SIFT) and some feature matching strategy to address directly detection and recognition [24,26,40]. Most commonly, a generic recognition task is addressed without considering the specificities of the concerned community of users, with the consequent need of identifying stringent requirements the user should adhere to in order to reach a reasonable quality level—see [19] for more details and references. Our approach stems from the classical bag-of-features (BoF) representations [10,12,25,35,39,41]. These strategies are very appropriate for image retrieval tasks, in that they favor the scalability to new visual concepts, while keeping the computational demand under control.

A specific domain that provides some interesting ad hoc object recognition methods is currency recognition, where we can find various task-specific and efficient approaches [17,26,40,43]. Among these methods, the iOS

LookTel Banknote reader is well known for the ability to recognize different currencies in real time, also in relatively bad acquisition conditions. The LookTel platform is one of the first apps proposed in the market of products for BVI users [43]. Experiments show that it is not as effective in terms of image-based recognition; it requires a precise framing to work properly, although it is also associated with a bar code reader which can be useful.

The ShelfScanner [48] is a system allowing for real-time detection of items in a grocery store, combining color information with SURF features. VizWiz:LocateIt [3] is a solution for assisting BVI users to locate objects in their environments where recognition is not performed automatically but with the support of a remote Mechanical Turk[3] service [6] (a remote sighted person). In a similar spirit, Tap Tap See[4] is a recent application that exploits a combination of computer vision and human crowdsourcing to describe the content of an image. The main drawback of this application is that it provides a precise, but relatively slow feedback, and, unless the framing is very careful, a feedback which is too "rich" and detailed (very long sentences describing different objects in the scene). We observe how the presence of anonymous humans in the tagging loop could raise privacy issues if the service is meant for daily use. The other two services rely on the presence of humans in the tagging loop: Be my Eyes,[5] and Aira Service[6]; these tools rely on a network of volunteers aware of the service they are providing and, in the case of the Aira service, volunteers also trained for the purpose. The Aira Service is based on images acquired by Google Glasses. Aipoly vision[7] has been recently released as a iOS app, which performs object recognition based on computer vision and a deep learning architecture running on the cloud. The app primarily addresses an image categorization problem, thus it is complementary to our perspective. Among these methods, only LookTel and Aipoly have a "teaching" mode where new classes can be inserted. A different but related problem is object finding [8,37], which aims at finding a specific missed object in the environment. This goal is rather different from object recognition, since there is need to cope with often severe scale variations and clutter. To date, the recognition performance is still too limited for the systems to be considered appealing to the users.

---

[3]https://www.mturk.com.
[4]http://taptapseeapp.com/.
[5]http://www.bemyeyes.org.
[6]http://aira.io.
[7]http://http://tech.aipoly.com/.

■ **FIGURE 4.1** The *Glassense system* prototype.

4.3 THE GLASSENSE SYSTEM

Architecture. The *Glassense system* prototype is shown in Fig. 4.1. As for the audio channel, the system includes two superdirective microphone arrays positioned on the temples, and two earphones sending back the processed signals to the ears. The arrays are composed of equally spaced elements over 0.10 m. Their working frequency band ranges from 400 Hz to 4 kHz and the sampling frequency is 16 kHz. The array is superdirective for frequencies up to 2300 Hz (i.e. more than two octaves). The directivity index (DI) has a mean value of 8.6 dB over the array working frequency band. Each microphone feeds a FIR filter (i.e. a digital filter) of order 127.

For the visual channel, the glasses are equipped with a Logitech Carl Zeiss Tessar HD 1080p webcam, and a tiny line of LED illuminators to be switched on when the ambient lighting is insufficient. The webcam is set to acquire video frames at resolution 640 × 480 pixels and at a rate of 25 fps.

An elaboration board is used to acquire and process the signals, and to deliver the appropriate output to the user. It consists of a MIYR Z-turn, a low-cost Linux-based development board based on the ARM processing system, and it is paired with a custom daughterboard.

Interaction modes. On the software side, the glasses offer functionalities based on [5,42,45], which provide interaction modalities that may be illustrated as follows.

How the Glassense system can help visually impaired people to recognize objects? Fig. 4.2 sketches this interaction modality. The user is observing an object lying in the peri-personal space, and grabs it to understand its category. To this purpose, he/she exploits other sensory channels, using e.g. touch to estimate size, shape, and weight, sound to obtain a feedback on the object content. Once the category is identified, the user may be interested in recognizing a specific instance belonging to the category, e.g. a favorite product brand, or a specific kind. The *Glassense system* provides this additional visual contribution, with an algorithmic pipeline organized in three main steps: (i) The user provides a vocal input to the system, pronouncing the name of the object category. (ii) The image retrieval engine is triggered

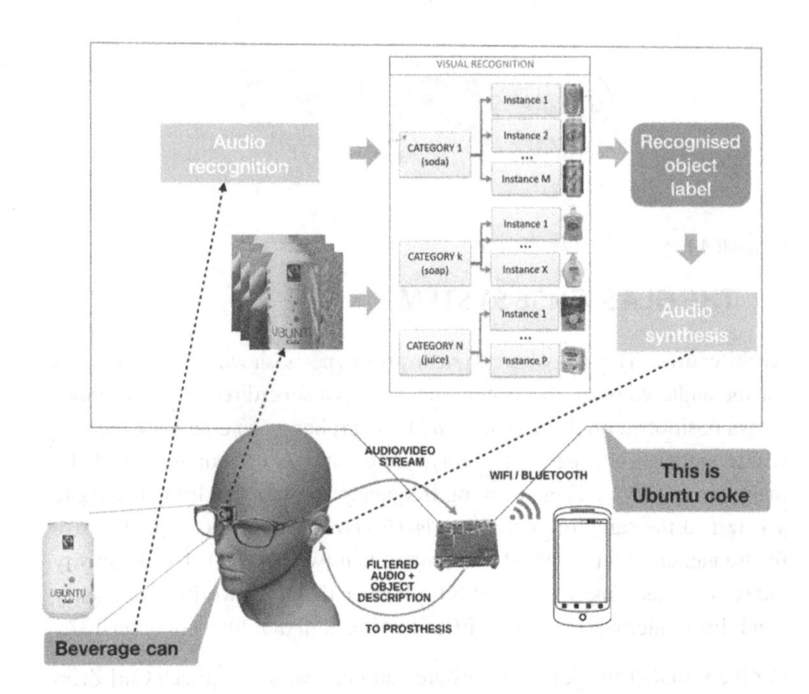

FIGURE 4.2 A visual representation of the interaction modality between the user and the *Glassense system* exploiting the visual recognition functionality.

to analyze the video stream and gather the object label, if any. (iii) The recognized label is provided by the system in the form of audio feedback.

This exemplified scenario suggests some properties of the desired architecture: (i) Since the object is placed in the peri-personal space of the user or it is hand-held, we can assume it occupies a large portion of the acquired image frames (no object detection is required). (ii) Since the objects in the category reflect the personal taste of users, we may assume the sets do not grow too much (of the order of tenths). On the other side, we need to consider some limitations due to the user's visual impairment, and as a consequence the method needs to be tolerant to viewpoint variations, changes in illumination, and to the presence of blur, motion blur, out of focus occurrences, and occlusions.

How does the Glassense system help a hearing impaired person to better understand speech in cocktail party noise? The peculiarity of our approach is that the microphone arrays in the *Glassense system* lie along the glasses temples. As shown in Fig. 4.3A, a listener (seen from above) stands among three talkers. The listener has a hearing disability at the right ear (in red). The voice intensity of the three speakers is the same and the cocktail party problem occurs. However, our audio filters in the Glassense system are designed

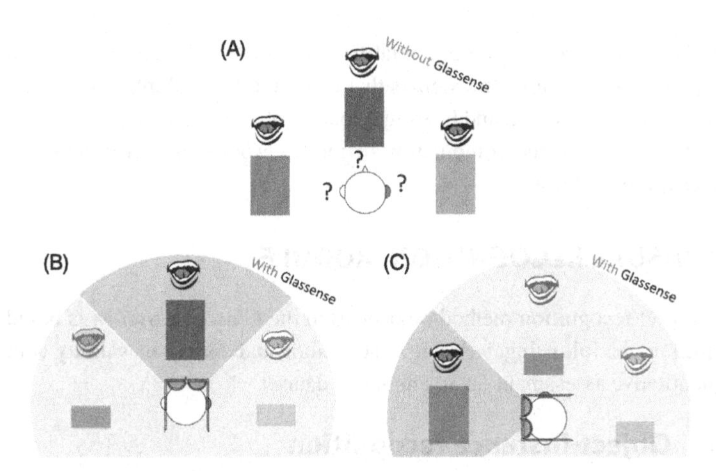

■ **FIGURE 4.3** A) Use case without audio beamforming. A listener is surrounded by three speakers, active at the same time (cocktail party) with equal loudness (colored rectangles). B) Use case with audio beamforming. The beamforming algorithm creates a sweet spot (yellow) where the loudness is doubled with respect to sound sources outside the beam (in gray background), which result attenuated (smaller rectangles). C) Use case same as B but with the listener actively choosing the source of interest by just turning the head.

so that the face of the listener, when looking at the speaker, puts him/her in the middle of a sound beam (subfigure B). When inside the beam, the speech frequency band is increased as opposed to sound sources outside the beam, which are attenuated. Therefore, (subfigure C) the listener chooses which speech sound sources have to be increased by naturally turning the head. Then, the enhanced speech can be sent either to a pair of headphones or to a pair of acoustic prostheses.

Solution for our user stories. With such interaction modes, Tina (US1) needs not to shake the boxes, as visual cues are captured by the camera. All she has to do is to make sure that the object is in the field of view of the camera, which can be easily done by holding the object in two hands or with some training. What can be gained from the interaction is **increased knowledge**. Tony (US2), as well, may not ask a sighted person for help, as the object recognition engine would suggest which type of biscuit he is holding. What can be gained from the interaction is **increased autonomy**. Lisa (US3), thanks to the directional audio, can now understand how beautiful the seaside is. All she needs to do is look at the person that she wants to understand better. What can be gained from the interaction is **more social interaction**. Andrew (US4) is now more confident with his glasses, since he can more clearly understanding his co-workers in noisy environments. What can be gained from the interaction is **more work efficacy and a better satisfaction from the workplace**. Finally, George (US5) benefits from both technologies, as lack of vision and hearing can be partially compensated by the *Glassense system*. As a potential benefit, not shown here, the camera can process many biometric signals that can serve for multiple disabilities, therefore both for visual and hearing impairment. What can be gained from the interaction and the combination of multisensory aids is **a potentially better quality of life**.

4.4 VISUAL RECOGNITION MODULE

The object recognition method embedded in the *Glassense system* is based on [42]. In the following, we review the method and discuss its validity with a quantitative assessment on an annotated dataset.

4.4.1 Object-instance recognition

Modeling phase. A Bag-of-Features (BoF) procedure is adopted to obtain a single global image vector from a set of local features, the SURF [2]). According to [21], let us define $\mathcal{X}$ a set of local features from an image I such that $\text{card}(\mathcal{X}) = n$. The embedding step ϕ: $\mathbb{R}^d \rightarrow \mathbb{R}^D$ encodes each feature $x \in \mathcal{X}$ into a new vector as

$$x \rightarrow \phi(x) \tag{4.1}$$

where D is the size (i.e. the number of visual words of a *visual dictionary* $\mathcal{D}$) used to encode the descriptors with respect to its visual words, i.e. local visual concepts, meaningful for recognition.

Let $\mathcal{C} = \{C_1, \ldots, C_M\}$ be a category with M category instances. $\mathcal{C}$ is modeled starting from a relatively small gallery $\mathcal{G}$ of annotated images of the different object instances (the *training set*):

$$\mathcal{G} = \left\{(I_1^1, 1), \ldots, (I_1^n, 1), (I_2^1, 2), \ldots, (I_2^m, 2), \ldots, (I_M^1, M), \ldots, (I_M^p, M)\right\}$$

where a input–output pair (I_i^j, i) is the jth image associated with the instance (or label) i. Thanks to the assumptions suggested by the application domain (see Section 4.3), the size of the gallery can be kept under control, while it is advisable to include images of the object from different but tactile ambiguous viewpoints. The gallery images should be acquired on a uniform background in order to only extract relevant information from the considered object.

We learn a visual dictionary $\mathcal{D}$ per category using K-means. Then, for each image we consider, we extract a sparse set of SURF and derive a BoF feature vector by quantizing the image features with respect to D (using L2-norm distance measure). Finally, we normalize the BoF feature vector $\mathbf{v} = (v_1, ..., v_D)$ by the power-law normalization [20], which has been shown to be very effective in applications. The updated vector v is then L2-normalized to reduce the impact of illumination changes.

At the end of the procedure the *category model* includes the dictionary of visual words and the gallery feature vectors

$$\mathcal{M}=\{(\mathbf{v}_1^1,1),\dots(\mathbf{v}_1^n,n),\dots,(\mathbf{v}_2^1,2),\dots(\mathbf{v}_2^m,2),\dots(\mathbf{v}_M^1,M),\dots(\mathbf{v}_M^p,M)\},$$

where $\mathbf{v}_i^j$ represents a feature vector describing the content of the jth gallery image of the ith object instance (image I_i^j) of the gallery.

Runtime analysis. In the *online* phase, each video frame I_t (the query image) is represented coherently with the category model selected by the user. We obtain a feature vector $\mathbf{v}_t$ which is processed with a retrieval procedure with respect to the gallery $\mathcal{G}$. The output of the retrieval step is a ranking of the gallery images based on their visual similarity with the query image. We then derive a recognition label from this ranking, using a voting procedure. We consider the first k positions in the ranking and incorporate a rejection rule as follows: if the similarity value between the query image and the kth image in the obtained ranking is lower than a threshold τ, the image is classified as a negative. Otherwise we associate the query image with the label which is majority in the first k positions. The whole recognition procedure is repeated for each available video frame and returns a feedback to the output module as soon as it accumulates at least Z consecutive coherent answers.

4.4.2 Experimental assessment

The goal of this section is to assess the visual recognition module with a quantitative experimental analysis on a set of images acquired for the task.

The *Glassense-Vision* dataset.[8] It is unique and different from other objects recognition benchmarks, since it provides a very specific observation point—the point of view of a user close to or manipulating the object. It has been acquired by taking into account all prior information available to a BVI user. The dataset includes seven object categories, which we will refer to as *recognition use cases*. These can be grouped in three main geometrical types: (i) *Flat objects*—which include the Euro banknote use case, which is a particularly meaningful flat, nonrigid, degradable type of object; (ii) *Boxes—Parallelepiped objects*—boxes, which include cereal-box and medicine-box use cases; (iii) *Cylindrical objects*—bottles and jars, which include water bottles, bean cans, tomato sauce jars and deodorants use

[8] The dataset is available for download at http://slipguru.unige.it.

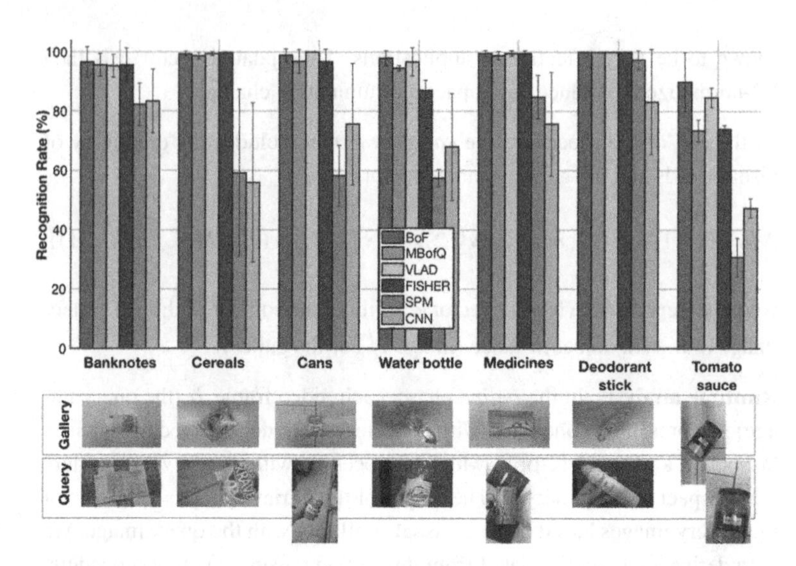

■ **FIGURE 4.4** Recognition rate of each descriptor considered in the experimental assessment averaged over all the query sets. Below, examples of gallery and query images are reported.

cases. Both gallery—used to train the models—and query—used for tests—images have been acquired. Gallery images have been acquired in a relatively uniform setting, while most query data have been acquired with the object hold in the user's hands and in a variety of conditions—background clutter, rotations, viewpoint changes, occlusions. We will refer to images reflecting these conditions as *query sets*. All images have been acquired by using two cellphones and stored at a resolution of 1182 × 665 pixels. Overall, three users participated to the acquisition. In Fig. 4.4, bottom, we report some examples of gallery and query images for each recognition use case.

Experiments. In this section we provide a quantitative assessment of the proposed method on the acquired dataset. More specifically, we compare our BoF representation with different global descriptors from the image retrieval literature: MBoFQ [41], VLAD [20], Fisher vector [34] and SPM [25], neural codes extracted from a pre-trained Convolutional Neural Network (CNN) [1]. We used the Computer Vision System Toolbox[9] from Matlab for extracting and describing SURF features, our implementation of the SPM, MBoFQ and BoF methods, and VLFeat Toolbox[10] for Fisher and VLAD vectors.

[9] http://www.mathworks.com/products/computer-vision/.

[10] http://www.vlfeat.org/.

For CNN-based image representations, we used a deep network trained on ILSVRC 2012 [36] from the *Caffe* framework.[11] We specifically use the output of Layer 6 as image vector, since it provides better semantic information of the image content (see [1] for further details). Layer 6 produces an output vector of size 4096.

In all experiments, the vector size of Fisher, VLAD, SPM and MBoFQ is equal to 20,000, while BoF is equal to 10,000. As for the choice of k, we experimentally observed that it is influenced by the number of images per instance in the gallery. In our experiments all the galleries have a comparable size with the exception of banknotes which is much larger. Thus we set $k = 7$ for the banknotes use case, and $k = 3$ for all the rest.

The rejection threshold τ is fixed automatically during the training phase by a leave-one-out procedure on the gallery, by fixing a maximum false positive rate to 10^{-2}.

Fig. 4.4 summarizes the performance that each descriptor obtained on average on the four query sets. For all the descriptors, the L2 measure is adopted for comparison.

Our BoF descriptor reached recognition rates higher than or comparable with the other descriptors, which overall achieve very good performance. An exception is the SPM descriptor, which is not rotation invariant. It can be observed that the performance of CNN-based image representations is in general lower than the rest of image descriptors. These results may be improved with a fine-tuning procedure[23], which however requires a large number of images and classes in order to reach a properly trained network, obstructing its use in scalable applications.

Table 4.1 reports a closer view on the BoF results, shown separately for each of the recognition use case and query sets. Overall, the recognition rates are remarkable. As a pointer to a different alternative, we also considered a different pipeline based on the use of state-of-the-art deep networks. In the last column of the table we report the average recognition rates obtained with the Caffe network [22] trained on ImageNet [36] and fine tuned on 5000 images acquired by our system for the purpose of specializing the descriptor to our case and to be able to use the last layer of the net as a single instance classifier. These images include background, rotation, viewpoint changes and partial occlusions of all the seven categories we are considering. We selected the model with the highest validation accuracy (98%). Although it is fair to mention that the network solves a more difficult classification problem—as it does not rely on an input from the user on the

[11] http://caffe.berkeleyvision.org/.

Table 4.1 Recognition rate of the BoF representation on each *Glassense-Vision* dataset use case

Use case	Background	Rotation	Viewpoint	Occlusion	Avg. RR	Caffe net Avg. RR (%)
Banknotes	100	100	97.22	89.15	97.57	75
Cereals	100	100	98.19	98.95	98.92	86.16
Cans	100	95.83	100	100	98.21	94.64
Water bottle	96.67	100	95	100	97.61	90.48
Medicines	98.14	100	100	100	99.49	90.91
Deodorant stick	100	100	100	100	100	87.88
Tomato sauces	95.45	95.45	75	93.75	92.10	93.42

intended category—it can be observed that the overall recognition rates are lower than the ones obtained by our system.

In preparation to the use of the method on the *Glassense* prototype, where the webcam has a resolution of 640×480 pixels, we performed an evaluation of the recognition rates with images resized to $\sim 60\%$ of the original dimensions. The performance remained, on average, above the 90%.

4.5 COMPLEMENTARY HEARING AID MODULE

In this section we describe our tool to assist people with partial hearing loss. Data-independent filter-and-sum beamforming [45] has been used to design the end-fire linear microphone arrays located in the temple of the *Glassense* prototype. The array has four equally spaced elements over 0.10 m. The working frequency band ranges from 400 Hz to 4 kHz, i.e. more than three octaves, and the frequency sampling is 16 kHz. The adopted technique allows one to attain a beamformer robust to array imperfections, like mismatches between microphones response and displacement errors in the element position.

4.5.1 Measurement of *Glassense* beam pattern

After the realization of the device some measurements have been performed in an anechoic boot to verify the consistency between the theoretical directional pattern of the synthesized filter and the physical implementation. Two different conditions have been observed in the positioning of the device: in the first one the microphone array has been suspended at about 1 m from the ground to avoid any possible reflection, while in the second one a mannequin built with biocompatible materials has been used as support for the glasses (see Fig. 4.5). The input signals were reproduced through a Behringer active loudspeaker positioned 70 cm away from the device. The measuring protocol (repeated twice, once with the single temple and

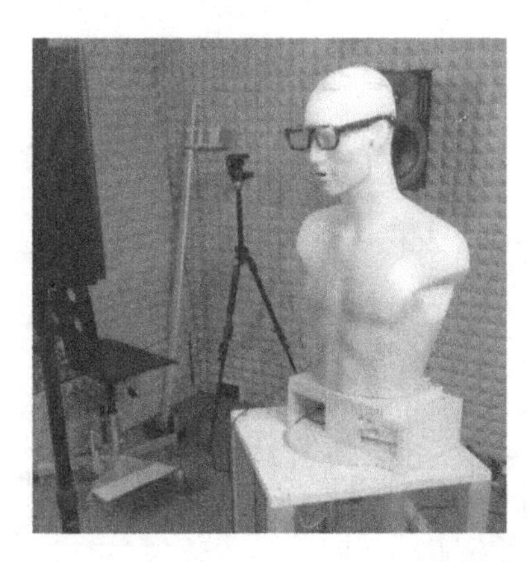

■ **FIGURE 4.5** The experimental setup, with the *Glassense* prototype positioned on the head of the mannequin.

once with the glasses on the mannequin) started with the device beam pointing exactly towards the speaker. Pure tones at various frequencies (between 250 Hz and 4500 Hz) were reproduced in sequence for some seconds each from the speaker, while being spatially filtered and recorded by the array. After the conclusion of the reproduction the devices were rotated by 15 degrees. The tones were repeated, filtered and recorded again for each position until we came back to the initial one.

At the end of the protocol, a recording of each pure tone in 24 different positions was stored and the level of spectral power of the signal at the tone frequency has been calculated and plotted, obtaining so-called beam power patterns.

4.5.2 **Analysis of measured beam pattern**

The beam power pattern shows the attenuation applied by the filter at the surrounding sounds depending on the frequency and the angle. The comparison between the theoretical expected pattern (Fig. 4.6), and the measured one (Fig. 4.7 on the left), shows that the latter one is pretty consistent with the former. Sounds coming from the back are attenuated between 5 dB and 15 dB, depending on the frequency. In most of the working frequencies of the filter it is possible to notice two main lobes, with the main one pointing towards the direction of interest and the other one directed symmetrically on the other side, which is the back. At 4 kHz sidelobes start to appear between the main and secondary lobes. Fig. 4.7 on the right shows the plots related to the binaural beam power pattern measured with the *Glassense* prototype positioned on the head of the mannequin. The black lines indicate the omnidirectional patterns, obtained recording one omnidirectional microphone on each temple, without any filtering. The red lines are the pattern of the

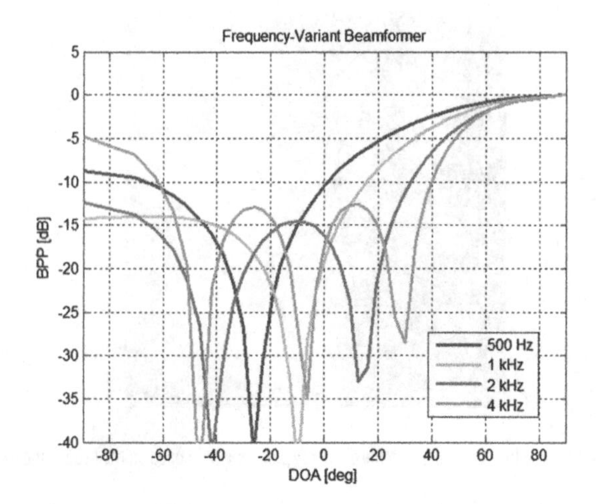

■ **FIGURE 4.6** Theoretical beam power pattern of *Glassense* prototype directional filter, plotted in Cartesian form.

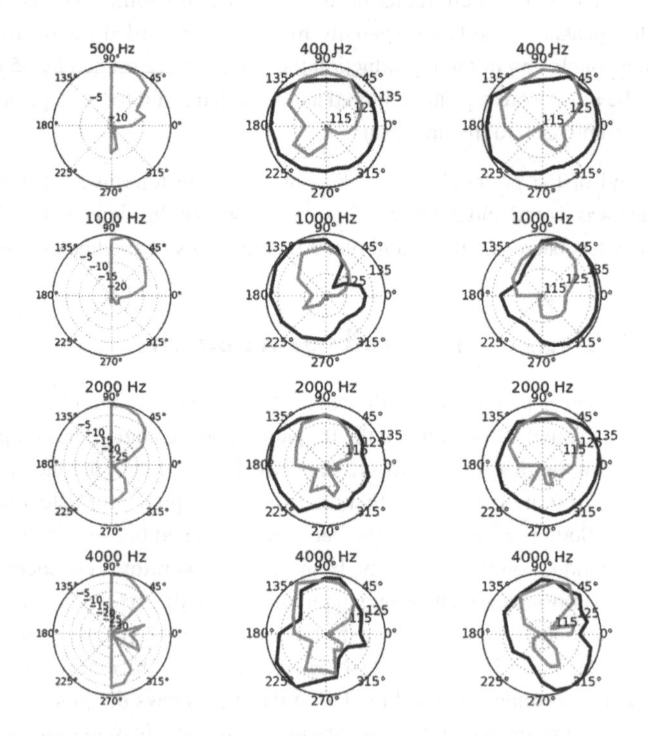

■ **FIGURE 4.7** (Left) Measured BPP, relative to a single temple array of the *Glassense* prototype, plotted at various frequencies in polar form. (Right) Measured binaural BPP of the *Glassense* prototype at various frequencies. The black line indicates the omnidirectional condition; the red line the directional one.

directional filter; differences between these lines and those on the left plots are a consequence of the presence of the head and torso. It is possible to notice that in this condition the patterns lose their symmetry and become more irregular.

4.6 ASSESSING USABILITY WITH IMPAIRED USERS

In this section we report the results of a testing session with BVI and hearing impaired users, involving the use of the *Glassense system*. The users represent the target of the project (older users and/or users with multiple disabilities and low familiarity with digital devices).

The *Glassense BVI* users group consists of 14 visually impaired individuals (selected from a larger group identified by the staff of *Istituto David Chiossone for the blind*.[12] The *Glassense Hearing Impaired* users group included six subjects impaired by light to severe sensorineural or mixed (sensorineural and transmissive) hearing loss. Three subjects have been included in both groups because they were affected by visual and hearing impairment. All subjects underwent a visual assessment tests (performed by Chiossone Institute staff), a hearing evaluation test (run from Linear audiometrist staff) and a cognitive assessment (performed by neurologists and neuropsychologists of the Fides Group[13]). The objective of the tests was to identify a set of subjects representing the profile of users which could profitably benefit from the proposed technology. The end users partners had identified the following *inclusion criteria* for participating to the users group: (i) No cognitive impairment, (ii) Visual impairment classified as fragile sight (visual acuity in the range [3/10–8/10] and binocular perimeter residual in the range [60%–80%]) or low vision (visual acuity in the range [1/20–3/10] and binocular perimeter residual in the range [10%–50%]), (iii) Sensorineural or mixed (sensorineural and transmissive) hearing loss in the range [21–80] dB. The 14 BVI subjects (age in the range [18–75]) met the above criteria for visual and cognitive impairment. They all have low vision: two of them are visually impaired since birth, two since early age due to an inherited condition, the rest from adult age, either caused by pathologies (diabetes, glaucoma), For five of them the impairment is age-related. The users were familiar with optical and electronic devices to assist them in daily life, in particular magnification devices. Five volunteers own a smart phone, and three of them use automatic OCR systems on a daily basis. None of them had used an object recognition app before. Two older volunteers admitted they are usually not comfortable with digital devices.

---

[12] http://www.chiossone.it.

[13] http://www.gruppofides.it.

The field tests have been performed at Istituto Chiossone, a familiar environment to the users, and at Linear s.r.l for the hearing part of the tests.

4.6.1 *Glassense* field tests with visually impaired

Tests have been performed one subject at a time, according to the following protocol:

- The subject wears the sensorized glasses and sits at a table.
- The system transmits the audio message "Say what category of objects you are interested in."
- The subject may select a category among the possible choices by pronouncing a simple keyword.
- The operator puts on the table a selection of objects belonging to the selected class.
- The user takes one object, holds it for a few seconds in front of the face. The users have been instructed to behave as naturally as possible and, in the case they do not receive feedback, to move the object in various poses, rotating it and moving it back and forth.
- When the system recognizes the object, it returns an audio feedback to the user through the glasses microphones.
- The recognition result is marked by the operator, then the user may select another object of the same class, request another class, or quit the test.

Users did not undergo a training session and had the opportunity to test the system as long as they wished and with the type of objects they preferred. All users tested the system with at least three categories. The overall recognition rate obtained was $\sim 98\%$. The users were positively impressed by the simplicity of use, the effectiveness of the answers, and the feedback speed and clarity.

4.6.2 *Glassense* field tests with binaural hearing loss

Setup. The experiment was held in a sound attenuated audiometric booth ($T60 = 0.2$ s). Each subject was surrounded by four speakers, located at a distance of 1 m from the head. They were positioned at 0° (directly ahead), 180° (back) and ±90° on either side of the midline. The subjects were instructed to look forward, without head constraints. The setup worn by the subjects consisted in the *Glassense* prototype plus a couple of completely-in-the-canal (CIC) hearing prostheses realized and fitted on their hearing loss by Linear s.r.l. Healthy hearing subjects also wore prostheses configured without hearing loss compensation. The glasses were connected in order to send audio signals to the prostheses by means of Near-Field Magnetic Induction (NFMI) technology through a relay device.

The speech stimuli used in the tests consisted of 20 lists of 10 bisyllabic Italian words as target and a four channel registration of a cocktail party speech as competing noise. During the execution the intensity of the target signal was fixed and the intensity of the noise was changed in order to explore different Signal-to-Noise Ratios (SNRs).

Procedure. For each subject, the first part of the tests consisted of free ear speech audiometry with the aim of verifying the intensity of the target speech signal at which they could understand 100% of the administered words. The chosen level of intensity was then fixed across all the trials. Each trial consisted of a listen and repeat task, where the subjects had to listen to the target words reproduced by the frontal speaker and repeat them, while the cocktail party noise was reproduced by all the four speakers.

Two independent variables affected the trials:

- The SNR value, which spanned between −15 and +15 dB.
- The listening modality, which had three possible values (randomized across subjects):
  1. **Hearing Aids**—The subject listened directly by the hearing aid.
  2. **Unfiltered**—The subject listened the audio recorded by the *Glassense system* using only one omnidirectional microphone for each temple and without any filtering algorithm.
  3. **Beamforming**—The subject listened to the audio recorded by the *Glassense system* applying beamforming filtering.

For each task the registered performance consisted of the percentage of correctly repeated words.

Data analysis and results. Each experimental condition produced a dependent variable consisting of the percentage of words the subject could correctly repeat. For each listening modality the variation of the SNR values produced similar performance profiles, characterized by a better performance associated to higher SNR values (i.e. lower noise intensity) and vice versa (see Fig. 4.8, left). We interpolated the performance over SNR values obtaining three logistic curves for each subject, related to the listening modalities. From each curve we obtained the SNR value corresponding to the standard threshold of 50% performance, known as the Speech Reception Threshold (SRT) [16].

The statistical analysis of the data consisted of a Friedman Test, followed by Wilcoxon–Mann–Whitney with Bonferroni correction between the SRT population of the beamforming listening condition against the two others. The results (Fig. 4.8, right) revealed a statistically significant improvement in the first: the beamforming modality mean SRT led to 2.4 dB lower re-

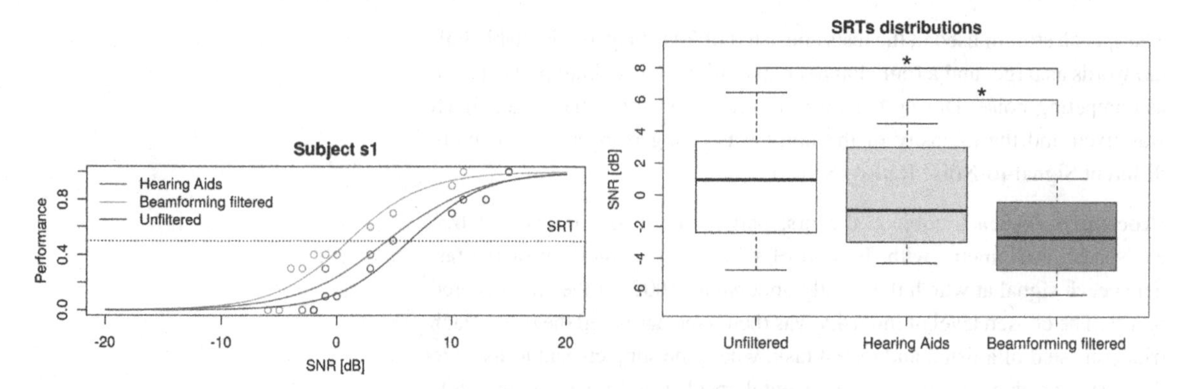

■ **FIGURE 4.8** Audiometric curves of a subject in the different listening modalities and distribution of the SRT values.

sults with respect to the hearing aid (p-value = 0.007813) and 3.3 dB lower results with respect to the unfiltered modality (p-value = 0.02344).

In conclusion, the results of the analysis show that the use of microphone arrays integrated in a pair of glasses to spatially filter sounds can improve the speech comprehension capabilities of hearing impaired people in noisy environments. Further studies will be conducted in order to verify the effectiveness of the *Glassense* prototype directional filter as a complementary device for modern hearing aids.

4.7 CONCLUSION

This chapter discussed the technologies devised in *Glassense—Wearable technologies for sensory supplementation*, a regional project whose aim was to develop a proof of concept prototype of a sensorized pair of glasses to assist users with limited technology skills and possibly multiple disabilities. The potential beneficiaries of this technology are primarily older users with low vision and/or hypoacusis.

For people with low vision, the glasses offer an object recognition functionality based on a collaborative principle: the user provides information on an object category of interest with the voice, the system then performs object-instance recognition and delivers an audio feedback, if any, to the user. The method is based on a content-based image retrieval engine, which guarantees fast processing and scalability. An experimental assessment of a batch dataset, the *Glassense-Vision* set, resembling the application domain of interest, shows the appropriateness of the strategy, with average recognition rate above 97% on varieties of scene conditions.

We have shown that it is possible to support partial hearing loss with a wearable multi-microphone device, able to partially isolate unwanted sound sources during speech understanding tasks. Specifically, we have shown that with our architecture the sounds coming from the back are attenuated between 5 dB and 15 dB. This results in improving the Speech Reception Threshold of about 2.4 dB when comparing the use of hearing aids alone to the use of the hearing aids coupled with a beamforming algorithm. This means that to maintain the same number of words that are understood in noise, a user can stand a significantly louder noise. We emphasize that our test have been carried with cocktail party noise, which is speech: while non-speech noises (fan, engines, alarms) can be isolated by state-of-the-art techniques that act on the spectral characteristics of sounds, this is not possible when the disturbing noise is speech itself. That is why *filtering in space* in addition to *filtering in frequency* is necessary.

The users feedback we obtained from a questionnaire we delivered to the volunteers is very encouraging and motivates future research. The users had to provide a score in the range 1 (low) to 5 (high) to a selection of questions, with the purpose of evaluating their experience and the proposed system. Overall, the users have been positively impressed by the prototype: they returned a high score to *"I would like to have it"*, *"easy to use"*, *"well integrated"*. They also confirmed they have high expectations, feeling the system would have a very positive impact to the quality of their lives. Overall, they do not feel their class of users would need too much technical support or training for using the glasses.

Future developments will include an easy-to-use interface for personalizing categories and object instances depending on the subject profile. In this phase the help of a sighted user will be needed only to guarantee a good quality of the gallery images.

The *Glassense system* is currently undergoing a major re-engineering of the prototype. With respect to its potential competitors in the market, such as Orcam glasses[14] and the platform based on Google glasses developed by Aira,[15] our prototype is meant as an open platform able to incorporate different sensors and software solutions to address different types of disabilities (including vision, hearing, and in the near future cognitive impairments).

REFERENCES

[1] A. Babenko, A. Slesarev, A. Chigorin, V. Lempitsky, Neural codes for image retrieval, in: ECCV, 2014, pp. 584–599.

[2] H. Bay, A. Ess, T. Tuytelaars, L. Van Gool, Speeded-up robust features (surf), Comput. Vis. Image Underst. 110 (3) (2008) 346–359.

[14] http://www.orcam.com.

[15] https://aira.io.

[3] J.P. Bigham, C. Jayan, A. Miller, B. White, T. Yeh, VizWiz:LocateIt-enabling blind people to locate objects in their environment, in: CVPR-W, 2010, pp. 65–72.
[4] M.M. Boone, Directivity measurements on a highly directive hearing aid: the hearing glasses, in: Audio Engineering Society Convention 120, 2006.
[5] L. Brayda, F. Traverso, L. Giuliani, F. Diotalevi, S. Repetto, S. Sansalone, A. Trucco, G. Sandini, Spatially selective binaural hearing aids, in: ACM Intern. Symp. on Wearable Computers, 2015, pp. 957–962.
[6] M. Buhrmester, T. Kwang, S.D. Gosling, Amazon's mechanical turk a new source of inexpensive, yet high-quality, data?, Perspect. Psychol. Sci. 6 (1) (2011) 3–5.
[7] M. Chessa, N. Noceti, F. Odone, F. Solari, J. Sosa-García, L. Zini, An integrated artificial vision framework for assisting visually impaired users, Comput. Vis. Image Underst. 149 (2016) 209–228.
[8] R. Chincha, Y. Tian, Finding objects for blind people based on surf features, in: IEEE BIBMW, 2011, pp. 526–527.
[9] J. Coughlan, R. Manduchi, Functional assessment of a camera phone-based wayfinding system operated by blind and visually impaired users, Int. J. Artif. Intell. Tools 18 (03) (2009) 379–397.
[10] G. Csurka, C. Dance, L. Fan, J. Willamowski, C. Bray, Visual categorization with bags of keypoints, in: SLCV-ECCV, vol. 1, 2004, p. 22.
[11] S. Doclo, M. Moonen, T. Van den Bogaert, J. Wouters, Reduced-bandwidth and distributed MWF-based noise reduction algorithms for binaural hearing aids, IEEE Trans. Audio Speech Lang. Process. 17 (1) (2009) 38–51.
[12] S.R. Fanello, N. Noceti, C. Ciliberto, G. Metta, F. Odone, Ask the image: supervised pooling to preserve feature locality, in: CVPR, 2014, pp. 851–858.
[13] J.F. Feuerstein, Monaural versus binaural hearing: ease of listening, word recognition, and attentional effort, Ear Hear. 13 (2) (1992) 80–86.
[14] M. Froehlich, K. Freels, T.A. Powers, Speech recognition benefit obtained from binaural beamforming hearing aids: comparison to omnidirectional and individuals with normal hearing, Audiology (Online) (2015).
[15] G. Fusco, E. Tekin, N.A. Giudice, J.M. Coughlan, Appliance displays: accessibility challenges and proposed solutions, in: SIGACCESS, ASSETS '15, 2015, pp. 405–406.
[16] L. Giuliani, L. Brayda, S. Sansalone, S. Repetto, M. Ricchetti, Evaluation of a complementary hearing aid for spatial sound segregation, in: IEEE ICASSP, 2017, pp. 221–225.
[17] F.M. Hasanuzzaman, X. Yang, Y. Tian, Robust and effective component-based banknote recognition for the blind, IEEE Trans. Syst. Man Cybern., Part C, Appl. Rev. 42 (6) (2012) 1021–1030.
[18] S. Haykin, Z. Chen, The cocktail party problem, Neural Comput. 17 (9) (2005) 1875–1902.
[19] R. Jafri, S.A. Ali, H.R. Arabnia, S. Fatima, Computer vision-based object recognition for the visually impaired in an indoors environment: a survey, Vis. Comput. 30 (11) (2014) 1197–1222.
[20] H. Jégou, F. Perronnin, M. Douze, C. Schmid, et al., Aggregating local image descriptors into compact codes, IEEE Trans. Pattern Anal. Mach. Intell. 34 (9) (2012) 1704–1716.
[21] H. Jégou, A. Zisserman, Triangulation embedding and democratic aggregation for image search, in: CVPR, 2014, pp. 3310–3317.
[22] Y. Jia, E. Shelhamer, J. Donahue, S. Karayev, J. Long, R. Girshick, S. Guadarrama, T. Darrell, Caffe: convolutional architecture for fast feature embedding, in: ACM Intern. Conf. on MM, 2014, pp. 675–678.
[23] A. Krizhevsky, I. Sutskever, G.E. Hinton, ImageNet classification with deep convolutional neural networks, in: Advances in Neural Information Processing Systems, 2012, pp. 1097–1105.

[24] R. Kumar, S. Meher, A novel method for visually impaired using object recognition, in: ICCSP, 2015, pp. 0772–0776.
[25] S. Lazebnik, C. Schmid, J. Ponce, Beyond bags of features: spatial pyramid matching for recognizing natural scene categories, in: CVPR, vol. 2, 2006, pp. 2169–2178.
[26] X. Liu, A camera phone based currency reader for the visually impaired, in: ACM SIGACCESS, 2008, pp. 305–306.
[27] R. Manduchi, J. Coughlan, (computer) vision without sight, Commun. ACM 55 (1) (2012) 96–104.
[28] M.L. Mekhalfi, F. Melgani, Y. Bazi, N. Alajlan, Toward an assisted indoor scene perception for blind people with image multilabeling strategies, Expert Syst. Appl. 42 (6) (2015) 2907–2918.
[29] M.M.L. Mekhalfi, F. Melgani, A. Zeggada, F.G.B. De Natale, M.A.-M. Salem, A. Khamis, Recovering the sight to blind people in indoor environments with smart technologies, Expert Syst. Appl. 46 (2016) 129–138.
[30] L.H.M. Mens, Speech understanding in noise with an eyeglass hearing aid: asymmetric fitting and the head shadow benefit of anterior microphones, Int. J. Audiol. 50 (1) (2011) 27–33.
[31] I. Léon, D.M. Merks, Binaural application of microphone arrays for improved speech intelligibility in a noisy environment, 2000.
[32] S. Nanayakkara, R. Shilkrot, K.P. Yeo, P. Maes, Eyering: a finger-worn input device for seamless interactions with our surroundings, in: ACM Augm. Human Intern. Conf., 2013, pp. 13–20.
[33] N. Ohnishi, T. Matsumoto, H. Kudo, Y. Takeuchi, A system helping blind people to get character information in their surrounding environment, in: ACM SIGACCESS, 2013, p. 34.
[34] F. Perronnin, Y. Liu, J. Sánchez, H. Poirier, Large-scale image retrieval with compressed fisher vectors, in: CVPR, 2010, pp. 3384–3391.
[35] J. Philbin, O. Chum, M. Isard, J. Sivic, A. Zisserman, Object retrieval with large vocabularies and fast spatial matching, in: CVPR, 2007, pp. 1–8.
[36] O. Russakovsky, J. Deng, H. Su, J. Krause, S. Satheesh, S. Ma, Z. Huang, A. Karpathy, A. Khosla, M. Bernstein, A.C. Berg, L. Fei-Fei, ImageNet Large Scale Visual Recognition Challenge, Int. J. Comput. Vis. 115 (3) (2015) 211–252.
[37] B. Schauerte, M. Martinez, A. Constantinescu, R. Stiefelhagen, An Assistive Vision System for the Blind that Helps Find Lost Things, Springer, 2012.
[38] R. Shilkrot, J. Huber, C. Liu, P. Maes, S.C. Nanayakkara, Fingerreader: a wearable device to support text reading on the go, in: ACM CHI, 2014, pp. 2359–2364.
[39] J. Sivic, A. Zisserman, Video Google: a text retrieval approach to object matching in videos, in: ICCV, 2003, pp. 1470–1477.
[40] Z. Solymár, A. Stubendek, M. Radványi, K. Karacs, Banknote recognition for visually impaired, in: ECCTD, 2011, pp. 841–844.
[41] J. Sosa-García, F. Odone, Mean BoF per quadrant – simple and effective way to embed spatial information in bag of features, in: VISAPP, 2015, pp. 297–304.
[42] J. Sosa-García, F. Odone, "Hands on" visual recognition for visually impaired users, ACM Trans. Access. Comput. 10 (3) (2017) 8.
[43] J. Sudol, O. Dialameh, C. Blanchard, T. Dorcey, LookTel—a comprehensive platform for computer-aided visual assistance, in: CVPR-W, 2010, pp. 73–80.
[44] Y. Tian, X. Yang, C. Yi, A. Arditi, Toward a computer vision-based wayfinding aid for blind persons to access unfamiliar indoor environments, Mach. Vis. Appl. 24 (3) (2013) 521–535.
[45] A. Trucco, F. Traverso, M. Crocco, Maximum constrained directivity of oversteered end-fire sensor arrays, Sensors 15 (6) (2015) 13477–13502.
[46] R.J.M. Van Hoesel, G.M. Clark, Evaluation of a portable two-microphone adaptive beamforming speech processor with cochlear implant patients, J. Acoust. Soc. Am. 97 (4) (1995) 2498–2503.

[47] B. Widrow, A microphone array for hearing aids, in: Adaptive Systems for Signal Processing, Communications, and Control Symposium 2000. AS-SPCC. The IEEE 2000, 2000, pp. 7–11.

[48] T. Winlock, E. Christiansen, S. Belongie, Toward real-time grocery detection for the visually impaired, in: CVPR-W, 2010, pp. 49–56.

[49] X. Yang, Y. Tian, C. Yi, A. Arditi, Context-based indoor object detection as an aid to blind persons accessing unfamiliar environments, in: ACM Int. Conf. on MM, 2010, pp. 1087–1090.

[50] H. Ye, M. Malu, U. Oh, L. Findlater, Current and future mobile and wearable device use by people with visual impairments, in: SIGCHI, 2014, pp. 3123–3132.

[51] C. Yi, Y. Tian, A. Arditi, Portable camera-based assistive text and product label reading from hand-held objects for blind persons, IEEE/ASME Trans. Mechatron. 19 (3) (2014) 808–817.

[52] A. Zeb, S. Ullah, I. Rabbi, Indoor vision-based auditory assistance for blind people in semi controlled environments, in: IPTA, 2014, pp. 1–6.

Chapter 5

Audio-visual learning for body-worn cameras

Andrea Cavallaro*, Alessio Brutti†
*Centre for Intelligent Sensing, Queen Mary University London, London, UK
†Center for Information Technology, Fondazione Bruno Kessler, Trento, Italy

CONTENTS

5.1 INTRODUCTION

A wearable camera is an audio-visual recording device that captures data from an ego-centric perspective. The camera can be worn on different parts of the body, such as the chest, a wrist, or the head (e.g. on a cap, on a helmet, or embedded in eyewear).

Body-worn cameras support applications such as life-logging and assisted living [30,50], security and safety [3,48], object and action recognition [28], physical activity classification of the wearer [21,27,38], and interactive augmented reality [33]. In particular, recognizing objects or people from an ego-centric perspective is an attractive capability for tasks such as diarization [46], summarization [17,30], interaction recognition [22], and person re-identification [9,23]. However, despite two pioneering multi-modal works presented already in 1999 and 2000 [14,13] and the availability of at least one microphone in wearable cameras [7], video-only solutions are still predominant even in applications that could benefit from multi-modal first-person observations [17,21–23,27,28,30,38]. While a few exceptions exist,

Multimodal Behavior Analysis in the Wild. https://doi.org/10.1016/B978-0-12-814601-9.00014-6
Copyright © 2019 Elsevier Ltd. All rights reserved.

■ **FIGURE 5.1** Sample challenging frames from a wearable camera.

multi-modal approaches combine audio or video with other motion-related sensors [25,45,47].

Wearable cameras can be used in a wide variety of scenarios and require new processing methods that adapt to different operational conditions, e.g. indoors vs. outdoors, well lit vs. poorly lit environments, sparse vs. crowded scenes. In this chapter, we focus on video and audio challenges.

Video processing challenges are related to the (unconventional) motion of a body-worn camera that leads to continuous and considerable pose changes of the objects being captured. Moreover, objects of interest are often partially occluded or outside the field of view. In addition, the rapid movements of body-worn cameras degrade the quality of the captured video due to motion blur and varying lighting conditions. Changes in lighting produce under or over exposed frames, especially when the camera is directed towards or away from a light source (see Fig. 5.1).

Audio processing challenges arise from the use of compact cases with (typically) low-quality microphones co-located with the imager. Moreover, plastic shields protecting the camera (e.g. GoPro Hero) may cover the microphones, attenuating and low-passing the incoming sound waves. Then, the variety of use conditions include changing and unfavorable acoustics and background noise, as well as interfering sound sources. Finally, wind noise and noise induced by movements of the microphone itself considerably reduce the signal quality.

These challenges prevent the use of off-the-shelf audio-visual processing algorithms and require specifically designed methods to effectively exploit the complementarity of the information provided by the auditory and visual sensors. In addition to these specific challenges, audio-visual processing for wearable cameras has to deal with more traditional issues such as asynchronism between video and audio streams, and different temporal sampling rates, which may increase uncertainty. Finally, an additional challenge for machine learning algorithms is the lack of large training datasets covering most operational conditions.

This chapter is organized as follows. Section 5.2 overviews state-of-the-art methods for audio-visual classification. Section 5.3 presents multi-modal

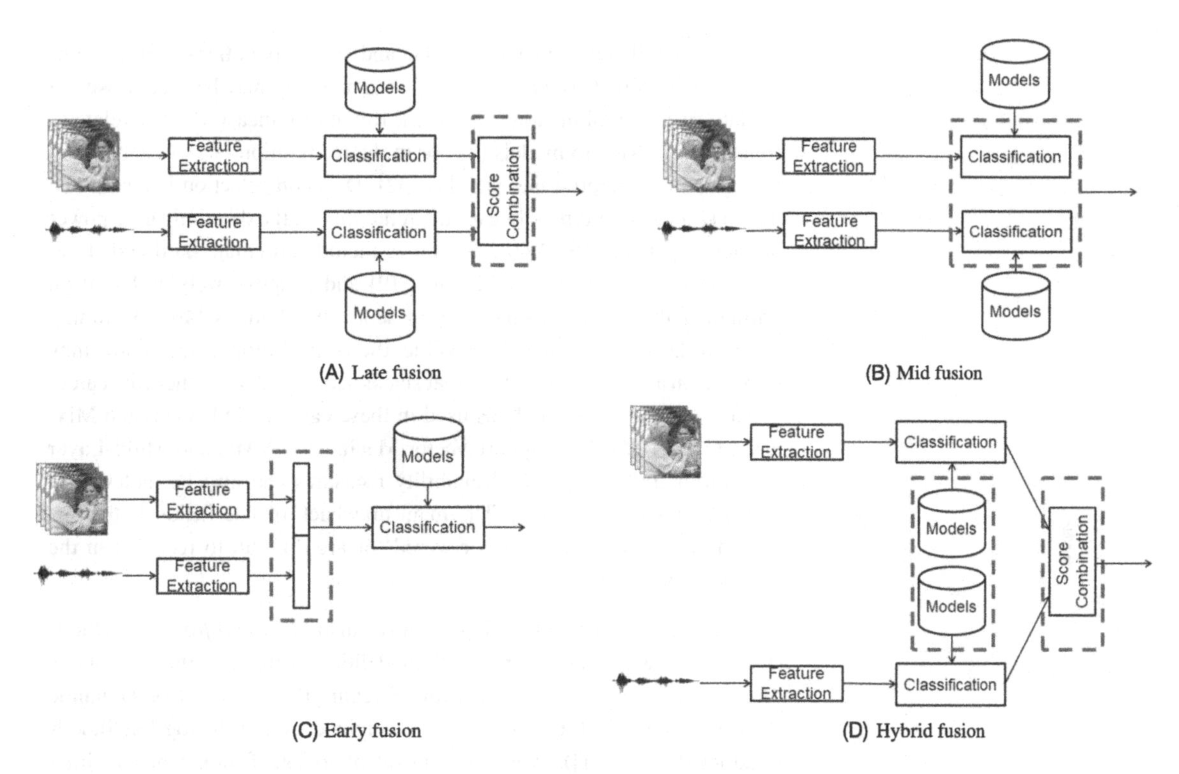

■ **FIGURE 5.2** Four strategies for multi-modal classification.

model adaptation as a strategy to exploit the complementarity of audio and visual information. Section 5.4 introduces person reidentification as use-case for model adaptation, Section 5.5 describes a body-worn camera dataset captured under four different conditions and Section 5.6 reports experimental results on this dataset. Finally, Section 5.7 summarizes the contributions of the chapter.

5.2 MULTI-MODAL CLASSIFICATION

The use of multiple modalities from wearable cameras may considerably help addressing the challenges presented in the previous section. We focus on how audio and visual signals that can be fused effectively in a machine learning context. Then we will discuss how this knowledge can help adapt to varying conditions under which body-worn cameras operates.

Audio-visual recognition can adopt late, early, mid or hybrid fusion strategies [9]. These four strategies for multi-modal fusions are illustrated in Fig. 5.2, and their main principles are discussed below.

Audio and video are processed independently in *late fusion* methods (Fig. 5.2A), which combine mono-modal scores (or decisions). These methods are generally efficient and modular, and other modalities or sub-systems can be easily added. The reliability of each modality may be used to weight the late fusion combination [18,24,32]. Reliability measures may relate to signals (e.g. SNR), to models [24] or to the recognition rate of each classifier by learning appropriate weights [32]. Decision selection is also a late fusion strategy that typically uses the reliability or the discriminative power of each expert [11,19,34,40]. Examples include articulate hard-coded decision cascades driven by reliability [19] and adaptive weights based on estimating the model mismatch from score distributions [40]. Assuming that a further training stage is possible, the combination of the scores may be based on a classification stage that stacks scores and treats them as feature vectors. Classification methods used in these cases include Gaussian Mixture Models (GMMs), Support Vector Machines (SVMs) and Multi-Layer Perceptrons (MLPs) [2,39]. Reliability measures can also be included to improve classification [1,6]. The main drawback of late fusion is that errors in one modality are not recovered (or are difficult to recover) in the classification stage.

Features are processed independently also by *mid-fusion* methods (Fig. 5.2B), which however merge modalities in the classification stage. These methods may employ a Multi-Stream HMM [31,52] or Dynamic Bayesian Networks [36]. In general, a weighted sum of the log-likelihoods is adopted (e.g. [31]), which is equivalent to late fusion when a time-independent classifier is used instead of Hidden Markov Models (HMMs). The stream with the highest posterior can be used in combination with an HMM for audio-visual speech recognition [41]. When the sampling rate of the modalities is different and misalignments between the modality occur, asynchronous HMM combinations are needed [52].

Audio and video features are combined before processing in a joint feature set in *early fusion* methods [13,44] (Fig. 5.2C). For example, stacking audio and video features for a HMM classifier outperforms processing the two modalities independently [35]. Early fusion exploits the correlation between feature sets but might lead to dimensionality problems. The dimensionality of the stacked feature vector can be reduced through Principal Component Analysis (PCA) and Principal Component Analysis (ICA) to remove redundancy across modalities [44].

Finally, *hybrid methods* perform fusion in at least two stages of their pipeline (Fig. 5.2D). In Co-EM [8], as an example, models are iteratively updated in the maximization step by exchanging labels between multiple views of the same dataset and minimizing disagreement between modalities. Examples of Co-EM in speech recognition or multi-modal interaction are referred to as co-training [29] and co-adaptation [12]. *Co-training* can be used for traffic analysis using multiple cameras [29]. The goal is to generate a larger training set for batch adaptation through unsupervised labeling of unseen training data. This labeling is performed based on the agreement between weak mono-modal classifiers trained on small labeled datasets. *Co-adaptation* for audio-visual speech recognition and gesture recognition jointly adapts audio and visual models using unseen unlabeled data of the new application domain by maximizing their agreement [12].

A limitation of most fusion methods is that a modality that deteriorates over time due to an increasing model mismatch is generally dropped without using mechanisms to limit its degradation. A solution to this problem is *cross-modal model adaptation*, which can be combined with the fusion strategies discussed above.

5.3 CROSS-MODAL ADAPTATION

Two major challenges for classification tasks with data from wearable cameras are rapidly varying capturing conditions and limited training data. To address these challenges, the complementarity of multiple modalities may be advantageous to prevent building poor models and to improve classification accuracy.

In this section we discuss cross-modal adaptation of models learned with deep learning approaches and, in particular, we use on-line person reidentification as application example. An additional challenge to address here is the discontinuous availability of visual or audio information, for example when a person goes outside the field of view or becomes silent. Audio-visual person reidentification methods typically use late fusion and combine classification scores (or decisions) produced by two independent mono-modal systems (see Section 5.2).

Let $\mathbf{x}_t^i$ be an observation (feature vector) from modality i and $t = 1, \ldots, T^i$ the temporal index. Let $\mathbf{x}^i = \{\mathbf{x}_1^i, \mathbf{x}_2^i, ..., \mathbf{x}_{T^i}^i\}$ be the T^i observations extracted from a segment to be classified. Let the identity of an enrolled target be denoted with one of S labels s in $\mathcal{S}$. Let $p(y_t = s|\mathbf{x}_t^i)$ be the probability that $\mathbf{x}_t^i$ is generated by target with identity $y_t = s$.

When a new person appears in front of the camera, a model for modality i is acquired during an unconstrained enrollment stage (which might involve some supervision from the user) and then employed to reidentify that person in future interactions, when the target is again visible and/or audible by the body-worn camera. The target identity estimated with modality i, $\hat{s}_t^i$, can be computed as

$$\hat{s}_t^i = \arg\max_{s \in \mathcal{S}} p(y_t = s|\mathbf{x}_t^i). \tag{5.1}$$

When modeled via deep learning architectures, $p(y_t|\mathbf{x}_t^i)$ can be expressed as $p(y_t = s|\mathbf{x}_t^i; \Theta^i)$, where Θ^i are the model parameters (weights w^i and bias b^i) of the neural network. The model parameters are estimated by maximizing the cross-entropy loss, $\mathcal{L}(\mathbf{x}^i|\Theta^i)$, on the training set:

$$\mathcal{L}(\mathbf{x}^i|\Theta^i) = -\sum_{t=1}^{T^i}\sum_{s=1}^{S} p(y_t = s|\mathbf{x}_t^i) \log p(y_t = s|\mathbf{x}_t^i, \Theta^i) + \lambda \|w^i\|^2, \tag{5.2}$$

where $\lambda \|w^i\|^2$ is a L2 regularization term on network weights. In multi-class classification problems, a hard alignment with training labels s_t^* is often used that models the posterior as $p(y_t|\mathbf{x}_t^i) = \delta(y_t = s_t^*)$. Therefore, Eq. (5.2) can be rewritten as a Negative Log-Likelihood (NLL):

$$\mathcal{L}(\mathbf{x}^i|\Theta^i) = -\sum_{t=1}^{T^i} \log p(y_t = s_t^*|\mathbf{x}_t^i, \Theta^i) + \lambda \|w^i\|^2. \tag{5.3}$$

The number of available samples is usually insufficient to train models that account for temporal changes in the environment and in target appearance. To handle this variability, models could be *continuously adapted* over time in an unsupervised manner. However, adapting models is challenging and requires integration of multi-modal information in a multi-modal model adaptation strategy.

Let Θ_t^i and Θ_t^j be time-dependent model parameters at time t for modality i and j, respectively. As soon as new observations are available, we adapt in an unsupervised fashion these model parameters by exploiting the complementarity of the modalities:

$$\Theta_t^i \leftarrow f\left(\Theta_{t-1}^i, \mathbf{x}_t^i, p(y_t|\mathbf{x}_t^j, \Theta_t^j)\right), \tag{5.4}$$

where $f(\cdot)$ is a model transformation (adaptation) function and $\mathbf{x}_t^i$ ($\mathbf{x}_t^j$) is the observation at time t from modality i (j). Model adaptation can be achieved by retraining the network using small learning rates when (supervised) labels are available. If (supervised) labels are unavailable, estimated *unsupervised labels* $\hat{s}_t^j$ provided by the other modality can be used. In this case the loss function is defined as:

$$\mathcal{L}(\Theta^i|\mathbf{x}^i) = -\sum_{t=1}^{T^i} \log p(y_t = \hat{s}_t^j|\mathbf{x}_t^i) + \lambda\|w^i\|^2, \tag{5.5}$$

To avoid over-fitting the unlabeled adaptation data, regularization can be used, for example with a term based on the Kullback–Leibler Divergence (KLD), as in unsupervised DNN adaptation for speech recognition [20,51]. In this case, the target distribution, $\tilde{p}(y_t|\mathbf{x}_t^i)$, is reformulated as linear combination of the actual distribution, $p(y_t|\mathbf{x}_t^i)$, and the original distribution, $\overline{p}(y_t|\mathbf{x}_t^i)$, i.e. the output of the network prior to adaptation:

$$\tilde{p}(y_t|\mathbf{x}_t^i) = (1-\rho^i)p(y_t|\mathbf{x}_t^i) + \rho^i\,\overline{p}(y_t|\mathbf{x}_t^i), \tag{5.6}$$

where the hyper-parameter ρ^i controls the amount of regularization and depends, for example, on the reliability of the unsupervised labels (see Section 5.4).

By replacing $p(y_t|\mathbf{x}_t^i)$ in Eq. (5.2) with $\tilde{p}(y_t|\mathbf{x}_t^i)$ from Eq. (5.6), the loss function becomes

$$\begin{aligned}\tilde{\mathcal{L}}(\Theta^i|\mathbf{x}^i) = &-\sum_{t=1}^{T^i}\sum_{s=1}^{S}\left[\left((1-\rho^i)p(y_t|\mathbf{x}_t^i) + \rho^i\,\overline{p}(y_t|\mathbf{x}_t^i)\right)\log p(y_t = s|\mathbf{x}_t^i)\right] \\ &+ \lambda\|w^i\|^2,\end{aligned} \tag{5.7}$$

which can be re-arranged as

$$\begin{aligned}\tilde{\mathcal{L}}(\Theta^i|\mathbf{x}^i) = &-(1-\rho^i)\sum_{t=1}^{T^i}\sum_{s=1}^{S} p(y_t|\mathbf{x}_t^i)\log p(y_t = s|\mathbf{x}_t^i) \\ &-\rho^i\sum_{t=1}^{T^i}\sum_{s=1}^{S}\overline{p}(y_t|\mathbf{x}_t^i)\log p(y_t = s|\mathbf{x}_t^i) \\ &+ \lambda\|w^i\|^2.\end{aligned} \tag{5.8}$$

Note that the first and the third terms on the right hand side of Eq. (5.8) are the right hand side of Eq. (5.2). Therefore the KLD-regularized loss function is [51]

$$\begin{aligned}&\tilde{\mathcal{L}}(\Theta^i|\mathbf{x}^i) \\ &= (1-\rho^i)\mathcal{L}(\Theta^i|\mathbf{x}^i) - \rho^i\sum_{t=1}^{T^i}\sum_{s=1}^{S}\overline{p}(y_t = s|\mathbf{x}_t^i)\log p(y_t = s|\mathbf{x}_t^i),\end{aligned} \tag{5.9}$$

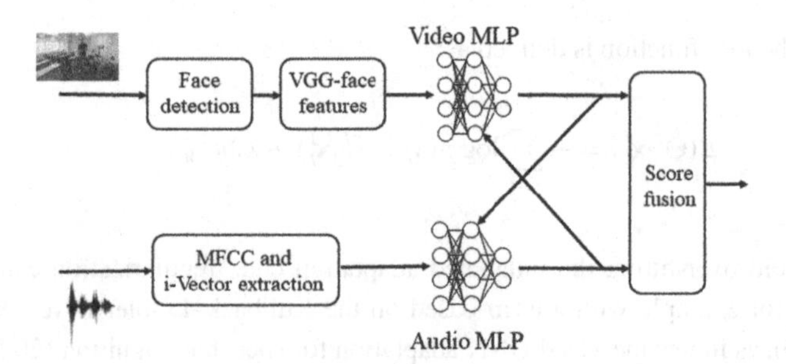

FIGURE 5.3 Example of audio-visual cross-modal adaptation strategy.

where hard alignment is not used for $\overline{p}(y_t|\mathbf{x}_t^i)$ since it has to replicate the output of the original network.

The amount of adaptation of the model can be controlled by resorting to the original model when the quality of the observations or the operational conditions do not allow an effective model adaptation. In particular, when $\rho^i = 0$, the model is adapted towards the new distribution, whereas when $\rho^i = 1$ the network is constrained to resemble the original distribution, prior to adaptation.

5.4 AUDIO-VISUAL REIDENTIFICATION

While both mono-modal and multi-modal approaches for person reidentification are moderately successful in controlled scenarios [5,26,48], their use on data from body-worn cameras requires the relaxation of constraints on the scene, target position or appearance, and speech signals.

Fig. 5.3 shows a specific instantiation for person reidentification of the cross-modal model adaptation approach presented in the previous section: the classification results of two mono-modal Multi-Layer Perceptrons (MLPs) undergo late score fusion. This method extends our previous work [10] with more advanced deep-features for the video modality.

We detect faces with an MXNet implementation in Python of light CNN[1] [49] and we extract VGG-face features[2] using the Oxford models and tools available for Keras [37]. A 4096-dimensional vector is extracted for each face by considering the 7th layer (i.e. "fc7"). Audio features are based on the Total Variability (TV) paradigm [15,16]: we generate 30-dimensional feature vectors by concatenating 15 Mel-Frequency Cepstral Coefficients

[1] https://github.com/tornadomeet/mxnet-face.

[2] VGG-face implements the same architecture as VGG16 [43] but it is trained to recognize faces.

(MFCC) with their first derivatives (energy is not used). We extract the coefficients from 20 ms windows, with 10 ms step. Starting from the MFCCs and using a TV feature extractor trained on the out-of-domain clean Italian APASCI dataset [4], we extract 200-dimensional i-vectors [15,16] for each utterance. Audio and video deep-features are classified frame by frame (i.e. for each face and each i-vector) using a modality-dependent MLP.

The goal is to identify the person present in each segment k (we assume that a single person is present in each segment) using T^i feature vectors for modality i and T^j feature vectors for modality j. We evaluate the maximum among the mean posteriors over each segment k:

$$\hat{s}_k^i = \arg\max_{s \in \mathcal{S}} \frac{1}{T^i} \sum_{t=1}^{T^i} p(y_t = s | \mathbf{x}_t^i), \tag{5.10}$$

and similarly for modality j. The KLD parameter, ρ_k^i, is segment-dependent and varies with the reliability of the observations. This parameter is a function of the average score of the label in that segment:

$$\rho_k^i = g^i \left[\frac{1}{T^j} \sum_{t=1}^{T^j} p(y_t = \hat{s}_k^j | \mathbf{x}_t^j) \right]. \tag{5.11}$$

Finally, score fusion is implemented as weighted average of the scores of the modalities [9].

5.5 REIDENTIFICATION DATASET

The *QM-GoPro* dataset contains 52 clips of 13 subjects talking for 1 minute to a person wearing a chest-mounted GoPro camera.[3] The dataset includes four conditions:

- C1 (indoors): in a lecture room;
- C2 (indoors): different clothes and possibly different room;
- C3 (outdoors): same clothes as in C1, quiet location;
- C4 (outdoors): same clothes as in C2, noisy location (near a busy road).

The video resolution is 1920 × 1080 pixels, at 25 frames per second. Audio streams are at 48 kHz, 16 bits. Speakers are up to a few meters away from the camera so the reidentification task is in distant talking conditions.

---

[3] The dataset is available at http://www.eecs.qmul.ac.uk/~andrea/adaptation.html and described in [9].

A noticeable illumination mismatch is present between the indoor and outdoor recordings. Conditions C1 and C2 are challenging for video processing as illumination conditions change considerably when the speaker moves, e.g. towards or away from the windows. Moreover, the person is often occluded by furniture thus challenging the person detector. The speaker-to-camera distance varies considerably: from so close that only a body part is visible to so far that the face is barely visible. The illumination conditions and the speaker-to-camera distance in C3 and C4 are instead rather constant.

The camera microphone is partially covered by a plastic shield thus resulting in a strong low-pass effect that limits the spectral content (even if audio is sampled at 48 kHz). Acoustics changes considerably between favorable conditions indoors (C1 and C2) and more challenging outdoors (C3 and C4). Although relatively quiet, a variety of interfering noise sources are present in C3 (e.g. wind blowing into the microphone, people passing by), which affect speaker reidentification performance. In addition, the lack of reverberation introduces a further mismatch with respect to C1 and C2. Finally, C4 has strong background noise, mostly due to road traffic, which at times makes the voice of the speaker inaudible.

5.6 REIDENTIFICATION RESULTS

In this section we discuss results on the *QM-GoPro* dataset (Section 5.5) using the reidentification pipeline of Section 5.4. We compare four strategies:

- intra (intra-modal): adaptation based on labels from the same modality;
- cross (cross-modal): adaptation based on labels from the other modality;
- KLD: KLD regularization added to cross;
- base (baseline): no adaptation.

We test both *matched conditions*, when target models are obtained under the same condition C_n as the test, and *mismatch conditions*, when target models are trained under condition C_n and tested under condition C_m, with $m \neq n$. Each clip is split in 5-second segments, excluding the first 8 and the last 5 s. Each segment ($k = 1, \dots, K$) consists of $T^V = 125$ frames (or VGG feature vectors) and $T^A = 1$ i-vectors. For training, the first 3 segments are used for each target under each condition. Moreover, $g^V[\cdot]$ is the identity function and $g^A[\cdot]$ is a sigmoid function (see [10] for details).

Fig. 5.4 shows the results under matched conditions. When matched models are available and video conditions are favorable (C3 and C4), the use of deep features, in combination with a deep neural classifier, results in a 100% accuracy (Fig. 5.4B). Similarly, i-vectors offers high performance indoors,

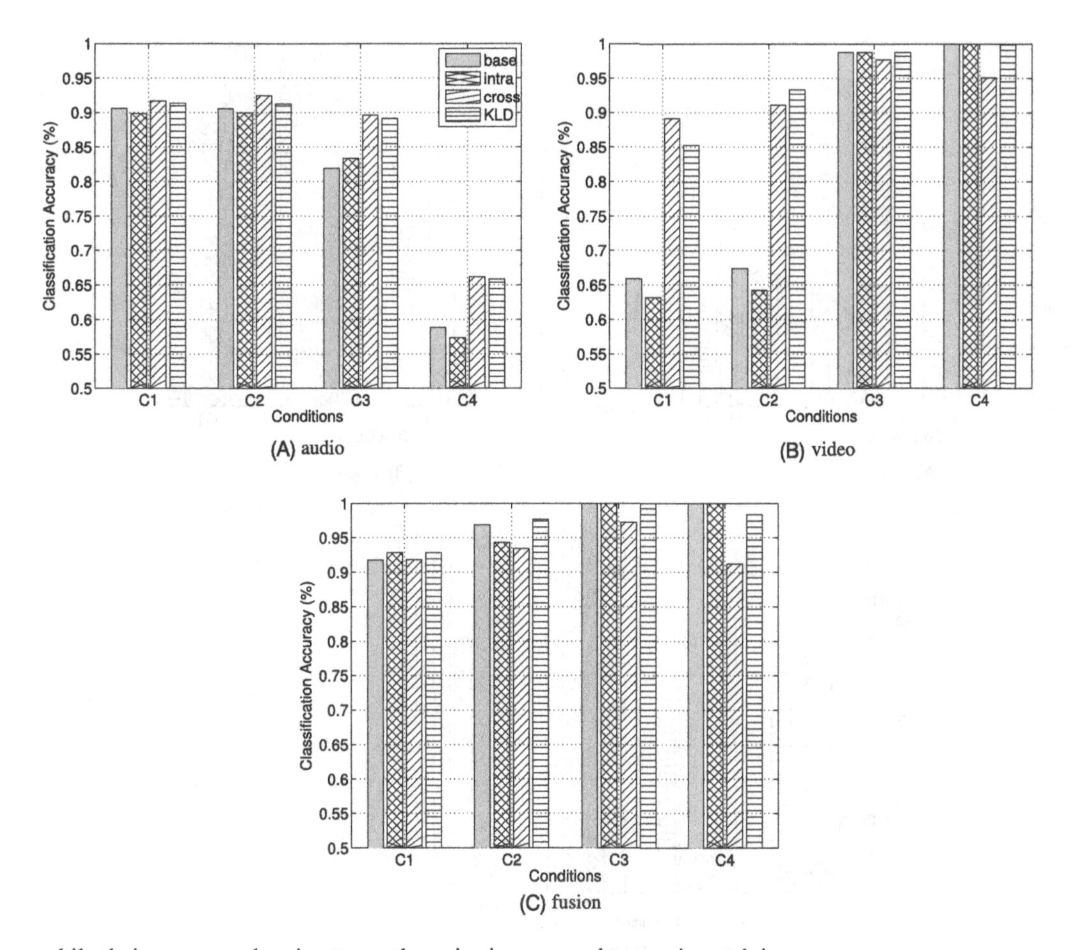

■ **FIGURE 5.4** Accuracy under matched conditions (*QM-GoPro* dataset).

while their accuracy deteriorates as the noise increases, due to mismatch in the i-vector extractor.

Cross-modal adaptation improves the performance of badly performing models but deteriorates in case of accurate video models (i.e. in C3 and C4). The KLD-based regularization improves poor models while preserving the performance of the good ones. Note that KLD limits the potential impact of cross-modal adaptation for audio, probably because it underestimates the reliability of the video models.

Fig. 5.4C reports the results after late score fusion. Note that this fusion strategy is not optimal for the system discussed here as it was developed in [9] to optimize the Equal Error Rate (ERR) on a different classifier. Note also that if an effective fusion is used, the overall accuracy is high and the benefits of model adaptation are less evident with respect to the baseline system, especially if the two modalities are always available (as in the majority

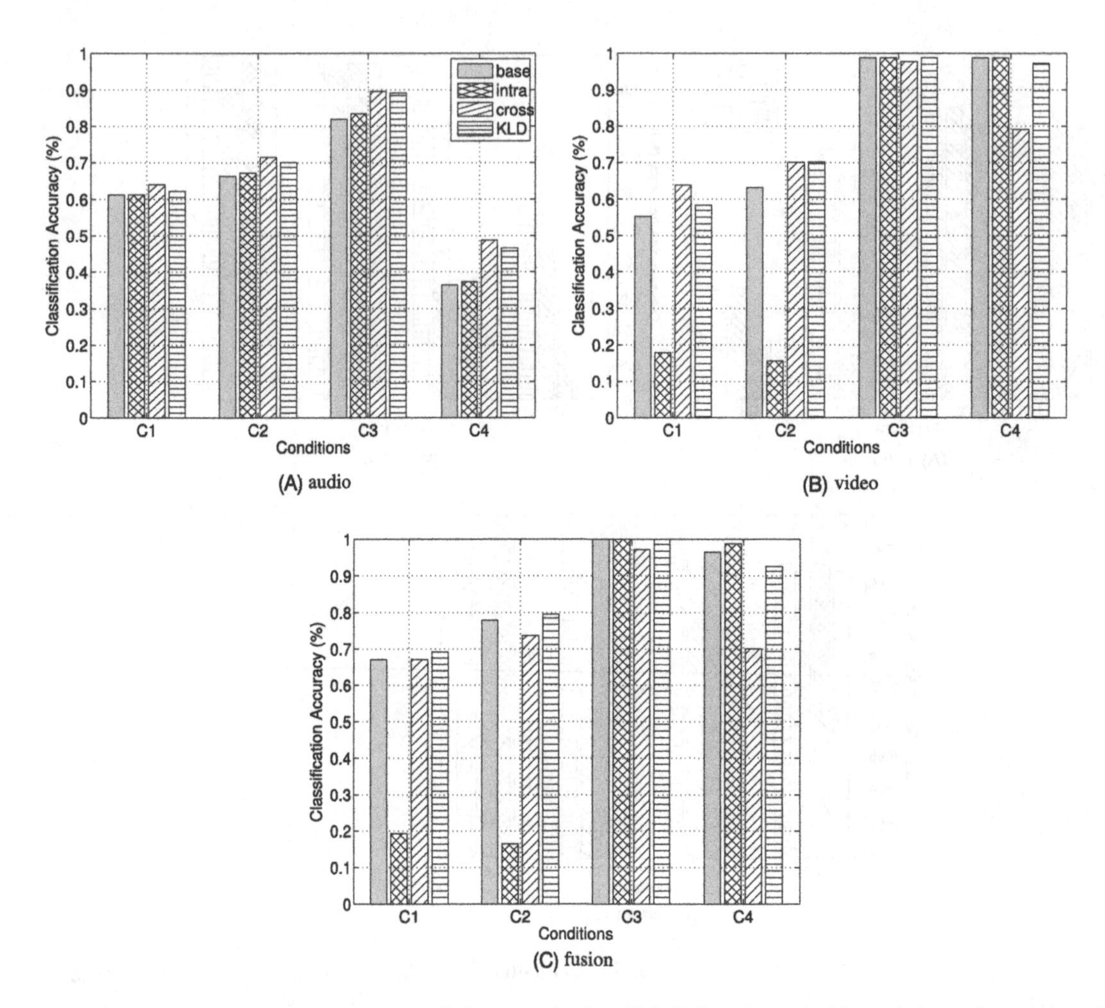

■ **FIGURE 5.5** Accuracy under "mismatched-C3" conditions (*QM-GoPro* dataset).

of the cases in the *QM-GoPro* dataset). Nevertheless, the results with the KLD adaptation are equivalent or superior to those of the baseline. A minor deterioration is observed in C4.

Figs. 5.5 and 5.6 show the classification accuracy under mismatched conditions: "mismatch-C3" (models are trained in C3) and "mismatch-C1" (models are trained in C1). As the behaviors of the methods are similar in the two cases, we will focus the analysis on "mismatch-C1" only.

In Fig. 5.6A the performance decreases from C1 to C4 as the amount of mismatch increases due to noise (outdoors). Note that the mismatch is not only in the speaker models but also in the i-vector extractor. Similar trends are observed in Fig. 5.5A: note however that despite some noise in the training set, C4 is still worse than C1 and C2. In terms of adaptation, the same behavior

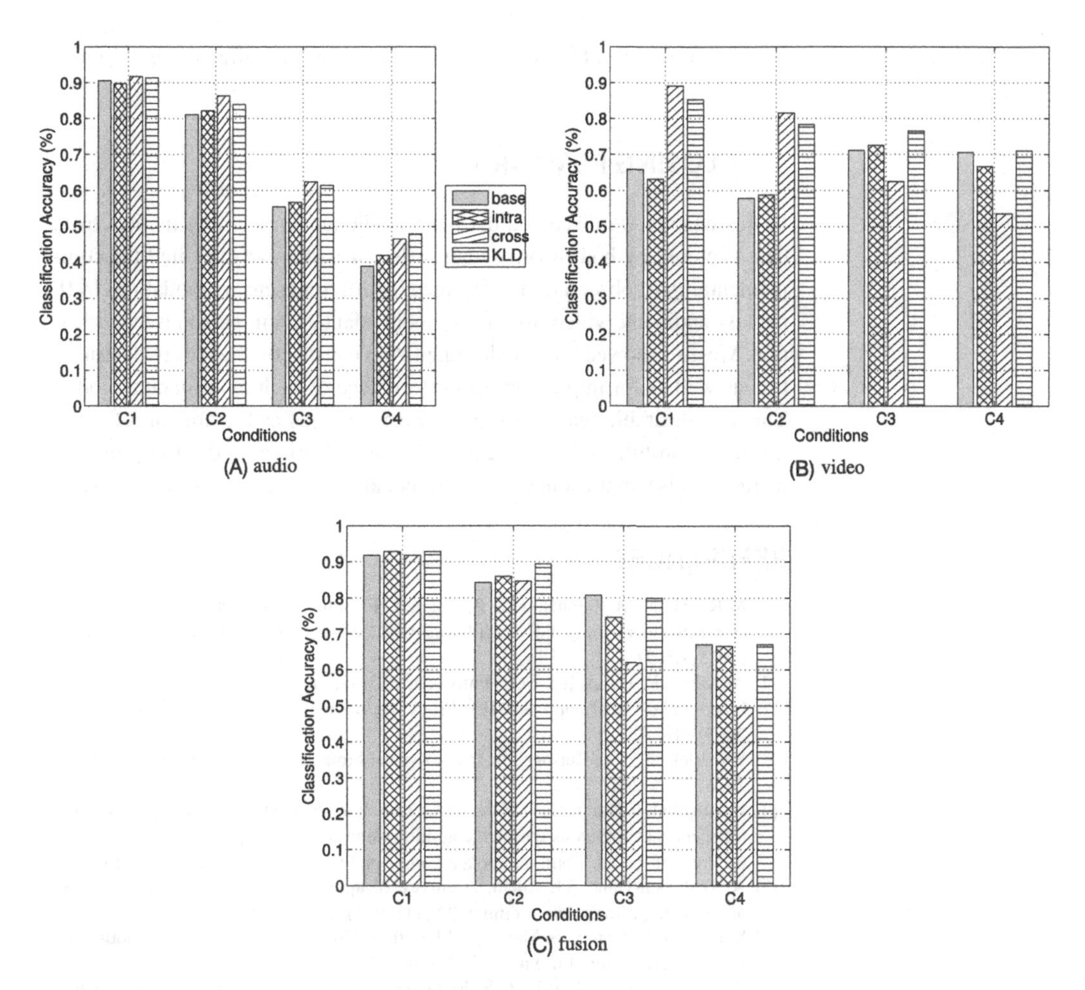

■ **FIGURE 5.6** Accuracy under "mismatched-C1" conditions (*QM-GoPro* dataset).

is observed here as in the matched case. Cross reduces the model mismatch and KLD, in some cases, limits the potential gain. The effect of model mismatch is less evident for the video modality as shown in Fig. 5.6B. In C3 and C4 the accuracy is higher than in C1 despite a high mismatch under light conditions. This is related to the robustness of VGG features in combination with the fact that in C3 and in C4 faces are visible without occlusions or over-exposure. This is also observed in Fig. 5.5B where models trained in C3 are suitable also in C4 (similar light conditions) but performance deteriorates in C1 and C2 to the same level as "mismatch-C1". Occlusions and over-exposure are the main reasons for the low performance indoors. Also for the video modality the KLD-regularized adaptation improves the performance over the baselines and avoids model deterioration introduced by cross adaptation in C3 and C4. Finally, after score fusion, the same trends are observed with KLD, improving baselines and not affecting good models.

5.7 CLOSING REMARKS

In this chapter we discussed the main challenges in learning models from data captured by body-worn cameras. In particular, we exploited the complementarity of observations generated by different sensing modalities [42], namely audio and video, for cross-modal adaptation in person reidentification. We also showed how model adaptation can be effective for reidentification as well as for improving the performance of each mono-modal model. This is a desirable feature for body-worn cameras as the amount of training (and enrollment) data is generally limited. Importantly, the proposed method is also applicable when each modality is used for a different task.

REFERENCES

[1] M.R. Alam, M. Bennamoun, R. Togneri, F. Sohel, A confidence-based late fusion framework for audio-visual biometric identification, Pattern Recognit. Lett. 52 (2015) 65–71.
[2] A. Albiol, L. Torres, E.J. Delp, Fully automatic face recognition system using a combined audio-visual approach, IEE Proc., Vis. Image Signal Process. 152 (3) (June 2005) 318–326.
[3] P.S. Aleksic, A.K. Katsaggelos, Audio-visual biometrics, Proc. IEEE 94 (11) (Nov. 2006) 2025–2044.
[4] B. Angelini, F. Brugnara, D. Falavigna, D. Giuliani, R. Gretter, M. Omologo, Speaker independent continuous speech recognition using an acoustic-phonetic Italian corpus, in: Proc. of the Int. Conf. on Spoken Language Processing, 1994, pp. 1391–1394.
[5] A. Bedagkar-Gala, S.K. Shah, A survey of approaches and trends in person re-identification, Image Vis. Comput. 32 (4) (2014) 270–286.
[6] S. Bengio, C. Marcel, S. Marcel, J. Mariéthoz, Confidence measures for multimodal identity verification, Inf. Fusion 3 (4) (2002) 267–276.
[7] A. Betancourt, P. Morerio, C.S. Regazzoni, M. Rauterberg, The evolution of first person vision methods: a survey, IEEE Trans. Circuits Syst. Video Technol. 25 (5) (May 2015) 744–760.
[8] S. Bickel, T. Scheffer, Estimation of mixture models using Co-EM, in: J. Gama, R. Camacho, P.B. Brazdil, A.M. Jorge, L. Torgo (Eds.), Machine Learning: ECML 2005, Springer, Berlin, Heidelberg, 2005, pp. 35–46.
[9] A. Brutti, A. Cavallaro, Online cross-modal adaptation for audio visual person identification with wearable cameras, IEEE Trans. Human-Mach. Syst. 47 (1) (Feb. 2017) 40–51.
[10] A. Brutti, A. Cavallaro, Unsupervised cross-modal deep-model adaptation for audio-visual re-identification with wearable cameras, in: Proc. of Int. Conf. on Computer Vision Workshop, Oct. 2017.
[11] U.V. Chaudhari, G.N. Ramaswamy, G. Potamianos, C. Neti, Audio-visual speaker recognition using time-varying stream reliability prediction, in: Proc. of IEEE Int. Conf. on Audio, Speech and Signal Processing, vol. 5, Apr. 2003, pp. 712–715.
[12] C.M. Christoudias, K. Saenko, L. Morency, T. Darrell, Co-adaptation of audio-visual speech and gesture classifiers, in: Proc. of the Int. Conf. on Multimodal Interfaces, Nov. 2006, pp. 84–91.

[13] B. Clarkson, K. Mase, A. Pentland, Recognizing user context via wearable sensors, in: Proc. of the Int. Symposium on Wearable Computers, Oct. 2000, pp. 69–75.
[14] B. Clarkson, A. Pentland, Unsupervised clustering of ambulatory audio and video, in: Proc. of IEEE Int. Conf. on Audio, Speech and Signal Processing, Mar. 1999, pp. 3037–3040.
[15] N. Dehak, R. Dehak, P. Kenny, N. Brümmer, P. Ouellet, P. Dumouchel, Support vector machines versus fast scoring in the low-dimensional total variability space for speaker verification, in: Proc. of Interspeech, Sep. 2009, pp. 1559–1562.
[16] N. Dehak, P. Kenny, R. Dehak, P. Dumouchel, P. Ouellet, Front-end factor analysis for speaker verification, IEEE Trans. Audio Speech Lang. Process. 19 (4) (May 2011) 788–798.
[17] A.G. del Molino, C. Tan, J.H. Lim, A.H. Tan, Summarization of egocentric videos: a comprehensive survey, IEEE Trans. Human-Mach. Syst. 47 (1) (Feb. 2017) 65–76.
[18] H.K. Ekenel, M. Fischer, Q. Jin, R. Stiefelhagen, Multi-modal person identification in a smart environment, in: Proc. of IEEE Int. Conf. on Computer Vision and Pattern Recognition, Jun. 2007, pp. 1–8.
[19] E. Erzin, Y. Yemez, A.M. Tekalp, Multimodal speaker identification using an adaptive classifier cascade based on modality reliability, IEEE Trans. Multimed. 7 (5) (Oct. 2005) 840–852.
[20] D. Falavigna, M. Matassoni, S. Jalalvand, M. Negri, M. Turchi, DNN adaptation by automatic quality estimation of ASR hypotheses, Comput. Speech Lang. 46 (Nov. 2017) 585–604.
[21] A. Fathi, A. Farhadi, J.M. Rehg, Understanding egocentric activities, in: Proc. of Int. Conf. on Computer Vision, Nov. 2011, pp. 407–414.
[22] A. Fathi, J.K. Hodgins, J.M. Rehg, Social interactions: a first-person perspective, in: Proc. of IEEE Int. Conf. on Computer Vision and Pattern Recognition, June 2012, pp. 1226–1233.
[23] F. Fergnani, S. Alletto, G. Serra, J.D. Mira, R. Cucchiara, Body part based re-identification from an egocentric perspective, in: Proc. of IEEE Int. Conf. on Computer Vision and Pattern Recognition Workshops, June 2016, pp. 355–360.
[24] N.A. Fox, R. Gross, J.F. Cohn, R.B. Reilly, Robust biometric person identification using automatic classifier fusion of speech, mouth, and face experts, IEEE Trans. Multimed. 9 (4) (June 2007) 701–714.
[25] N. Kern, B. Schiele, H. Junker, P. Lukowicz, G. Troster, Wearable sensing to annotate meeting recordings, in: Proc. of the Int. Symposium on Wearable Computers, June 2002, pp. 186–193.
[26] T. Kinnunen, H. Li, An overview of text-independent speaker recognition: from features to supervectors, Speech Commun. 52 (1) (Jan. 2010) 12–40.
[27] K.M. Kitani, T. Okabe, Y. Sato, A. Sugimoto, Fast unsupervised ego-action learning for first-person sports videos, in: Proc. of IEEE Int. Conf. on Computer Vision and Pattern Recognition, Nov. 2011, pp. 3241–3248.
[28] Y.J. Lee, J. Ghosh, K. Grauman, Discovering important people and objects for egocentric video summarization, in: Proc. of IEEE Int. Conf. on Computer Vision and Pattern Recognition, June 2012, pp. 1346–1353.
[29] A. Levin, P. Viola, Y. Freund, Unsupervised improvement of visual detectors using co-training, in: Proc. of Int. Conf. on Computer Vision, vol. 1, Oct. 2003, pp. 626–633.
[30] Z. Lu, K. Grauman, Story-driven summarization for egocentric video, in: Proc. of IEEE Int. Conf. on Computer Vision and Pattern Recognition, June 2013, pp. 2714–2721.
[31] S. Lucey, T. Chen, S. Sridharan, V. Chandran, Integration strategies for audio-visual speech processing: applied to text-dependent speaker recognition, IEEE Trans. Multimed. 7 (3) (June 2005).

[32] J. Luque, R. Morros, A. Garde, J. Anguita, M. Farrus, D. Macho, F. Marqués, C. Martínez, V. Vilaplana, J. Hernando, Audio, video and multimodal person identification in a smart room, in: Rainer Stiefelhagen, John Garofolo (Eds.), Multimodal Technologies for Perception of Humans: First International Evaluation Workshop on Classification of Events, Activities and Relationships, CLEAR 2006, 2007.
[33] Z. Lv, A. Halawani, S. Feng, S. Réhman, H. Li, Touch-less interactive augmented reality game on vision-based wearable device, Pers. Ubiquitous Comput. 19 (3) (July 2015) 551–567.
[34] C. McCool, S. Marcel, A. Hadid, M. Pietikainen, P. Matejka, J. Cernocky, N. Poh, J. Kittler, A. Larcher, C. Levy, D. Matrouf, J. Bonastre, P. Tresadern, T. Cootes, Bi-modal person recognition on a mobile phone: using mobile phone data, in: Proc. of ICME Workshop on Hot Topics in Mobile Multimedia, July 2012, pp. 635–640.
[35] I. McCowan, S. Bengio, D. Gatica-Perez, G. Lathoud, F. Monay, D. Moore, P. Wellner, H. Bourlard, Modeling human interaction in meetings, in: Proc. of IEEE Int. Conf. on Audio, Speech and Signal Processing, vol. 4, Apr. 2003, pp. 748–751.
[36] A. Noulas, G. Englebienne, B.J.A. Krose, Multimodal speaker diarization, IEEE Trans. Pattern Anal. Mach. Intell. 34 (1) (Jan. 2012).
[37] O.M. Parkhi, A. Vedaldi, A. Zisserman, Deep face recognition, in: British Machine Vision Conference, 2015.
[38] H. Pirsiavash, D. Ramanan, Detecting activities of daily living in first-person camera views, in: Proc. of IEEE Int. Conf. on Computer Vision and Pattern Recognition, June 2012, pp. 2847–2854.
[39] C. Sanderson, S. Bengio, H. Bourlard, J. Mariethoz, R. Collobert, M.F. BenZeghiba, F. Cardinaux, S. Marcel, Speech face based biometric authentication at IDIAP, in: Proc. of the Int. Conf. on Multimedia and Expo, vol. 3, July 2003, pp. 1–4.
[40] C. Sanderson, K.K. Paliwal, Noise compensation in a person verification system using face and multiple speech features, Pattern Recognit. 36 (2) (2003) 293–302.
[41] R. Seymour, J. Ming, S.D. Stewart, A new posterior based audio-visual integration method for robust speech recognition, in: Proc. of Interspeech, Sep. 2005, pp. 1229–1232.
[42] S.T. Shivappa, M.M. Trivedi, B.D. Rao, Audiovisual information fusion in human computer interfaces and intelligent environments: a survey, Proc. IEEE 98 (10) (Oct. 2010) 1692–1715.
[43] K. Simonyan, A. Zisserman, Very deep convolutional networks for large-scale image recognition, CoRR, abs/1409.1556, 2014.
[44] P. Smaragdis, M. Casey, Audio/visual independent components, in: Proc. of the Int. Symposium on Independent Component Analysis and Blind Source Separation, Apr. 2003, pp. 709–714.
[45] S. Song, V. Chandrasekhar, B. Mandal, L. Li, J.H. Lim, G.S. Babu, P.P. San, N.M. Cheung, Multimodal multi-stream deep learning for egocentric activity recognition, in: Proc. of IEEE Int. Conf. on Computer Vision and Pattern Recognition Workshops, June 2016, pp. 378–385.
[46] R. Stiefelhagen, K. Bernardin, R. Bowers, R. Rose, Martial Michel, J. Garofolo, The CLEAR 2007 evaluation multimodal technologies for perception of humans, in: Rainer Stiefelhagen, Rachel Bowers, Jonathan Fiscus (Eds.), Multimodal Technologies for Perception of Humans, in: Lect. Notes Comput. Sci., Springer, Berlin, Heidelberg, 2008, chapter 1.
[47] A. Subramanya, A. Raj, Recognizing activities and spatial context using wearable sensors, in: Proc. of the Conf. on Uncertainty in Artificial Intelligence, July 2006.
[48] R. Vezzani, D. Baltieri, R. Cucchiara, People reidentification in surveillance and forensics: a survey, ACM Comput. Surv. 46 (2) (Dec. 2013) 29:1–29:37.
[49] X. Wu, R. He, Z. Sun, A lightened CNN for deep face representation, arXiv preprint arXiv:1511.02683, 2015.

[50] Y. Yan, E. Ricci, G. Liu, N. Sebe, Egocentric daily activity recognition via multitask clustering, IEEE Trans. Image Process. 24 (10) (2015) 2984–2995.

[51] D. Yu, K. Yao, H. Su, G. Li, F. Seide, KL-Divergence regularized deep neural network adaptation for improved large vocabulary speech recognition, in: Proc. of IEEE Int. Conf. on Audio, Speech and Signal Processing, May 2013, pp. 7893–7897.

[52] D. Zhang, D. Gatica-Perez, S. Bengio, I. McCowan, Modeling individual and group actions in meetings with layered HMMs, IEEE Trans. Multimed. 8 (3) (June 2006).

Chapter 6

Activity recognition from visual lifelogs: State of the art and future challenges

Mariella Dimiccoli, Alejandro Cartas, Petia Radeva
University of Barcelona, Department of Mathematics and Computer Science, Barcelona, Spain

CONTENTS

6.1 INTRODUCTION

The last decade has witnessed impressive advances in the field of wearable sensor technologies [19]. From small magnetic tracer secured on the tongue to smart watches, wearable sensors are ubiquitous. Consequently, they are specially suited to capture human activity signals under free living conditions, paving the road to novel opportunities in the field of assistive technologies and tele-rehabilitation systems [15]. In recent years, a great amount of efforts have been made to come up with systems able to convert sensory information into human-readable signals that could be directly used for medical purposes. Among wearable sensors, wearable cameras are of special interest since they allow one to record what the user is doing and seeing from a first-person perspective, hence providing rich contextual information about the activities being performed [20]. Indeed, wearable cameras could potentially provide information not only about what the person is doing, but also about how the person is doing something, in which

Multimodal Behavior Analysis in the Wild. https://doi.org/10.1016/B978-0-12-814601-9.00017-1
Copyright © 2019 Elsevier Ltd. All rights reserved.

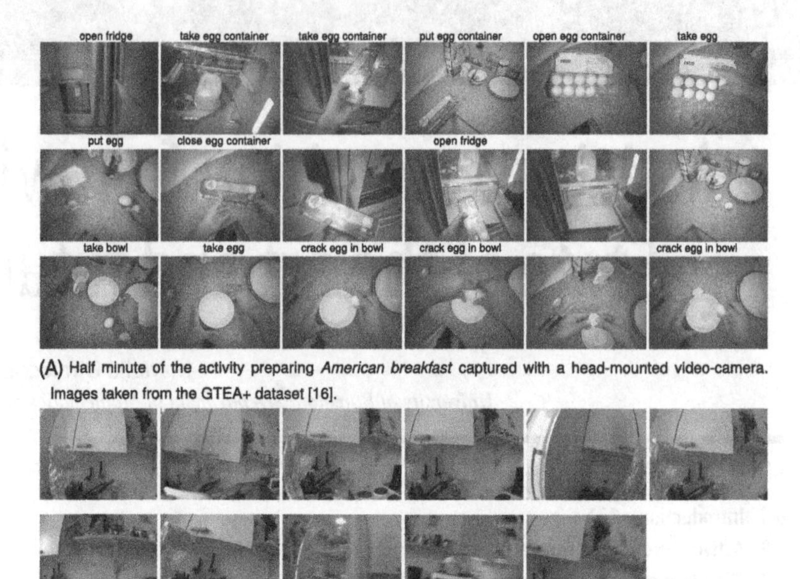

(A) Half minute of the activity preparing *American breakfast* captured with a head-mounted video-camera. Images taken from the GTEA+ dataset [16].

(B) Five minutes of the activity *cooking* captured by a chest-mounted photo-camera. Images taken from the NTCIR 2016 dataset [13].

FIGURE 6.1 Example of consecutive frames from an egocentric video and a visual lifelog.

environment, with whom, etc. However, privacy concerns are limiting the widespread adoption of wearable cameras for activity recognition applications. For this reason, the recognition of human activities from a first-person (egocentric) point of view and the solution of related privacy issues are both active areas of research [6,7,10,14,25,28].

So far the main focus of the computer vision community has been in recognizing activities from at most a few hours of egocentric videos [1–3,12, 16–18,21,23,24,26,29] more than on the analysis of photo-streams covering the full day [6–8], also called *visual lifelog*. As extensively discussed in [4], visual lifelogs provide a rich source of information for several health applications, including lifestyle characterization and monitoring, and cognitive training based on digital memories, among others. However, in their analyses there arise several challenges with respect to conventional and egocentric videos. As a result of the free motion of the camera and its very low frame (2–3 fpm), temporally adjacent images present abrupt appearance changes so that motion features cannot be reliably estimated. Additionally, some actions crucial to identifying the activity being performed can be missed, and due to the non-intentional nature of the photos they may capture non-informative regions such as wall or skin, or present severe occlusions of the objects being interacted with. Fig. 6.1 shows some examples of activities recorded with a video-camera (top) and a photo-camera (bottom). As it can be observed, a half-minute video of the activity *Preparing American breakfast*, recorded with a wearable video-camera, includes the complete sequence of actions performed by the wearer such as *taking an egg container*, *putting down the egg container*, etc. On the contrary, a five minute photo-stream of the activity *cooking*, captured by a wearable photo-camera, does not include the complete sequence of actions performed and even the visible ones are difficult to identify, because often the objects of interest are occluded or only partially visible. Consequently, while the focus in activity recognition from egocentric videos has been on classifying actions such as *taking bread*, *putting honey*, in visual lifelogs the focus has been on classifying activities at a larger temporal scale such as *eating*, *driving*, etc. The intrinsically different nature of the image data on the one hand, and the diversity in the scope on the other hand, make available methods for egocentric activity recognition from video inadequate to the visual lifelog domain.

In this chapter, we restrict our analysis to the recognition of daily activities captured in a naturalistic setting through a wearable photo-camera. In Sections 6.2 and 6.3, we provide a detailed discussion of state of the art methods acting at image and event level, respectively. In Section 6.4, we report an experimental survey of state of the art methods and in view of their analysis we give recommendations for future research. Finally, in Section 6.5, we summarize the main contributions and conclude this chapter.

6.2 ACTIVITY RECOGNITION FROM EGOCENTRIC IMAGES

The first attempt to address the problem of egocentric activity recognition from images captured at regular intervals was made by Castro et al. [8]. The authors proposed a Convolutional Neural Network (CNN) Late Fusion Ensemble (LFE) method that incorporates relevant contextual information to increase the classification accuracy of a fine-tuned CNN. More specifically, the output probabilities of a CNN are merged, through a random forest, with contextual information provided by a color histogram and time metadata such as the day of the week and the time of the day. This particular choice of contextual information was justified by the fact that the dataset used for validation was acquired by a single person having a routine life. For example, the user used to spend time with dogs almost always in the same park at about the same time of the day so that both color histogram and time metadata were useful to improve the classification accuracy of the activity "spending time with dogs". However, for a dataset acquired by several users, the use of time metadata and a color histogram would hardly help

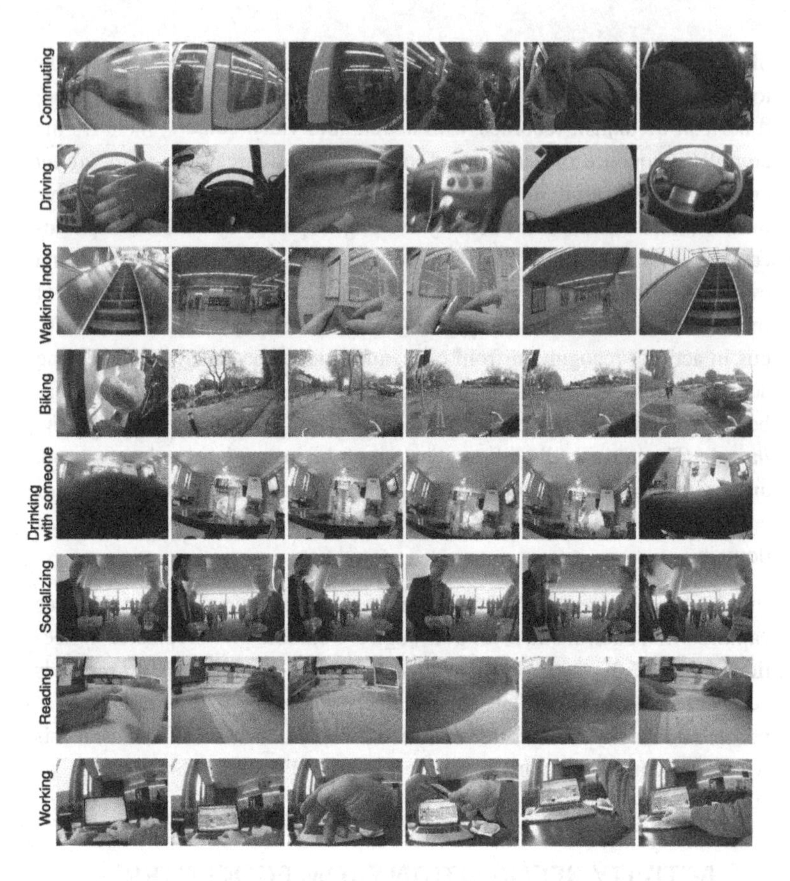

■ **FIGURE 6.2** Example of short photo-streams capturing different activities.

to increase the classification accuracy, since it is unlikely that different users perform the same activity at the same time of the day and day of the week, and in places having similar color distributions. For instance, the *cooking* activity of a given user will have very different appearance and time scale than for other users depending on their working hours and the style of their kitchen. With the aim of generalizing the LFE method to multiple users, Cartas et al. [7] proposed to use the output of a fully connected layer of a CNN as contextual information. The insight beyond the latter approach is that a fully connected layer provides more discriminative and complete features than a color histogram to characterize the typical scene where a given activity takes place across different users.

However, both these approaches treated images captured at regular intervals as contextually unrelated, while it has been shown that, in photo-streams, the temporal coherence of concepts is preserved despite the lack of temporal coherence [5]. Fig. 6.2 shows some photo-streams corresponding to different activities that well illustrate how images corresponding to the same activity share the same concept vocabulary. For instance, for the activity *biking* the concept vocabulary includes the words *car, tree, sky,* etc. Additionally, it is worth to observe how in many images the object being interacted with is occluded and consequently these images can be annotated correctly only taking into account their temporal neighborhood. Finally, photo-streams may contain images representing different sub-events of a particular activity, and it is not known which of these images are responsible for the label of the activity. Consequently, treating images as contextually unrelated when they have been annotated in batches is unfair. In the next section, we review state of the art approaches that take as input a batch of images instead of a single one.

6.3 ACTIVITY RECOGNITION FROM EGOCENTRIC PHOTO-STREAMS

Following the success of convolutional neural networks for image classification, during the last few years there have been several efforts aiming to exploit Long Short Term Memory (LSTM) recurrent neural networks to improve the classification accuracy in videos. The classical approach, firstly proposed in [30], is to use a LSTM network on top of a CNN to learn long-range temporal relationships. Since the video is down-sampled at 1 fps to reduce the computational cost, the lost of implicit motion information is compensated for by explicit motion information that is computed on the original video and fed to the CNN. Experimental results reported by the authors suggested that LSTMs are able to capture the temporal evolution of convolutional features, and that this conveys useful information to classify videos. However, in photo-streams explicit motion information is not available, because the frame rate is originally very low. Additionally, shot boundaries are typically unknown and this hardens the classification problem. In fact, in order to learn how features change over time while the user is performing a given activity, all LSTM cells of a shot, whose frames have the same label, must be connected. Fig. 6.3 shows the ground truth labels of ten annotated photo-streams. It can be observed that, in spite of the sparseness of the observations, temporally adjacent images are more likely to have the same label. This is specially true when the user is performing activities such as *working* or *attending a seminar* that typically lasts from one to a few hours.

To overcome these problems, Cartas et al. [6] proposed a batch-based approach for egocentric activity recognition that does not rely on the knowledge of shot boundaries but works at batch level. More precisely, the approach consists in training in an end-to-end fashion a LSTM recurrent neural

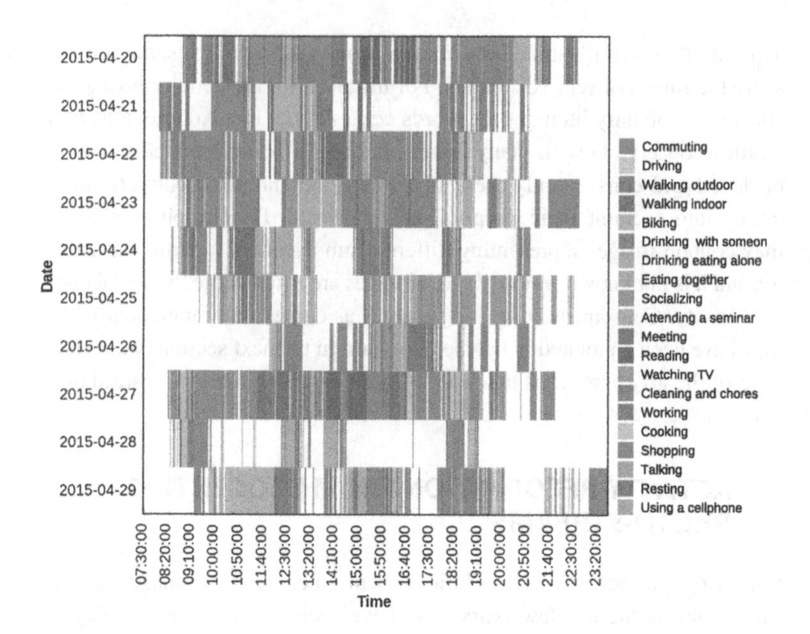

FIGURE 6.3 Ten day sequences of annotated images from one person.

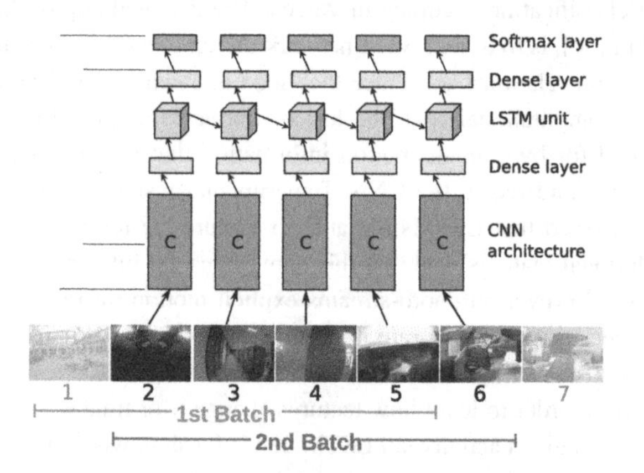

FIGURE 6.4 Illustration of the batch-based training process proposed in [6].

network on top of a CNN by using overlapping batches of N consecutive frames, with a sliding window of size w, with $w < N$. For the sake of clarity, the training process is illustrated in Fig. 6.4. This work has shown firstly that LSTM recurrent neural networks improve the classification accuracy of photo-streams for which motion information is not available. Secondly, it has shown that the knowledge of shot boundaries, which is in general not available, is not needed, since a sliding window strategy allows one to parse the photo-stream within and across event boundaries.

■ **FIGURE 6.5** Examples of activity categories from the dataset.

Lately, Cartas et al. [7] extended the work in [7] by proposing an approach that takes into account both the contextual correlation and temporal coherence of concepts in photo-streams. More specifically, the authors proposed a two-stages approach that first computes the activity probabilities by using the LFE method proposed in [7], and then apply a strategy similar to [6] for training an LSTM recurrent neural network. The LSTM is able to learn long-term feature relationships by processing overlapping sequences of consecutive frames. This work shows that the LFE method provides better features than those obtained in an end-to-end fashion as in [6].

6.4 EXPERIMENTAL RESULTS

In this section, we provide an experimental survey of the techniques described in this chapter for activity recognition from visual lifelogs.

6.4.1 Experimental setup

All methods have been tested on a dataset acquired by three people during several consecutive days, under free living conditions, for a total of 44,902 images almost equally divided among the three wearers. The images were acquired by an OMG Autographer wearable camera worn on the chest and tuned to capture 2 frames per minute. Examples of activity categories are shown in Fig. 6.5. As it can be appreciated by looking at the visual dataset summary shown in Fig. 6.6, the dataset is highly unbalanced, and

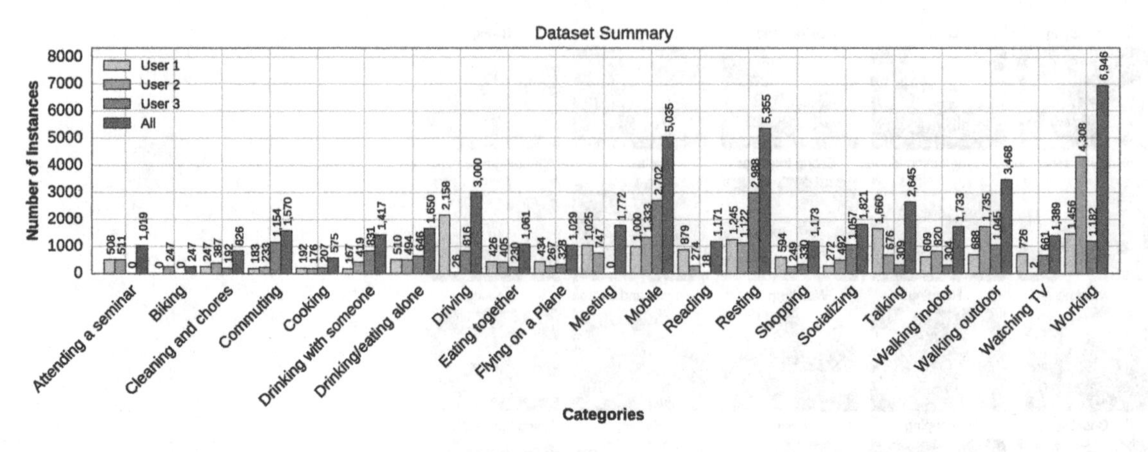

■ FIGURE 6.6 Distribution of annotated activity categories in our dataset.

this reflects the fact that in our daily lives we spend most of the time doing activities like *working* or *writing* more than others like *resting* or *cleaning*. Additionally, it can be observed that the three users have different lifestyles and the activity *biking* is performed just by one user, and other activities as *reading*, *watching TV*, and *driving* are performed only by two users.

The annotated images are part of the NTCIR 2016 dataset [13] and were annotated in batches by using the annotation tool described in [7].[1]

Since the dataset is highly unbalanced, in addition to the accuracy, we measured also macro-metrics such as macro-precision, macro-recall and macro-F1-score.

6.4.2 Implementation

In this subsection we detail the training setup used for obtaining the results reported in Section 6.4.3 for each activity recognition algorithm described in this chapter.

6.4.2.1 Activity recognition at image level

VGG-16. The VGG-16 CNN was fine-tuned by using a SGD optimization strategy during 14 epochs. The first 10 epochs were aimed at updating only the weights of the fully connected layers using a learning rate $\alpha = 1 \times 10^{-5}$, a batch size of 1, a momentum $\mu = 0.9$, and a weight decay equal to 5×10^{-6}. During the last four epochs, also the weights of the last three convolutional layers were fine-tuned and the learning rate changed to $\alpha = 1 \times 10^{-5}$.

[1]The annotation tool is available at https://github.com/hermetico/TFG/tree/master/annotation-tool.

VGG-16+FR. Three different LFE methods were considered that use a Random Forest (RF) with 500 trees to combine the softmax probabilities of the VGG-16 firstly with the output of the first convolutional layer (FC1) of the VGG-16; secondly, with time metadata and color information the day of the week, the time of the day, and a 10-bin size histogram for each color channel as proposed in [8]; and thirdly with the output of the FC1 layer having more time metadata.

6.4.2.2 Activity recognition at batch level

VGG-16+LSTM. For this architecture, a LSTM layer of 256 units was added after the first fully-connected layer of the VGG-16 network and was followed by a fully-connected layer. During training, the weights of the first four blocks of the convolutional layers were frozen and SGD was used as optimization algorithm. The training was performed in batches of 15 frames, a value that showed better performances in [6]. These batches were sampled using a sliding window from each sequence. As initial weights, the ones obtained for the base model were used. This configuration was trained for two epochs with a learning rate $\alpha = 1 \times 10^{-4}$, a momentum $\mu = 0.9$, and a weight decay equal to 5×10^{-6}.

VGG-16+RF+LSTM. For this configuration, a LSTM recurrent neural network was trained using the predictions from the ensemble of VGG-16 plus the RF on FC1. The LSTM had 32 units and its output was connected to a fully-connected layer. The training was performed in batches of 15 frames. Dropout layers were added between the input and output of the LSTM layer as a way to perform regularization.

In both architectures, the sliding window approach also acted as a data augmentation strategy without which the convergence of the network would have been impossible. In fact, the use of batches instead of single images drastically reduces the number of available training examples.

6.4.3 Results and discussion

In Table 6.1, we report the activity recognition performance in terms of accuracy and macro-metrics, obtained by applying state of the art methods to the dataset described in the previous section. As expected, the fine-tuned VGG-16 baseline is outperformed by all considered approaches. However, surprisingly, these results show that the use of time metadata still helps in improving performance. We believe this is likely due to the fact that certain activities such as *biking* are performed just by one user, and some other activities such as *working*, performed by two or more users, occur at regular times on the considered dataset. Therefore these results shed light on the

Table 6.1 Summary of the performance of activity recognition methods

Method	Accuracy	Macro precision	Macro recall	Macro F1-score
VGG-16	75.97	68.50	67.49	66.80
VGG-16+RF on FC1+Softmax [7]	76.64	71.41	66.43	67.02
VGG-16+RF on Softmax+ Date&Time+Color [8]	77.39	69.66	67.79	66.99
VGG-16+RF on FC1+ Softmax+Date&Time	76.91	72.07	66.77	67.27
VGG-16+LSTM [6]	81.73	76.68	74.04	74.16
VGG-16+RF+LSTM	**87.07**	**81.05**	**83.48**	**81.77**

need of a standard benchmark for activity recognition from visual lifelogs, which includes activities captured by a very large number of users having different lifestyles, instead of just three persons. Such a benchmark is difficult to build mainly for three reasons. Firstly, because for building a visual lifelogs datasets each person has to wear a camera during a long period of time. Secondly, because of the huge amount of images captured that need to be annotated. Thirdly, because visual lifelogs necessarily include anyone who comes into the view of a camera, with or without consent, and privacy concerns have to be taken into account.

The use of a LSTM neural recurrent network for taking advantage of the temporal coherence of concepts in photo-streams as a result was successful both when using contextual cues through a random forest and when training in an end-to-end fashion. In Fig. 6.7, we visualized, for three one day sequences (top to bottom panel), activity predictions with respect to the ground truth (top rows) when using the VGG-16 baseline (second rows) and when using LSTM on its output (third rows) in an end-to-end fashion (red/blue correspond to correct/wrong predictions, respectively). The improvement obtained when using LSTM is quite evident when a single activity during a continuous period of time is split into subintervals. For instance, for the large light green bloc "eating together" in panel one, VGG-16 yields mostly wrong predictions while the accuracy is significantly improved when combined with LSTM.

Although there results were achieved without relying on the knowledge of shot boundaries, we assume that segmenting a day sequence into different event subsequences and using them as batches during training could further improve the activity recognition performance. The problem of event segmentation in egocentric videos has indeed received relatively much attention in the past decade [9,11,22,27], and state of the art solutions for visual lifelog segmentation are robust enough for enabling the understand-

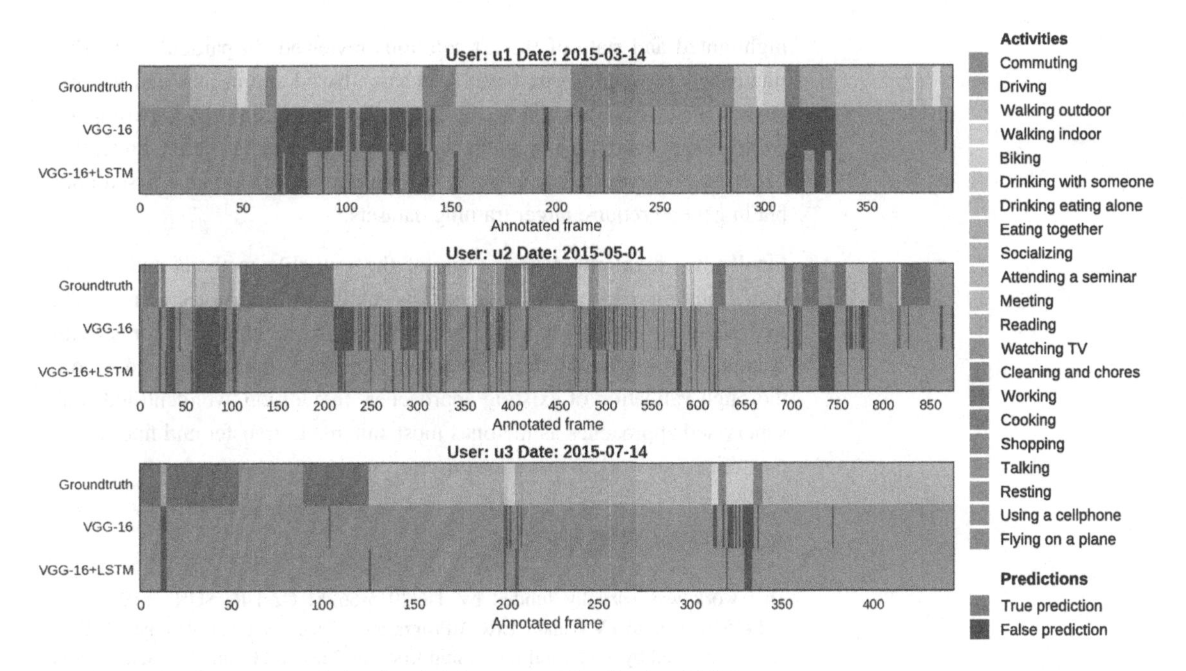

FIGURE 6.7 Prediction comparison on three day sequences.

ing of the role of event boundaries for activity recognition in visual lifelogs. In particular, instead of using overlapping batches of fixed size to train a LSTM recurrent neural network, we could use consecutive batches of different lengths corresponding to the semantically homogeneous segments extracted by a temporal segmentation algorithm.

Although the 21 activities considered in this chapter cover a good number of basic universal activities such as *eating*, *walking*, etc., they are by no means able to cover the richness of human activities. Consequently, a fully supervised approach could possibly generate misleading results in real world applications when an unseen user performs activities not covered by these 21 ones. These limitations suggest going beyond supervised frameworks to explore unsupervised and self-supervised approaches. On the one hand, unsupervised approaches seem to be more suitable to cope with both the variation of human activities in a real world setting and the availability of unlabeled data. On the other hand, self-supervised approaches would allow one to easily fine-tune the system for a given user, which is a desirable property in the context of personalized activity recognition.

6.5 CONCLUSION

This chapter has focused on the problem of activity recognition from visual lifelogs. Both challenges and opportunities of this domain have been highlighted and state of the art solutions reviewed. In particular, we distinguished between image-based and batch-based approaches and we discussed their strengths and weaknesses also in the light of experimental comparisons on a publicly available dataset. These results have shown that batch-based approaches are in general more robust than image-based ones but in general require larger training datasets.

Finally, we highlighted limitations of the available methods and benchmarks, and in view of our analysis, we outlined future lines of research. In particular, we pointed out the need of a standard benchmark for activity recognition from visual lifelogs that could enable further studies and a more thorough validation of existing approaches. In addition, we identified self-supervised approaches as the ones most suitable to transfer and fine-tune a pre-trained network to an unseen user.

ACKNOWLEDGMENTS

This work was partially funded by TIN2015-66951-C2-1-R, SGR 1219, Grant 20141510 (Marató TV3) and CERCA Programme/Generalitat de Catalunya. A. Cartas is supported by a doctoral fellowship from the Mexican Council of Science and Technology (CONACYT) (grant-no. 366596). M. Dimiccoli and P. Radeva are partially supported by ICREA Academia 2014. The funders had no role in the study design, data collection, analysis, and preparation of the manuscript.

REFERENCES

[1] Girmaw Abebe, Andrea Cavallaro, Xavier Parra, Robust multi-dimensional motion features for first-person vision activity recognition, Comput. Vis. Image Underst. 149 (2016) 229–248.

[2] Sven Bambach, Stefan Lee, David J. Crandall, Chen Yu, Lending a hand: detecting hands and recognizing activities in complex egocentric interactions, in: Proceedings of the IEEE International Conference on Computer Vision, 2015, pp. 1949–1957.

[3] Ardhendu Behera, David C. Hogg, Anthony G. Cohn, Egocentric activity monitoring and recovery, in: Asian Conference on Computer Vision, Springer, 2012, pp. 519–532.

[4] Marc Bolanos, Mariella Dimiccoli, Petia Radeva, Toward storytelling from visual lifelogging: an overview, IEEE Trans. Human-Mach. Syst. 47 (1) (2017) 77–90.

[5] Daragh Byrne, Aiden R. Doherty, Cees G.M. Snoek, Gareth J.F. Jones, Alan F. Smeaton, Everyday concept detection in visual lifelogs: validation, relationships and trends, Multimed. Tools Appl. 49 (1) (2010) 119–144.

[6] Alejandro Cartas, Mariella Dimiccoli, Petia Radeva, Batch-based activity recognition from egocentric photo-streams, in: Proceedings on the International Conference in Computer Vision (ICCV), 2nd International Workshop on Egocentric Perception, Interaction and Computing, Venice, Italy, 2017.

[7] Alejandro Cartas, Juan Marín, Petia Radeva, Mariella Dimiccoli, Recognizing activities of daily living from egocentric images, in: Iberian Conference on Pattern Recognition and Image Analysis, Springer, 2017, pp. 87–95.

[8] Daniel Castro, Steven Hickson, Vinay Bettadapura, Edison Thomaz, Gregory Abowd, Henrik Christensen, Irfan Essa, Predicting daily activities from egocentric images using deep learning, in: Proceedings of the ACM International Symposium on Wearable Computers, 2015, pp. 75–82.

[9] Mariella Dimiccoli, Marc Bolaños, Estefania Talavera, Maedeh Aghaei, Stavri G. Nikolov, Petia Radeva, Sr-clustering: semantic regularized clustering for egocentric photo streams segmentation, Comput. Vis. Image Underst. 155 (2016) 55–69.

[10] Mariella Dimiccoli, Juan Marín, Edison Thomaz, Mitigating bystander privacy concerns in egocentric activity recognition with deep learning and intentional image degradation, Proc. ACM Interact. Mob. Wearable Ubiquitous Technol. 1 (4) (2018) 132.

[11] A.R. Doherty, A.F. Smeaton, Automatically segmenting lifelog data into events, in: Proceedings of the 2008 Ninth International Workshop on Image Analysis for Multimedia Interactive Services, 2008, pp. 20–23.

[12] Alireza Fathi, Ali Farhadi, James M. Rehg, Understanding egocentric activities, in: 2011 International Conference on Computer Vision, IEEE, 2011, pp. 407–414.

[13] Cathal Gurrin, Hideo Joho, Frank Hopfgartner, Liting Zhou, Rami Albatal, Overview of NTCIR-12 lifelog task, in: NTCIR-12, National Institute of Informatics (NII), 2016.

[14] Paul Kelly, Simon J. Marshall, Hannah Badland, Jacqueline Kerr, Melody Oliver, Aiden R. Doherty, Charlie Foster, An ethical framework for automated, wearable cameras in health behavior research, Am. J. Prev. Med. 44 (3) (March 2013) 314–319.

[15] Oscar D. Lara, Miguel A. Labrador, A survey on human activity recognition using wearable sensors, IEEE Commun. Surv. Tutor. 15 (3) (2013) 1192–1209.

[16] Yin Li, Zhefan Ye, James M. Rehg, Delving into egocentric actions, in: Proceedings of the IEEE Conference on Computer Vision and Pattern Recognition, 2015, pp. 287–295.

[17] Minghuang Ma, Haoqi Fan, Kris M. Kitani, Going deeper into first-person activity recognition, in: Proceedings of the IEEE Conference on Computer Vision and Pattern Recognition, 2016, pp. 1894–1903.

[18] Tomas McCandless, Kristen Grauman, Object-centric spatio-temporal pyramids for egocentric activity recognition, in: BMVC, vol. 2, 2013, p. 3.

[19] Subhas Chandra Mukhopadhyay, Wearable sensors for human activity monitoring: a review, IEEE Sens. J. 15 (3) (2015) 1321–1330.

[20] Thi-Hoa-Cuc Nguyen, Jean-Christophe Nebel, Francisco Florez-Revuelta, Recognition of activities of daily living with egocentric vision: a review, Sensors (Basel) 16 (1) (Jan 2016) 72, sensors-16-00072 [PII].

[21] Hamed Pirsiavash, Deva Ramanan, Detecting activities of daily living in first-person camera views, in: IEEE Conference on Computer Vision and Pattern Recognition (CVPR), 2012, pp. 2847–2854.

[22] Y. Poleg, C. Arora, S. Peleg, Temporal segmentation of egocentric videos, in: Proceedings of the IEEE Conference on Computer Vision and Pattern Recognition, 2014, pp. 2537–2544.

[23] Yair Poleg, Ariel Ephrat, Shmuel Peleg, Chetan Arora, Compact cnn for indexing egocentric videos, in: IEEE Winter Conference on Applications of Computer Vision (WACV), 2016, pp. 1–9.

[24] Michael S. Ryoo, Larry Matthies, First-person activity recognition: what are they doing to me?, in: Proceedings of the IEEE Conference on Computer Vision and Pattern Recognition, 2013, pp. 2730–2737.

[25] Michael S. Ryoo, Brandon Rothrock, Charles Fleming, Hyun Jong Yang, Privacy-preserving human activity recognition from extreme low resolution, arXiv preprint, arXiv:1604.03196, 2016.

[26] M.S. Ryoo, Larry Matthies, First-person activity recognition: feature, temporal structure, and prediction, Int. J. Comput. Vis. 119 (3) (2016) 307–328.
[27] Estefania Talavera, Mariella Dimiccoli, Marc Bolaños, Maedeh Aghaei, Petia Radeva, R-clustering for egocentric video segmentation, in: Pattern Recognition and Image Analysis, Springer, 2015, pp. 327–336.
[28] Edison Thomaz, Aman Parnami, Jonathan Bidwell, Irfan Essa, Gregory D. Abowd, Technological approaches for addressing privacy concerns when recognizing eating behaviors with wearable cameras, in: Proceedings of the 2013 ACM International Joint Conference on Pervasive and Ubiquitous Computing, ACM, 2013, pp. 739–748.
[29] Yan Yan, Elisa Ricci, Gaowen Liu, Nicu Sebe, Egocentric daily activity recognition via multitask clustering, IEEE Trans. Image Process. 24 (10) (2015) 2984–2995.
[30] Joe Yue-Hei Ng, Matthew Hausknecht, Sudheendra Vijayanarasimhan, Oriol Vinyals, Rajat Monga, George Toderici, Beyond short snippets: deep networks for video classification, in: Proceedings of the IEEE Conference on Computer Vision and Pattern Recognition, 2015, pp. 4694–4702.

Chapter 7

Lifelog retrieval for memory stimulation of people with memory impairment

Gabriel Oliveira-Barra*, Marc Bolaños*,†, Estefania Talavera*,†,§, Olga Gelonch‡, Maite Garolera‡, Petia Radeva*

*University of Barcelona, Department of Mathematics and Computer Science, Barcelona, Spain †Computer Vision Center, Barcelona, Spain ‡Consorci Sanitari de Terrassa, Barcelona, Spain §University of Groningen, Intelligent Systems Group, AG Groningen, the Netherlands

CONTENTS

7.1 INTRODUCTION

Lifelogging is the capturing of a person's everyday activities, typically via wearable sensors including accelerometers, GPS, heart rate monitors, tem-

Multimodal Behavior Analysis in the Wild. https://doi.org/10.1016/B978-0-12-814601-9.00016-X
Copyright © 2019 Elsevier Ltd. All rights reserved.

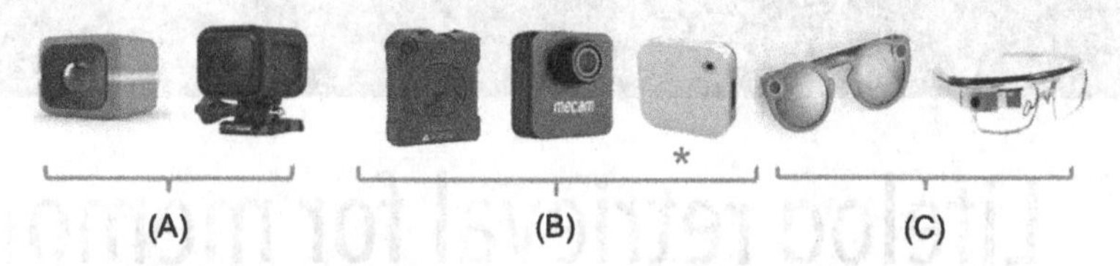

FIGURE 7.1 Egocentric cameras. Some egocentric vision wearable cameras currently available on the market, having different purposes and specifications, and worn in different ways: (A) Head and torso cameras used mostly for sports, leisure and entertainment [Polaroid Cube and GoPro Hero5 Session]. (B) Torso mounted, devices used for visual lifelogs creation and security [Axon Body 2, MeCam HD, Narrative Clip 2]. (C) Glass mounted, used for social media and augmented reality [Spectacles and Google Glass]. The camera that we use to record egocentric photo-streams is the Narrative Clip 2, the camera on the right within the group (B).

perature, audio, cameras, among others. When the recorded signals are obtained through a wearable camera, it is referred to as visual lifelogging [9]. The egocentric vision field refers to addressing and exploring computer vision problems on egocentric images. Thanks to advances in wearable technologies, the trend is rapidly increasing over recent years. Lifelogging is presented as a potential tool for user life-style analysis, with the aim of improving quality of life, memory, or even safety. By recording the person own view of the world, lifelogging opens new questions and leads to the desired and personalized analysis of the camera wearer's lifestyle. Egocentric information is used in order to have a better insight of a person's pattern of behavior; daily habits, such as eating habits through food recognition [12], social interactions [3], indoor or outdoor activities and actions [14,43]. Information is extracted from the recorded lifestyle data in the egocentric images.

Users and specialists are both interested in gaining an overview and finding relevant events from lifelogging data. Some of the possible applications could be the analysis and summarization of this data into narratives, offering a memory aid to people with memory impairments due to brain injury or neurodegenerative diseases by reactivating their memory capabilities [16]. Another application is to extract the camera wearer's activity diary based on with whom the user performs a specific activity; what it is, and where it is done.

Egocentric photo-streams are usually automatically recorded by wearable cameras (see Fig. 7.1). The obtained data represent a first-person view of the world, describing daily activities and providing information about the interaction of the user with the surrounding environment. Its popularity has been increasing in the last years due to the easy of access, simplicity and reduced price of wearable technology.

Table 7.1 Comparison of some popular wearable cameras

	Main use	Temporal resolution (FPS)	Mount	Weight (g)
GoPro Hero5	Entertainment	High and low	Head and torso	73
Google Glass	Augmented reality	High	Head	36
Spectacles	Social networks	High	Head	25
Narrative Clip 2	Lifelogging	High and low	Torso	19
Axon Body 2	Security	High and low	Torso	141

GoPro and Google Glasses are the well-known devices in the current wearable cameras market. Both cameras have high temporal resolution; in other words, they are able to capture many Frames Per Second (FPS). These devices are able to record videos during certain moments of time with high quality resolution. On the other side, there are low temporal resolution cameras which record images with a certain time-lapse. This operation mode, sometimes known as *story mode*, takes a new picture every t seconds, with t usually ranging between 0.5–30 s. These cameras create a daily life log of their user. Table 7.1 describes some of the most popular wearable cameras currently available on the market designed with different goals in mind.

They are small devices that can be worn all day and automatically record the everyday activities of the wearer in a passive fashion, from a first-person point of view. Their usability, last of battery, and possibility to capture in images the user's whole day make their use attractive and useful to extract behavioral patterns of the wearer from long periods of time (days, weeks, months).

Lifelogging-oriented wearable cameras with low temporal resolution usually record up to 2000 egocentric images per day with a $t = 30$ s time span, i.e. up to 70,000 per month. The camera is commonly attached to the user's chest, giving a first view of the user's environment and actions through the recorded images. If the user performs static activities such as working with a laptop, watching TV, etc., it will lead to redundancies within the sequences. The summarization of egocentric photo-streams appeared as a solution for the problem of selecting relevant information when many pictures are taken among a long time interval [34]. The aim is the selection of images for the creation of a summary which can recapitulate the user's day. Image selection is usually based on classifying the semantically rich ones [42], social interactions [3], and event representation [35,28]. However, extracting and locating semantic content from a huge collection of personal data triggers major challenges that strongly limit their utility and usability in practice. In consequence, lifelogging requires efficient retrieval techniques in combination with other intelligent filtering, selection and classification solutions to efficiently process, store and retrieve semantic information out of the huge amount of egocentric data generated on a daily basis.

7.2 RELATED WORK

Lifelogging. The visual lifelogging process allows one to record a rich set of data that captures the wearer's day in a holistic way: the events the person participated in, environments he/she visited, persons he/she interacted with, activities she/he did, etc. Here, we comment on several approaches that, by analyzing egocentric photo-streams, address different computer vision problems. In [22], a method was introduced to passively recognize human activities through recorded lifelogs. Social interaction analysis is also addressed in [3,2,1]. In [1], the authors proposed a face tracking method with the aim of finding correspondences along the egocentric sequence for each detected face. Moreover, in [54,55], the sentiment evoked by the images was analyzed in order to select the positive moments within the lifelog sequence.

Summarization. When dealing with lifelogging data, we can usually think of the terms *routine* or *repetition*. Most of the data acquired describe very repetitive information that should be summarized in order to both improve its comprehension and avoid having to review the whole amount of data. These methods are probably even more necessary when dealing with lifelogging images, because we need to avoid that the user overviews hundreds or thousands of pictures of the same scene again and again. For this reason, several researchers have focused on proposing useful summarization methods. From the egocentric video point of view, most of the recently proposed works take into account semantics for producing the resulting summaries. In [34], and later in [38], a new method is proposed where the video is chunked into independent units. These units are considered to belong to different scenes or subshots based on color features. The most relevant frames of each subshot are detected based on the people and objects appearing. More recently, [36] proposed a method to highlight detection based on a structured SVM that also selects the most relevant video subshots. A different set of methods, which are called fast-forward summarization methods, intend to find the most relevant parts and instead of removing the parts that are less important they fast-forward them on the video. A good example of this is the work from [48], which uses a semantic selection method. The problem of these methods is that they are only suitable for video visualization and are not intended to provide a selection of images.

Focusing on the egocentric photo-streams summarization methods, [28] used a nearest neighbor ratio strategy for selecting the best key-frame in each temporal scene. The work in [10] proposed a random walk algorithm based on image similarity. More recently, [35] worked on a method specific for applying a key-frame ranking based on already segmented temporal frames [21]. The authors employed several semantic detection techniques that intend to capture relevant faces, objects and highly salient frames in order to generate an initial relevance ranking. Afterwards, they proposed applying a novelty-based strategy that, taking into account the CNN-based features extracted for each frame, calculates the similarities between all relevant key-frames and re-ranks them considering both their relevance and novelty.

Retrieval. Conventional instance-level retrieval approaches rely on handcrafted features like SIFT descriptors, which are usually encoded into bag-of-words histograms [53]. Encoding techniques, such as the Fisher vector [45] or the Vector Locally Aggregated Descriptor (VLAD) [29], obtain a compact representation that could replace bag-of-words histograms, achieving good results, while requiring less storage. These methods combined with compression [46,27,47] produce global descriptors that scale to larger databases at the cost of reduced accuracy. All these methods can also be combined with other post-processing techniques such as query expansion [17,6]. Major improvements generated by CNNs in computer vision tasks [50,44,23,58] show a growing interest in investigating possible uses of CNNs for image retrieval tasks. A pioneering attempt was made in [50] with OverFeat [52]. However, the performance of CNN features extracted from the final layers of AlexNet lagged behind the ones of SIFT-based methods with BoW and VLAD encoding. Only by incorporating spatial information they achieved similar results. In [19], CNN features learned with various augmentation and pooling schemes are applied to painting retrieval and achieved very satisfactory results. Reference [25] introduced Multi-scale Orderless Pooling (MOP) to aggregate CNN activations from higher layers with VLAD, where these activations are extracted by a sliding window with multiple scales.

The work in [57] conducted a study on applying CNN features to image retrieval with model retraining and similarity learning. In this direction, [7] retrained a CNN model on a landmark dataset that is similar to the images at query time.

Encouraging experimental results showed that CNN features are effective in bridging the semantic gap between low-level visual features and high-level concepts. Recently, the work in [49] conducted extensive experiments on different instance retrieval datasets and obtained excellent results by using a spatial search with CNN features. Some efforts have been made trying to benefit from widely used, robust text-based information retrieval libraries based on an inverted index for visual instance retrieval tasks. For instance, the authors of [39] introduced an open source library for content-based image retrieval supporting a variety of handcrafted image descriptors. In [40], an extension of this library is proposed and in [41], the authors benefited from the same library as a complement for a lifelogging moment retrieval engine with visual instance retrieval support. In [4], the authors proposed a technique for generic approximate similarity searching, based on the use of inverted files, and in [24] the authors used CNN features to support content-based image retrieval on large-scale image datasets by converting the CNN features into a textual form. Current work such as [41] proposed a retrieval engine oriented to egocentric images; the work of [24] described a method for working with CNN features as textual features, and thus proposed a valuable technique to be applied for lifelog moment retrieval for memory improvement.

Memory improvement based on reviewing autobiographical images. One of the first applications of lifelogging data analysis was to improve autobiographical memory in people with memory impairments, such as amnesic patients due to brain injury or people with dementia [33]. The proposed method aimed to extract consisting cues from the lifelog, later to be reviewed by the persons with episodic memory impairment.

In order to re-live the events reviewing the images, several studies have shown that reviewing the images taken by the wearable camera can be effective for memory rehabilitation. Thus, it helps to reinstate previous thoughts, feelings and sensory information, thus stimulating recollection. In memory impairment, self-initiated processes are impaired, and reviewing images of wearable camera plays an important environmental support for memory retrieval. This device facilitates access to cues and information that are pertinent for the patient, thus providing effective recall cues which can facilitate recollection. It is a very important attribute, so this capacity to re-live an event with the majority of its initial richness (for example, to re-live the emotional feelings of the event) is what allows a person to maintain a coherent self-identity. This device should not only allow access to images, but also affect memory by re-initiating thoughts, feelings and (non-visual) sensorial input linked to the images [32]. Studies have shown that reviewing images taken by the wearable camera enhances recall of personality experienced events that had been recorded and retrieved, both in healthy and memory impaired populations [37]. It has been shown that it is efficacious in the retrieval for autobiographical memory and personal semantic memory, but it also acts as a cognitive stimulant in daily life for a healthy population [5]. To

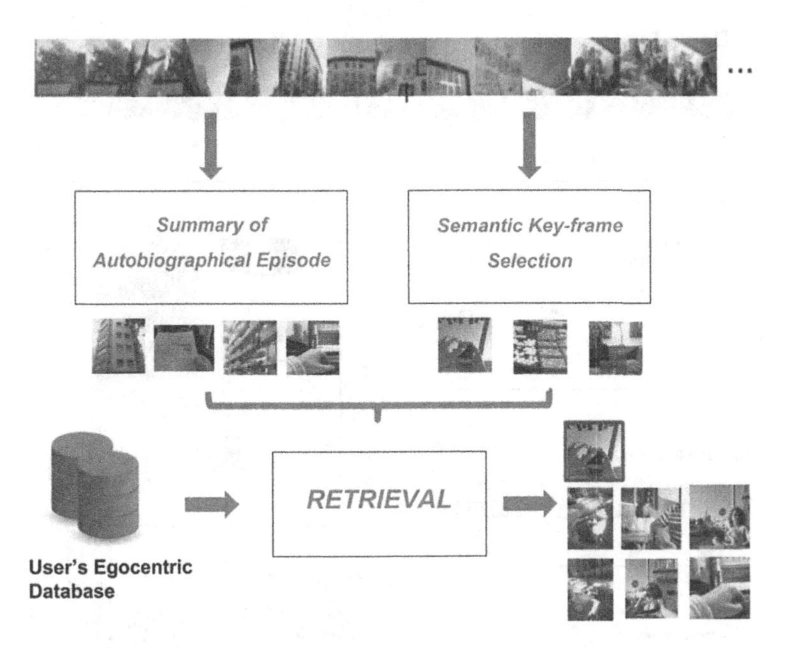

■ FIGURE 7.2 **Egocentric photo-stream pipeline**. Sketch of the proposed approach for egocentric photo-streams retrieval. On the one hand, we semantically summarize and extract rich images from the egocentric photo-stream. On the other hand, we use these selected images and retrieve similar ones from the user's egocentric database. Some results are shown after the retrieval process, where the query image is shown with a blue border.

apply a memory stimulation procedure, we need efficient egocentric vision techniques to retrieve special moments from episodic images and prepare a set of semantically rich enough non-redundant data to be used by the experts in neurodegenerative diseases. See Fig. 7.2.

The methodology of egocentric retrieval is detailed in the next section. A description of the performed run trials and quantitative results are detailed in Section 7.4. Finally, Section 7.5 provides concluding remarks.

7.3 RETRIEVAL BASED ON KEY-FRAME SEMANTIC SELECTION

In our work, we propose a complete methodology for egocentric image selection based on semantic characteristics. Our method provides a smart key-frame selection based on summarization techniques, people and semantically-rich scenes. Although applicable to other problems, the main purpose of our approach is to provide a selection of a subset of egocentric images in order to obtain better results when applying retrieval techniques on the generated daily lifelogging data. More precisely, these techniques

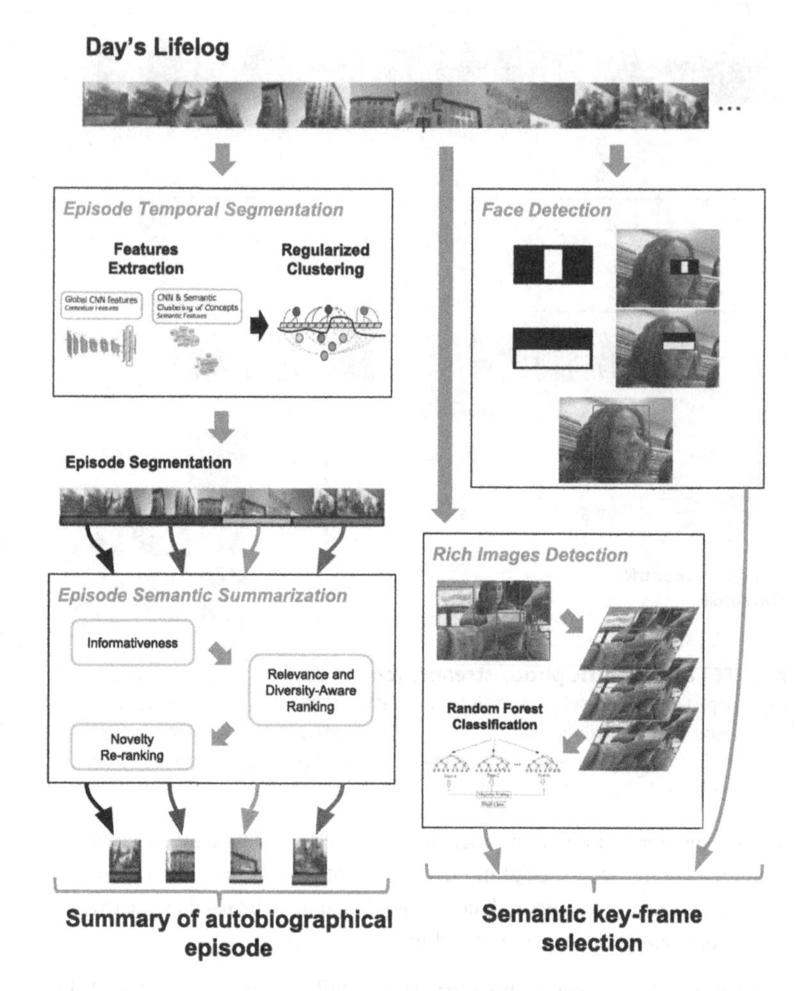

■ FIGURE 7.3 **Photo-streams for memory aid pipeline.** Sketch of the proposed model for egocentric photo-streams analysis. 1) Summary of autobiographical episodes: First, a temporal segmentation is applied over the egocentric photo-stream. Later, events semantic summarization is performed to extract the informative images. 2) Image retrieval for memory stimulation program: On the one hand, we extract the faces of the people who played a role in our day, and on the other hand, we seek for *rich images*, selected based on informativeness, relevance and novelty estimators.

can serve as an image selection process for developing memory training tools (i.e. image retrieval, among others) for neurodegenerative patients.

In order to apply the image selection process, this work proposes two steps that can be applied in parallel (see Fig. 7.3 for a detailed pipeline of our proposed approach). On the one hand, we apply key-frame selection based on semantic segmentation and summarization of the complete day by considering the joint relationship of images. On the other hand, we apply two image selection processes based simply on the appearance of faces and of rich semantic information for each image separately.

We use text-based, inverted index retrieval techniques in combination with the feature vectors obtained from state-of-the-art convolutional neural networks for very fast content-based image retrieval with low memory footprint. We use this approach as a technique for memory improvement based on lifelogging images, evaluating its performance on the EDUB2015 egocentric dataset [11].

7.3.1 Summarization of autobiographical episodes

This part of the model intends to provide a summarization of certain relevant and memorable episodes from the life of the user. By applying a summarization of the whole day of the user, we gain several advantages: 1) we select only the most relevant moments, 2) we remove all redundant and repeated information, 3) we remove non-informative data, and 4) we save computational time for the next processes and space needed for the data.

From a clinical practice point-of-view of patients with memory impairments, it is clear they have major problems to recall certain moments of their life. Showing them scenes and people of their daily life can help them create memory links and allow them to retrieve autobiographic events. The main problem of providing visual cues to these patients is when these cues are not clear enough (blurred or dark), since when this happens they need an extra effort for keeping their attention when reviewing the images. For these reasons, it is important to provide useful summarization techniques for egocentric data. By developing these methods, users with dementia could enhance their cognitive capabilities through a reinforcement of the positive emotions drawn from it. At the same time, both the removal of non-informative data and the reduction of exposure to images would allow for a higher degree of concentration by the users.

Our proposal for summarization of the autobiographical episodes is composed of two main algorithms. The first one, named SR-clustering [21], is in charge of temporally segmenting the complete day in relevant events and separate units of information. In order to do so, it extracts semantic and global features for each frame and applies an energy minimization technique by joining two different clustering methods. The second one, named semantic summarization [35], treats each of the previously extracted segments separately. For each of them, it applies an informativeness filtering, extracts a set of different semantic indicators like faces, objects and saliency, and performs a relevance and novelty ranking in order to select a subset of key-frames.

7.3.1.1 Episode temporal segmentation

Given an egocentric photo-stream, we obtain the events experienced by the user along the day by applying the temporal segmentation SR-clustering algorithm [21]. This algorithm starts by extracting a combined semantic and global representation of the sequence of frames. The global visual features are extracted by means of a convolutional neural network trained for image classification [31]. The resulting feature vector serves as a general low-level encoding of the information contained in the frame. The semantic features are computed by:

1. detecting high-level concepts on each image separately,
2. applying a semantic concept clustering for defining an ontology of concepts specific to that day, and
3. building a feature vector that encodes the certainty of appearance of each of the clustered concepts on each image. After concatenating both the global and semantic feature vectors, we apply a regularized clustering (R-clustering) technique. This technique consists in applying two clustering techniques separately on the photo-stream of images: Agglomerative clustering, which usually over-segments the data, and ADWIN [8], which tends to under-segment it. After this, the output of both techniques are combined by applying an energy minimization method that obtains the final segmentation result.

7.3.1.2 Episode semantic summarization

A day photo-stream is normally composed by thousands of images, often repetitive and not relevant. With the aim of semantically summarizing the day, we apply the method introduced in [35]. In order to apply this technique, we propose using each of the segments extracted in the previous process as independent input samples. Taking this into account, for each input segment we select the subset of key-frames that better summarize the event. The general scheme of the method can be described in the following steps: 1) select images based on their informativeness, removing any non-informative frame; 2) calculate their relevance considering several semantic concepts like the appearance of faces and objects and the level of saliency; and 3) compute a final ranking of images based on relevance and novelty estimators. This approach was validated by medical experts in the field of memory training [35].

7.3.2 Semantic key-frame selection

Unlike the summarization of autobiographical episodes, which selects images considering groups of sets that describe separate units of time, the semantic key-frames we propose aim to select images based solely on their individual appearance.

The main factor for providing a meaningful image selection algorithm for memory stimulation is that images should provoke a sentiment enhancement and cognitive mechanisms reactivation. Considering the free-motion and non-intentionality of the pictures taken by wearable cameras [9], it is very important to provide a robust method for image selection.

Two of the most important and basic factors that determine the memorability of an image [30,13] can be described as: 1) the appearance of human faces, and 2) the appearance of characteristic and recognizable objects. In order to satisfy these criteria, we propose using, in parallel, two different computer vision algorithms. On the one hand, a face detection algorithm is applied that allows to select images where a person is present; and on the other hand, a rich image detection algorithm [42] which intends to detect images with a high number of objects and variability avoiding at the same time images with low semantical content.

7.3.2.1 With whom was I? Face detection

We humans are naturally drawn towards focusing on faces when looking at images. This is an important salient feature that at the same time can easily be associated to certain emotions and sentiments, allowing to easily associate a certain memory to their appearance. In order to detect images containing faces, we propose using a cascade learning algorithm based on Haar-like features [56], which has proven to be fast and robust towards faces detection.

The algorithm consists in applying a cascade of classifiers, which consists of a series of stages that at each step increases its strength in classification, allowing it to rapidly discard most negative examples. In our case, the classifiers are trained on Haar-like features, which are a series of masks that after being applied on the input sample provide a positive or negative response depending on whether they match an expected spatial pattern or not.

7.3.2.2 What did I see? Rich image detection

Another important characteristic that allows one to identify and remember a certain environment are the objects and the elements that it contains. Following this premise, we propose using an algorithm for rich image detection [42], understanding as rich any image that is neither blurred, nor dark and that contains clearly visible non-occluded objects.

The algorithm for rich images detection consists in: 1) multiple object detection, where the neural network named YOLO9000 [51] is applied in order

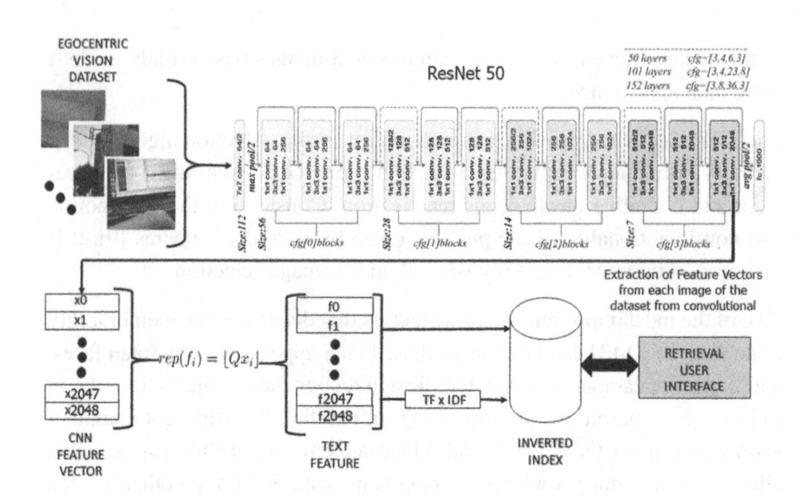

■ **FIGURE 7.4 Overview of the proposed system.** First, CNN features are extracted from the fully-connected layers of the CNN and later the feature vector is converted to text and used to index the image in a retrieval user interface.

to detect any existent object in the images and its associated confidence c_i, 2) extraction of a set of richness-related features, and 3) a random forest classifier [18] trained to distinguish the difference between rich or non-rich images.

7.3.3 Egocentric image retrieval based on CNNs and inverted index search

In the following, we detail the two major steps of the proposed methodology: the extraction and the textual representation of CNN features and their embedding in a web-based search platform running on top of a Lucene core. Our approach is summarized in Fig. 7.4.

7.3.3.1 *Extraction of CNN features and their textual representation*

To benefit from a text retrieval engine and to perform efficient image search through similarity, we combine the richness of the representation of CNNs features with the scalability of the text-based search engine approach. First, we need to extract the CNN feature vectors out from any given CNN and then proceed to re-write them as text, ensuring direct proportionality between the feature components and the term frequencies.

■ FIGURE 7.5 **EDUB2015 Dataset overview**. Sample tagged categories from EDUB2015 dataset: paper, person bottle, building, glass, hand, chair, sign, face, glass, TVmonitor, window, mobile phone. There might be more than one labeled category per image.

We extract CNN features from the EDUB2015 dataset (see Fig. 7.4) by using the Tensorflow[1] framework and the pre-trained ResNet50 [26] model over the ImageNet [20] dataset. Following the recommendations from [59], we choose to conduct our experiments by extracting the feature vectors from the last hidden layer of the ResNet50 of dimension 2048. Once we have all the features extracted from the EDUB2015 egocentric dataset (Fig. 7.5), we quantify and map each component of the CNN feature vector, $e_i, i = 1, \ldots, 2048$, with a unique alphanumeric term, T_i [24]. Each term has the prefix 'f' for each feature followed by the numeric values corresponding to the index i of the i-th CNN feature vector component. We form the textual representation of e_i by repeating the term T_i for the non-zero components a number of times directly proportional to its values. The number of repetitions $rep(f_i)$ of the term f_i corresponding to the component e_i is given by

$$rep(f_i) = \lfloor Qx_i \rfloor,$$

where $\lfloor\rfloor$ denotes the floor function and Q is an integer multiplication factor > 1 that works, as explained, as a quantification factor. For instance, if we fix $Q = 2$, for $x_i < 0.5$, $rep(f_i) = 0$, while for $x_i \geqslant 0.5$, $rep(f_i) = 1$. In this way, the weights of the feature vectors describing an image, are translated into term frequencies for the text-based search engine. With this procedure, we are now able to give textual descriptions of images with a dictionary size as long as the length of the feature vector. Although this process introduces a quantification error due to the representation of float components as integers, the error introduced is negligible.

Since an average of 25% of the CNN features is greater than zero (in our specific case of the last layer extracted from ResNet50), the size of its cor-

[1] https://www.tensorflow.org/.

responding textual representation will have a small fraction of the unique terms present in the whole dictionary composed of 2048 terms. The average document contains about 1000 terms from which about $1/4$ are unique representations, since the quantification procedure applied before sets feature components below $1/Q$ to 0. Another issue we face is that, in order to process searches, if no kind of pre-processing is done, the engine should handle unusual query sizes. These abnormal long queries have negligible effect on the response time, if the inverted index contains few thousands items, but have great impact if the index contains million of terms. A way to overcome this issue is to reduce the size of the query by exploiting the knowledge of the $TF \cdot IDF$ (Term Frequency · Inverse Document Frequency) statistics of the documents and the index, respectively. By keeping the top-10 elements of a query that exhibit higher $TF \cdot IDF$ scores from a given image, we can query the index using these 10 unique terms instead of the average 1000 terms, achieving a query reduction time $> 95\%$ in most cases. experiments, these top-10 features are stored in the index for evaluation purposes, but they could be extracted on-the-fly from any query image.

7.3.3.2 Text based index and retrieval of CNN features with inverted index

There are a wide variety of text-based retrieval engines currently available on the market, most of them are based on an inverted index fashion for its fast query time and low memory footprint. The basic idea is that, in order to perform a similarity search over CNN features, the features extracted from a given query image should be compared with all the features extracted from the images of the dataset, and the nearest images according to a predefined (e.g. $L2$) distance should be selected and presented, sorted by their similarity, i.e. results should be presented from more similar to less similar ones. However, if the database is large, this approach becomes resource-demanding and time-consuming. Taking advantage of the observed sparsity of CNN features, which are densely populated with zeros (about 75%), we aim to efficiently perform searches without affecting the performance. Typically, CNN feature vectors exhibit better results if they are $L2$-normalized [15,50]. Two vectors x and y are normalized as in our case, hence, the $L2$ distance between two feature vectors x and y, say $d_2(x, y)$, can be expressed by means of the inner product: $x \cdot y$, that is, $d_2(x, y)^2 = 2(1 - x \cdot y)$. Formulating the $L2$ distance in terms of an inner product allows one to efficiently benefit from the sparsity of the vectors, moving the non-zeros entries to the vector x and the non-zeros entries to the vector y.

A common, standard similarity measure among different search engines to compare documents is the cosine similarity, that is,

$$similarity = \cos(\theta) = \frac{\mathbf{x} \cdot \mathbf{y}}{\|\mathbf{x}\|\,\|\mathbf{y}\|}, \tag{7.1}$$

where θ is the angle between $\mathbf{x}$ and $\mathbf{y}$. We exploit this property so that the cosine similarity equals the dot product. The goal now is to adapt a text-based inverted index retrieval engine to populate its inverted index with the CNN feature vectors. By doing so, when the engine running the index receives a query that is an adapted CNN feature vector, the results are a ranked list of documents, images in our case, sorted by decreasing values of the cosine similarity or, what is the same, by increasing values of the $L2$ distance. For memory reasons, we do not store float numbers in the inverted index. Instead, we store the term frequencies represented as integers. To keep the integrity of the system, we must ensure that the inverted index will maintain the proportionality between numeric values representing image features and the float values obtained from the CNN features.

7.4 EXPERIMENTS

7.4.1 Dataset

Given that there are no public egocentric datasets designed for retrieval purposes up to the moment, for the evaluation of the effectiveness of the proposed approach, we used the public EDUB2015 dataset [11]. The mentioned dataset is composed of a total of 4912 images acquired by the wearable camera Narrative,[2] configured to capture images in a passive way every 30 s. The dataset is divided in eight different days which capture daily life activities like shopping, eating, riding a bike, working, etc. It has been acquired by four different subjects who captured two different full days each. The dataset includes both the images and the ground truth object segmentation for each of the images. It includes a total of 11,281 different object instances and 21 different classes in total. Fig. 7.6 represents an insight of how these 11,281 tags are distributed among the dataset. Considering the characteristics of this dataset, it is a natural candidate to apply our proposed techniques for its key elements:

1. Images considering complete days of the users.
2. Redundancies naturally present in routine environments that are not necessary for describing the complete days, and therefore could be summarized.

[2] http://getnarrative.com/.

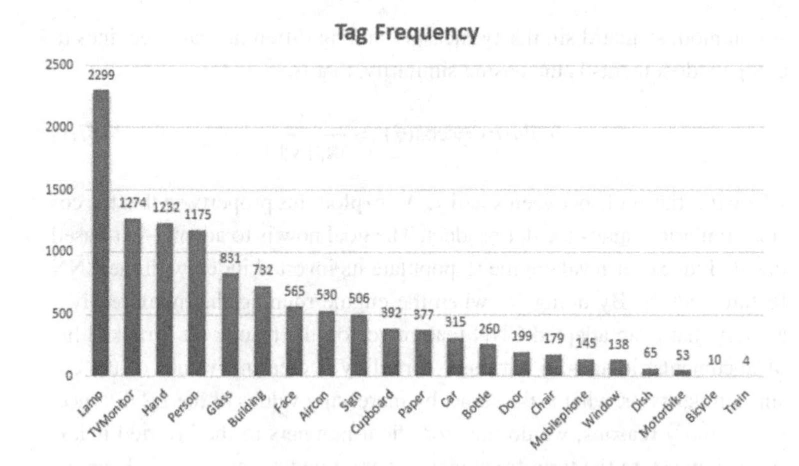

FIGURE 7.6 EDUB2015 ground-truth overview. Distribution of the 11,281 labels composing the ground-truth of the EDUB2015 dataset among 21 different categories present on the ground truth of the 4912 images.

3. Social interactions and people.
4. Objects related to daily life that can serve as a way of evaluating the retrieval performance.

7.4.2 Experimental setup

As a starting point for indexing and retrieval, we follow the setup described at previous works such as [39] and [24], where the authors propose an open source Java library based on the Apache Lucene core[3] for content-based image retrieval supporting a variety of handcrafted image descriptors. Specifically, in [41], the authors benefited from the same library as a complement for a lifelogging moment retrieval engine with visual instance retrieval support. We opted for Solr,[4] an open source, inverted index, text-based retrieval engine based on this Lucene core. We used a Tensorflow implementation of ResNet50 for extracting the features from the last layer of the neural network and Scikit-learn[5] python library for calculating the $TF \cdot IDF$ of the textual representation obtained from each image following the methodology described at 7.3.3. The Q factor has a strong effect on the system and the approximation directly depends on it. There is a trade-off between the size of Q and the memory footprint of the inverted index. For large-scale data, high Q values offer higher retrieval resolution, but on the other hand,

[3]http://lucene.apache.org/.
[4]http://lucene.apache.org/solr/.
[5]http://scikit-learn.org/.

Table 7.2 Retrieval settings for query and index

	Query	Index
$A \rightarrow A$	Full dataset	Full dataset
$B \rightarrow A$	Semantically rich images	Full dataset
$B \rightarrow B$	Semantically rich images	Semantically rich images

they require larger memory. For our experiments we choose $Q = 5$ since we find a good balance between the effectiveness of the retrieval system and the resources it demands.

Once $TF \cdot IDF$ is calculated for each image with $Q = 5$, we prepare three different retrieval setups to validate semantic key-frame selection as shown in Table 7.2. Finally, we used Curl[6] to upload the index and to perform the queries into the Solr engine.

7.4.3 Evaluation measures

The quality of the retrieved images is typically evaluated by means of *precision* and *recall* measures. We combined this information by means of the mean Average Precision (mAP), which represents the area below the precision and recall curve. In addition, we will use the Precision at n (P@n), that is the proportion of the top-n documents that are relevant in a ranked list of retrieved images.

7.4.4 Results

We present our results in Tables 7.3 and 7.4, where we show the P@1 and mAP@10 respectively, with $Q = 5$, using as query vector the top-10 $TF \cdot IDF$ of each image.

As seen in Fig. 7.7, the categories *bicycle* and *train* have too few samples on the dataset (10 and 4, respectively) to experiment with them so we do not take these categories in consideration when performing evaluations. Therefore, we perform our experiments on the remaining 19 categories. We observe a slight increase on the retrieval accuracy when queries are performed among rich images alone. Specifically, 4.8% with P@1 and 2.3% with mAP@10. From the high score obtained at P@1 we observe that it is enough with only 10 terms out of a vocabulary composed by 2048 terms to describe and retrieve a specific image.

The memory footprint of the inverted index of the system when using the feature with $Q = 5$ to index the 4192 images of the EDUB2015 dataset is

[6] https://curl.haxx.se/.

Table 7.3 P@1 obtained for $Q = 5$ with the EDUB2015 ground-truth for the setups described at 7.4.2. Queries are performed with the top10 $TF \cdot IDF$ of each image

	Aircon	Bottle	Building	Car	Chair	Cupboard	Dish	Door	Face	Glass	Hand
$A \to A$	0.782	0.867	0.883	0.929	0.857	0.839	0.982	0.946	0.909	0.942	0.86
$B \to A$	0.67	0.743	0.757	0.796	0.734	0.719	0.841	0.81	0.779	0.807	0.738
$B \to B$	0.80	0.866	0.942	0.945	0.926	0.926	0.983	0.978	0.949	0.967	0.916

	Lamp	Mobilephone	Motorbike	Paper	Person	Sign	TVmonitor	Window	Average
$A \to A$	0.819	0.85	0.82	0.915	0.927	0.844	0.876	0.931	**0.883**
$B \to A$	0.702	0.728	0.703	0.784	0.795	0.723	0.751	0.798	**0.762**
$B \to B$	0.986	0.95	0.837	0.968	0.968	0.904	0.928	0.959	**0.931**

Table 7.4 mAP@10 obtained for $Q = 5$ with the EDUB2015 ground-truth for the setups described at 7.4.2. Queries are performed with the top 10 $TF \cdot IDF$ of each image

	Aircon	Bottle	Building	Car	Chair	Cupboard	Dish	Door	Face	Glass	Hand
$A \to A$	0.749	0.615	0.752	0.631	0.685	0.778	0.695	0.546	0.693	0.757	0.61
$B \to A$	0.535	0.604	0.503	0.561	0.35	0.549	0.733	0.418	0.497	0.702	0.399
$B \to B$	0.944	0.665	0.715	0.701	0.615	0.896	0.751	0.552	0.696	0.823	0.599

	Lamp	Mobilephone	Motorbike	Paper	Person	Sign	TVmonitor	Window	Average
$A \to A$	0.867	0.535	0.376	0.719	0.717	0.436	0.814	0.504	**0.656**
$B \to A$	0.501	0.422	0.354	0.418	0.522	0.374	0.587	0.253	**0.488**
$B \to B$	0.864	0.552	0.454	0.627	0.736	0.516	0.876	0.326	**0.679**

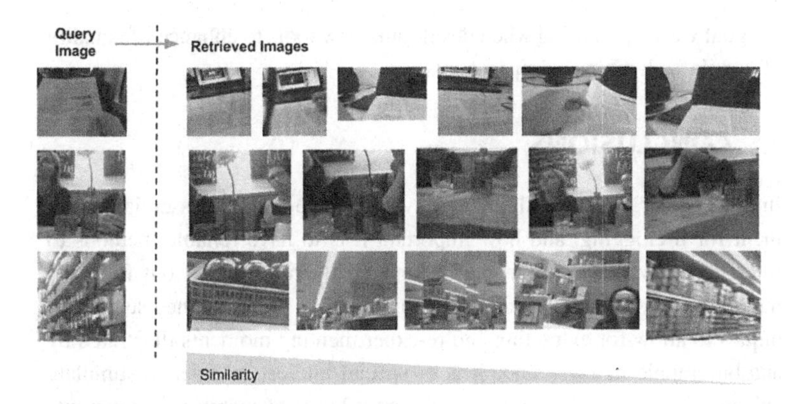

■ **FIGURE 7.7 Some image retrieval examples**. For each of the query images on the left, the retrieval engine returns ranked results based on similarity.

25.12 MB and 12.9 MB if the index is populated with the rich images alone. The average query time is only 0.5 ms per query. The average terms per image document with $Q = 5$ is 1183.

7.4.5 Discussions

It is not a simple task to describe a semantically rich image with few words. In the same way, it is difficult to illustrate and quantify the resulting image set obtained from applying the method for filtering semantically rich images out of the whole dataset where many images had very little to no semantical context (i.e. roof and wall pictures, occluded pictures, blurry, dark, etc.). Big amounts of images have many more activations at the last layers of ResNet50, therefore producing a larger variety of relevant feature words and making it harder to reduce the query vector to only 10 feature words. This is a plausible explanation for why, when querying the full dataset with only the semantically rich images, the results obtained were significantly lower than when querying with the filtered semantically rich ones or when the index was populated with the semantically rich images instead of the full dataset. We also found it interesting to mention the significant reduction of the image set when the non-semantically rich images are filtered out. Specifically, from the 4912 images conforming the EDUB2015 dataset, 1594 (32.4%) were classified as semantically rich. This indicates that an important part of the huge amount of pictures obtained on a daily basis by lifelogging devices are irrelevant or semantically poor. This shows the need for methods such as the ones presented in this chapter. Nevertheless, we observe that semantically rich images slightly enhance retrieval results quantitatively, and qualitatively, the quality of the pictures returned to the user are generally much nicer, more colorful and with more objects present, which is after all the goal we keep in mind when developing new tools to enhance life quality of people with memory impairments.

7.5 CONCLUSIONS

In this chapter, we showed how retrieval techniques are an essential instrument for lifelogging, and how important it is to have reliable methods to filter the relevant, non-redundant, semantically rich images out from the huge amount of images obtained on a daily basis. We combined such techniques to allow for extracting and re-experimenting moments illustrated by autobiographic images, since it is of special interest in order to stimulate episodic memory of patients with different kinds of memory impairments (Alzheimer, mild cognitive impairment, among others). We used state-of-the-art retrieval methods based on convolutional neural network features obtained from the last layer of the ResNet50 pre-trained on ImageNet. Through representing CNN features in a textual form, we used the features obtained for the purpose of the retrieval in combination with Solr, an efficient, widely used open source text-based retrieval engine. The approach allows one to tune the query costs versus the quality of the approximation by specifying the length of the query, without the need of maintaining the original features for reordering the resulting set. We evaluated the strategy of textual representation in order to optimize the index occupation and the query response time in order to show that this approach is able to handle large datasets, and we obtained a memory footprint of only 12.9 MB and an average query response time of 0.5 ms. We evaluated the deployed system with filtered semantically rich images for lifelogging purposes with the EDUB2015 dataset and obtained a mAP@10 of 67.9% and an average P@1 of 93.1%, which is an improvement of 2.3% and 4.8%, respectively, from the case where the unfiltered whole dataset was used.

The final system proved to be a generic, straightforward and efficient way for indexing and retrieval. Our experiments showed that the features of the ResNet50 CNN have a strong semantic content since same-category objects are frequently described by the same words.

The final developed system is a memory stimulation program for patients with mild cognitive impairment.

ACKNOWLEDGMENTS

This work was partially funded by TIN2015-66951-C2-1-R, SGR 1219, Grant 20141510 (Marató TV3) and CERCA Programme / Generalitat de Catalunya. M. Dimiccoli and P. Radeva are partially supported by ICREA Academia 2014. The funders had no role in the study design, data collection, analysis, and preparation of the manuscript.

REFERENCES

[1] Maedeh Aghaei, Mariella Dimiccoli, Petia Radeva, Multi-face tracking by extended bag-of-tracklets in egocentric photo-streams, Comput. Vis. Image Underst. 149 (2016) 146–156.
[2] Maedeh Aghaei, Mariella Dimiccoli, Petia Radeva, With whom do I interact? Detecting social interactions in egocentric photo-streams, arXiv:1605.04129, 2016.
[3] Maedeh Aghaei, Mariella Dimiccoli, Petia Radeva, All the people around me: face discovery in egocentric photo-streams, in: 2017 IEEE International Conference on Image Processing (ICIP), 2017.
[4] Giuseppe Amato, Claudio Gennaro, Pasquale Savino, MI-File: using inverted files for scalable approximate similarity search, Multimed. Tools Appl. 71 (3) (August 2014) 1333–1362.
[5] A.R. Silva, M.S. Pinho, L. Macedo, C.J.A. Moulin, A critical review of the effects of wearable cameras on memory, Neuropsychol. Rehabil. (2016) 1–25.
[6] R. Arandjelović, A. Zisserman, Three things everyone should know to improve object retrieval, in: IEEE Conference on Computer Vision and Pattern Recognition, June 2012, pp. 2911–2918.
[7] A. Babenko, A. Slesarev, A. Chigorin, V. Lempitsky, Neural codes for image retrieval, in: Lect. Notes Comput. Sci., vol. 8689, 2014, pp. 584–599.
[8] A. Bifet, R. Gavalda, Learning from time-changing data with adaptive windowing, in: Proceedings of SIAM International Conference on Data Mining, 2007.
[9] Marc Bolaños, Mariella Dimiccoli, Petia Radeva, Toward storytelling from visual lifelogging: an overview, IEEE Trans. Human-Mach. Syst. 47 (1) (2017) 77–90.
[10] Marc Bolaños, Ricard Mestre, Estefanía Talavera, Xavier Giró-i Nieto, Petia Radeva, Visual summary of egocentric photostreams by representative keyframes, in: 2015 IEEE International Conference on Multimedia & Expo Workshops (ICMEW), IEEE, 2015, pp. 1–6.
[11] M. Bolaños, P. Radeva, Ego-object discovery, arXiv:1504.01639, Apr. 2015, pp. 1–9.
[12] Marc Bolaños, Petia Radeva, Simultaneous food localization and recognition, in: 23rd International Conference on Pattern Recognition, ICPR, 2016, Cancún, Mexico, December 4–8, 2016, 2016, pp. 3140–3145.
[13] Marc Carné, Xavier Giro-i Nieto, Petia Radeva, and Cathal Gurrin. Egomemnet: Visual memorability adaptation to egocentric images, 2016.
[14] Alejandro Cartas, Juan Marín, Petia Radeva, Mariella Dimiccoli, Recognizing activities of daily living from egocentric images, in: Pattern Recognition and Image Analysis – 8th Iberian Conference, IbPRIA 2017, Faro, Portugal, June 20–23, 2017, Proceedings, 2017, pp. 87–95.
[15] K. Chatfield, K. Simonyan, A. Vedaldi, A. Zisserman, Return of the devil in the details: delving deep into convolutional nets, CoRR, arXiv:1405.3531, 2014.
[16] Tiffany E. Chow, Jesse Rissman, Neurocognitive mechanisms of real-world autobiographical memory retrieval: insights from studies using wearable camera technology, Ann. N.Y. Acad. Sci. 1396 (1) (2017) 202–221.
[17] O. Chum, J. Philbin, J. Sivic, M. Isard, A. Zisserman, Total recall: automatic query expansion with a generative feature model for object retrieval, in: IEEE 11th International Conference on Computer Vision, Oct. 2007, pp. 1–8.
[18] Antonio Criminisi, Jamie Shotton, Ender Konukoglu, et al., Decision forests: a unified framework for classification, regression, density estimation, manifold learning and semi-supervised learning, Found. Trends Comput. Graph. Vis. 7 (2–3) (2012) 81–227.

[19] E.J. Crowley, A. Zisserman, In search of art, in: Workshop on Computer Vision for Art Analysis, ECCV, 2014.

[20] J. Deng, W. Dong, R. Socher, L.-J. Li, K. Li, L. Fei-Fei, ImageNet: a large-scale hierarchical image database, in: CVPR, 2009.

[21] Mariella Dimiccoli, Marc Bolaños, Estefania Talavera, Maedeh Aghaei, Stavri G. Nikolov, Petia Radeva, Sr-clustering: semantic regularized clustering for egocentric photo streams segmentation, Comput. Vis. Image Underst. (2016).

[22] Aiden R. Doherty, Niamh Caprani, Ciarán í Conaire, Vaiva Kalnikaite, Cathal Gurrin, Alan F. Smeaton, Noel E. O'Connor, Passively recognising human activities through lifelogging, Comput. Hum. Behav. 27 (5) (September 2011) 1948–1958.

[23] J. Donahue, Y. Jia, O. Vinyals, J. Hoffman, N. Zhang, E. Tzeng, T. Darrell, Decaf: a deep convolutional activation feature for generic visual recognition, CoRR, arXiv:1310.1531, 2013.

[24] C. Gennaro, Large scale deep convolutional neural network features search with lucene, arXiv:1603.09687, 2016, pp. 1–10.

[25] Yunchao Gong, Liwei Wang, Ruiqi Guo, Svetlana Lazebnik, Multi-scale orderless pooling of deep convolutional activation features, CoRR, arXiv:1403.1840, 2014.

[26] Kaiming He, Xiangyu Zhang, Shaoqing Ren, Jian Sun, Deep residual learning for image recognition, in: Proceedings of the IEEE Conference on Computer Vision and Pattern Recognition, 2016, pp. 770–778.

[27] Hervé Jégou, Ondřej Chum, Negative evidences and co-occurrences in image retrieval: the benefit of PCA and whitening, in: Proceedings of the 12th European Conference on Computer Vision – Part II, Springer-Verlag, Berlin, Heidelberg, 2012, pp. 774–787.

[28] Amornched Jinda-Apiraksa, Jana Machajdik, Robert Sablatnig, A keyframe selection of lifelog image sequences, in: MVA, 2013, pp. 33–36.

[29] H. Jégou, F. Perronnin, M. Douze, J. Sánchez, P. Pérez, C. Schmid, Aggregating local image descriptors into compact codes, IEEE Trans. Pattern Anal. Mach. Intell. 34 (9) (Sept. 2012) 1704–1716.

[30] Aditya Khosla, Akhil S. Raju, Antonio Torralba, Aude Oliva, Understanding and predicting image memorability at a large scale, in: Proceedings of the IEEE International Conference on Computer Vision, 2015, pp. 2390–2398.

[31] Alex Krizhevsky, Ilya Sutskever, Geoffrey E. Hinton, ImageNet classification with deep convolutional neural networks, in: Advances in Neural Information Processing Systems, 2012, pp. 1097–1105.

[32] L. Dubourg, A.R. Silva, C. Fitamen, C.J.A. Moulin, C. Souchay, Sensecam: a new tool for memory rehabilitation?, Rev. Neurol. (Paris) 172 (12) (2016) 735–747.

[33] Matthew L. Lee, Anind K. Dey, Lifelogging memory appliance for people with episodic memory impairment, in: Proceedings of the 10th International Conference on Ubiquitous Computing, UbiComp '08, ACM, New York, NY, USA, 2008, pp. 44–53.

[34] Yong Jae Lee, Joydeep Ghosh, Kristen Grauman, Discovering important people and objects for egocentric video summarization, in: 2012 IEEE Conference on Computer Vision and Pattern Recognition (CVPR), IEEE, 2012, pp. 1346–1353.

[35] Aniol Lidon, Marc Bolaños, Mariella Dimiccoli, Petia Radeva, Maite Garolera, Xavier Giró-i Nieto, Semantic summarization of egocentric photo stream events, arXiv preprint, arXiv:1511.00438, 2015.

[36] Yen-Liang Lin, Vlad I. Morariu, Winston Hsu, Summarizing while recording: context-based highlight detection for egocentric videos, in: Proceedings of the IEEE International Conference on Computer Vision Workshops, 2015, pp. 51–59.

[37] M.A. Conway, C. Loveday, Using SenseCam with an amnesic patient: accessing inaccessible everyday memories, Memory 19 (2011) 697–704.

[38] Zheng Lu, Kristen Grauman, Story-driven summarization for egocentric video, in: Proceedings of the IEEE Conference on Computer Vision and Pattern Recognition (CVPR), 2013.

[39] M. Lux, O. Marques, Visual Information Retrieval Using Java and LIRE, vol. 5, 2013.

[40] G. Oliveira-Barra, LIvRE: a Video Extension to the LIRE Content-Based Image Retrieval System, Master Thesis, Universitat Politecnica de Catalunya, 2015.

[41] G. Oliveira-Barra, X. Giró-i Nieto, A. Cartas-Ayala, P. Radeva, LEMoRe: a lifelog engine for moments retrieval at the NTCIR-lifelog LSAT task, in: 12th NTCIR Conference, Tokyo, Japan, NII, 2016.

[42] Gabriel Oliveira-Barra, Marc Bolaños, Estefania Talavera, Adrián Duenas, Olga Gelonch, Maite Garolera, Serious games application for memory training using egocentric images, arXiv preprint, arXiv:1707.08821, 2017.

[43] Gabriel Oliveira-Barra, Mariella Dimiccoli, Petia Radeva, Leveraging activity indexing for egocentric image retrieval, in: Iberian Conference on Pattern Recognition and Image Analysis, Springer, 2017, pp. 295–303.

[44] M. Oquab, L. Bottou, I. Laptev, J. Sivic, Learning and transferring mid-level image representations using convolutional neural networks, in: Proc. 2014 IEEE, CVPR, 2014, pp. 1717–1724.

[45] F. Perronnin, C. Dance, Fisher kernels on visual vocabularies for image categorization, in: IEEE Conference on Computer Vision and Pattern Recognition, June 2007, pp. 1–8.

[46] F. Perronnin, Y. Liu, J. Sánchez, H. Poirier, Large-scale image retrieval with compressed fisher vectors, in: IEEE Conference on Computer Vision and Pattern Recognition, June 2010, pp. 3384–3391.

[47] Filip Radenovic, Hervé Jégou, Ondrej Chum, Multiple measurements and joint dimensionality reduction for large scale image search with short vectors – extended version, CoRR, arXiv:1504.03285, 2015.

[48] Washington L.S. Ramos, Michel M. Silva, Mario F.M. Campos, Erickson R. Nascimento, Fast-forward video based on semantic extraction, in: 2016 IEEE International Conference on Image Processing (ICIP), IEEE, 2016, pp. 3334–3338.

[49] A. Razavian, J. Sullivan, A. Maki, S. Carlsson, Visual instance retrieval with deep convolutional networks, CoRR, arXiv:1412.6574, 2014.

[50] A.S. Razavian, H. Azizpour, J. Sullivan, S. Carlsson, CNN features off-the-shelf: an astounding baseline for recognition, in: IEEE CVPR Workshops, 2014, pp. 512–519.

[51] Joseph Redmon, Ali Farhadi, YOLO9000: better, faster, stronger, arXiv preprint, arXiv:1612.08242, 2016.

[52] P. Sermanet, D. Eigen, X. Zhang, M. Mathieu, R. Fergus, Y. LeCun, OverFeat: integrated recognition, localization and detection using convolutional networks, CoRR, arXiv:1312.6229, 2013.

[53] Josef Sivic, Andrew Zisserman, Video google: a text retrieval approach to object matching in videos, in: Proceedings of the Ninth IEEE International Conference on Computer Vision, vol. 2, IEEE Computer Society, Washington, DC, USA, 2003, p. 1470.

[54] Estefania Talavera, Petia Radeva, Nicolai Petkov, Towards egocentric sentiment analysis, in: 16th International Conference on Computer Aided Systems Theory, 2017.

[55] Estefanía Talavera, Nicola Strisciuglio, Nicolai Petkov, Petia Radeva, Sentiment recognition in egocentric photostreams, in: Pattern Recognition and Image Analysis – 8th Iberian Conference, IbPRIA 2017, Faro, Portugal, June 20–23, 2017, Proceedings, 2017, pp. 471–479.

[56] Paul Viola, Michael J. Jones, Robust real-time face detection, Int. J. Comput. Vis. 57 (2) (2004) 137–154.

[57] J. Wan, D. Wang, S. Hoi, H. Chu, P. Wu, J. Zhu, Y. Zhang, J. Li, Deep learning for content-based image retrieval: a comprehensive study, in: Proceedings of the 22nd International Conference on Multimedia, MM'14, ACM, New York, NY, USA, 2014, pp. 157–166.
[58] M.D. Zeiler, R. Fergus, Visualizing and understanding convolutional networks, CoRR, arXiv:1311.2901, 2013.
[59] L. Zheng, Y. Zhao, S. Wang, J. Wang, Q. Tian, Good Practice in CNN Feature Transfer, arXiv:1604.00133, April 2016.

Chapter 8

Integrating signals for reasoning about visitors' behavior in cultural heritage

Tsvi Kuflik, Eyal Dim
The University of Haifa, Israel

CONTENTS

8.1 INTRODUCTION

Tourists' behavior and especially museum visitors' behavior has triggered research attention for decades, if not more, as can be seen in the following non-representative list of examples: [27,25,7,12,14,26,11,5,15,21,22,8,10, 20,23,1,2,9,17,18,28,30]. They are studied in order to learn about their behavior, characteristics, needs and preferences, so museums can better serve them. Recently such studies were done even in order to promote sustainable tourism [6]. Over the years a plethora of studies of museum visitors and their behavior were published by numerous researchers. One of the best knowns is by John Falk, who worked on studying museum visitors for about four decades [12–15]. Falk studied their behavior and suggested five generic stereotypes of visitors' characters:

1. the explorer, whose visit to the museum is motivated by curiosity or general interest in discovering more about the subject matter introduced by the museum,
2. the experience seeker, often a tourist, whose visit is typically motivated by the main attraction that the museum is known to offer,
3. the professional/hobbyist, who is interested in specific topics out of the full collection of the museum visit,
4. the recharger, who comes to the museum to reflect and rejuvenate, or to relax and absorb the atmosphere,

Multimodal Behavior Analysis in the Wild. https://doi.org/10.1016/B978-0-12-814601-9.00021-3
Copyright © 2019 Elsevier Ltd. All rights reserved.

■ **FIGURE 8.1** Illustrative animals representing visitors' types. 1. https://freeclipartimage.com//storage/upload/grasshopper-clip-art/grasshopper-clip-art-23.jpeg 2. https://freeclipartimage.com//storage/upload/fish-clip-art/fish-clip-art-60.jpg 3. https://freeclipartimage.com//storage/upload/butterfly-clip-art/butterfly-clip-art-88.jpg 4. http://www.allthingsclipart.com/07/ant.10.jpg.

5. the facilitator, who visits the museum to satisfy the needs and desires of someone she or he cares about rather than just of her- or himself.

A well-known early study is the work of Véron and Levasseur [29], which tracked visitors and identified four distinctive types of them, and assigned animal types to describe them:

1. an ant which walks slowly, examines each and every exhibit, following an order,
2. a grasshopper, which jumps around, examines carefully only exhibits of interest,
3. a butterfly that hovers around, in an unordered manner and briefly explores exhibits of interest,
4. a fish that enters an exhibition area, scans it, gets the overall idea and leaves.

The above attractive illustration of visitors' behavior, maybe due to its simplicity and intuitiveness (see Fig. 8.1 for an illustrative representation), inspired later studies that adopted the animals' metaphor of visitors' behavior, aiming to identify these types using technological means rather than by ethnographic research.

Naturally, early studies were based on observations and tracking of visitors while they visit museums and other cultural heritage (CH) sites. The researchers followed, tracked and recorded their behavior. While tracking visitors, researchers recorded their paths through the museum – places visited, time spent, interaction among them, and more. As a physical environment, spatial and temporal aspects are among the most informative ones when it comes to reasoning about visitors' behavior. With the evolution of Information and Communication Technology (ICT), it was only natural that researchers explored the potential of novel technology to enhance the services provided to visitors to CH sites. When tracking visitors in CH sites, places visited, and time spent there were among the first to me measured, together with visitors' interaction with technology. As additional sensors became available, new opportunities opened.

8.2 USING TECHNOLOGY FOR REASONING ABOUT VISITORS' BEHAVIOR

With the evolvement of ICT, various attempts were made to experiment with its potential in enhancing the museum experience. At first, the focus was on enriching the visit experience by providing access to rich information, beyond the limited textual description that traditionally accompanies museum exhibits. This is when mobile audio guides and later on multimedia guides started to appear (see Ardissono et al. [3] for a review of the state of the art in mobile guide technology). At first, with early versions of mobile guides, visitors' paths, places visited and the order of the visit, as well as time spent at Points of Interest (PoIs) and the interaction with the device were measured. Inspired by previous ethnographic studies, attempts were made to replicate them and identify visitors' types with technological means. Zancanaro et al. [31] and Kuflik et al. [19], inspired by the work of Véron and Levasseur [29], analyzed museum visit logs, where a visit was represented by the time spent, the order and completeness of the visit and the interaction with the mobile device and clustered the logs into four clusters, representing the "Ant", "Fish", "Butterfly" and "Grasshopper", showing a similar general behavior in their study.

As technology progressed and mobile devices that are rich with mobile sensors started to appear, new opportunities opened for researchers and practitioners to reason about visitors to CH sites, as part of the evolution of context-aware systems – systems that reasoned about the contextual aspects of their users by fusing signals measured by sensors embedded in the user's mobile devices and the environment (see Baldauf et al. [4] for a survey of context-aware systems at that time). Nowadays it is the era of the "Internet of Things" – IoT (see Gil et al. [16] for a recent review), and again, IoT is being adopted for supporting visitors to CH – making the exhibits smart and context aware, so they can deliver the right information to the right users, at the right time, taking into account their current context.

So how is it done in practice?

How can a system/the environment reason about its inhabitants – the visitors to the CH site? Let us consider an illustrative scenario: A middle age couple and their teenage son are visiting an archeological museum. The couple is interested in the exhibits, they walk slowly throughout the exhibition, look carefully at the exhibits and discuss them. The teenager is a bit bored. He is wondering around, looking occasionally at exhibits, going back and forth between his parents and various exhibits. He is not interested. It is clear from the description that it will not take long until the bored teenager will start complaining and eventually drag his parents out of that boring place... As

■ **FIGURE 8.2** Two pictures, taken less than half a minute apart, depicting a group of three visiting the museum: a middle-aged couple, engaged with exhibits, and a bored teenager, wandering around, yawning.

in any group leisure activity, if one group member is not happy, the whole group will not be happy.

Our question here is if and how present-days technology can help in identifying such a situation. Assuming we identify the group situation correctly, a smart museum visitor guide system may be able to suggest an engaging activity to the bored teenager. However, reasoning about a solution is beyond the scope of this chapter, as the current focus is on identifying the situation automatically.

Fig. 8.2 actually depicts the above described scenario. It presents two frames out of a video taken at a museum, showing a group of three visitors that entered as a group to the museum. The two middle-aged visitors are engaged looking at exhibits, while the teenager is a bit further away (Fig. 8.2 left), then, a few seconds later, the teenager moves to the other side of the corridor, yawning (right). Obviously, he is bored while the others are engaged with the exhibits.

The video of which two frames are presented in Fig. 8.2 was taken as part of a study at the Hecht museum,[1] which experimented with the ability to infer group context using simple signals measured by sensors worn by the visitors during the visit. The system included a wearable (matchbox size) set of sensors that included a proximity sensor (able to identify similar sensors), a compass, an accelerometer, a microphone and a ZgBee transmitter. Similar sensors were installed at various PoIs in the museum. The mobile sensor reported every half a second its ID, IDs of nearby mobile/stationary devices, acceleration, orientation and vice versa activity data to the systems servers, see Fig. 8.3 for details. It is worth noting that the mobile sensors are able to identify similar sensors in close proximity and only when they face each

[1] http://mushecht.haifa.ac.il/Default_eng.aspx.

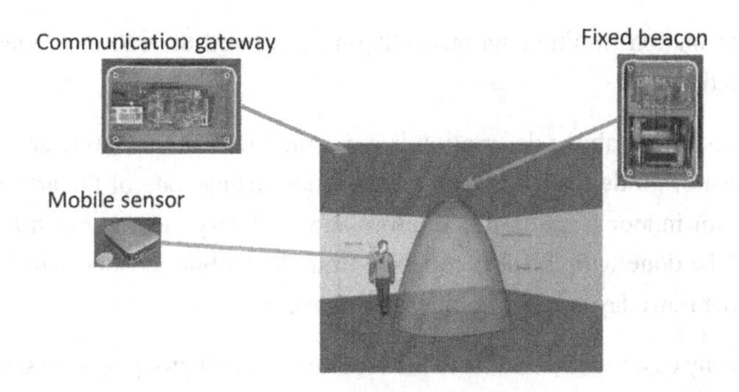

■ **FIGURE 8.3** The system sensors and basic communication. A matchbox size mobile sensor set, detectable by other mobile sensors, stationary beacons – detectable by the mobile sensors. Information was transmitted from the mobile sensors to communication gateways in every room that transferred the data over Ethernet to a server.

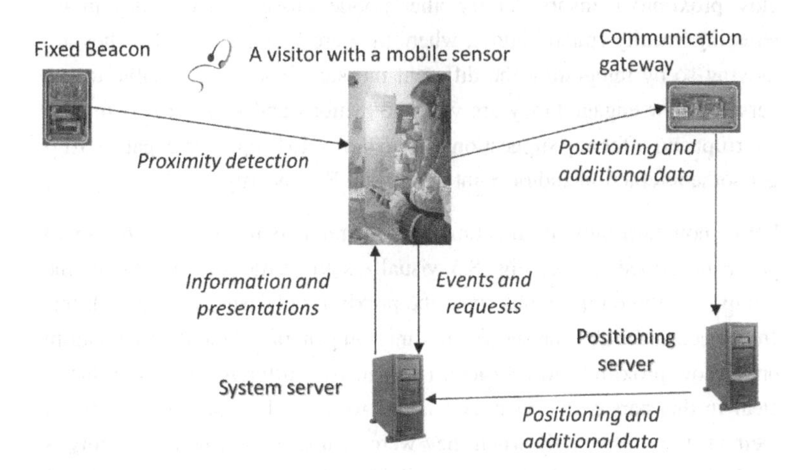

■ **FIGURE 8.4** The system architecture: Information from the mobile sensors set is reported to the positioning server through a communication gateway every half a second, including the proximity to stationary/mobile devices. The data, including PoI, is reported to the system server, which offers relevant multimedia presentation to the visitor.

other (so when visitors stand side to side or face each other, but not when one is turning the back to another).

The museum visitors guide system itself (see Fig. 8.4) was composed of two servers – a positioning server and a system server. The positioning server received the information from the mobile sensors every half a second, analyzed it and reported to the system server whether a mobile device was in proximity of any PoI, as well as additional signals measured by the mobile device. The system server reasoned about the positioning data and whenever a visitor was in the vicinity of an exhibit as regards which the system had information (multimedia presentation), it offered the visitor to view a presentation.

While so far the above description is concerned with a context-aware museum visitor guide, which is a nice application, using state of the art technology for indoor location-based information delivery there is much more that can be done with the data collected from the mobile sensors (which, in fact, exist nowadays in our mobile smartphones).

As in many cases visitors come to the museum in small groups (as presented above), technology enables us to reason about the overall group context and act accordingly, to mitigate the scenario described earlier.

Using the set of sensors it is possible to understand whether visitors are in close proximity (sensors identify other mobile sensors, whether they are interacting vocally (microphone), where they are looking, how fast they are moving so by integrating the different measurements it is possible to understand how engaged they are with the exhibits and whether it is time to interrupt and offer any suggestion/information – to keep them engaged, trigger some interaction and/or point to relevant PoIs nearby.

Let us now turn into visualization of measurements recorded for the small group described above. Fig. 8.5 visualizes the data collected about the group. The three top graphs show the proximity of visitors in three different aspects – the top one shows proximity in general, while the two bottom ones show proximity to a beacon (PoI) in two different regions. What is clear in this part is that visitors 1 and 2 were in close proximity most of their visit and especially when they were near exhibits (their proximity is marked by a + symbol). Contrarily, visitor 3 is occasionally near visitor 1 (∗) or 3 (a small circle). It is quite clear that the third visitor, the teenager, is separated from the others.

Then we can turn our attention to voice interaction (fourth graph from the top at Fig. 8.5), and see that the microphones of visitors 1 and 2 are active most of the time (there was no voice recording, only detection of voice activity).

Finally, we look at orientation (compass) data (the bottom graph). We can see that visitors 1 and 2 are nicely oriented – they walk together and look in the same directions. Visitor 3 is wandering around, as can be seen from the data (the small circle, at the top) for a reference, there is also a random orientation which is around 0 degrees, which resembles the overall behavior of visitor 3.

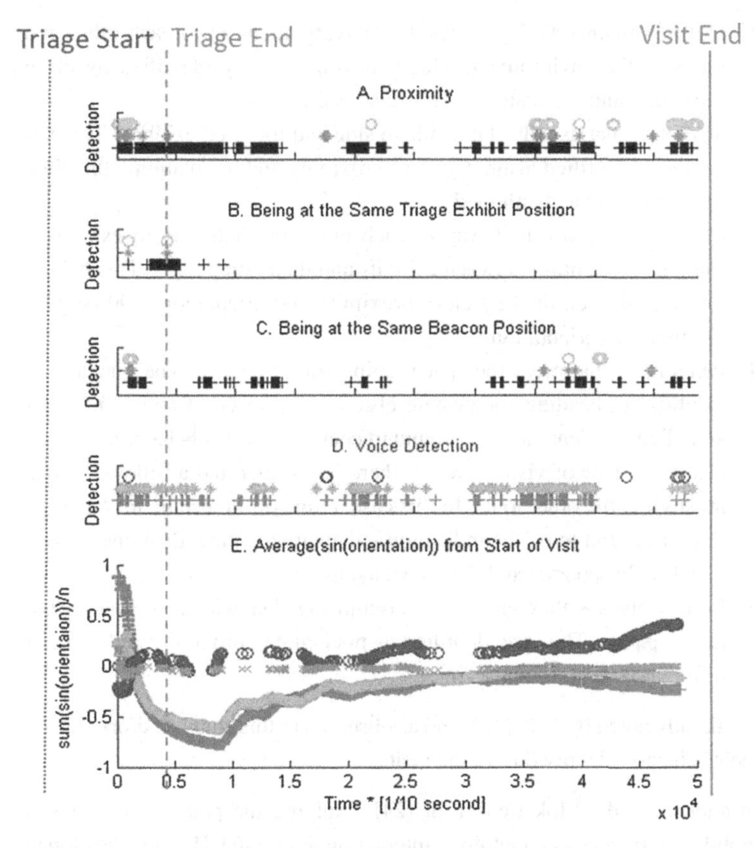

FIGURE 8.5 Social interaction in CH: visitors' proximity and proximity to exhibits, voice activity and orientation. Top three plots: "+": visitor 1&2; "∗" visitor 1&3; "o" visitors 2&3. Two bottom plots: "+": visitor 1; "∗": visitor 2; "o" visitor 3; "x": random.

Given the above signals we can integrate them and given our CH scenario, and reason about what we learn. The above was an illustrative example of how different signals measured from mobile devices carried by visitors may enable reasoning about their interests and needs.

Still inspired by the work of Véron and Levasseur [29] and the work of Falk [15], Dim and Kuflik [10] studied the behavior of pairs of visitors that visited the Hecht museum. They used the infrastructure describe above and measured and analyzed proximity, orientation and speed. Analysis of visit logs combined with analysis of videos taken at the museum enabled them to identify six different behaviors, which were named after different animals and included:

1. Doves – mainly facing each other, paying attention to each other and ignoring the environment. This behavior is easily identified by close proximity and orientation – 180 degrees difference.[2]
2. Mirkats – that usually stand side to side and focus on exhibits. This behavior is identified again, by close proximity and orientation – this time orientation is nicely aligned.
3. Parrots – that turn half way to each other and half way to exhibits – share their attention between exhibits and their companion. This behavior is again identified by close proximity and orientation – 90 degrees difference in orientation.
4. Penguins – that cross the space, going somewhere, maybe glancing at exhibits but heading somewhere else. Here a speed of above 0.25 m/s and aligned orientation was found to characterize this behavior.
5. Geese – a pair of visitors where there is a leader and a follower – one moves and the other joins. In this case a time difference in arrival at the PoI was found to be the indication – they are together, then one moves, they briefly separate and then meet again.
6. Lone wolves – they enter the museum together with companions and then separate. They are identified by not being in proximity and being at different PoIs at the same time.

As already noted by Falk [15], and as observed in this study, visitors change their behavior/identity during the visit.

In another study, Mokatren et al. [24], explored the potential of using a mobile eye tracker as an intuitive interaction device in CH. They developed a system that, by using image-based positioning and temporal aspects, was able to identify when a visitor expresses interest in a specific exhibit and then provided audio explanation of the exhibit of interest.

8.3 DISCUSSION

As we can see from the above examples, as technology evolves, becoming inexpensive, widespread and pervasive, it can easily be used for reasoning about the behavior of users in a smart environment. When we consider current state of the art technology – with the Internet of Things, cloud computing and smart environments, we are all surrounded by sensors and soon we will be wearing them. The variability and wide spread of sensors make it easy (and also a must) to integrate their data and reason about it.

[2]Note that proximity is not detected when one of the visitors turns his back to the other – only in face to face or side by side – when there is a line of sight between the devices.

Every domain has its specific characteristics. CH sites are places for leisure activity – visitors come as individuals or groups (whether small – family or friends or large – guided groups) to explore them, learn and have a meaningful good experience. Current and future sensors enable us to reason about individuals and group behavior and infer from the behavior about their interests, the group cohesion, whether they are getting bored/tired/hungry (the well-known phenomenon of "museum fatigue" needs to be addressed). Being able to analyze all that from where the visitors are, their visit path and pace, exhibits of interest and focus of attention, their engagement and interaction, provides the basic means needed for identifying the right actions to take in order to ensure an enjoyable experience – such as trying to engage the bored teenager by pointing out exhibits of interest or offering a quest or an engaging game, to keep him interested, while leaving the couple alone as they are doing fine as they are, without any need for any support.

Each and every sensor may contribute specific aspects that may be taken into account. We already discussed indoor positioning and proximity to PoIs, orientation, speed, voice activity and fixation (measured by eye tracker). Wearable sensors, measuring heart bit, blood pressure and skin conductivity may provide hints about excitement. Further analysis of the eye tracker's data, like pupil dilation and saccades, may provide additional information as regards the focus of attention and interests.

All this is becoming a reality as our mobile devices and environments get smarter. Still, while new sensor technology opens new opportunities for reasoning about the users' needs (and this is not limited only to CH), there is also a cost and it is the loss of privacy that we already witness when apps are offering us location-based services we never asked for... So careful thought needs to be given so as to find a balance between the desire to provide better services and the intrusiveness of the invisible technology.

8.4 CONCLUSIONS

Sensor technology is becoming widespread. It is embedded in our mobile devices and soon will also be worn. The individual signals measured by the different sensors can be integrated and reasoned about and given meaning using specific domain knowledge for understanding the preferences and needs of users. In CH, these novel technologies will enable better engaging visitors, providing them with relevant information and services, helping them enjoy the sites both as individuals and groups – the technology may accompany them and support them unobtrusively – and monitor and react when an opportunity presents itself.

REFERENCES

[1] A. Antoniou, G. Lepouras, Adaptation to visitors' visiting and cognitive style, in: Proceedings of the 3rd International Conference of Museology and ICOM-AVICOM Annual Conference, 2006.

[2] A. Antoniou, G. Lepouras, Using bluetooth technology for personalised visitor information, in: Proceedings of the IADIS International Conference on Mobile Learning, 2005, pp. 28–30.

[3] L. Ardissono, T. Kuflik, D. Petrelli, Personalization in cultural heritage: the road travelled and the one ahead, User Model. User-Adapt. Interact. 22 (1–2) (2012) 73–99.

[4] M. Baldauf, S. Dustdar, F. Rosenberg, A survey on context-aware systems, Int. J. Ad Hoc Ubiq. Comput. 2 (4) (2007) 263–277.

[5] F. Bohnert, I. Zukerman, S. Berkovsky, T. Baldwin, L. Sonenberg, Using interest and transition models to predict visitor locations in museums, AI Commun. 21 (2–3) (2008) 195–202.

[6] P. Buonincontri, A. Marasco, H. Ramkissoon, Visitors' experience, place attachment and sustainable behaviour at cultural heritage sites: a conceptual framework, Sustainability 9 (7) (2017) 1112.

[7] C.A. Cone, K. Kendall, Space, time, and family interaction: visitor behavior at the Science Museum of Minnesota, Curator: Mus. J. 21 (3) (1978) 245–258.

[8] S. Cuomo, P. De Michele, A. Galletti, G. Ponti, Classify visitor behaviours in a cultural heritage exhibition, in: International Conference on Data Management Technologies and Applications, Springer International Publishing, 2015, pp. 17–28.

[9] C. De Rojas, C. Camarero, Visitors' experience, mood and satisfaction in a heritage context: evidence from an interpretation center, Tour. Manag. 29 (3) (2008) 525–537.

[10] E. Dim, T. Kuflik, Automatic detection of social behavior of museum visitor pairs, ACM Trans. Interact. Intell. Syst. 4 (4) (2015) 17.

[11] N.G. Espelt, J.A.D. Benito, Visitors' behavior in heritage cities: the case of Girona, J. Travel Res. 44 (4) (2006) 442–448.

[12] J.H. Falk, The use of time as a measure of visitor behavior and exhibit effectiveness, Roundtable Rep. 7 (4) (1982) 10–13.

[13] J.H. Falk, Time and behavior as predictors of learning, Sci. Educ. 67 (2) (1983) 267–276.

[14] J.H. Falk, J.J. Koran, L.D. Dierking, L. Dreblow, Predicting visitor behavior, Curator: Mus. J. 28 (4) (1985) 249–258.

[15] J.H. Falk, Identity and the Museum Visitor Experience, Left Coast, 2009.

[16] D. Gil, A. Ferrández, H. Mora-Mora, J. Peral, Internet of things: a review of surveys based on context aware intelligent services, Sensors 16 (7) (2016) 1069.

[17] S. Hsi, H. Fait, RFID enhances visitors' museum experience at the exploratorium, Commun. ACM 48 (9) (2005) 60–65.

[18] H.J. Klein, Tracking visitor circulation in museum settings, Environ. Behav. 25 (6) (1993) 782–800.

[19] T. Kuflik, Z. Boger, M. Zancanaro, Analysis and prediction of museum visitors' behavioral pattern types, in: Ubiquitous Display Environments, Springer, Berlin, Heidelberg, 2012, pp. 161–176.

[20] T. Kuflik, A.J. Wecker, J. Lanir, O. Stock, An integrative framework for extending the boundaries of the museum visit experience: linking the pre, during and post visit phases, Inf. Technol. Tour. 15 (1) (2015) 17–47.

[21] J. Lanir, T. Kuflik, E. Dim, A.J. Wecker, O. Stock, The influence of a location-aware mobile guide on museum visitors' behavior, Interact. Comput. 25 (6) (2013) 443–460.

[22] J. Lanir, P. Bak, T. Kuflik, Visualizing proximity-based spatiotemporal behavior of museum visitors using tangram diagrams, Comput. Graph. Forum 33 (3) (2014) 261–270.

[23] J. Lanir, T. Kuflik, J. Sheidin, N. Yavin, K. Leiderman, M. Segal, Visualizing museum visitors' behavior: where do they go and what do they do there?, Pers. Ubiquitous Comput. 21 (2) (2017) 313–326.
[24] M. Mokatren, T. Kuflik, I. Shimshoni, Exploring the potential of a mobile eye tracker as an intuitive indoor pointing device: a case study in cultural heritage, Future Gener. Comput. Syst. 81 (2018) 528–541.
[25] L.C. Nielsen, A technique for studying the behavior of museum visitors, J. Educ. Psychol. 37 (2) (1946) 103.
[26] C. Sandifer, Time-based behaviors at an interactive science museum: exploring the differences between weekday/weekend and family/nonfamily visitors, Sci. Educ. 81 (6) (1997) 689–701.
[27] E.S. Robinson, The Behavior of the Museum Visitor, 1928.
[28] F. Sparacino, The Museum Wearable: Real-Time Sensor-Driven Understanding of Visitors' Interests for Personalized Visually-Augmented Museum Experiences, 2002.
[29] E. Véron, M. Levasseur, Ethnographie de l'exposition: l'espace, le corps et le sens, Centre Georges Pompidou, Bibliothèque publique d'information, 1989.
[30] Y. Yoshimura, S. Sobolevsky, C. Ratti, F. Girardin, J.P. Carrascal, J. Blat, R. Sinatra, An analysis of visitors' behavior in the Louvre museum: a study using bluetooth data, Environ. Plan. B, Plan. Des. 41 (6) (2014) 1113–1131.
[31] M. Zancanaro, T. Kuflik, Z. Boger, D. Goren-Bar, D. Goldwasser, Analyzing museum visitors' behavior patterns, in: International Conference on User Modeling, Springer, Berlin, Heidelberg, 2007, pp. 238–246.

Chapter 9

Wearable systems for improving tourist experience

Lorenzo Seidenari, Claudio Baecchi, Tiberio Uricchio, Andrea Ferracani, Marco Bertini, Alberto Del Bimbo
University of Florence, Firenze, Italy

CONTENTS

9.1 INTRODUCTION

Natural and adaptive interfaces exploiting recent progress in mobile, multimedia and computer-vision technologies can effectively improve user experience during a visit of a cultural heritage location or a museum. Multimodal data coming from mobile and wearable device sensors can be processed, analyzed and fused on-board in real time in order to automatically understand the context of the visit, user's needs, behavior and interests, e.g. what the visitor is looking at, for how long, and if other events occur during the visit itself.

In particular, visual object classification can help in disambiguating where user's attention is directed by observing what he is looking at, as well as understanding if other people are occluding his view. Data from localization and motion sensors, such as GPS, gyroscope and accelerometer, can give insights about the user position with respect to Point-Of-Interests (POIs). His movement, both direction and velocity, can help to understand if the visitor is just wandering through the museum, or if he is looking at an exhibit that really interests him. Finally, voice activity detection (VAD) can reveal if the visitor has friends that accompany him during the visit, and if he is busy

Multimodal Behavior Analysis in the Wild. https://doi.org/10.1016/B978-0-12-814601-9.00020-1
Copyright © 2019 Elsevier Ltd. All rights reserved.

in a conversation. These results are particularly well suited to the intended use of smart audio-guides that should accompany and help the user, as a real museum guide, to satisfy curiosities and respond to personal needs [6].

However, approaches for inferring the context and detecting user behavior exploiting device sensors must necessarily be different in the case of indoor or outdoor scenarios, given the extreme diversity of these situations. In the outdoor scenario, GPS provides a fairly reliable localization which is hardly available indoor. Conversely, audio sensing is extremely limited outdoor, due to the poor quality of the acquired signal. Considering these problems, in the following we present two versions of the audio-guide: one designed for the indoor scenario and the other for the outdoor. These two versions have the same human–computer interface for a fully automatic use: the system is worn and the output is provided through speech synthesis and presents a slight difference in the case that some user interaction is needed.

9.2 RELATED WORK

Personalized museum experience

Recent statistics from the US National Travel and Tourism Office reveal a new record of tourism-related activities.[1] Despite their simplicity, the audio-guides is still one of the most used interactive devices in a museum visit [5]. In recent years there has been a growing interest in improving user experience using smart digital guides. In fact, although classic human guides are still an available option, the modern tourist expects new personal augmented experiences, founded on individual preferences, context awareness and ease of use, possibly using their personal device.

Much work in the literature proposes the use of mobile systems to have an augmented personalized experience in the context of the visit to cultural heritage sites. The need of personalization and user adaptation has recently become of great interest in research [6]; a review of the different types of smart mobile guides addressing the problem of personalization has been provided in [3]. The work of Abowd et al. [1] is one of the first on this subject, marking the difference between systems which provide improved experiences in indoor (e.g. museums) and outdoor scenarios (e.g. monuments visible on traveling in a city). This difference, from a system-implementation perspective, is mainly that in the indoor case there is more control on the dataset of artworks to be evaluated with respect to the outdoor scenario where this control is missing. The distinction, as detailed later in this chapter, is especially important in the case that the system must scale in performing

[1] http://tinet.ita.doc.gov/tinews/archive/tinews2017/20170413.asp.

non-trivial tasks hosted on-board, such as object classification and artwork recognition—tasks that may rely on the availability of an adequate training dataset.

Using sensors and cameras it is possible to localize visitors in the proximity of artworks, thus inferring an implicit interest. In [44] the Cultural Heritage Information Personalization (CHIP) system has been proposed, where a personalized visit could be created through a web interface. The tour can be downloaded to a mobile device. A track of the artworks visited by the user in the museum is obtained using RFIDs. Information is then synced back on the server side in order to update the user profile and provide a better personalization for the next tours. In [48] and [26] the behavioral patterns of museum visitors is analyzed and predicted according to four main patterns, which emerged from ethnographic observations by [13]. The work shows that these patterns can be identified exploiting features such as average time spent on each artwork, percentage of observed artworks, etc. Reference [21] presents a non-intrusive computer-vision system based on person re-identification of museum visitors observed through surveillance cameras. The system tracks the visitor's behavior in the museum and measures the time spent in looking at each artwork. In this way a personalized user profile of interests is built. The user profile is then used at the end of the visit to adapt an interface and recommend targeted multimedia content on a wide interactive table located at the exit of the exhibit. In this way the system can provide more information on the artworks that attracted the visitor the most and suggest additional targeted tours outdoor.

Regarding outdoor systems and the more broad context of city tourism, several notable works propose context aware mobile guides that can provide information depending on the user position and contextual cues [10,15,22], such as time, user profile, device profile or network conditions. In [38], one of the first software systems for a tourist guide was presented. It shows areas of interest on a map as well as available public utility points, by also exploiting the GPS location of the user. Takeuchi and Sugimoto developed the CityVoyager system [39], which uses GPS to detect user location, understand visited places such as shops and suggest new items based on user preferences. To reduce the cognitive effort needed to use the system, it features a simple "metal detector" interface that beeps faster when recommended shops are nearby. In [24], a mobile agent that automatically selects attractions and plans individual tours following the tourist's specific behavior was proposed. The focus is on planning and group behavior analysis. A wearable device, sharing information with a hand-held device, was used in [23] to develop an un-intrusive mobile tour guide with real-time updates.

Furthermore, thanks to the recent developments in mobile computing, some advanced solutions for mobile guides with embedded processing capabilities have been proposed. Augmented reality solutions have been exploited in specific domains. For instance, in [19], a computer-vision-based mobile guide, able to automatically recognize and enrich artworks by overlaying information, was developed for religious museums. However, this class of solutions are distracting and highly cumbersome, since they require the user to constantly interact with the device, interrupting the experiential flow. RFIDs have been applied to tags placed near artworks and scanned by mobile phones [8,43]. Nonetheless, this approach is still limited to controlled environments; it scales poorly and is unsuitable for outdoor applications. Wearable systems have been used to interact with artworks [4]. This approach is the less intrusive and more immersive, since it adopts a first person perspective, in the sense that a wearable system equipped with a camera and pointed in the field of view can process images exactly as the user sees the scene.

Object detection and recognition

Recent object detection methods are all based on modern deep convolutional neural networks [25]. To avoid the computation of a CNN forward pass on multiple sub-windows [18], recent approaches applied the object classification and bounding box regression stage directly on the last convolutional feature map of the CNN [17,36]. An even faster strategy is to directly generate a set of class-labeled bounding boxes with a single pass of a convolutional network. Redmon et al. [35] argue that "You should Only Look Once" (YOLO) at frames. This idea treats the task of object detection as a regression problem, generating location and scores directly. Liu et al. [28] proposed an approach named Single-Shot Detection (SSD), which is very similar to YOLO with the difference that multiple convolutional maps are used in order to evaluate more windows, at multiple scales.

The availability of multi-core CPUs and GPUs on mobile devices has recently allowed one to implement CNNs on smartphones. In [47] an analysis of the best CNN architectures for mobile devices has been performed, evaluating the effect of BLAS routines and NEON SIMD instructions of the ARM CPUs. The use of weight quantization is employed to deal with reduced memory capability [30,40,46]. In [20] a framework to execute deep learning algorithms on mobile devices has been presented, using OpenCL to exploit the GPUs. In [27] an open source framework for GPU-accelerated CNNs on Android devices has been presented that parallelizes more computationally intensive types of layers on GPUs and the execution of the others on CPUs using SIMD instructions.

Content-based retrieval for cultural heritage

Over the years different methods and applications of content-based image retrieval (CBIR) techniques have been proposed for the cultural heritage domain. Retrieving 3D artworks, e.g. statues, using salient SIFTs has been proposed in [11]; mutual information is used to filter background features. Fisher vectors are employed to recognize style and author in [2]. In [34] has been presented a comparison of different techniques for image classification and retrieval in cultural heritage archives, using engineered and learned features. This work highlights two issues when applying current state-of-the-art CBIR techniques for cultural heritage: *(i)* the need to account both for micro properties, e.g. brush strokes, and macro properties, e.g. scene layout, in the design of similarity metrics; *(ii)* datasets that are relatively small, with few images for each item, thus hampering methods requiring large scale training datasets.

Voice activity detection

Voice activity detection (VAD) is the process of detecting human speech in an audio stream; this task is essential to improve further processing like automatic speech recognition or audio compression. Early approaches to this problem were based on heuristics and simple energy modeling, by thresholding or observing zero-crossing rate rules [45]. These methods work well when no background noise is present.

Successive methods have addressed this limitation using autoregressive models and line spectral frequencies [32] to observe signal statistics in current frame and comparing it with the estimated noise statistics. Most of these algorithms assume that noise statistics are stationary over long periods of time. Therefore, given the extreme diversity and rapid changes of noise in different environments, they cannot detect the occasional presence of speech.

More recent data-driven approaches avoided to make an assumption on the noise distribution. They usually used a classifier trained to predict speech vs. non-speech given some acoustic features [14,31]. However, their performance degraded when the background noise resembled that of speech. Current state-of-the-art methods exploit long-span context features learned through the use of recurrent neural networks [12,16,42] to adapt the classification on the basis of the previous frames.

9.3 BEHAVIOR ANALYSIS FOR SMART GUIDES

Audio-guides come into play in different scenarios. In our analysis there is one key difference that can be made, also leading to slightly different choices in sensor integration and recognition algorithms. We differentiate between indoor and outdoor scenarios. In both cases we aim to provide information meeting the user want of knowledge and predicting correctly user attention towards elements of the environment. The only sensor that can be used reliably in both scenarios is the camera.

User location cannot easily be inferred indoor, unless a WiFi positioning system is in place. On the other hand guides to be used indoor have usually a fixed set of objects to be recognized and content can be provided in advance. In the outdoor scenario it is hard to pre-load content related to all possible landmarks a user may encounter. Nonetheless GPS and A-GPS are often reliable enough to provide an approximate location of the user, limiting the amount of objects of interest.

Recognition of objects of interests can be obtained with a simple object detection/recognition framework, using a CNN trained for artwork detection and the very same features for recognition. This framework assumes objects to be *detectable* in frames, meaning they have a well-defined contour so that we can unambiguously understand user attention.

Outdoor landmark recognition faces a few more challenges with respect to indoor artwork recognition. First of all landmarks have a more diverse distribution, including buildings, statues and archaeological sites, to name a few. Linked to this diverse set of data are different fruition strategies. While looking at a statue in a museum or in a public square follow similar behavioral patterns, visiting a vast ruin site or observing a large cathedral gives rise to imagery which is very far apart. Buildings have a well-defined physical limit but they are hardly visible as a whole during a visit. In the case sites are to be considered, global views are never the case and visitors often stroll around an area which is pretty distinguishable. In both these emblematic outdoor use cases, unfortunately a framework based on object detection and recognition is hard to deploy. For the outdoor use case we advocate the use of local features. This paradigm allows for matching of partial views of objects and perform geometric verification. Interestingly, the matching database can be built online, exploiting even noisy geo-localizations.

9.4 THE INDOOR SYSTEM

As mentioned before, the indoor context has different constraints from the outdoor one. In this case, several components can be used jointly to under-

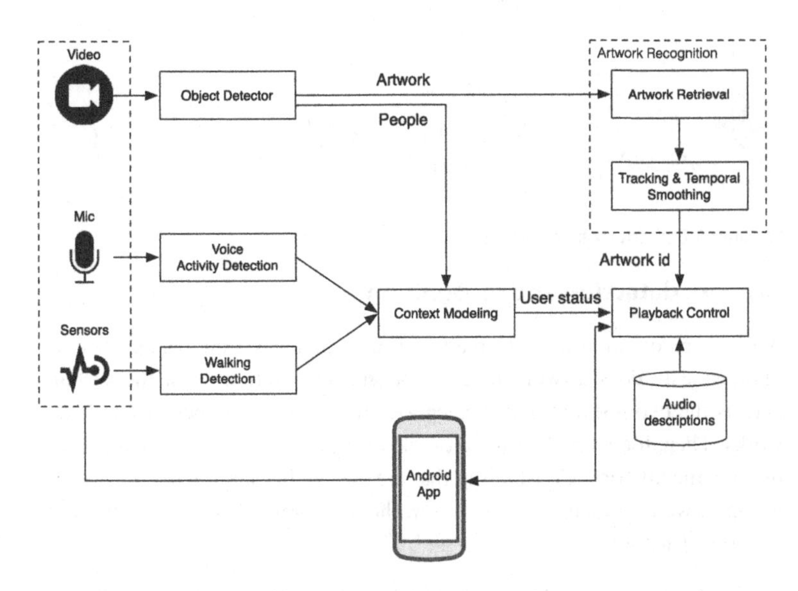

■ **FIGURE 9.1** The overall system architecture.

stand the context. Fig. 9.1 shows the architectural diagram illustrating the main submodules of the system.

The system may be divided into two main parts that are responsible for the primary functions of the system. The first one is responsible to model the *User status* while the other one is responsible for recognizing artworks, i.e. provide the *Artwork id*. Together they work with the *playback control* component by providing input signals that enable the system to play descriptions at appropriate time.

Inputs to the systems are the camera, the microphone and the movement sensors, accessed through an Android app that also operates as front-end of the system. A computer-vision system looks at the scene through the camera, detects objects (*object detector*) and recognizes what artwork the user is looking at (*artwork recognition*). Artwork recognition is performed with a two-step process: first a database of known artworks is used to retrieve the most similar artwork, then tracking is performed to filter out wrong predictions.

In order to generate the *user status* signal, the *context modeling* module exploits the three inputs to generate behavioral signals. The microphone is a source to the *voice activity detection* module, the movement sensors to perform walking detection and the camera to detect *people*. They are all used to understand if the user is actively looking at an artwork or if he is engaging in other activities.

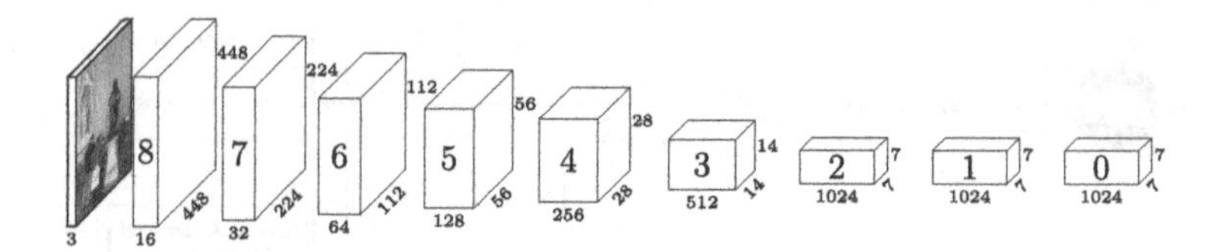

■ **FIGURE 9.2** Our network architecture, with tensor size and layer numbering.

Artwork detection and recognition

Artwork recognition is performed with a computer-vision system that simultaneously does artwork localization and recognition. First the system detects the two relevant object categories for our task, i.e. persons and artworks. Then, for every detected artwork, the system performs recognition of the specific artwork. Considering that the system has a sequence of frames in input, we also take advantage from the temporal coherence to filter out spurious detections.

We base our object detector on YOLO [35], using its *tiny net* version, fine-tuned on our artworks dataset. Using a small CNN is mandatory for real-time requirements of our mobile platform. We train it to detect *people* and *artworks*. People detection may be used to improve context understanding and it is also fundamental to avoid false positive detections on people, since artworks present in our training set are often statues depicting humans.

The detection is performed as follows. For each frame, we generate scored boxes for the two required categories. After splitting the original image in 7×7 blocks, each of the 49 regions (which have a size of $5 \times 2 \times |C|$) encodes the box predictions as a tuple $\langle x, y, h, w, s \rangle$. Then, maximal suppression is used to remove multiple predictions for the same objects. The confidence s represents the probability of that class being present inside the bounding box.

Artwork recognition. After detecting an artwork, the system performs recognition by re-using the computed activations of convolutional layers in the detection phase as features. A low dimensional descriptor of a region is obtained by applying a global max-pooling over feature activation maps of two convolutional layers. We selected features from layers 3 and 4 by experimental evaluation, as detailed in Section 9.4 (see Fig. 9.2).

Recognition is performed with a simple 1-NN classifier. Using non-parametric classifiers allows the artwork dataset to grow without the need of re-training.

Even in the case of very few wrong recognitions, user experience would be compromised since the audio-guide would present different information

■ **FIGURE 9.3** Shape-based filtering: artwork in yellow (left) is not considered for recognition (too small) while the other is recognized as "marzocco" (the heraldic lion symbol of Florence).

■ **FIGURE 9.4** Example of artwork tracking. Only after a stable recognition over M frames ($M = 15$ in this example) the system labels the artwork.

with respect to the observed artwork. To avoid this we use three temporal processing strategies. As regards *distance*: we discard recognitions from artwork with a small pixel area, which may be erroneous (see Fig. 9.3).

As regards *consistency*: we consider a recognition correct only after a certain amount of consistent labels (see Fig. 9.4). Finally for *persistence*: it is unlikely that the user moves quickly from an artwork to another in just a few frames. So, after the system recognizes an artwork, it continuously outputs its label proportionally to the time elapsed since the recognition.

Context modeling

Embedded devices can be easily packed with different sensors. A smart guide should act as an autonomous agent, automatically understanding the right time to engage users and when to avoid audio feedback. Independently from the acquired visual information we must understand if the user behavior is changing, e.g. due to interaction with another person.

Detecting conversations. We want audio feedback to pause in the case a conversation is happening. Hearing multiple superposed speeches is certainly a cause of disturbance, not delivering the right information and, moreover, degrading the user experience. This mechanism should come into play even if the user is standing in front of an artwork, thus regardless of the visual feedback. In the indoor scenario, we assume a high signal to noise ratio and exploit the device microphone to detect nearby speech. Although museums are typically quiet environments, in some cases there can be background music or some other environmental noise. We use a Voice Activity Detection (VAD) system, based on the implementation and model available in the OpenSMILE framework.[2] The method is based on LSTM modeling a long-term signal relationship and it is able to model environmental noise [16].

Once another voice is detected, the audio playback will stop, therefore in the case of a false positive the user experience will be compromised. To improve the classifier quality we process an entire second of audio sampling every 0.01 s and averaging all predictions for the final conversation detection score.

Sensors for walking detection. User movements are an extremely important hint for context understanding. The acts of standing still, walking and staying seated give important cues to interpret the attention of an user towards an artwork. In fact:

- A fast walking behavior can be safely interpreted as a sign of no interest in the visible artworks, therefore we can avoid giving feedback in such situations.
- In the case the user is standing still and is listening to a description, we must avoid interrupting the audio, even in the case visual cues get lost, e.g. due to temporary occlusions.

We detect walking using accelerometer data. Specifically we estimate mean and standard deviation and, after subtracting the means, we detect peaks

---

[2] http://audeering.com/research/opensmile/.

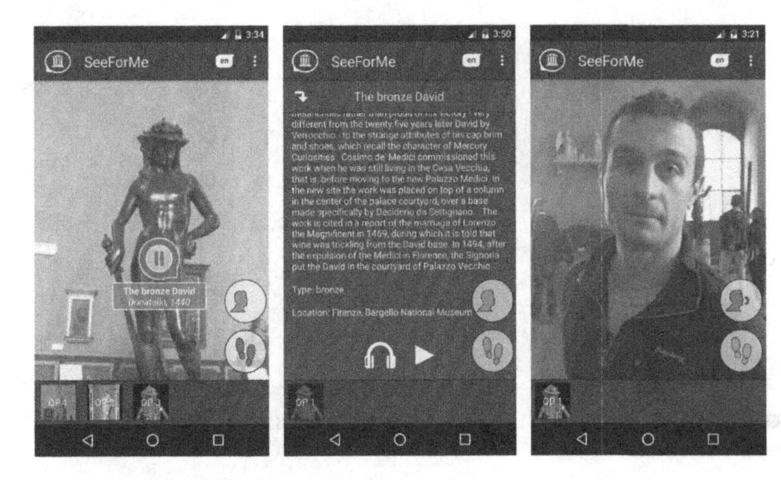

■ **FIGURE 9.5** *(left)* The user is listening to the description of the artwork, *(center)* the user is reviewing an item in the history, *(right)* the user is speaking with someone not focusing on any artwork.

above the standard deviation considering them as steps. If at least a step is made in the last 1 s window, we consider the person walking.

Orientation change can be detected using gyroscope data. After averaging the orientation vector on the same 1 s window we consider the facing direction changed if differing for more than 45°.

User interface. SeeForMe has been designed with the aim to handle three different scenarios: *(i)* The user puts the device in a front pocket with the camera facing forward in order to use the system in a fully automated mode. Recognition is performed by the system on the camera stream. The user can start/stop audio reproduction through voice commands. *(ii)* The user exploits the application in a semi-automated way pointing the camera to the artworks of interest. *(iii)* The user revisits his experience once the visit is completed through a visual history of the tour in the form of a carousel.

In Fig. 9.5 two use cases of the application interface in semi-automatic mode are shown. In Fig. 9.5*(left)* the GUI presents recognized artworks through overlaid green icons that provide information on the artwork and the possibility to start/pause audio playback. If the visitor gets away from the artwork the audio played will fade out, avoiding an abrupt interruption. In the case the visitor approaches back the same artwork the audio feedback is resumed automatically from where it stopped.

In Fig. 9.5*(right)* the app is used in fully automatic mode: the user is speaking with a person. The app understands the situation detecting the voices through VAD and consequently stops audio playback. Once the conversation

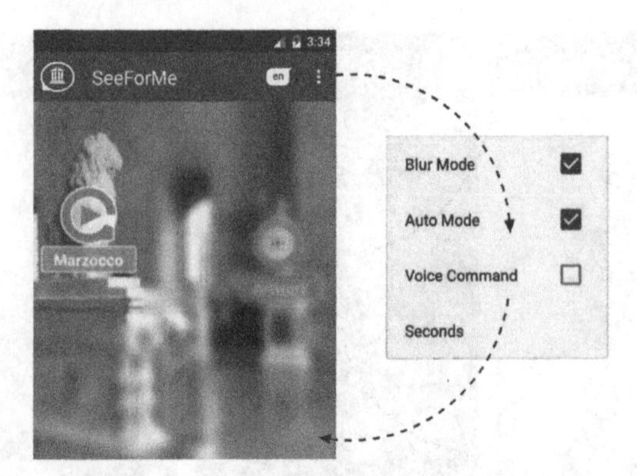

■ **FIGURE 9.6** The contextual menus to configure the app properties. Here is shown the appearance of the interface in blur mode.

is finished and the attention of the visitor turns to the artwork, the system automatically resumes the playback.

Some configuration properties are available to users to manage the user interface as shown in Fig. 9.6.

The mobile app has been developed using the Android SDK. The interface follows the guidelines of Material Design proposed by Google. SQLite is exploited to persist the information on the local storage of the device. A Java Native Interface (JNI) is used to establish the communication between the vision module and the application. Data-interchange is achieved through JSON messages which contain data about artwork detected and recognized, persons in the field of view of the camera, the presence of speech and user movements found through visual information.

Experimental results

Dataset. To train and evaluate our object detector described in Section 9.1, and later evaluate the full recognition system, we collected a dataset of artworks situated in the Bargello Museum in Florence. Artwork data have been collected in the form of 1237 images taken from live footage of the museum under diverse illumination and viewpoint conditions. Changes in these conditions produce a significant alteration of artwork appearance mainly due to the sensitivity of the camera sensor and the diverse behavior of light reflection of different surfaces.

To comply with the object detection task, we annotated a subset of images with information about the pictured artworks and split them into training and testing sets. To balance the artwork and person classes the latter has been augmented with images from PASCAL VOC2007, resulting in ~ 300

person and ~ 300 artwork images. Data augmentation techniques have also been applied, such as image flipping.

We evaluate our recognition system with an additional set of images annotated with the correct artwork id. To produce this set a tool has been developed that uses our pipeline by employing both the artwork detector to generate bounding boxes and a tracker, to aggregate boxes belonging to the same sequence. Such sequences have then been manually annotated, assigning each one to a new or existing artwork id. As already discussed in Section 9.4, thanks to the non-parametric nature of the recognition system, this process can be repeated multiple times to further enrich the dataset. Special attention has been put to include examples with multiple artworks.

The resulting dataset consists of sequences accounting for 8820 frames, where each visible artwork is annotated with its corresponding label and bounding box coordinates. All the sequences account for a total of ~ 250 s of video with 7956 detections.

Artwork detection. We evaluate the performance of the artwork detection system as a first experiment. After fine-tuning the network on our dataset we measure the detector average precision on the test set. As described in Section 9.4, since we are interested in detecting the artwork in front of the user, we limit the results to those detections for which the detected area, normalized to the dimension of the frame, is at least T, and report the average precision when varying its value. The experiment shows that increasing the minimum area of a box yields an increase of the average precision, which reaches its peak of 0.9 at 40% of the area. This means that the classifier is more effective at recognizing artworks close to the user. Increasing the minimum area is also not a guarantee of better results; in fact, while far detections are prone to errors due to small scale, errors can also arise at close distance due to the blur effect of the camera.

A good value for T must therefore be a trade-off between good precision and how close an artwork must be to the user. We choose a value of $T = 0.1$, which provides a significant improvement in precision with respect to the bare detector at a distance of ~ 5 m. Fig. 9.8 reports the precision-recall curve relative to the chosen T, showing very good precision at high recall rates and exhibiting only a small decrease in precision until ~ 0.8 recall. Note that, according to the results reported in Fig. 9.7 for the selected threshold, a higher recall cannot be obtained, and therefore the curve in Fig. 9.8 is truncated at that point.

Artwork recognition: nearest neighbor evaluation. With the following experiment we aim at first evaluating the descriptor fusion approach described in Section 9.4 to find the best pair of layers, and then evaluate how

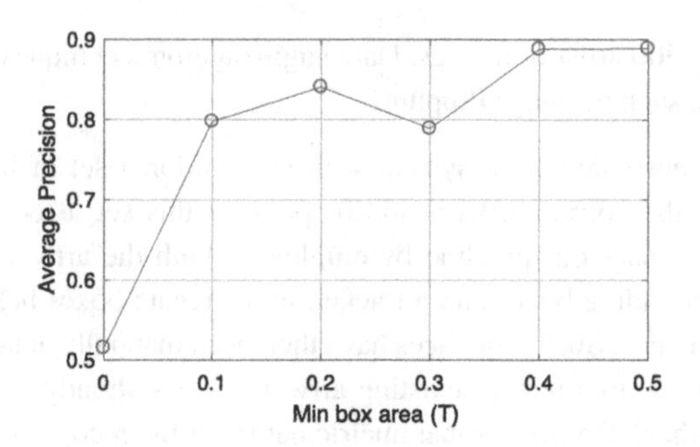

■ **FIGURE 9.7** Average precision of artwork detection varying the minimum box area T.

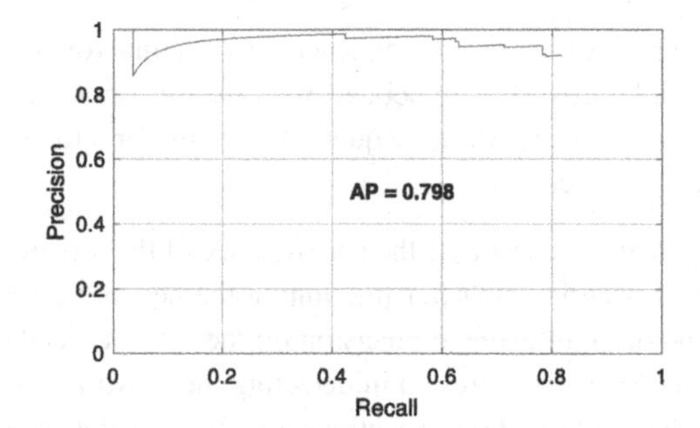

■ **FIGURE 9.8** Precision-recall curve for artwork detection using a threshold $T = 0.1$.

varying the number of nearest neighbors affects in terms of precision the artwork recognition system when using the best configuration. Fig. 9.9 reports the recognition accuracy of the combination of layer 3 and 4 with all the other convolutional layers, showing that fusing the third and fourth layers yields the best results. Using this configuration we then vary the number of neighbors and report in Fig. 9.10 the variation in accuracy, observing that using just one nearest neighbor provides the best performance in recognizing an artwork. Using more neighbors to vote the correct artwork id does not improve the accuracy. The reason for this behavior is that for each artwork we acquired multiple poses in different lighting conditions, so for each query there will be only a small number of samples in a similar pose with a similar lighting condition. Increasing the amount of neighbors simply adds noisy data to the results.

Temporal processing evaluation. Here we test the effectiveness of the three temporal processing strategies, i.e. *consistency*, *distance* and *persistence*, described in Section 9.4, by experimenting with several of their combinations. Using a simulation of the system where we set $T = 0.1$, $M = 15$ and with a persistence of artwork detection of $P = 20$ recognitions, we eval-

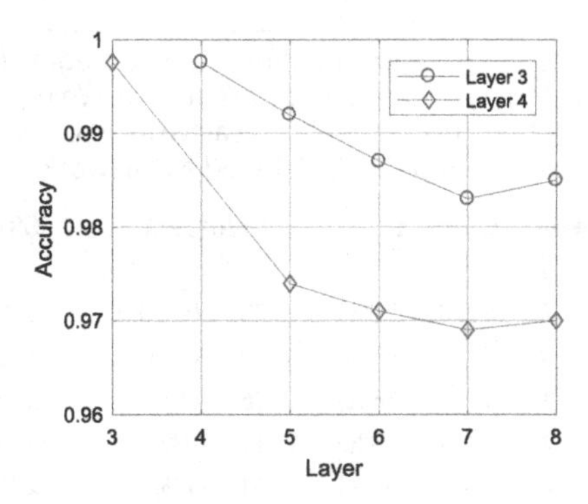

FIGURE 9.9 Recognition accuracy of combinations of layer 3 and 4 with layers $[3, \ldots, 8]$.

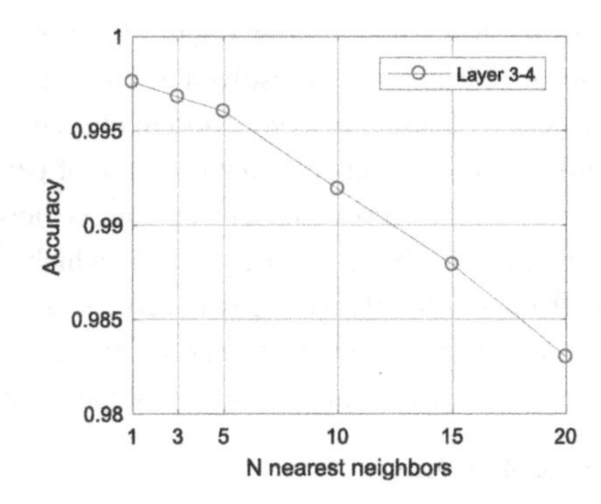

FIGURE 9.10 Recognition accuracy of the best layer combination 3+4, varying number of NN.

uate the annotated video sequences tracking every combination of output bounding box and label, comparing the result with the ground truth. The number of correctly, incorrectly and generically labeled "artwork" detections is used as performance measure and reported in Table 9.1.

We perform six tests labeled from T1 to T6. In T1, where we do not apply any strategy to provide a baseline for the other tests, we can observe that only 70% of the labels are correctly classified, leaving 30% of incorrect labels. By applying the *distance* criterion in T2 we halve the number of incorrect labels but at the expense of assigning a generic label to the other half, confirming that a large number of errors is made on distant, and therefore difficult to recognize, artworks. In T3 we evaluate the *consistency* strategy which shifts almost all incorrect labels and around 15% of the correct ones to the generic label. This is due to the video system not providing a stable output label among consecutive frames.

In T4 we combine *distance* and *consistency* criteria, noting that, while *consistency* is able to almost nullify the incorrect recognitions, it is not robust

Table 9.1 Performance by applying the three strategies for temporal smoothing: consistency, distance and persistence. We report the number of detections where, respectively, the artwork was: correctly recognized, misclassified, and generically labeled as "artwork"

Test	Strategy			Correct	Incorrect	Skipped
	C	*D*	*P*			
T1	✗	✗	✗	5598 (~70%)	2358 (~30%)	0 (0%)
T2	✗	✓	✗	5334 (~67%)	1267 (~16%)	1355 (~17%)
T3	✓	✗	✗	4475 (~56%)	36 (~0%)	3445 (~43%)
T4	✓	✓	✗	4363 (~55%)	11 (~0%)	3582 (~45%)
T5	✓	✗	✓	5141 (~65%)	61 (~1%)	2754 (~35%)
T6	✓	✓	✓	4966 (~62%)	22 (~0%)	2968 (~37%)

to sparse errors, causing the system to often bounce from the correct recognition to the generic label. This issue is resolved in T5 where *consistency* is combined with *persistence*, leading to a reduction in misclassified labels, an increase in the number of correct labels and a decrease of generic ones. Finally in T6 we combine all three criteria. Here we can see how using all the strategies together produces only 22 incorrect labels, which approximately corresponds to less than one cumulative second every ~ 5 minutes of video, at the expense of a reasonable number of generically classified artworks. This also confirms our intuition about the efficacy of the three strategies.

Voice detection evaluation

Here we evaluate two strategies for voice activity detection: *(i)* sample, we evaluate one audio sample per second; and *(ii)* mean, we evaluate the mean sample computed over a second of audio. To minimize the number of false positives we measure the classifier performance varying the false positive threshold.

In Fig. 9.11 we report the Receiver Operating Characteristic (ROC) curve. Although both strategies have a high Area Under the Curve (AUC) and correctly classify voice most of the time, the mean strategy performs always better than sample.

9.5 THE OUTDOOR SYSTEM

The system, for which the architectural diagram is shown in Fig. 9.13, is composed of three interacting modules: *(i)* a *location module* to provide current location and nearby interest points; *(ii)* a *content provider* responsible for fetching textual information; and *(iii)* a *vision module* constantly comparing the user point of view to a set of expected interest point appearance images.

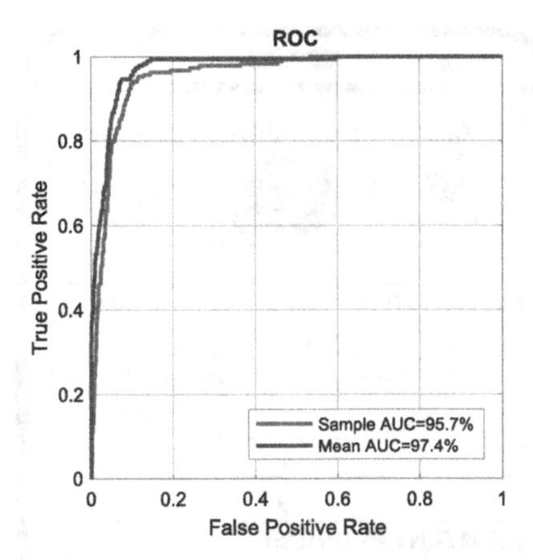

■ **FIGURE 9.11** Receiver operating characteristic curve of the tested voice activity classifiers.

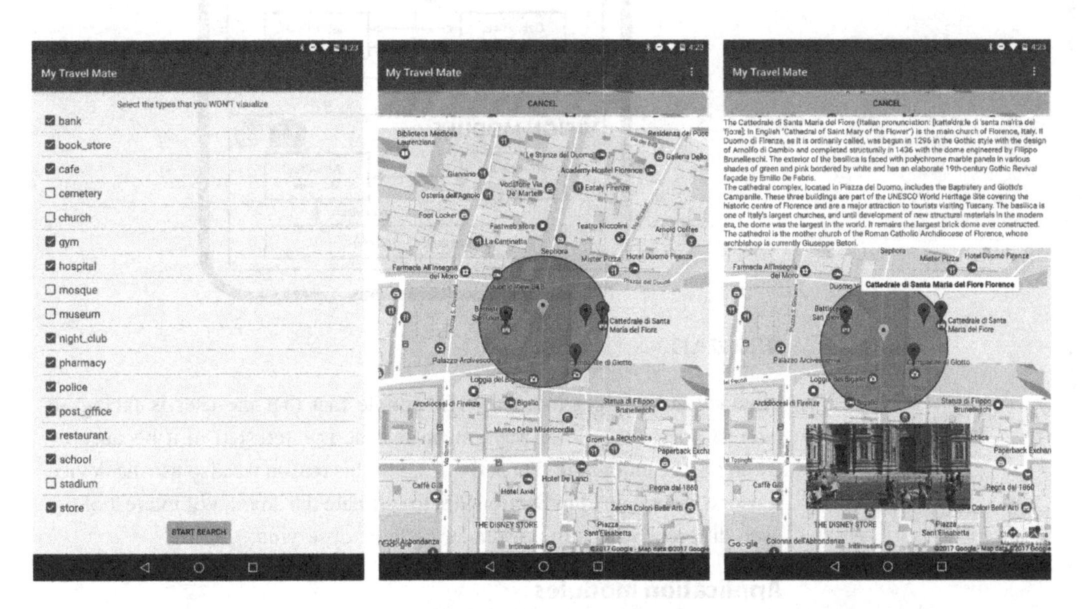

■ **FIGURE 9.12** *(left)* The application is asking the user the interest point types he does not want to be included in the search, *(center)* the application is showing the interest point around the user, *(right)* the application shows the textual description for the selected point.

The system has been developed as an android application and makes use of the following sensors: a *GPS receiver* to perceive the user position, and a *camera* to validate the artwork he faces. The application is responsible of coordinating the three main modules and to present the final information to the user (see Fig. 9.12).

Context awareness

For a fully autonomous audio-guide, understanding the wearer context must be considered of paramount importance if meaningful information is to be

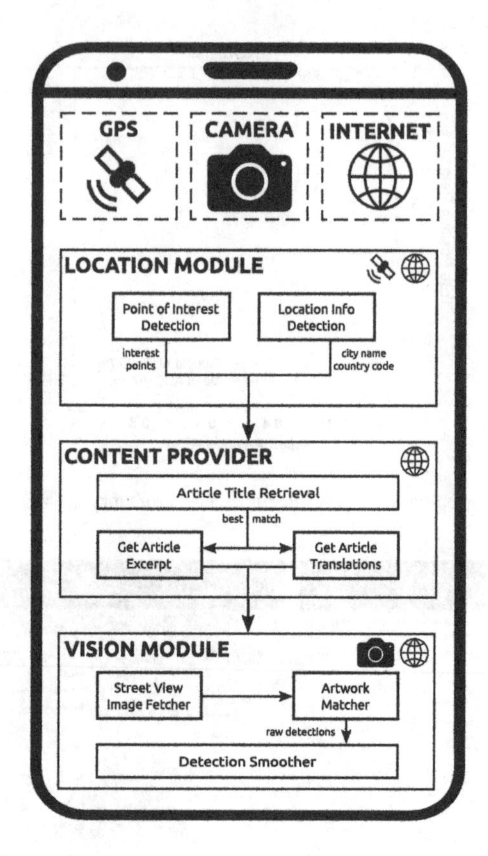

FIGURE 9.13 System architecture.

provided at the right moment. The simple fact that the user is facing an interest point does not always mean that he is interested in it because he could also be enjoying other activities. For this reason we also use the device sensors, as explained in Section 9.5, to estimate the amount of interest of the user, differentiating simple wandering from observing.

Application modules

Location module. This module is responsible for providing interest points by exploiting the current user position and querying the Google Place Search API[3] for a given radius. The response is a list of 60 interest points each one annotated with one or more `types`,[4] such as `art_gallery` or `bank`. The user can personalize the application by specifying which type he is interested in Fig. 9.12(*left*). The interest points are then passed to the *content*

[3]https://developers.google.com/places/web-service/search.
[4]https://developers.google.com/places/supported_types.

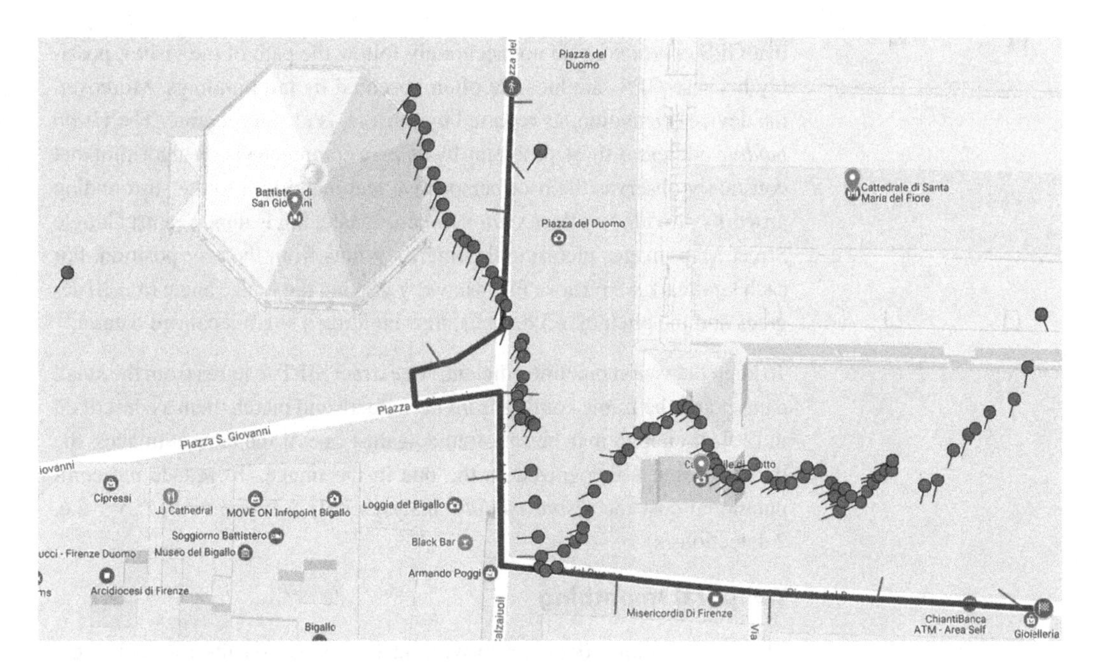

■ **FIGURE 9.14** Map of the Duomo Square of Florence. The three points of interest (red markers) are highlighted together with the user location points obtained from GPS (red circles) and the real tourist itinerary (green line). The real tourist orientation and the one obtained from GPS are marked, respectively, with green lines along the itinerary and black lines on the red circles.

provider to retrieve textual descriptions, together with the name of the city the user is visiting which is obtained by performing a Geocoding[5] query. This is necessary to discriminate interest points with the same name but located elsewhere.

Content provider. This module fetches textual information for interest points found by the *location module* by taking the first result of a Wikipedia[6] search in the local language for articles containing both the city name and the interest point name. The description is then obtained with another Wikipedia[7] search, limiting the description text to the page summary. To provide translations into other languages the module performs an additional query to Wikipedia requesting the Interlanguage links[8] for the retrieved page and downloading the extracts for the desired languages. If an interest point name does not match any Wikipedia article or the content is not available for the desired language, then the point is discarded.

Vision module. Understanding when the user is correctly facing a landmark is not trivial. As shown in Fig. 9.14, GPS position and device orientation are not reliable information and cannot be used directly. We can observe

[5] https://developers.google.com/maps/documentation/geocoding.
[6] https://www.mediawiki.org/wiki/API:Search.
[7] https://www.mediawiki.org/wiki/API:Query.
[8] https://en.wikipedia.org/wiki/Help:Interlanguage_links.

that GPS coordinates do not accurately follow the path of the visitor, possibly because GPS satellites are often obscured by tall buildings. Moreover, the device orientation, as reported by sensors, is also inaccurate. The *vision module* addresses these problems by using a computer-vision algorithm that constantly observes the user perspective and matches it to the surrounding artworks provided by the *location module*. Matching is done against Google Street Map images picturing the interest points from the user position. For each landmark we retrieve four views by varying the image angle of ± 10 degrees and the pitch of ± 5 degrees, thus building a small localized dataset.

To keep the vision pipeline efficient, we extract SIFT features from the small dataset and the frame coming from the camera and match them as described in [29]. If enough matches are found against one of the dataset images, the interest point is recognized as the one in the image. To reduce the computational cost the *vision module* analyzes only a frame every 0.5 s (i.e. 2 detections/s).

Temporal smoothing

The *vision module* does not provide instantaneous results for each input frame; instead, it needs to process a sequence of frames before outputting a decision. This allows us to perform post-processing over single outputs and present smoothed output to the user. The module evaluates each frame and considers the output *valid* if it persists for at least five frames; moreover, a Time To Live (TTL) of five frames is also applied to tolerate sporadic incorrect results. The TTL is decremented whenever a decision differs from the valid one and is reset in the case of a matching decision. If TTL reaches zero all invalid frames are considered incorrect, otherwise they are all assigned to the artwork of the valid frames. This behavior allows the module to tolerate small classification noise. In addition, to enforce the production of long lasting sequences, the module reports valid output only if it lasts at least for 10 frames.

Exploiting sensors for modeling behavior

In an outdoor scenario where artworks are often whole buildings, moving does not necessarily imply disinterest. We monitor GPS data and compute user distance from the artwork to understand if he moves but stays close to the artwork, thus continuing to be interested in it, or if he moves away.

This spacial information is used to make the audio persist when the *vision module* stops detecting an artwork. This is crucial in outdoor scenarios where occlusion is a very frequent event, so once the artwork is recognized the description is provided to the user as long as he stays close to it or a different artwork is recognized by the *vision module*. If he walks away and no other artwork is recognized, then audio is stopped.

System implementation

As for the indoor system, the vision system introduced in Section 9.5 has been developed and tested using an NVIDIA Jetson TX1 board. The audio-guide application has been deployed on an NVIDIA Shield Tablet K1 which has the same hardware as the TX1 board but runs Android 6.0 instead of Linux. The mobile application has been developed using the Android SDK and makes use of the Java Native Interface (JNI) to communicate with the *vision module*. Data between Java and Native Code is exchanged using JSON messages providing information on the detected artworks.

Application use cases

The application has been developed with the intention of letting the user enjoy his visit without needing to interact with the guide. Although this is the primary behavior, the application can also work in a non-automated way, giving the user the ability to virtually explore his surrounding. These modalities define two different use cases: *(i)* a *fully automated* mode and *(ii)* an *interactive* mode.

Fully automated. In this scenario the user walks freely with the device positioned in a front pocket to let the camera look at what is in front of him. Meanwhile, the application monitors the user GPS position and looks for interest points through the *location module*. When the user walks in a zone where interest points are available, the *content provider* fetches information about them and the *vision module* analyzes the camera stream to understand if the visitor is facing one of them. In the case of a positive match, the application delivers the downloaded information to the user as audio description.

Interactive. The interactive modality is available to the user to let him virtually explore his surroundings. The user can move to any position on the map, as shown in Fig. 9.12, to discover interest points. By touching one of them the application will show the user the relative textual description, which can then be transposed to audio if desired. In this modality the *vision module* is paused.

Experimental results

Dataset. To evaluate the application vision system we collected a small dataset of video footage of the city center of Florence. The city of Florence has a rich architectural and artistic history and its center hosts a great variety of artworks. The footage records the visit of three buildings, filmed from different points of view and different lighting conditions, for a total of 5 min

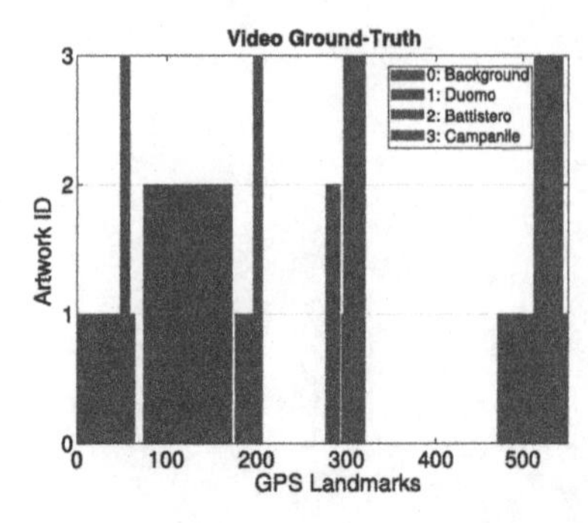

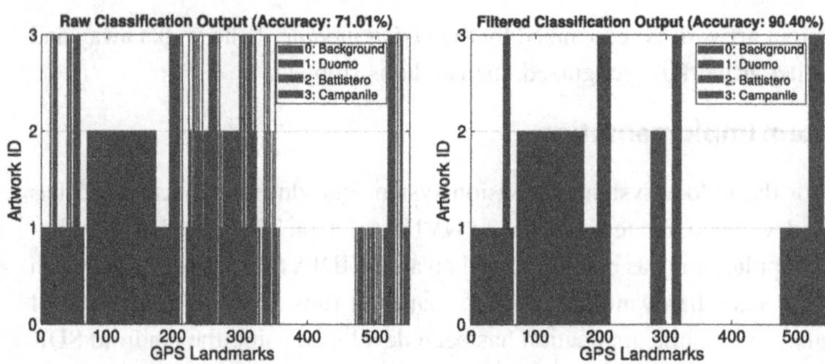

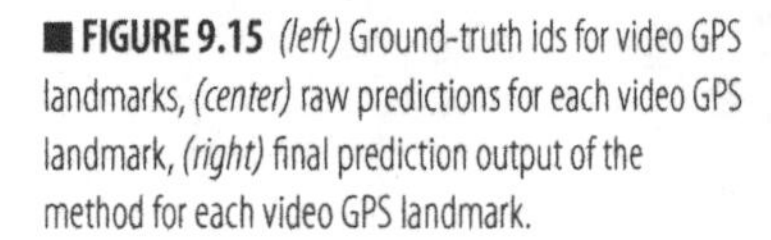

■ **FIGURE 9.15** *(left)* Ground-truth ids for video GPS landmarks, *(center)* raw predictions for each video GPS landmark, *(right)* final prediction output of the method for each video GPS landmark.

of video. Every 0.5 s we annotated the current frame with time of the visit and the GPS position, resulting in 522 annotated frames.

The evaluation is carried by selecting four points during the visit and for each of them we queried the Google Street View API to download images of the three interest points and obtain their appearances from those positions. To account for sensor inaccuracy we slightly varied the *heading* and the *pitch* of $\pm 10°$ and $\pm 5°$, respectively, obtaining four different images of the same artwork.

Method evaluation. We evaluate the system as follows: for each of the 522 GPS landmarks we extract SIFT descriptors from the relative frame and match them to the ones extracted from the reference images in the dataset. If enough matches are found between two images then we label the current frame as containing the artwork pictured in the matching reference image. After this process we obtain a sequence of 522 descriptors labeled either with one of the artworks or as background.

Performance is evaluated in terms of accuracy, comparing the predicted label series to the ground truth. In Fig. 9.15 we report the ground truth (*left*) and the results, showing both the raw output (*center*) of the module and the smoothed one (*right*) as explained in Section 9.5. The center image confirms the erratic behavior of the results when no filtering is applied, producing a quite low accuracy of 70%. Applying filtering, accuracy increases considerably by 20%, proving that the smoothing is necessary. It can be noted that filtered results and ground truth differ slightly only on the starting and ending points of the detections. This means that we always give the correct output label but we make a small error in detecting the exact starting and ending landmarks. It is worth noting that perfect timing is not essential for an audio-guide application; in fact it is preferable to give the correct audio description with a short delay than giving a wrong description at the correct time.

Table 9.2 Functions comparison of our guide with respect to human and traditional audio guides

Type	Instrumental	Social	Interactional	Communicative
Human guide	* * *	* * *	* * *	* * *
Audio guide	*	–	–	* *
Our system	* *	* *	* *	* *

User experience evaluation

Human guide roles have been connoted in [9] through a distinction between their functions: instrumental (guide), social (animator), interactional (leader) and communicative (intermediator). Instrumental functions convey information such as artwork localization and user routing. Interactional aspects provide users with means of getting information. Sociality involves user engagement, also supplying collaborative experiences. Communicative functions facilitate personalization and easy access to targeted content. Although humans are the best in carrying out these functions, recent progresses in technology have given the opportunity to achieve close results. In Table 9.2 a qualitative comparison is shown between functions provided by human and traditional audio-guides with those achievable by our system.

Our audioguide presents appreciable improvements especially as regards instrumental, interactional and socialization aspects like: 1) it allows the least cognitive effort to obtain information, which is provided automatically allowing at the same time for playback control; 2) it senses the context and understands user behavior adapting accordingly (e.g. stopping when the user loses attention), avoiding the 'isolated experiences' provided by traditional audio-guides. The overall experience achieved by the system in a real environment was measured through an evaluation of its usability both for the outdoor and the indoor scenario.

The popular Standard Usability Scale (SUS) [7] was used, following a user-centered approach. Different groups of users were asked to fulfill some tasks in both the scenarios and then to respond to a 10 points questionnaire. Answers to questions are based on a Likert scale [41] from 1 to 5, expressing a range between "Strongly Disagree" and "Strongly Agree" with opposite meaning, a strategy proved effective in minimizing acquiescence bias, and alternating positive with negative sentences so that users have to read carefully. As stated in [33], it is sufficient to collect five polls to detect the 85% of design errors of a user interface. For both the indoor and the outdoor system a supervised and an unsupervised scenario (i.e. receiving explanations or not on the app functionalities and on the recognition system) were evaluated. Users had to get information on one or more artworks/landmarks of interest. The indoor system was tested by 12 users and obtained an average SUS of 74.0 for the unsupervised scenario and 79.5 for the supervised scenario; the outdoor system was tested by 10 users with an average SUS of 69.5 for the unsupervised scenario and 71.8 for the supervised scenario. Results, quite close to 80 for the first scenario, denote an interface providing a very good experience; while for the second, which reached a score between the reference values of 60 and 80, the user experience was above average [37]. The small difference in supervised and unsupervised scores suggests that the system is easy to use also without previous training in both scenarios. Feedback, gained through open-ended questions on system usability, which, as stated by Nielsen, allow you 'to find out more than you anticipate' are summarized as follows: 11 users indicated the automatic recognition of artworks (i.e. 6 users) and the consequent start/stop of the audio reproduction (i.e. 5 users) as the best feature. Issues were highlighted by 2 users: they are relative to the use of a system as a mobile app in assisted mode and regard the fatigue caused by having to constantly hold the device in front of the artworks with the hands.

9.6 CONCLUSIONS

New devices featuring powerful GPUs for building and deploying embedded systems applications, like the NVIDIA Jetson TX1 and the NVIDIA Shield Tablet K1, give the possibility to design and implement new tools, such as the smart audio-guide presented in this chapter, which can considerably improve and change consolidated paradigms for the fruition of multimedia insights in the context of museum and cultural experiences.

The presented audio-guide system allows one to profile the visitor in real time, understanding his interests and the context of the visit, and to provide a natural mechanism of interaction and feedback for the attainment and the semi-automatic control of multimedia insights. Audio and sensor data are exploited to improve the user experience, reducing the difficulty of use and the cumbersome approach of a traditional audio-guide. In proposing our method we discuss the opportunity to adopt different techniques for performing artwork recognition in the two main scenarios of an outdoor and indoor visit (i.e. CNN and SIFT). Furthermore, the proposed solution shows some good practices for the fusion of data (visual, acoustic, motion and localization) coming from different types of sensors on wearable mobile devices and suggests simple but effective temporal smoothing and consistency strategies. These strategies allow one to obtain a lower error rate in automatic artwork recognition and an overall improved human–computer interaction.

The smart audio-guides have been implemented as Android apps providing the users with the possibility to switch between a fully automated experience to a semi-automatic, more interactive mode where the system is not worn but used as a traditional application on a mobile device through an *ad hoc* configurable interface.

Usability testing, performed with the commonly used SUS (System Usability Scale), has given good results for both scenarios, proving that understanding the context can help to provide an improved experience to the user.

REFERENCES

[1] Gregory D. Abowd, Christopher G. Atkeson, Jason Hong, Sue Long, Rob Kooper, Mike Pinkerton, Cyberguide: a mobile context-aware tour guide, Wirel. Netw. 3 (5) (1997) 421–433.

[2] Rao Muhammad Anwer, Fahad Shahbaz Khan, Joost van de Weijer, Jorma Laaksonen, Combining holistic and part-based deep representations for computational painting categorization, in: Proc. of ICMR, 2016.

[3] Liliana Ardissono, Tsvi Kuflik, Daniela Petrelli, Personalization in cultural heritage: the road travelled and the one ahead, User Model. User-Adapt. Interact. 22 (1) (2012) 73–99.

[4] Lorenzo Baraldi, Francesco Paci, Giuseppe Serra, Luca Benini, Rita Cucchiara, Gesture recognition using wearable vision sensors to enhance visitors' museum experiences, IEEE Sens. J. 15 (5) (2015) 2705–2714.

[5] Jonathan Bowen, James Bradburne, Alexandra Burch, Lynn Dierking, John Falk, Silvia Filippini Fantoni, Ben Gammon, Ellen Giusti, Halina Gottlieb, Sherry Hsi, et al., Digital Technologies and the Museum Experience: Handheld Guides and Other Media, Rowman, Altamira, 2008.

[6] Jonathan P. Bowen, Silvia Filippini-Fantoni, Personalization and the web from a museum perspective, in: Proc. of Museums and the Web (MW), 2004.

[7] John Brooke, SUS – a quick and dirty usability scale, in: Usability Evaluation in Industry, 1996, pp. 189–194.

[8] Luca Caviglione, Mauro Coccoli, Alberto Grosso, A framework for the delivery of contents in RFID-driven smart environments, in: Proc. of RFID-TA, 2011.

[9] Erik Cohen, The tourist guide: the origins, structure and dynamics of a role, Ann. Tour. Res. 12 (1) (1985) 5–29.

[10] Alberto Del Bimbo, Andrea Ferracani, Daniele Pezzatini, Federico D'Amato, Martina Sereni, LiveCities: revealing the pulse of cities by location-based social networks venues and users analysis, in: Proc. of WWW, 2014.

[11] Alberto Del Bimbo, Walter Nunziati, Pietro Pala David, Discriminant analysis for verification of monuments in image data, in: Proc. of ICME, 2009.

[12] Thomas Drugman, Yannis Stylianou, Yusuke Kida, Masami Akamine, Voice activity detection: merging source and filter-based information, IEEE Signal Process. Lett. 23 (2) (2016) 252–256.

[13] Veron Eliseo, Levasseur Martine, Ethnographie de l'exposition, in: Études et recherche, Centre Georges Pompidou, Bibliothèque publique d'information, 1991.

[14] Benjamin Elizalde, Gerald Friedland, Lost in segmentation: three approaches for speech/non-speech detection in consumer-produced videos, in: Proc. of ICME, 2013.

[15] Christos Emmanouilidis, Remous-Aris Koutsiamanis, Aimilia Tasidou, Mobile guides: taxonomy of architectures, context awareness, technologies and applications, J. Netw. Comput. Appl. 36 (1) (2013) 103–125.

[16] Florian Eyben, Felix Weninger, Stefano Squartini, Björn Schuller, Real-life voice activity detection with LSTM recurrent neural networks and an application to Hollywood movies, in: Proc. of ICASSP, 2013.
[17] Ross Girshick, Fast R-CNN, in: Proc. of ICCV, 2015.
[18] Ross Girshick, Jeff Donahue, Trevor Darrell, Jagannath Malik, Rich feature hierarchies for accurate object detection and semantic segmentation, in: Proc. of CVPR, 2014.
[19] Luca Greci, An augmented reality guide for religious museum, in: Proc. of AVR, 2016.
[20] Loc Nguyen Huynh, Rajesh Krishna Balan, Youngki Lee, DeepSense: a GPU-based deep convolutional neural network framework on commodity mobile devices, in: Proc. of WearSys, 2016.
[21] Svebor Karaman, Andrew D. Bagdanov, Lea Landucci, Gianpaolo D'Amico, Andrea Ferracani, Daniele Pezzatini, Alberto Del Bimbo, Personalized multimedia content delivery on an interactive table by passive observation of museum visitors, Multimed. Tools Appl. 75 (7) (2016) 3787–3811.
[22] Michael Kenteris, Damianos Gavalas, Daphne Economou, Electronic mobile guides: a survey, Pers. Ubiquitous Comput. 15 (1) (2011) 97–111.
[23] Doyeon Kim, Daeil Seo, Byounghyun Yoo, Heedong Ko, Development and evaluation of mobile tour guide using wearable and hand-held devices, in: Proc. of HCI, 2016.
[24] Ronny Kramer, Marko Modsching, Klaus Ten Hagen, A city guide agent creating and adapting individual sightseeing tours based on field trial results, Int. J. Comput. Intell. Res. 2 (2) (2006) 191–206.
[25] Alex Krizhevsky, Ilya Sutskever, Geoffrey E. Hinton, ImageNet classification with deep convolutional neural networks, in: Proc. of NIPS, 2012.
[26] Tsvi Kuflik, Zvi Boger, Massimo Zancanaro, Analysis and Prediction of Museum Visitors' Behavioral Pattern Types, Springer, Berlin Heidelberg, 2012, pp. 161–176.
[27] Seyyed Salar Latifi Oskouei, Hossein Golestani, Matin Hashemi, Soheil Ghiasi, CNN-droid: GPU-accelerated execution of trained deep convolutional neural networks on android, in: Proc. of ACM MM, 2016.
[28] Wei Liu, Dragomir Anguelov, Dumitru Erhan, Christian Szegedy, Scott Reed, Cheng-Yang Fu, Alexander C. Berg, SSD: single shot multibox detector, in: Proc. of ECCV, 2016.
[29] David G. Lowe, Distinctive image features from scale-invariant keypoints, Int. J. Comput. Vis. 60 (2) (2004) 91–110.
[30] Austin Meyers, Nick Johnston, Vivek Rathod, Anoop Korattikara, Alex Gorban, Nathan Silberman, Sergio Guadarrama, George Papandreou, Jonathan Huang, Kevin P. Murphy, Im2Calories: towards an automated mobile vision food diary, in: Proc. of ICCV, 2015.
[31] Ananya Misra, Speech/nonspeech segmentation in web videos, in: Proc. of Interspeech, 2012.
[32] Saman Mousazadeh, Israel Cohen, AR-GARCH in presence of noise: parameter estimation and its application to voice activity detection, IEEE Trans. Audio Speech Lang. Process. 19 (4) (2011) 916–926.
[33] Jakob Nielsen, Rolf Molich, Heuristic evaluation of user interfaces, in: Proc. of CHI, 1990.
[34] David Picard, Philippe-Henri Gosselin, Marie-Claude Gaspard, Challenges in content-based image indexing of cultural heritage collections, IEEE Signal Process. Mag. 32 (4) (2015) 95–102.
[35] Joseph Redmon, Santosh Divvala, Ross Girshick, Ali Farhadi, You only look once: unified, real-time object detection, in: Proc. of CVPR, 2016.

[36] Shaoqing Ren, Kaiming He, Ross Girshick, Jian Sun, Faster R-CNN: towards real-time object detection with region proposal networks, in: Proc. of NIPS, 2015.
[37] Jeff Sauro, James R. Lewis, Quantifying the User Experience: Practical Statistics for User Research, Morgan Kaufmann, 2012.
[38] Todd Simcock, Stephen Peter Hillenbrand, Bruce H. Thomas, Developing a location based tourist guide application, in: Proc. of ACSW Frontiers, 2003.
[39] Yuichiro Takeuchi, Masanori Sugimoto, A user-adaptive city guide system with an unobtrusive navigation interface, Pers. Ubiquitous Comput. 13 (2) (2009) 119–132.
[40] Ryosuke Tanno, Koichi Okamoto, Keiji Yanai, DeepFoodCam: a DCNN-based real-time mobile food recognition system, in: Proc. of MADiMa, 2016.
[41] William M. Trochim, et al., Likert scaling, in: Research Methods Knowledge Base, 2nd edition, 2006.
[42] Fabio Vesperini, Paolo Vecchiotti, Emanuele Principi, Stefano Squartini, Francesco Piazza, Deep neural networks for multi-room voice activity detection: advancements and comparative evaluation, in: Proc. of IJCNN, 2016.
[43] Yafang Wang, Chenglei Yang, Shijun Liu, Rui Wang, Xiangxu Meng, A RFID & handheld device-based museum guide system, in: Proc. of ICPCA, 2007.
[44] Yiwen Wang, Natalia Stash, Rody Sambeek, Yuri Schuurmans, Lora Aroyo, Guus Schreiber, Peter Gorgels, Cultivating personalized museum tours online and on-site, Interdiscip. Sci. Rev. 34 (2–3) (2009) 139–153.
[45] Kyoung-Ho Woo, Tae-Young Yang, Kun-Jung Park, Chungyong Lee, Robust voice activity detection algorithm for estimating noise spectrum, Electron. Lett. 36 (2) (2000) 180–181.
[46] Jiaxiang Wu, Cong Leng, Yuhang Wang, Qinghao Hu, Jian Cheng, Quantized convolutional neural networks for mobile devices, in: Proc. of CVPR, 2016.
[47] Keiji Yanai, Ryosuke Tanno, Koichi Okamoto, Efficient mobile implementation of a CNN-based object recognition system, in: Proc. of ACM MM, 2016.
[48] Massimo Zancanaro, Tsvi Kuflik, Zvi Boger, Dina Goren-Bar, Dan Goldwasser, Analyzing museum visitors' behavior patterns, in: Proc. of UM, 2007.

Chapter 10

Recognizing social relationships from an egocentric vision perspective

Stefano Alletto*, Marcella Cornia*, Lorenzo Baraldi*, Giuseppe Serra†, Rita Cucchiara*

*University of Modena and Reggio Emilia, Department of Engineering "Enzo Ferrari", Modena, Italy †University of Udine, Department of Mathematics, Computer Science and Physics, Udine, Italy

CONTENTS

10.1 INTRODUCTION

Humans are inherently good at understanding and categorizing social formations. A clear signal of this cue is that we respond to different social situations with different behaviors: while we accept to stand in close proximity to strangers when we attend some kind of public event, we would feel uncomfortable in having people we do not know close to us when we take a coffee. This processing happens so naturally in our brains that we

Multimodal Behavior Analysis in the Wild. https://doi.org/10.1016/B978-0-12-814601-9.00015-8
Copyright © 2019 Elsevier Ltd. All rights reserved.

rarely stop wondering who is interacting with whom or which people form a group and which do not. Nevertheless, understanding social relationships and social groups is a complex task which can hardly be transferred to a fully automatic system.

Initial work has started to address the task of social interaction analysis from the videosurveillance perspective [24,33]. Fixed cameras, however, lack the ability to immerse in the social environment, effectively losing an extremely significant portion of the information about what is happening. Wearable cameras, on the contrary, put the research on this matter in a new and unique perspective. A video taken from the egocentric perspective provides a more meaningful insight in the social interaction, given that the recording is performed by a member of the group itself with a clear view of the social formation.

This privileged perspective lets researchers use wearable cameras for acquiring and processing the same visual stimuli that humans acquire and process. In this regard, first-person vision (or egocentric vision) assumes the broader meaning of understanding what a person sees calling for similar learning, perception and reasoning paradigms of humans. While this approach carries exceptional benefits, it also features several problems: the camera continuously follows the wearer's movements, resulting in severe camera motion, steep lighting transitions, background clutter and severe occlusions. These situations are required to be properly tackled in order to process the video automatically and extract higher level information.

There are significant cues which can be captured by an egocentric camera and which can help the automatic understanding of the social formation the user is involved in. First, when we are interacting with each other we naturally tend to place ourselves in determined positions to avoid occlusions in our group, stand close to the ones we interact with and orientate our head so as to place the focus on the subjects of our interest. Moreover, when engaged in a conversation we naturally tend to look at the people we are interacting with, and to ignore others, so eye fixations are an important cue to determine the strength of a social relationship between people. Distances between individuals and mutual orientations also assume clear significance and must be interpreted according to the situation. *F-formation* theory [18] describes patterns that humans naturally tend to create when interacting with each other and can be used to understand whether an ensemble of people forms a group or not, based on the mutual distances and orientations of the subjects in the scene. *F-formations* have recently been successfully applied in videosurveillance, with fixed cameras, in studies aimed at social interaction analysis showing great promise [9,14].

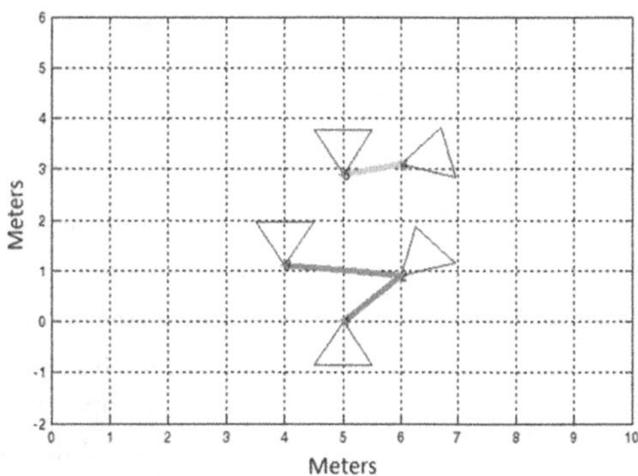

■ **FIGURE 10.1** An example of our method output. In the left image: different colors in bounding box indicate their belonging to different groups. The red dot represents the first person wearing the camera. In the right image: the bird's eye view model where each triangle represents a person and links among them represent the groups.

Following these cues, we adopt distance and orientation information and use them to build a pairwise feature vector capable of describing how two people relate. Since orientations and distances differ as regards importance and meaning in different situations, we use a supervised correlation clustering framework to learn about social groups. Once social groups are detected, we estimate the gaze of the camera wearer by using a saliency prediction approach. This lets us recover important information, which the camera cannot record, and predict the importance of the social relation between the user and the people involved in his social group. While head orientations and distances can be inferred for the people the user can see (e.g. *the others*), saliency estimates a component of the behavior of the wearer himself. An example of the output of our method is shown in Fig. 10.1.

In the rest of this chapter, we will discuss the following issues:

- The definition of a novel head pose estimation approach which can cope with the challenges of the egocentric vision scenario: using a combination of facial landmarks and shape descriptors, our head pose method is robust to steep poses, low resolutions and background clutter.
- The formulation of a 3D ego-vision people localization method capable of estimating the position of a person without relying on calibration. Camera calibration is a process that cannot be automatically performed on different devices and would cause a loss in generality for our method. We use instead random regression forests that employ facial landmarks and the head bounding box as features, resulting in a robust pose-independent distance estimation of the head.
- Modeling of a supervised correlation clustering algorithm using structural SVM to learn how to weight each component of the feature vector depending on the social situation it is applied to. This is due to the fact that humans perform differently in different social situations and the way groups are formed can greatly differ.

- The estimation of the degree of social interaction between the camera wearer and the other people involved in his social group, through the definition of a social strength score derived from a supervised saliency prediction model.

The proposed method is evaluated on publicly available datasets, and by comparing it with several recent algorithms. Each component of the framework is extensively discussed. While experimental results highlight some open problems, they show a new way for computer vision to deal with the complexity of unconstrained scenarios such as egocentric vision and human social interactions.

10.2 RELATED WORK

In this section, we review the literature related to head pose estimation and social interactions with particular attention to egocentric vision approaches.

10.2.1 Head pose estimation

The problem of estimating the head pose has been widely studied in computer vision. Existing methods can be roughly divided into two main categories, regardless of whether their aim is to assess the head pose on still images or video sequences.

Considering the most important solutions for head pose estimation in still images, We et al. [31] proposed a two-level classification framework based on Gabor wavelets in which the first level has the objective of deriving a good estimate of the pose within some uncertainty, while the second level aims at minimizing this uncertainty by analyzing finer structural details captured by bunch graphs. Ma et al. [21] presented a multiview face representation based on local Gabor binary patterns extracted on different sub-regions of the images. Despite these methods performing well on different publicly available datasets, they have significant performance losses when applied to less constrained environments, as egocentric vision contexts.

In [36], an unified model for face detection, pose estimation and landmark localization for "in the wild" images is presented. In particular, the proposed model is based on mixtures of trees with a shared pool of parts in which every facial landmark is modeled as a part and global mixtures are used to capture topological changes due to varying the viewpoint. A different approach is introduced by Li et al. [20] who developed a central profile-based 3D face pose estimation method. The central profile is a 3D curve that divides the face and has the characteristic of having its points lying on the symmetry plane. By relying on the Hough transform to determine the symmetry plane, Li et al. estimate the head pose using the normal vectors of the central profile which are parallel to the symmetry plane. For a comprehensive study summarizing about 90 of the most innovative head pose estimation methods, we refer to the survey presented in [22].

Moving to the video domain, several existing approaches exploit 3D information to estimate the head pose [23,25]. [23] who However, this kind of information can be hardly employed in an egocentric vision environment in which wearable devices, being aimed at more general purpose users and being on a mid-low price tier, usually lack the ability to capture 3D feature points. Moreover, due to the unpredictable motion of both camera and object, a robust 3D model is often hard to recover from multiple images. Instead of using a 3D model, Huang et al. [12] developed a computational framework capable of performing detection, tracking and pose estimation of faces captured by video arrays. To estimate the face orientation, they compared through extensive experiments a Kalman filtering based tracker and multistate continuous density hidden Markov models. Orozco et al. [26] presented a head pose estimation technique based on mean appearance templates and multiclass SVM, and effectively applied to low-resolution video frames representing crowded public spaces under poor light conditions.

10.2.2 Social interactions

There is a growing interest in understanding social interactions and human behavior of individuals present in video frames using computer vision techniques. However, the majority of these methods are based on the video surveillance setting [24,33,35], which presents significant differences with respect to the first-person perspective. One of the first attempts of studying social interactions in the egocentric vision domain is that presented by Fathi et al. [11] who aim at recognizing five different social situations (monologue, dialog, discussion, walking dialog, walking discussion). By using day-long videos recorded from an egocentric perspective in an amusement park, they extract three categories of features: location of faces around the first person, patterns of attention and roles take by individuals and patterns of first-person head movement. These features are then used in a framework that explores the temporal dependency over time to detect the types of social interactions.

Yonetani et al. [34] presented a method to understand the dynamics of social interactions between two people by recognizing their actions and reactions using a head-mounted camera. In particular, to recognize micro-level actions and reactions, such as slight shifts in attention, subtle nodding, or small hand actions, they proposed to use paired egocentric videos recorded by two interacting people. In [1], instead, a new pipeline for automatic social

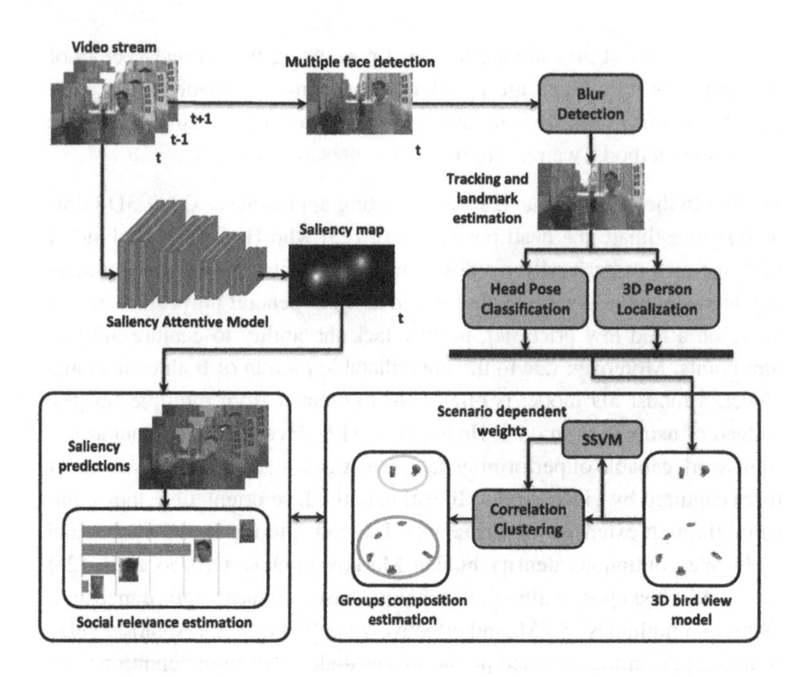

■ FIGURE 10.2 Schematization of the proposed approach.

pattern characterization of a wearable photo-camera user is presented. The proposed pipeline first of all studies a wider set of features for social interaction detection and second categorizes the detected social interactions into two broad categories of meetings (i.e. formal and informal). Even though all these methods provide interesting insights for understanding social interactions in the egocentric vision domain, none of them takes into account the group dynamics and the social relations within the group as presented in this work.

10.3 UNDERSTANDING PEOPLE INTERACTIONS

To deal with the complexity of understanding people interactions and detecting groups in real and unconstrained egocentric vision scenarios, our method relies on several components (see Fig. 10.2). We start with an initial face detection and then track the head to follow the subjects between frames. Head pose and 3D people locations are estimated to build a "bird view" model that is the input of the supervised correlation clustering in order to detect groups in different contexts based on the estimation of pairwise relations of their members. To further analyze the social dynamics, we estimate the social relevance of each subject by means of a saliency prediction model based on deep neural networks.

10.3.1 Face detection and tracking

A typical egocentric vision feature is a steep motion of the camera wearer. This can happen, for example, when a camera wearer is looking around for something. In this case, this person has not focused his attention on some point of interest and hence those frames are likely to be not interesting. Therefore, a first step to the face tracking procedure is to recognize whether tracking itself should be performed or not.

From a technical point of view, the steep motion can significantly increase the blur effect in the video sequence. If not addressed properly, this situation can degrade the tracking to the point that it may not be possible to resume it when the attention of the subject stabilizes again. To deal with this issue, at each frame we compute the amount of blurriness and decide whether to proceed with the tracking or to skip it. The idea of our approach is to evaluate the gradient intensity in the frame and to learn a threshold that can recognize a fast head movement from the normal blur caused by motion of objects, people or background. We define a simple blur function which recognizes the blur degree in a frame F, according to a threshold θ_B:

$$Blur(F, \theta_B) = \sum_F \sqrt{\nabla S_x^2(F) + \nabla S_y^2(F)}, \tag{10.1}$$

where $\nabla S_x^2(F)$ and $\nabla S_y^2(F)$ are the x and y components of Sobel's gradient in the frame and θ_B is the threshold under which the frame is discarded due to excessive motion blurriness, a parameter which can be learned by computing the average amount of gradient in a sequence.

This preprocessing step, which can be performed in real-time, effectively allows us to discard frames that could lead the tracker to adapt its model to a situation where gradient features cannot be reliably computed. To robustly track people in our scenario we adopt the tracker TLD [17] as it is able to deal with fast camera motion and occlusions which often occur between members of different groups.

10.3.2 Head pose estimation

To estimate an accurate head pose our approach is based on two different techniques: facial landmarks and shape-based head pose estimation.

With the facial landmarks approach, head pose can be accurately estimated if the face resolution is high enough and the yaw, pitch and roll angles of the head are not excessively steep. However, when these conditions are not satisfied and the first strategy fails, our method relies on shape-based head pose estimation and uses HOG features and a classification framework composed of SVM followed by HMM.

The facial landmark estimator is the first component of our solution: if this can be computed, the head pose can be reliably inferred and no further processing is needed.

To estimate facial landmarks, we employ the method proposed by Smith et al. [32]. We fix the number of landmarks at 49 as it is the minimum number of points for a semantic face description [28]. To obtain the head pose we perform a face alignment procedure by applying the supervised gradient descent method, which minimizes the following function over $\Delta\mathbf{x}$:

$$f(\mathbf{x}_0 + \Delta\mathbf{x}) = \|\mathbf{h}(\boldsymbol{\lambda}(\mathbf{x}_0 + \Delta\mathbf{x})) - \phi_*\|_2^2, \tag{10.2}$$

where $\mathbf{x}_0$ is the initial configuration of landmarks, $\boldsymbol{\lambda}(\mathbf{x})$ is the function that indexes the N landmarks in the image and $\mathbf{h}$ is a non-linear feature extraction function; in this case the SIFT operator. $\phi_* = \mathbf{h}(\boldsymbol{\lambda}(\mathbf{x}_*))$ represents the SIFT descriptors computed over manually annotated landmarks in the image. Finally, the obtained pose is quantized over five classes, representing the intervals $[-90, -60)$, $[-60, -30)$, $[-30, 30]$, $(30, 60]$ and $(60, 90]$.

If the landmark estimator fails, we combine it with a second component based on the shape of the subject's head. Before computing the head descriptor, which will be used in the pose classification step, a preprocess step is required in an unconstrained scenario as egocentric vision: background removal inside the bounding boxes of the tracked faces. We use an adaptation of the segmentation algorithm GrabCut [27], which minimizes the following energy function:

$$\hat{a} = \arg\min_{\alpha} U(\alpha, \mathbf{k}, \theta, \mathbf{z}) + V(\alpha, \mathbf{z}), \tag{10.3}$$

where $\mathbf{z}$ is the image; α is the segmentation mask with $\alpha_i \in \{\pm 1\}$. θ is the set of parameters of the K components of a GMM and $\mathbf{k}$, $k_n \in \{1, \ldots, K\}$ is the vector assigning each pixel to a unique GMM. The U term encodes the likelihood of each color that exploits the GMM models, and V is the term describing the coherence between neighborhood pixels (see [28] for more details). Intuitively, the key aspect of the GrabCut algorithm is its usage of GMMs to model pixels belonging to background or foreground in the term U. These models represent the distribution of color and are used to assign a label α to each pixel. Using the standard GrabCut, we manually initialize both foreground and background region T_F and T_B to build the respective GMMs.

Exploiting the high-frame rate of egocentric videos it is possible to assume that only slight changes in the foreground and background mixtures will occur between two subsequent frames. This allows us at time t to build a GMM_t based on GMM_{t-1} instead of reinitializing the models. This is equivalent to soft assigning pixels that would end up in the T_U region, which is sensitive to noise.

In our preliminary experimental evaluation we observe that this initialization is necessary, because a segmentation on a bounding box, resulting from the tracking phase, without any assumptions yields poor results. This is due to the fact that small portions of background pixels are often included in the tracked bounding box. When those elements do not appear outside the target region, $p \in T_U$, $p \notin T_B$ (where T_U is the region of pixels marked as unknown), they cannot be correctly assigned to the background by the algorithm and they produce a noisy segmentation.

Once a precise head segmentation is obtained, the resulting image is resized to a fixed size 100×100 (to ensure invariance to scale), converted to grayscale and equalized. On this image a dense HOG descriptor is extracted using 64 cells and 16 bins per cell. To obtain the final feature vector, a power normalization technique has been applied. Using these features, the head pose is then predicted using a multiclass linear SVM classifier following the same quantization used in the landmark based estimation.

In a social scenario where three or more subjects' activity revolves around a discussion or any kind of similar social interaction, orientation transitions are temporally smooth and abrupt changes are avoided as changes tend not to occur when one subject is talking.

To enforce temporal coherence that derives from a video sequence, a stateful hidden Markov model technique is employed. The HMM is a first order Markov chain built upon a set of time-varying unobserved variables/states $\mathbf{z}_t$ and a set of observations $\mathbf{o}_t$. In our case, we set the latent variables to coincide with the possible head poses, while the observed variables are the input images. In practice, we set in the state transition matrix $\mathbf{A}$ a high probability of remaining in the same state, a lower probability for a transition to adjacent states and a very low probability for a transition to the non-adjacent states. This leads our approach to have continuous transitions between adjacent poses, removing impulsive errors that are due to wrong segmentation. This translates into a smooth transition among possible poses, which is what conventionally happens during social interaction among people in egocentric vision settings.

Our method combines the likelihood $p(\mathbf{z}_t|\mathbf{o}_t)$ of a measure $\mathbf{o}_t$ to belong to a pose $\mathbf{z}_t$ provided by the SVM classifier with the previous state $\mathbf{z}_{t-1}$ and the transition matrix **A** derived from the HMM, obtaining the predicted pose likelihood, which is the final output.

To calibrate a confidence level to a probability in a SVM classifier, so that it can be used as a observation for our HMM, we trained a set of Venn Predictors (VPs) [19], on the SVM training set. We have the training set in the form $S = \{s_i\}_{i=1\dots n-1}$ where s_i is the input-class pair $(\mathbf{x}_i, y_i)$. Venn predictors aim to estimate the probability of a new element $\mathbf{x}_n$ belonging to each class $Y_j \in \{Y_1 \dots Y_c\}$. The prediction is performed by assigning each one of the possible classification Y_j to the element $\mathbf{x}_n$ and dividing all the samples $\{(\mathbf{x}_1, y_1) \dots (\mathbf{x}_n, Y_j)\}$ into a number of categories based on a taxonomy. A taxonomy is a sequence $Q_n, n = 1, \dots, N$ of finite partitions of the space $S^{(n)} \times S$, where $S^{(n)}$ is the set of multisets of S of length n. In the case of multiclass SVM the taxonomy is based on the largest SVM score; therefore each example is categorized using the SVM classification in one of the c classes.

After partitioning the element using the taxonomy, the empirical probability of each classification Y_k in the category τ_{new} that contains (x_n, Y_j) is

$$p^{Y_j}(Y_k) = \frac{|\{(\mathbf{x}^*, y^*) \in \tau_{\text{new}} : y^* = Y_k\}|}{|\tau_{\text{new}}|}. \tag{10.4}$$

This is the pdf for the class of $\mathbf{x}_n$ but after assigning all possible classifications to it we get

$$P_n = \{p^{Y_j} : Y_j \in \{Y_1, \dots, Y_c\}\}, \tag{10.5}$$

which is the well-calibrated set of multiprobability predictions of the VP used in the HMM to compute the final output.

10.3.3 3D people localization

To deal with any egocentric camera, we decided not to use any calibration technique in estimating the distance of a subject from the camera wearer. The challenges posed by this decision are somewhat mitigated by the fact that, aiming to detect groups in a scene, the reconstruction of the exact distance is not needed and small errors are lost in the quantization step. We have a depth measure which preserves the positional relations between individual suffices.

With that in mind, we assume that all the heads in the image lie in a plane, so the only two significant dimensions of our 3D reconstruction are (x, z), resulting in a "bird's eye view" model. To estimate the distance from the person wearing the camera, we first use the facial landmarks computed in

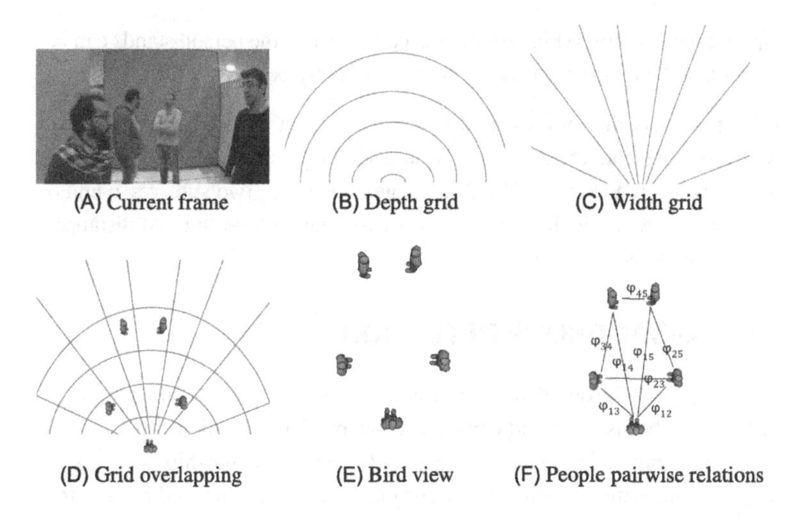

■ **FIGURE 10.3** Steps used in our distance estimation process.

the head pose estimation phase. Letting N be the number of landmarks, we build the feature vector

$$\mathbf{d} = \{d_i = \|l_i, l_{i+1}\|, i = 1, \ldots, N-1, l_i \in L\}, \quad (10.6)$$

where $\|\cdot\|$ is the standard euclidean distance. This feature vector is used in a random regression forest [3] trained using the ground truth depth data obtained from a Kinect sensor. To reduce the impact on the distance of a wrong set of landmarks, we apply over a 100 frame window a robust local regression smoothing (RLOESS) based on the LOWESS method [6].

This solution provides a good estimation of the distance between a subject and the camera wearer dealing with the topological deformations that are due to changes in pose and with the non-linearity of the problem.

In the case where the facial landmarks estimator fails, we compute the distance by using a random regression forest trained on the tracked bounding box used as feature. The estimation accuracy of this approach is less than the landmark solution, but it makes our approach more robust in unconstrained scenarios. To estimate the location of a person accounting for the projective deformation in the image, we build a grid with variable cells sizes. The distance allows us to locate the subject with one degree of freedom (x) (Fig. 10.3B): the semicircle in which the person stands is decided based on the distance computed previously, resulting in a quantization of the distance. Using the x position of the person in the image plane and employing a grid capable of accounting for the projective deformation (Fig. 10.3C), it is now possible to place the person with one further degree of freedom z. By overlapping the two grids (Fig. 10.3D) the cell in which the person stands can be decided and the bird's eye view model can finally be created (Fig. 10.3E).

Each person is then represented by its location in the 3D space (x, z, o), where o represents the estimated head orientation, and a graph connecting people is created (Fig. 10.3F). Each edge connecting two subjects p and q has a weight ϕ_{pq} which is the feature vector that includes mutual distances and orientations.

10.4 SOCIAL GROUP DETECTION

To deal with the group detection problem, head pose and 3D people information can be used, introducing the concept of the relationship between two individuals. Given two people **p** and **q**, their relationship ϕ_{pq} can be described in terms of their mutual distance, the rotation needed by the first to look at the second and vice versa $\phi_{pq} = (d, o_{pq}, o_{qp})$.

Notice that the distance d is by definition symmetric, while the orientations o_{pq} and o_{qp} are not. An example is given by the situation where two people have the same orientation resulting in **p** looking at **q**'s back; they will have $o_{pq} = 0$ and $o_{qp} = \pi$, so we need two separate features.

Practically it can be hard to fix this definition of relationship and use it in any scenario. In fact, in some contexts people in the group are looking at the same object/scene and none of them looks at any other. Therefore, the need of an algorithm is obvious that is able to differentiate social contexts and learn how to weight distance and orientation features.

10.4.1 Correlation clustering via structural SVM

To detect social groups based on the pairwise relations of their members we use the correlation clustering algorithm [2]. Let **x** be a set of people in the video sequence, their pairwise relations can be encoded by an affinity matrix W, where for $W_{\mathbf{pq}} > 0$ two people **p** and **q** are in the same group with certainty $|W_{\mathbf{pq}}|$ and for $W_{\mathbf{pq}} < 0$ p and q belong to different clusters. The correlation clustering **y** of a set of people **x** is then the partition that maximizes the sum of affinities for item pairs in the same cluster:

$$\arg\max_{\mathbf{y}} \sum_{y \in \mathbf{y}} \sum_{r \neq t \in y} W_{\mathbf{rt}}, \tag{10.7}$$

where the affinity between two people **p** and **q**, $W_{\mathbf{pq}}$, is represented as a linear combination of the pairwise features of orientation and distance over a temporal window. To obtain the best partition of social groups in different social contexts, our experiments showed that the weight vector $\mathbf{w}$ should not be fixed but, instead, learned directly from the data.

Given an input $\mathbf{x}_i$, a set of distance and orientation features of a set of people, and $\mathbf{y}_i$, their clustering solution, we can observe that a graph describing connections between members suits better the social dimension of the group interaction. This leads to an inherently structured output that is required to be treated accordingly. Structural SVM [29] offers a framework to learn structured outputs. This classifier, given a sample of input–output pairs $S = \{(\mathbf{x}_1, \mathbf{y}_1), \ldots, (\mathbf{x}_n, \mathbf{y}_n)\}$, learns the function mapping an input space $\mathcal{X}$ to the structured output space $\mathcal{Y}$.

A discriminant function $F : \mathcal{X} \times \mathcal{Y} \to \Re$ is defined over the joint input–output space. Hence, $F(\mathbf{x}, \mathbf{y})$ can be interpreted as measuring the compatibility of an input $\mathbf{x}$ and an output $\mathbf{y}$. As a consequence, the prediction function f results:

$$f(\mathbf{x}) = \arg\max_{\mathbf{y} \in \mathcal{Y}} F(\mathbf{x}, \mathbf{y}, \mathbf{w}) \tag{10.8}$$

where the solution of the inference problem is the maximizer over the label space $\mathcal{Y}$, which is the predicted label. Given the parametric definition of correlation clustering in Eq. (10.7), the compatibility of an input–output pair can be defined by

$$F(\mathbf{x}, \mathbf{y}, \mathbf{w}) = \mathbf{w}^T \Psi(\mathbf{x}, \mathbf{y}) = \mathbf{w}^T \sum_{y \in \mathbf{y}} \sum_{r \neq t \in y} \phi_{pq} \tag{10.9}$$

where ϕ_{pq} is the pairwise feature vector of elements p and q. This problem of learning in structured and interdependent output spaces can be formulated as a maximum-margin problem. We adopt the n-slack, margin-rescaling formulation of [29]:

$$\begin{aligned} \min_{\mathbf{w}, \xi} \quad & \frac{1}{2}\|w\|^2 + \frac{C}{n}\sum_{i=1}^{n} \xi_i \\ \text{s.t.} \quad & \forall i : \xi_i \geq 0, \\ & \forall i, \forall \mathbf{y} \in \mathcal{Y} \backslash \mathbf{y}_i : \mathbf{w}^T \delta \Psi_i(\mathbf{y}) \geq \Delta(\mathbf{y}, \mathbf{y}_i) - \xi_i . \end{aligned} \tag{10.10}$$

Here, $\delta\Psi_i(\mathbf{y}) = \Psi(\mathbf{x}_i, \mathbf{y}_i) - \Psi(\mathbf{x}_i, \mathbf{y})$, ξ_i are the slack variables introduced to accommodate for margin violations and $\Delta(\mathbf{y}, \mathbf{y}_i)$ is the loss function. In this case, the margin should be maximized in order to jointly guarantee that, for a given input, every possible output result is considered worst than the correct one by at least a margin of $\Delta(\mathbf{y}_i, \mathbf{y}) - \xi_i$, where $\Delta(\mathbf{y}_i, \mathbf{y})$ is bigger when the two predictions are known to be more different. We rely on the cutting plane algorithm in which we start with no constraints, and iteratively we find the most violated one and re-optimize it until convergence.

The problem of group detection is similar to the noun-coreference problem [5] in NLP, where nouns have to be clustered according to who they refer to. For this problem, recently a suitable scoring measure has been proposed: the MITRE loss function [30]. It, or formally $\Delta_M(\mathbf{y}, \bar{\mathbf{y}})$, is based on the understanding that, instead of representing each subject's links towards every other person, connected components are sufficient to describe dynamic groups and thus spanning trees can be used to represent clusters.

10.5 SOCIAL RELEVANCE ESTIMATION

Group detection estimates how people interact with each other by analyzing the geometry of their social formations. What cannot be unveiled by detecting social groups are all the properties of the social formation which depend on the particular observer, like the importance attached by an observer to each person in a group. We name this subjective property *social relevance*. By providing complementary information to what is provided by group estimation, we argue that social relevance enables a better understanding of the social dynamics from an egocentric prospective. Clearly, social relevance cannot be fully estimated from features like head pose and distance. To some extent, the camera wearer could give more importance to a distant person than to a closer one, or even to somebody who is turned away.

Sensors that can objectively measure the relevance of a person from the point of view of an observer, like eye-tracking glasses, are expensive and more uncomfortable to wear in public than a tiny camera. Therefore, to estimate social relevance relying only on the frames captured by a wearable camera, we choose to rely on saliency prediction [7,8,13,15]. Saliency prediction architectures predict the distribution of eye fixation points on a given image, and are trained on data captured from eye-tracking devices. By providing a distribution of eye fixations over an image, this lets us estimate the amount of fixations each person on the scene would receive from the wearer, and by extension, the social relevance of each person. More in detail, given a video, we first extract the saliency maps for all video frames. We then define the social relevance of each person as the accumulation of the saliency values inside the person's bounding box summed over time. In this way, for each subject appearing in the video frames, we can obtain a measure of his individual relevance to the social interaction.

To compute saliency maps, we employ the Saliency Attentive Model (SAM[1]) presented in [8], which has shown state-of-the-art results on popular saliency benchmarks, such as the MIT Saliency Benchmark [4] and the SALICON dataset [16]. This model is composed of three main components. First, a Convolutional Neural Network (CNN) extracts a set of feature maps from the original image. Because of the presence of spatial pooling operations, which compute the maximum activation over a sliding window, this would largely downscale the activations of the last layers with respect to the original image size. As is easy to see, a scaled output reduces the accuracy of predictions and their localization accuracy, which is fundamental in the context of predicting the saliency on faces that might occupy only a small portion of the frame. To control this phenomenon, we employ dilated convolutions. In short, dilation increases the spatial support of convolutions, by enlarging the kernel size, but keeps the number of parameters constant, setting the pixels of the kernel to zero at evenly spaced locations. The feature maps coming from the CNN are then fed through a recurrent layer which, thanks to the incorporation of attentive mechanisms, selectively attends to different regions of the input. In particular, we use a Long Short-Term Memory network (LSTM) with convolutional operations to sequentially refine and enhance the input feature maps. Predictions are finally combined with multiple prior maps directly learned by the network, thus effectively incorporating the center bias present in human eye fixations.

10.6 EXPERIMENTAL RESULTS

To provide an evaluation of the proposed social groups detection method and its main components, we rely on two publicly available datasets, namely EGO-HPE for evaluating the head pose estimation component and EGO-GROUP to assess the performance of group detection, distance estimation and social relevance.

The EGO-HPE dataset[2] is used to test the proposed head pose estimation method. It features more than 3400 frames where bounding boxes and head pose labels are provided for every person in the frame. Aiming at specifically evaluating the head pose estimation in egocentric vision, this dataset features the typical challenges of body-worn cameras, such as background clutter, different lighting conditions and motion blur.

---

[1] Source code available at http://github.com/marcellacornia/sam. The method has also won the LSUN Saliency Challenge in 2017.

[2] http://imagelab.ing.unimore.it/files/EGO-HPE.zip.

■ FIGURE 10.4 Example sequences from the EGO-GROUP dataset.

On the other hand, the EGO-GROUP dataset[3] features 18 video sequences in the five different scenarios represented in Fig. 10.4: a coffee break scenario with very poor lighting and random backgrounds (first row), a laboratory setting with limited background clutter and fixed lighting conditions (second row), an outdoor scenario (third row), a festive moment with a crowded environment (fourth row), and a conference room setting where people's movement and positioning is tied to seats (fifth row). A total of 23 different subjects appear in the videos.

10.6.1 Head pose estimation

Among the different features employed in the social group detection, head pose estimation is arguably one of the most important. In fact, errors in the head pose create a strong bias in the features used in the group estimation.

To provide the best results, our head pose estimation relies on two steps: landmark estimation and HOG-based pose classification. Both approaches have different characteristics: facial landmarks are very accurate and fast but their performance drops quickly when facing more extreme head poses, making them suitable for near frontal images but unreliable under steep pose angles. On the other hand, shape features such as histogram of gradients ex-

[3]http://imagelab.ing.unimore.it/files/EGO-GROUP.zip.

Table 10.1 Comparison between different techniques. In the different methods, PN indicates the usage of power normalization, HMM indicates the use of HMM-based temporal smoothing

Method	EGO-HPE1	EGO-HPE2	EGO-HPE3	EGO-HPE4
HOG+PN	0.710	0.645	0.384	0.753
HOG+PN+HMM	0.729	0.649	0.444	0.808
Landmarks	0.537	0.685	0.401	0.704
Landmarks+HOG	0.750	**0.731**	0.601	0.821
Landmarks+HOG+HMM	**0.784**	0.727	**0.635**	**0.821**

cel at discriminating steep head poses and can complete the estimation when landmarks cannot be reliably computed. Furthermore, HOG descriptors are much less sensitive to scale, which can be helpful when dealing with subjects in the background.

Table 10.1 compares the two approaches, clearly showing that the combination of landmarks and HOG features achieves the best results.

To show how egocentric vision's unique perspective can affect the results of an approach if not explicitly taken into account, we tested our egocentric head pose estimation method against other recent methods over the EGO-HPE dataset. The first method we compared to is proposed by Zhu et al. [36]: by building a mixture of trees with a shared pool of parts, where each part represents a facial landmark, they use a global mixture in order to capture topological changes in the face due to the viewpoint, effectively estimating the head pose. To achieve a fair comparison in terms of required time, we used their fastest pretrained model and reduced the number of levels per octave to one. This method, while being far from real-time, provides extremely precise head pose estimations even in egocentric vision scenarios when it can overcome detection difficulties. The second method used in our comparison is [10]. This method provides real-time head pose estimations by using facial landmark features and a regression forest trained with examples from five different head poses. Table 10.2 shows the results in terms of the accuracy of this comparison.

10.6.2 **Distance estimation**

To assess the quality of our distance estimation method, by keeping the regression architecture unvaried, we test two commonly employed techniques. The first relies on using the dimensions of the head bounding box as features, while the second one uses the area of the segmented face. Table 10.3 shows this comparison in terms of absolute error. The *Bounding Box* method employs the TLD tracker in order to estimate the subject's bounding box,

Table 10.2 Comparison of our head pose estimation and two recent methods on EGO-HPE dataset

	Our method	Zhu et al. [36]	Dantone et al. [10]
EGO-HPE1	**0.784**	0.685	0.418
EGO-HPE2	**0.727**	0.585	0.326
EGO-HPE3	**0.635**	0.315	0.330
EGO-HPE4	**0.821**	0.771	0.634

Table 10.3 Comparison between different distance estimation approaches

Method	Abs. error
Area (baseline)	12.67
Bounding Box	5.59
Landmarks	1.91
Landmarks + Moving Average	1.72
Landmarks + LOESS	1.68
Landmarks + RLOESS	**1.60**

while the *Area* method relies on the segmentation of the face surface. The results show that using biologically-inspired features such as the ratio between different facial landmarks can greatly improve the results when compared to other methods.

Aiming to improve our results, we apply to our distance sequence a smoothing filter. As Table 10.3 shows, using a moving average filter can improve the results by 9,95%, while LOESS and RLOESS smoothing methods yield, respectively, an error reduction of 12.04% and 16.23%. In both the LOESS and the RLOESS methods the span has been set to 10% of the data.

10.6.3 Groups estimation

A critical issue when training the social group detection method is how to deal with data from different scenarios. In fact, depending on the social situation, distance and pose features can have different significance, e.g. when there is not much space available, people cluster together regardless of their will to form a coherent group. For this reason, training should be *context dependent*. Table 10.4 reports the performance of our method on the EGO-GROUP dataset, where the training is repeated on the first video of each scenario. Furthermore, results obtained by training over the union of the individual training sets are also reported. To the best of our knowledge, no other methods deal with the estimation of social groups in egocentric scenarios.

Table 10.4 Comparison between training variations on our method. The table shows how different training choices can deeply impact the performances. All tests have been performed using a window size of eight frames

Test scenario	**Training: Laboratory**			**Training: Coffee**			**Training: Party**		
	Error	***Precision***	***Recall***	***Error***	***Precision***	***Recall***	***Error***	***Precision***	***Recall***
Coffee	10.74	83.04	97.29	9.23	82.67	100.00	18.04	68.76	100.00
Party	9.33	100.00	83.63	0.00	100.00	100.00	0.00	100.00	100.00
Laboratory	11.91	91.68	85.79	14.75	74.67	99.43	14.43	74.81	100.00
Outdoor	11.47	87.88	95.11	10.22	82.09	98.27	11.30	81.17	100.00
Conference	16.27	75.24	93.32	14.56	73.94	95.15	18.97	75.58	95.28
Test scenario	**Training: Outdoor**			**Training: Conference**			**Training: All**		
	Error	***Precision***	***Recall***	***Error***	***Precision***	***Recall***	***Error***	***Precision***	***Recall***
Coffee	6.80	92.54	94.92	13.88	79.99	88.41	8.11	85.50	99.60
Party	10.92	100.00	80.34	7.11	90.12	95.42	3.15	96.27	98.05
Laboratory	27.75	72.60	72.81	12.02	90.75	87.22	19.97	74.32	88.05
Outdoor	16.22	81.11	90.24	16.71	74.92	94.81	16.24	84.33	88.67
Conference	14.46	74.09	95.20	13.95	74.67	95.10	17.07	74.04	93.73

Table 10.5 Comparison between training the correlation clustering weights using SSVM and performing clustering without training (fixed weights). The window size used in the experiment is 8

Method		Coffee	Party	Laboratory	Outdoor	Conference
CC	Error	12.75	0.00	14.28	17.13	15.54
	Precision	74.86	100.00	73.12	71.81	74.43
	Recall	96.29	100.00	97.55	97.98	91.39
CC+SSVM	Error	9.23	0.00	11.91	16.22	13.95
	Precision	82.67	100.00	91.68	81.11	74.67
	Recall	100.00	100.00	85.79	90.24	95.10

From the data reported in Table 10.4, it can be noticed how training in specific scenarios can result in overfitting. For example, the weights learned by training on the *outdoor* sequences provide better results when testing on *coffee* than the ones trained on *coffee* itself. This is due to overfitting on a particular dynamic present in both scenarios but, unsurprisingly, it provides poor performances when testing on videos from other scenarios. In order to have an estimate of how different training methods perform, the standard deviation over the absolute error can be computed. It emerges that the *laboratory* setting is the more general training solution with an average error of 11.94 and a standard deviation of 2.61, while training over the *party* sequence, although it can achieve impeccable results over its own scenario and an average error of 12.55, presents a much higher deviation (7.65). Training over the set given by the union of each training set from the different scenarios results in a standard deviation of 7.01 over a mean error of 12.91, showing how this solution, while maintaining the overall error rates, does not provide a gain in generality. This confirms that different social situations call for different feature weights and that context-dependent training is needed to adapt to how humans change their behavior based on the situation.

To stress the importance of domain specific training of the feature weights, we report results of trainingless clustering. That is, all the feature weights are fixed at the same value, effectively assigning the same importance to distance and orientation features. Table 10.5 reports these results: it can be noticed how the algorithm is often biased towards placing all the subjects into one single group. This is showed by the high recall and the lower precision: the MITRE loss function penalizes precision for each person put in the wrong group, while the recall stays high. Placing every person in the same group hence results in an average error due to the fact that, not leaving any subject out of a group provides a high recall. In our experiments, we

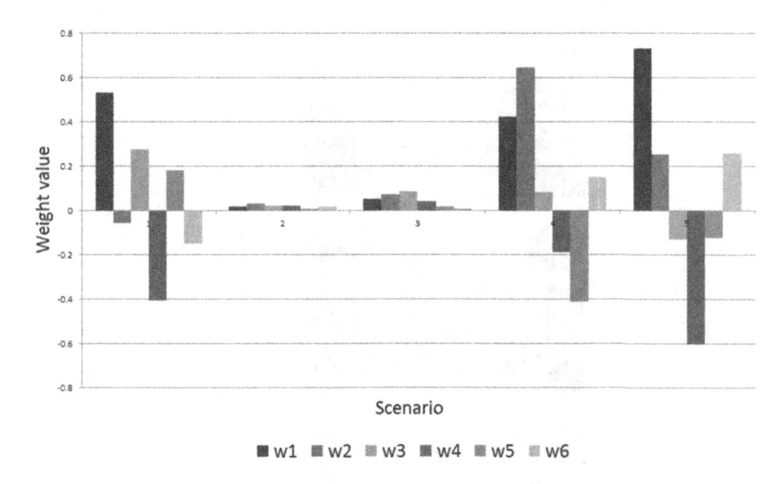

■ **FIGURE 10.5** Weight values in the five different training scenarios. The scenarios are: *1) laboratory, 2) party, 3) conference, 4) coffee, 5) outdoor.*

set the clustering window size to eight frames, a value that our preliminary experiments showed to achieve a good compromise between robustness to noise and fine grained responsiveness to group changes.

To further evaluate our approach, we discuss how the clustering weights vary in different scenarios. Fig. 10.5 shows the comparison between the different components of the weight vectors. As can be noticed, performing the training over different scenarios yields significantly different results. For example, clustering a sequence in the fourth scenario gives more importance to the second feature (the orientation of subject 1 towards subject 2), slightly less importance to the spatial distance between the two and very little importance to the orientation of 2 towards 1. In scenario 5, the outdoor sequence, the most important feature is recognized to be the distance, correctly reflecting the human behavior where, being outdoor, different groups tend to increase the distance between each other thanks to the high availability of space (Fig. 10.6).

A negative weight models the fact that, during the training, our approach has learned that the feature that weight relates to can decrease the affinity of a pair. A typical example of such situation is when a person is giving us the back: while our orientation can have a high similarity value towards that person, that feature will probably lead the system to wrongly put us in the same group. Our approach learns that there are situations where some features can produce wrong clustering results and assign a negative weight to them.

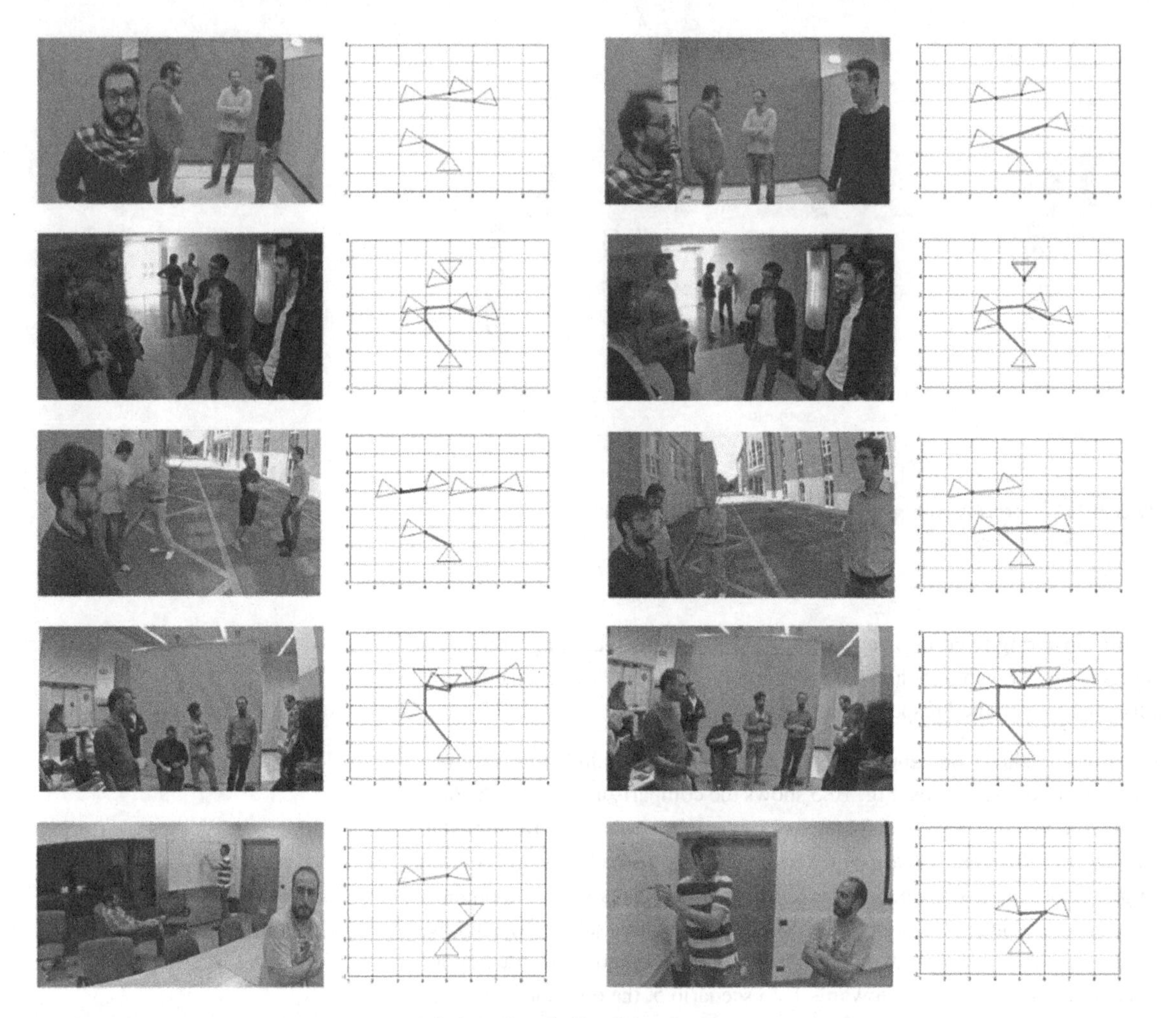

■ **FIGURE 10.6** Examples of the results obtained by our method. Different groups are shown by different link colors.

10.6.4 Social relevance

While the group estimation evaluated in the previous section describes how people interact with each other, we also evaluate the individual relevance of the subjects that partake to the social interaction. In particular, we argue that while head pose and distance information are instrumental in estimating social groups, visual saliency can be used to understand who are the most relevant subjects according to what is recorded by the egocentric camera.

Here, we rely on the method described in Section 10.5 to compute saliency maps of the EGO-GROUP dataset sequences. The overall saliency value of

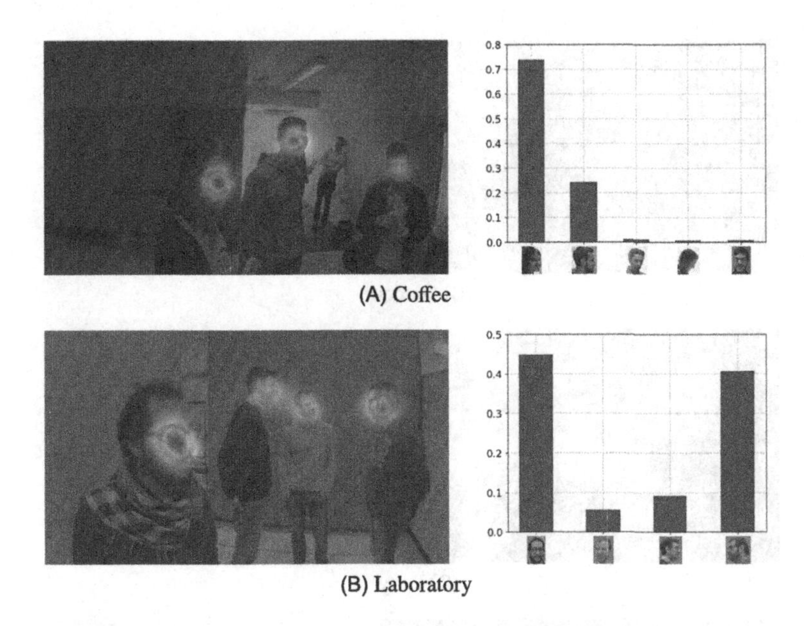

FIGURE 10.7 Sample results on the social relevance of each subject. Left: saliency map overlay. Right: social relevance of the participants to each considered scenario.

a person is then obtained by accumulating the raw saliency contained in the person's bounding box and summing it over the temporal dimension. The insight behind this is that, while group estimation captures the overall group dynamics, not all the members of a group may have the same importance for the person wearing the camera and analyzing the social relevance through the saliency estimation can provide information complementary to what is provided by the group estimation.

Fig. 10.7 provides qualitative results of this analysis on five different scenarios. On the left, a sample frame with the saliency map overlay is provided, while on the right side a plot of the individual saliency scores of the participants is provided. It may be noticed how the information provided by evaluating the saliency of individual participants is complementary to the group data. In fact, taking Fig. 10.7A as an example, the saliency estimation provides remarkable cues on who are the most interesting members of the foreground group, while it agrees with the group estimation in assigning low relevance to the people forming the background group. Similarly, in the *party* scenario (Fig. 10.7D) there is only one big group, and evaluating the individual saliency values can provide a better insight on the intra-group dynamics.

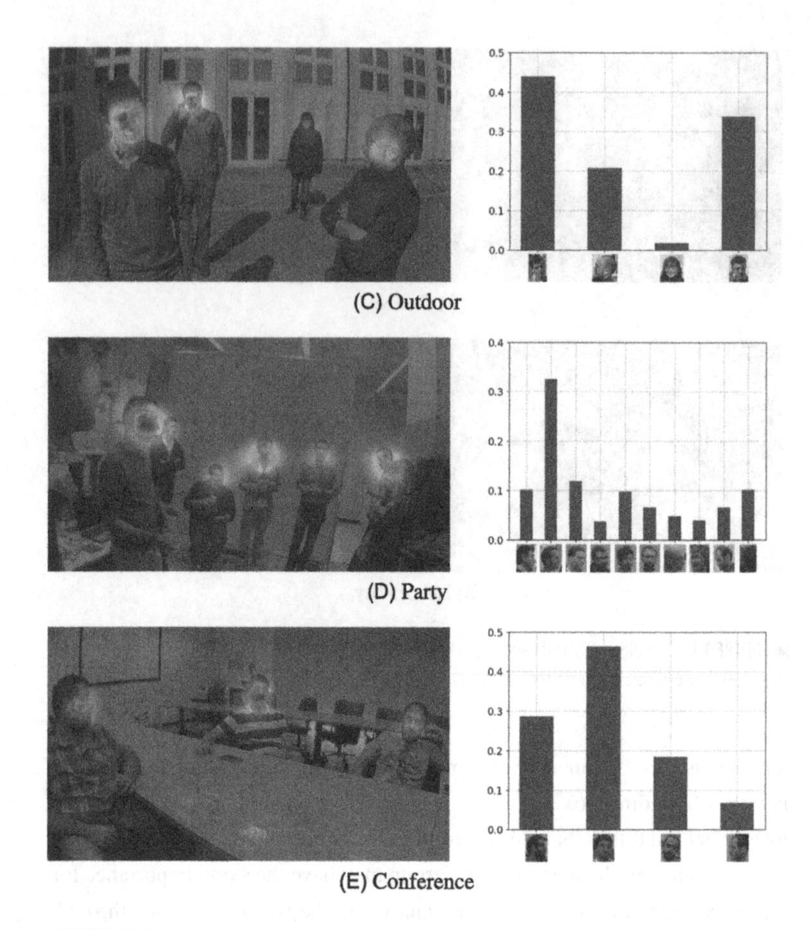

■ FIGURE 10.7 *(continued)*

10.7 CONCLUSIONS

In this chapter we presented a novel approach to detecting social groups using a head-mounted camera. The proposal relies on a head pose classification technique combining landmarks and shape descriptors in a temporally smoothed HMM framework. Furthermore, 3D location estimation of the people without the need of camera calibration is also presented. Using this information, the approach is able to build a "bird's eye view" model that is the input of the supervised correlation clustering in order to detect the group in different contexts. An extensive experimental evaluation shows competitive performance on two publicly available egocentric vision datasets, recorded in real and challenging scenarios.

REFERENCES

[1] Maedeh Aghaei, Mariella Dimiccoli, Cristian Canton Ferrer, Petia Radeva, Towards social pattern characterization in egocentric photo-streams, arXiv preprint, arXiv:1709.01424, 2017.

[2] Nikhil Bansal, Avrim Blum, Shuchi Chawla, Correlation clustering, Mach. Learn. 56 (2004) 89–113.

[3] Leo Breiman, Random forests, Mach. Learn. 45 (1) (2001) 5–32.

[4] Zoya Bylinskii, Tilke Judd, Ali Borji, Laurent Itti, Frédo Durand, Aude Oliva, Antonio Torralba, MIT saliency benchmark, http://saliency.mit.edu/, 2017.

[5] Claire Cardie, Kiri Wagstaff, Noun phrase coreference as clustering, in: Joint SIGDAT Conference on Empirical Methods in Natural Language Processing and Very Large Corpora, 1999.

[6] William S. Cleveland, Robust locally weighted regression and smoothing scatterplots, J. Am. Stat. Assoc. 74 (368) (1979) 829–836.

[7] Marcella Cornia, Lorenzo Baraldi, Giuseppe Serra, Rita Cucchiara, A deep multi-level network for saliency prediction, in: International Conference on Pattern Recognition, 2016.

[8] Marcella Cornia, Lorenzo Baraldi, Giuseppe Serra, Rita Cucchiara, Predicting human eye fixations via an LSTM-based saliency attentive model, IEEE Trans. Image Process. 27 (10) (2018) 5142–5154.

[9] Marco Cristani, Loris Bazzani, Giulia Paggetti, Andrea Fossati, Diego Tosato, Alessio Del Bue, Gloria Menegaz, Vittorio Murino, Social interaction discovery by statistical analysis of F-formations, in: British Machine Vision Conference, 2011.

[10] Matthias Dantone, Juergen Gall, Gabriele Fanelli, Luc Van Gool, Real-time facial feature detection using conditional regression forests, in: IEEE International Conference on Computer Vision and Pattern Recognition, 2012.

[11] Alircza Fathi, Jessica K. Hodgins, James M. Rehg, Social interactions: a first-person perspective, in: IEEE International Conference on Computer Vision and Pattern Recognition, 2012.

[12] Kohsia S. Huang, Mohan M. Trivedi, Robust real-time detection, tracking and pose estimation of faces in video streams, in: International Conference on Pattern Recognition, 2004.

[13] Xun Huang, Chengyao Shen, Xavier Boix, Qi Zhao, SALICON: reducing the semantic gap in saliency prediction by adapting deep neural networks, in: IEEE International Conference on Computer Vision, 2015.

[14] Hayley Hung, Ben Kröse, Detecting F-formations as dominant sets, in: ACM International Conference on Multimodal Interaction, 2011.

[15] Laurent Itti, Christof Koch, Ernst Niebur, et al., A model of saliency-based visual attention for rapid scene analysis, IEEE Trans. Pattern Anal. Mach. Intell. 20 (11) (1998) 1254–1259.

[16] Ming Jiang, Shengsheng Huang, Juanyong Duan, Qi Zhao, SALICON: saliency in context, in: IEEE International Conference on Computer Vision and Pattern Recognition, 2015.

[17] Zdenek Kalal, Krystian Mikolajczyk, Jiri Matas, Tracking-learning-detection, IEEE Trans. Pattern Anal. Mach. Intell. 34 (7) (2012) 1409–1422.

[18] Adam Kendon, Studies in the Behavior of Social Interaction, vol. 6, Humanities Press Intl, 1977.

[19] Antonis Lambrou, Harris Papadopoulos, Ilia Nouretdinov, Alexander Gammerman, Reliable probability estimates based on support vector machines for large multi-class datasets, in: Artificial Intelligence Applications and Innovations, vol. 382, 2012, pp. 182–191.

[20] Deqiang Li, Witold Pedrycz, A central profile-based 3d face pose estimation, Pattern Recognit. 47 (2) (2014) 525–534.
[21] Bingpeng Ma, Wenchao Zhang, Shiguang Shan, Xilin Chen, Wen Gao, Robust head pose estimation using LGBP, in: International Conference on Pattern Recognition, 2006.
[22] Erik Murphy-Chutorian, Mohan Manubhai Trivedi, Head pose estimation in computer vision: a survey, IEEE Trans. Pattern Anal. Mach. Intell. 31 (4) (2008) 607–626.
[23] Rhys Newman, Yoshio Matsumoto, Sebastien Rougeaux, Alexander Zelinsky, Real-time stereo tracking for head pose and gaze estimation, in: International Conference on Automatic Face and Gesture Recognition, 2000.
[24] Nicoletta Noceti, Francesca Odone, Humans in groups: the importance of contextual information for understanding collective activities, Pattern Recognit. 47 (11) (2014) 3535–3551.
[25] Shay Ohayon, Ehud Rivlin, Robust 3D head tracking using camera pose estimation, in: International Conference on Pattern Recognition, 2006.
[26] Javier Orozco, Shaogang Gong, Tao Xiang, Head pose classification in crowded scenes, in: British Machine Vision Conference, 2009.
[27] Carsten Rother, Vladimir Kolmogorov, Andrew Blake, "GrabCut": interactive foreground extraction using iterated graph cuts, ACM Trans. Graph. 23 (3) (2004) 309–314.
[28] Brandon M. Smith, Jonathan Brandt, Zhe Lin, Li Zhang, Nonparametric context modeling of local appearance for pose- and expression-robust facial landmark localization, in: IEEE International Conference on Computer Vision and Pattern Recognition, 2014.
[29] Ioannis Tsochantaridis, Thomas Hofmann, Thorsten Joachims, Yasemin Altun, Support vector machine learning for interdependent and structured output spaces, in: International Conference on Machine Learning, 2004.
[30] Marc Vilain, John Burger, John Aberdeen, Dennis Connolly, Lynette Hirschman, A model-theoretic coreference scoring scheme, in: Message Understanding Conference, 1995.
[31] Junwen Wu, Jens M. Pedersen, D. Putthividhya, Daniel Norgaard, Mohan M. Trivedi, A two-level pose estimation framework using majority voting of gabor wavelets and bunch graph analysis, in: International Conference on Pattern Recognition Workshops, 2004.
[32] Xuehan Xiong, Fernando De la Torre, Supervised descent method and its applications to face alignment, in: IEEE International Conference on Computer Vision and Pattern Recognition, 2013.
[33] Xu Yan, Ioannis A. Kakadiaris, Shishir K. Shah, Modeling local behavior for predicting social interactions towards human tracking, Pattern Recognit. 47 (4) (2014) 1626–1641.
[34] Ryo Yonetani, Kris M. Kitani, Yoichi Sato, Recognizing micro-actions and reactions from paired egocentric videos, in: IEEE International Conference on Computer Vision and Pattern Recognition, 2016.
[35] Ting Yu, Ser-Nam Lim, Kedar Patwardhan, Nils Krahnstoever, Monitoring, recognizing and discovering social networks, in: IEEE International Conference on Computer Vision and Pattern Recognition, 2009.
[36] Xiangxin Zhu, Deva Ramanan, Face detection, pose estimation and landmark localization in the wild, in: IEEE International Conference on Computer Vision and Pattern Recognition, 2012.

Chapter 11

Complex conversational scene analysis using wearable sensors

Hayley Hung*,†, Ekin Gedik*, Laura Cabrera Quiros*,‡

*Delft University of Technology, Intelligent Systems, the Netherlands †CWI, Distributed and Interactive Systems, the Netherlands ‡Instituto Tecnológico de Costa Rica, Electronic Engineering Department, Costa Rica

CONTENTS

11.1 INTRODUCTION

In recent years, researchers have shown that it is possible to automatically detect complex social phenomena, often using predominantly or only nonverbal behavior; from dominance [15,20,21,26,39,43], functional roles [12,28,54], deception [7,40], cohesion [24], attraction [29,30,41,49], interest [16,36,48,52], influence [38,44], rapport [19,46], to friendship [53].

Most research on social behavior analysis has focused on the analysis of conversational scenes in predominantly pre-arranged settings of typically no more than 6 people. Little is known about the function and nature of

Multimodal Behavior Analysis in the Wild. https://doi.org/10.1016/B978-0-12-814601-9.00019-5
Copyright © 2019 Elsevier Ltd. All rights reserved.

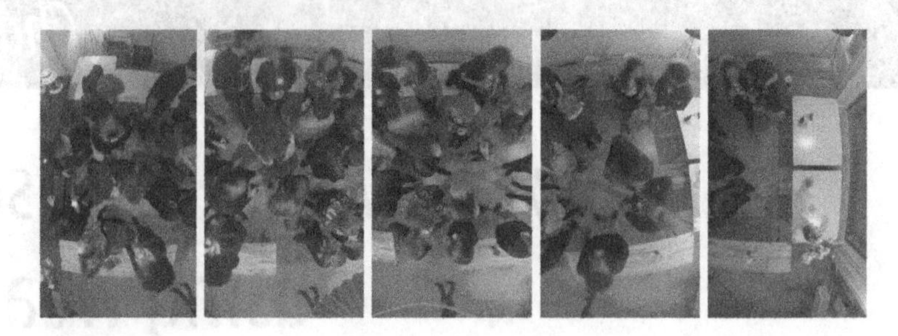

■ **FIGURE 11.1** Example snapshots of a mingling event. Taken from [27].

social interactions during complex conversational scenes that is, mingling behavior that occurs during social networking events (e.g. see Fig. 11.1). However attending such face-to-face mingling events has been linked to career and personal success [51]. The long term goal of such an analysis of complex conversational scenes is threefold:

- to be able to generate a social network describing the relationships between attendees of an event and how cohesive the group is;
- to quantify the interaction by considering how the social signals expressed and coordinated during the conversations can be indicative of the experience of a person and therefore how much potential influence there is within the network. This goes beyond more commonly used proxies of social relationship such as the frequency of interaction, where interactions are considered binary values [10,11,13,14];
- or ultimately to predict future behavior such as the chance that someone starts a personal or business relationship with people they first encountered at a particular social event.

One of the reasons that it is challenging to observe these types of events is that they require the coordination of scores of people together to capture such phenomena. Observing and analyzing such behavior is both hard for social scientists and computer scientists because the setting is so complex; What aspects of the behavior indicate how involved a person is in the conversation or that they are interested or engaged in a conversation? What are the patterns of behavior that determine when or how groups split or merge?

Usually when analyzing smaller groups such as for meeting analysis, findings from social science provide a framework for further analysis from computer science. For instance, we can use measures proposed in social science to build automated methods to perceive social phenomena (e.g. estimating dominance using the visual dominance ratio [21]). However, in the case of complex conversational scenes, little is known about how people behave because it has been too difficult to observe at a scale appropriate for social scientists, the technical challenges for automatically analyzing behavior in these scenes are also difficult to handle. For instance, since the scenes are so crowded, relying on video alone is unwise due to the high level of occlusion. Amalgamating this with wearable sensors can be one way to mitigate this problem, such as for improving head and body pose estimates using audio and infrared sensing from a body worn sensing device [1,2].

Concretely, complex conversational scene understanding has to overcome the following challenges.

- Scene noise: Crowding causes frequent visual occlusions, audio is contaminated by surrounding chatter, the sheer density of human bodies reflect and distort proximity and localization signals (e.g. from radio or wifi).
- Uncontrolled scenario: People move dynamically from group to group based on their own individual goals. Conversational groups or the length of an interaction are not prearranged. Multiple conversations occur simultaneously and conversation partners can change dynamically.
- Multi-sensor fusion: exploiting wearable sensors allows us to mitigate the data association problem by linking all digital sensor information to the wearer. However, we multiply the sensor problem from one of capturing the entire social scene in a single camera to requiring one device per person in the scene.

In this chapter, we propose a break from conventional approaches to the first step in doing such a behavior analysis by detecting social actions (e.g. speaking). Detecting individual speaking status is important for generating derived turn-taking features that are foundational for further analysis of more complex social constructs such as dominance [21,25,26], cohesion [24,35], and attraction [49].

11.2 DEFINING 'IN THE WILD' AND ECOLOGICAL VALIDITY

Let us first discuss some definitions. The term 'in the wild' has been much used in recent years to refer to research analyzing human behavior in uncontrolled settings. It has been used frequently in relation to a video-based analysis of facial expressions recorded outside of laboratory conditions. The expectation is that the facial behavior will be unposed and therefore more spontaneous and true to real life. One might question where the 'wild' aspect of this example sits. Conceptually, we could say that it lives within the conditions in which the data is collected outside of the lab, so presumably in uncontrolled/uneven lighting conditions, varying pixel resolution and frame rates. Aside from the recording conditions, there needs to be truthfulness to the behavior of the person being recorded; that is, the behavior should not be posed or fake.

When we consider human social behavior, we cannot avoid being influenced by social science. And when experimental psychologists conduct experiments, one of the issues they consider is ecological validity. The reason for introducing such a term in this chapter is perhaps most evident in its definition. Ecological validity describes the extent to which an experimental setting and task is true to real life. That is, it tells us the extent to which the location and the task fit with our expectations of what occurs in the expected ecology of someone's everyday life.

We can further understand the idea of ecological validity by considering a typical example. Suppose that an experiment is conducted to study whether people are able to follow emergency instructions provided in a leaflet if a building was burning down. One could approach this task by asking people to fill out a survey where they are shown the instruction leaflet and then asked to imagine what they would do in such a situation. This setting has low ecological validity as the respondents have to imagine what they might do and the task of reading the instruction leaflet is not done in the setting it was intended to be.

The experimenters could improve the ecological validity by using a state-of-the-art virtual reality system to simulate the burning building and then observe how participants react. In this case, the ecological validity may be higher but one might still doubt how realistic the experiment really is. In the optimal case, one might consider actually measuring the behavior of people as a real building is burning down. However, it is unlikely that such an experiment would get ethical approval and this would be so realistic that perhaps the sensing required to actually measure the genuine responses would either be unavailable or too noisy to be useful.

11.3 ECOLOGICAL VALIDITY VS. EXPERIMENTAL CONTROL

In both instantiations of the experiment, we see that there is an inherent trade-off between ecological validity and experimental control; while we want to allow the participants to carry out the task in as realistic a scenario as possible, we need to be able to measure the resulting behavior as accurately as possible too.

This is where social science and computer science diverge. For social science, at least traditionally speaking, to have good experimental control means that it is necessary to have accurate measures of behavior and also good quality survey responses. This is easier to achieve with laboratory based experiments. In computer science, the whole premise behind 'in the wild' perception is that this trade-off between experimental control and ecological validity can be tuned more precisely to the experimenter's requirements. That is, traditionally speaking, in multimedia tasks we no longer expect uniform lighting, no background noise, and the sample rate or frame rate, bitrate or pixel resolution may all be unknown or may vary. Thus the methods to interpret the data need to be robust to this situation, opening up interesting research challenges for computer science.

11.4 ECOLOGICAL VALIDITY VS. ROBUST AUTOMATED PERCEPTION

The notion of 'in the wild' can be taken even one step further using the scenario of mingling that we examine in this chapter by considering whether unconventional sensing modalities can act as a proxy for more traditional sensing modalities. Here, we address this problem for social behavior analysis from wearables. Traditionally, if we want to observe social behavior, extracting turn-taking features has shown to have great discriminative power for a number of tasks related to the analysis of small group behavior [17]. Typically, we would expect to measure this from audio of the speakers recorded with microphones. However, recording audio 'in the wild' can have considerable consequences both from a privacy and from an automated analysis perspective.

Privacy concerns relate to recording unwilling participants accidentally as one person's microphone can easily pick up sounds from the surroundings and other people. Moreover, people may not be willing to have their voices recorded at all, leading to a further sample bias when identifying volunteers (one might expect more sample bias the more experimental control is required). As the scene becomes denser, the background noise can become so great that it hinders robust audio processing of the speech signal. This is where wearables in the wild can address these problems by using accelerometer signals as a proxy for speech and also social behavior. The reason why this unconventional method is still deemed a feasible approach is that we know from social science that when people converse, they gesture and move their bodies [34]. By leveraging these body movements, we hypothesize that we can estimate when someone is performing a social action.

This pushes the boundaries of 'in the wild' processing while trading off ecological validity for the following reasons. First, the wearable sensor we propose to use is an ID badge that is hung around the neck; similar to what one might wear during a conference or festival. Second, we do not record audio which could make the wearer self-conscious of what they say when interacting with others. In this respect, the sensing is ecologically valid. It is 'in the wild' because the setting does not control for exactly what actions people need to perform when they are socializing. Their social behavior is genuine within the context of the situation.

Finally, a key question one might ask is whether the trade-off in spending so much effort on ensuring an ecologically valid solution is worth it. We can point to a prior study by Ekin and Hung [18] that demonstrated this point clearly. For the task of detecting speaking from body movements, data collected in the lab with acted social behavior yielded easily discriminable features. However, evaluating the same method on data recorded in a more ecologically valid setting led to significantly worse results. Our conclusion here is that laboratory data can lead to an underestimation of the difficulty of a task when transferred to real-life settings.

11.5 THIN VS. THICK SLICES OF ANALYSIS

Much work on wearables in the wild has been conducted using smart phones or wearable sensors for analyzing social phenomena on a large scale [37]. Analyzing on a large scale and accumulating observations over weeks or months has the benefit that sensor data such as the frequency of proximity as estimated from blue tooth readings can be used directly as a proxy for the quality of a social relationship. In this chapter however, we aim not to take the sensor data at face value but to investigate how signal processing and machine learning techniques can be used to squeeze out more meaning from noisy sensor data at shorter times scales of minutes or even seconds. The reason why this could even be considered possible is based on the thin slice theory proposed by psychologists Ambady and Rosenthal [3] who discovered that short observations (of typically just a few seconds) of social behavior were often enough to reliably assess some social situations.

11.6 COLLECTING DATA OF SOCIAL BEHAVIOR

In pushing to more 'in the wild' sensing, one needs to consider what is appropriate as data to investigate this phenomenon. One can imagine that the act of collecting data exists on a continuum in terms of the research question being addressed. For instance, one might have a specific research question that needs to be answered and so therefore the research is more inductive—the data acts to validate some hypotheses. In other cases, a more deductive approach may be used where the data collection acts as a vehicle to investigate currently unknown patterns of behavior. It is vitally important to consider this when using data to analyze social behavior 'in the wild'.

An individual data sample may exist in an 'in the wild' setting. However, when multiple data samples are aggregated to make an entire corpus, could there be selection bias at this stage? In this chapter, based on the research goals listed in Section 11.1, we focus on data collection using a deductive approach, considering how this further impacts the machine perception process at the end. Note in this case that due to the realistic nature of the data, class labels can have high levels of imbalance.

11.6.1 Practical concerns when collecting data during social events

In this section, we describe an approach to collect data in what may be considered an extremely uncontrolled and ecologically valid setting where many research challenges lie. Here we focus on mingle scenarios or free-standing conversational gatherings. That is, we address crowded social settings where people come together purely to socialize.

Very few data sets exist to investigate the machine perception of nonverbal social behavior in mingle scenarios with wearables [1,27], although little work analyzing social behavior with wearables does exist [5,18,32, 33]. There have been made more efforts in the computer vision community [8,9,22,55], which provide many insights for addressing this problem from a multimodal perspective. Usually researchers who focus on multimedia analysis problems are not likely to have practical experience of wearable sensor system deployment. Therefore we provide a primer on some key issues to consider when collecting data in such settings. Here we focus on lessons learned from capturing our own data set [27] with respect to the use of wearable sensors that capture acceleration and proximity, as well as of cameras that are able to validate the behaviors captured. In moving outside of the lab while still wanting to maintain some experimental control, we encounter important logistic issues that need to be taken into account. While many of these adhere to common sense practices, with so many elements to keep in balance, it can be easy to overlook some aspects. This can have severe negative consequences for the data collection such as lost or unusable data.

Requirements of the hardware and software

The selection of adequate sensor or wearable devices is an important aspect of the analysis of social interactions 'in the wild'. And this is strictly related to the event and behavior that is going to be analyzed. Aside from this, the method of analysis will greatly affect the requirements. For example, will the data be analyzed offline or is wireless communication necessary (e.g. for live or realtime processing)?

Moreover, even when each module is independently capable of fulfilling the requirements of the event, all devices (as individual units) have certain restrictions given their hardware (memory, CPU, size of registers) and software (execution time, interruptions and deadlocks, delays). One could, for example, try to use the maximum sample rate of the accelerometer (sensor) but one wonders: can all this information be stored locally in the device or sent wirelessly to a storing unit without critical package loss? And is the software/firmware used capable of handling this rate?

Thus, a balance must be sought between factors such as sensing configurations (e.g. sample rates, sensitivity, mode of sensing, sleep states), storage space, and power consumption. To do so, good practices from the embedded systems community such as efficient programming (e.g. use of idle states and proper scheduling) are advised, so that factors such as power consumption of the final badges are optimal [31].

Ease of use

For the participants and experimenters, we prefer to have a wearable that is 'grab and go'. That is, no additional registering is required once consent forms have been signed and the sensing device has been fitted. This allows for easier scaling of a data collection event to more and more participants. This has the drawback that associating the correct video data with the sensor requires manual work. However, alternative approaches with automated methods of associating the data together are showing promise [6,47,50].

For the researcher, having a system or firmware that is easily adapted to slight variations in research questions may also be important e.g. adjusting the bit rate, sample rate, wireless sensitivity. This is particularly important if one wants to examine hardware trade-offs e.g. high sample rate vs. long battery life. The sociometer is an example of a custom-made wearable device that could be useful for mingle events. It has several well-selected sensors that comply with the nature of mingle scenarios. However, as a commercial product, the selection and visualization of the raw sensor data are restricted. Reconfiguration of the device in terms of the trade-offs listed above is not possible. Sadly, there are few commercial products that give full access to the device's capabilities for a researcher's needs. For such cases, perhaps an open-source solution such as a platform-based device (e.g. raspberry pi or arduino) can be adapted for use. In our case, custom hardware was used.

Technical pilot test

Finally, it is vitally important that a technical pilot is carried out, preferably in the venue itself; it can often be the case that even commercial sensors have significant problems in delivering the functionality that they are designed for. This enables the entire sensor set-up to be tested before participants are involved. Note, however, that in our experience, carrying out a pilot test without participants has one drawback as the participants themselves can cause disruption to sensor signals. For instance, we found in prior data collections that a high density of people led to system performance degradation for an indoor localization system where all people in the same conversing group were detected to be located at exactly the same location in the ground plane, despite having a much better localization resolution when fewer people were present.

It is generally safe to assume that something will break and that all sensors will need to be continuously monitored to ensure that they are recording. This is also a key period to verify that the data across all modalities is correctly synchronized. Even if the primary analysis is carried out using the wearable devices, correctly synchronized video as well as correct data association (knowing where person A wearing sensor X is located in a video) is vital for data labeling.

Issues during the data collection event

As with any experimental setting, informed consent needs to be sought and ethical approval from an institutional committee needs to be obtained. Since we are dealing with a short term event where multiple people need to attend, overbooking participants is recommended. A financial incentive can be given to those who are turned away to minimize disappointment.

Other practical conditions include pre-defining a clear procedure for the event and if this is particularly intense, rest breaks may need to be planned for participants if there is some level of experimental control necessary. If participants are free to come and go as they please, a mechanism needs to be in place to ensure that they do not leave with the sensor (unless this was planned originally), and it is possible to find out who left (in terms of wearable sensor ID) and when.

The sensors themselves must be linked to an individual if survey responses are required from the participants. To ensure anonymization of the person's identify from their data, usually a participant is given a number. Their associated sensor then needs to be logged. If sensors and participants are already assigned before the start of the experiment, this can be problematic if there is sensor failure and a replacement needs to be brought in and a new sensor number logged.

For the data collection itself, having at least one person responsible per sensor type or modality ensures that multiple simultaneous failures can be handled relatively quickly.

11.7 ANALYZING SOCIAL ACTIONS WITH A SINGLE BODY WORN ACCELEROMETER

In this section, we will present a case study of social behavior analysis, focusing on automatic social action detection in complex conversational scenes. Various (social) actions will be discussed with respect to their physical manifestation and their connection to the worn sensing device, the required approaches, and available data size. In our case the sensing device we focus on is a single tri-axial accelerometer which is embedded in an ID badge hung around the neck.

When analyzing human behavior, past analysis has tended to assume that this is more or less person-independent. Throughout the text, 'person-independent' will be used for settings where data for training a model comes from different sources (people in our case) compared to the test data. Much work has been done on estimating daily activities such as walking or running from accelerometer data, showing promising results with a person-independent setup [4,42]. There is a direct connection between the sensing medium and the physical manifestation of the behavior so that actions such as walking and stepping result in acceleration readings that are easy to discriminate directly from the magnitude of the signal. This makes a person-independent setup for discriminating such behavior quite easy to implement.

However, some of the actions observed in crowded social settings tend to be much more person specific and the connection between the existence of these actions and the accelerometer readings is more ambiguous. In our case, the physical manifestation of speaking comes from vibration of the vocal chords, so unless the subject has a very sensitive accelerometer attached tightly to the body (e.g. the chest [32,33]), there will not be a direct connection between the action and the sensing. However, speaking also has a physical gestural aspect, and it has been shown in previous work that the connection between body movements and speech can still be exploited for detecting if someone is speaking or not [18,23]. Actions like speaking, which are loosely connected with the sensing medium, are expected to be harder to detect and may require specialized approaches.

To examine this, we conducted a number of experiments on a dataset that is collected from a real-life 'in the wild' event. The dataset is comprised of mingling events from three separate evenings where each evening includes data from approximately 32 people. Each participant wore a sensor hung around the neck that records individual tri-axial acceleration at 20 Hz. Note that the sample rate is not high enough to detect vocal chord vibration. However, it is high enough to capture body movements such as

Table 11.1 AUC scores for various actions

	AUC (%)	Std (±)	Annotator agreement
Stepping	76.0	10.5	0.51
Speaking	69.5	8.3	0.55
Hand gestures	70.4	9.1	0.61
Head gestures	64.4	7.4	0.25
Laughter	67.8	12.5	0.39

gestures. Different social actions are manually labeled by trained annotators for 30 min of the mingling sessions. For more information about the dataset, please refer to [27]. We have focused on the mingling session from the first day for the experiments presented in this section.

11.7.1 Feature extraction and classification

We have extracted features for each of the 26 subjects with valid accelerometer data. Statistical and spectral features are extracted from each axis of raw and absolute values of the acceleration and the magnitude of the acceleration, using 3 s windows with 1.5 s overlap. As the statistical features, mean, and variance values are calculated. The spectral features consist of the power spectral density binned into eight components with logarithmic spacing between 0–8 Hz.

We have used the L2 penalized Logistic Regressor as the classifier. Performance evaluation is done with leave-one-subject-out cross-validation. Hyperparameter optimization for regularization is carried out with nested cross-validation. Stepping, speaking, hand and head gestures, and laughter are selected as the target actions. Since the class distributions for each participant are different, we have chosen the AUC (area under the ROC curve) as the performance metric. Performances obtained with the aforementioned setup are presented in Table 11.1. We also present the mean annotator agreement for each action using Fleiss'–Kappa for three annotators. Values higher than 0.4 are considered to be of moderate agreement.

We can see that the results presented in Table 11.1 support the claim that actions that are loosely connected to the physical manifestation of the behavior are harder to detect. Stepping, as expected, has the highest performance of all. We also see that the performance tends to drop as the connection between the physical manifestation of the action itself and the acceleration reduces. For example, head gesture labels in the dataset, the social action with the lowest detection rate, include many subtle nods which are harder to capture via acceleration, compared to a step or hand gesture.

It should be noted that there might be a second factor at play here. In real-life events, it is generally harder to obtain annotations. Thus, the annotations must be made later manually. This of course introduces some differences in annotator agreement with respect to the type of action. Table 11.1 shows the annotator agreements as reported on a subset of the data taken from [27]. It can be seen that the lowest annotator agreement values are for the head gestures, followed by laughter. Variation in agreement (due to behavioral ambiguity or visual occlusion of the person being annotated) in the labels might have also contributed to the low performance of these actions, in addition to the nature of the connection between the action and the sensing medium. Thus, noisy labels, at least for some actions, are a reality of data collection in the wild, which needs to be taken into account when evaluating the perception performance. A further discussion of the trade-offs between using crowd sourced annotations compared to onsite annotators are also discussed in [27].

11.7.2 Performance vs. sample size

In the former experiment, 30 min of data from each participant was used. The results obtained showed that 30 min was enough to capture a variety of actions with various different situational contexts (i.e. differing conversing partners with different levels of conversational involvement), obtaining acceptable performance even for more subtle actions. But what is the minimum required amount of data for acceptable performance? Will the patterns be similar if we had less data? Since it is not guaranteed that we would have a continuous stream of 30 min of data, we conducted another experiment, where we used the earlier setup but with gradually increasing amounts of data for each participant, starting from five samples to a total of 1198 that covers the entire 30 min period.

As mentioned in the former section, each sample is extracted with a sliding window of 3 s with 1.5 s shift. Thus, we can say that five samples roughly corresponds to 7.5 s of data, 40 samples correspond to 1 min, and so on. We still used a leave-one-subject setup where for each fold, all the data from one participant corresponded to the test set. However, the training set is formed randomly by selecting n samples from each of the other participant's data. Since the selection is random, the process is repeated m times, which was also dependent on the number of samples selected. For computational reasons, we gradually reduced the number of repetitions from 150 to 15, resulting in 5 to 1000 samples for each repetition. We have selected two relatively well performing actions, stepping and speaking. These actions have different characteristics as described earlier with stepping being

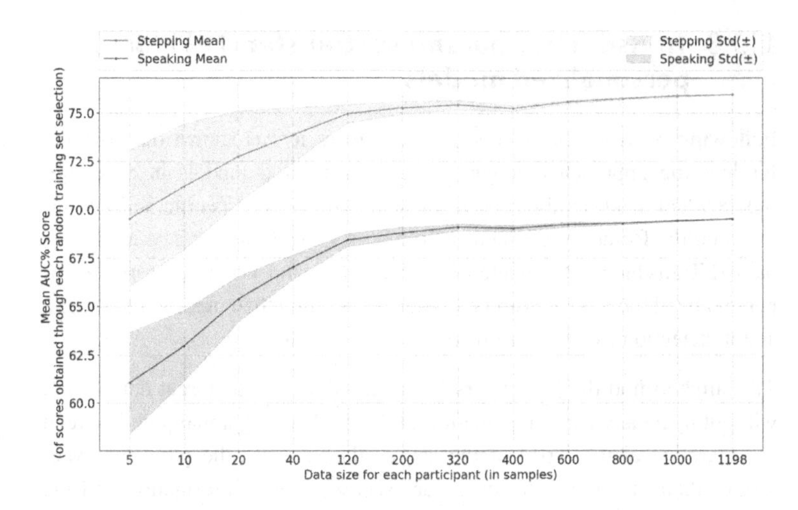

■ **FIGURE 11.2** AUC scores of stepping and speaking with respect to data size.

more closely connected to the physical manifestation of the behavior compared to speaking, which relies on detecting bodily gestures that are related to speech. In addition, this selection is based on former studies that showed that the connection between speech and acceleration is highly person specific, compared to stepping–walking [18]. The mean of the AUC scores of all repetitions, with increasing data size, are shown in Fig. 11.2 with standard deviation.

First, from Fig. 11.2 we observe the higher standard deviation for smaller sample sizes. This is related to the decreasing number of repetitions but we argue that is not the only factor. We believe there are parts of the event that are less informative than others and if the selected samples are coming from such intervals the performance tends to be low, and therefore fails to generalize over the whole event. This issue will be discussed further later in this section where we will present results of an experiment where the samples are not randomly sampled but selected chronologically. We also observe that the standard deviation for both actions converges to small values with increasing sample size.

We can see from Fig. 11.2 that the pattern for the two actions are quite similar. Performances for the actions increase with a steep curve in the beginning and after 120 samples the increase gets smaller. This suggests that 3 min of data from each person is enough to cover the variations in each type of action in such an event. The question then becomes if it is possible to provide a specialized solution which can guarantee better results even if the number of samples is relatively low.

11.7.3 Transductive parameter transfer (TPT) for personalized models

Following on from the results of [18], where it was shown that a transfer learning approach that guarantees personalized models in a person-independent setup tends to perform better for person specific actions, we repeated the former experiment with a personalized model. The method is named Transductive Parameter Transfer (TPT) and was first proposed for personalized facial expression recognition [45] and then modified for social action detection from a body worn accelerometer in [18].

TPT aims to find the parameters of the classifier for the target dataset X^t, without using any label information of X^t, by learning a mapping between the marginal distributions of the source datasets and the parameter vectors of their classifiers. N source datasets with label information and the unlabeled target dataset are defined as $D_1^s, ..., D_N^s$, $D_i^s = \{x_j^s, y_j^s\}_{j=1}^{n_i^s}$ and $X^t = \{x_j^t\}_{j=1}^{n_t}$, respectively. The main steps of the TPT are shown below (for a detailed explanation, please refer to [18]).

1. Compute $\{\boldsymbol{\theta}_i = (\boldsymbol{w}_i, c_i)\}_{i=1}^N$ using L2 penalized logistic regression.
2. Create the training set $\tau = \{X_i^s, \boldsymbol{\theta}_i\}_{i=1}^N$.
3. Compute the kernel matrix $\boldsymbol{K}$ that defines the distances between distributions where $\boldsymbol{K}_{ij} = \kappa(X_i^s, X_j^s)$.
4. Given $\boldsymbol{K}$ and τ, compute $\hat{f}(.)$ with kernel ridge regression.
5. Compute $(\boldsymbol{w}_t, c_t) = \hat{f}(X^t)$ using the mapping obtained in former step.

We conducted the performance vs. sample size experiment explained in the former section, with the addition of TPT. TPT is also used in a person-independent setup, where data from other participants are treated as source datasets with label information whereas the data to be classified is the target dataset. Although [18] suggests the use of an Earth Mover's Distance (EMD) kernel for computing the distance between distributions, we employed a density estimate kernel [45], since it is computationally less complex and more suitable for many random repetitions. The resulting AUC scores are plotted in Fig. 11.3.

According to Fig. 11.3, TPT outperforms a traditional person-independent setup when using small sample sizes for both actions. It seems to generalize better even with a small amount of data. For speaking, with the increasing data size, the gap between the two methodologies starts to close, showing that the single logistic regressor in the person-independent setup has seen enough diverse cases to generalize better. A one tailed paired t-test between AUC scores showed that, up until 320 samples, TPT provides significantly better performance ($p < 0.05$ for 40 samples and $p < 0.01$ for the rest).

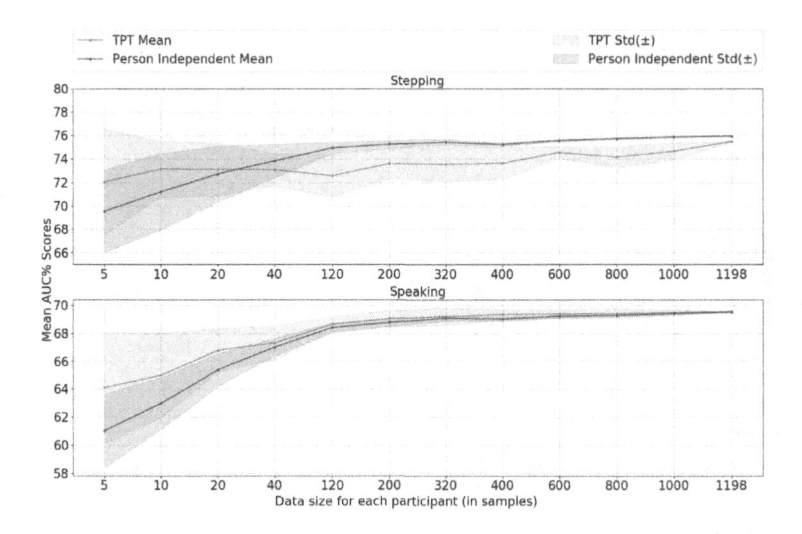

■ **FIGURE 11.3** AUC scores of stepping and speaking with respect to data size.

After that point, the mean scores provided by TPT seemed to be still higher than the person-independent setup but the significance is not guaranteed (some results such as those at 400 and 600 samples are still significant, though). We can say that with the increasing data size, the two methods converge to similar performances. However, especially for smaller sample sizes, we can still conclude that for estimating an action in a person specific manner, TPT is more robust.

For stepping, the trend shown is different. For extremely small amounts of data of 5, 10 and 20 samples, TPT outperforms the traditional person-independent setup (significantly for 5 and 10 samples). With increasing data sizes, the person-independent setup clearly outperforms TPT. It can be argued that this is related to the nature of the action. Stepping is less person specific than speaking and the connection between the sensor and the physical manifestation of the action is more direct. Thus, it can be expected that the representations of such an action should not vary too much between participants. With the increasing number of samples, the person-independent classifier will see more samples and since samples from different participants can be expected to be equally informative for all, a more optimal and general decision boundary can be obtained, unlike for speaking. So although we can advocate the use of TPT for really small sample sizes, a traditional person-independent setup seems to be a more robust selection for less person-specific actions.

Now, we want to go back to our claim that some parts of the event are more informative than others. The first parts of the dataset correspond to the be-

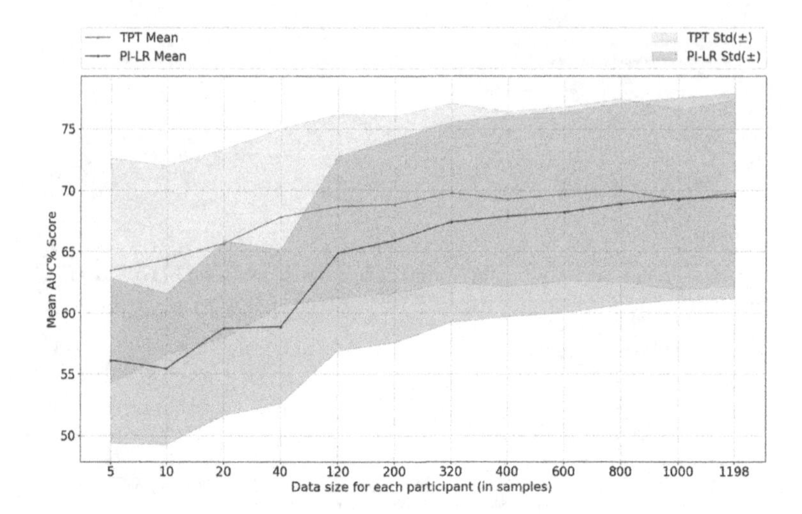

■ **FIGURE 11.4** AUC scores of speaking with temporally increasing data size.

ginning of the event, when groups are just starting to be formed. We might expect people to be less involved in the conversation as the discussions are not yet in full flow. This might result in samples that are not representative of all variations of actions that can occur in a real-life event, throughout time. So, we did a follow-up experiment where we compared the performances of TPT and the traditional person-independent setup for speaking detection. However, this time for each participant in the training set, we increased the number of samples in chronological order. Thus, n samples for a participant correspond to the first n samples in time. Since there are no repetitions, the means and the standard deviations are computed on the individual performances of all participants. The results of this experiment are shown in Fig. 11.4.

The first thing we observe from Fig. 11.4 is how the performances of the person-independent method is lower compared to those from Fig. 11.3. Using random selection of the samples throughout the event, the person-independent method was providing an AUC of roughly 61% for 5 samples. However, in the temporally increasing setup, the performance for the same number of samples is roughly 56%. The pattern is similar for the following sample sizes and the performance of temporally increasing selection is only able to reach the level of random selection if at least 320 samples are used for training. TPT on the other hand still provides similar results to the random selection method and provides relatively satisfactory results even with samples that were less informative for a traditional person-independent approach.

One other interesting observation is the relatively high standard deviations for both methods, even with an increasing number of samples. This shows that, for some participants, classifying the action is harder compared to others regardless of the sample size, further showing the person specific characteristics of speaking. These results further strengthens the claim that TPT should be considered for person specific and indirect actions such as speaking.

11.7.4 Discussion

With the presented perception analysis results, a few issues emerge that are all related to the 'in the wild' nature of the experiment. When collecting data from real-life events, many challenges arise. Some of these restrictions and difficulties come from the unrestricted nature of the event: the variety and frequency of actions might cause some cases to be under or over-represented making detection harder. The difficulty of the annotation process (either due to the ambiguity of the behavior or occlusion) can also result in label noise. Thus, when designing and conducting experiments on real-life data, a researcher should always first consider how these issues will affect the machine perception problem to be solved.

Specifically, for the case study presented in this chapter, when focusing on the detection of actions through wearables, there are some important points to consider. First, one should understand the connection between the physical manifestation of the action, and the sensing medium they are using. This is required for the valid selection of features and models that will be used for classification. In real-life scenarios, it is not guaranteed to have each action perfectly represented in all its possible variations for each participant. This is particularly true because natural 'in the wild' behavior samples only come into being as the result of the dynamics of a conversation as it unfolds over time. That is, a monologue in a group would yield more positive examples of speaking for the speaker of the group but no speaking samples for the members of the group who are just listening. So, the experimental setup and methodology chosen should encapsulate this together with the physical nature of the action. The experiments presented in this section are good examples of this, where two approaches for the detection of two actions tend to perform differently, because of the physical nature of the actions in relation to the sample sizes.

11.8 CHAPTER SUMMARY

This chapter has introduced some basic concepts of how to perform social behavior analysis 'in the wild' and specifically in the case of analyzing complex conversational scenes. In conducting research in this area, we have discussed two conceptual concerns: how to consider the relationship between ecological validity and 'in the wild' automated perception. Next, we provided concrete guidelines on how to collect data in such settings, differentiating between inductive versus deductive research practices and how this influences the data collection process. Finally, we provide some experiments on doing social action detection during complex conversational scenes using just accelerometer data recorded from a body worn sensor pack. Within this setting, we address challenging questions related to recording data in a deductive setting; When is there enough data? Does the learning model change depending on the amount of data available? How does the amount of data and the learning model vary with respect to the level of physical connection between the social behavior being detected? All these investigations provide an initial glimpse of what could be further investigated when considering automated social behavior analysis 'in the wild'. We have presented behavior analysis from the perspective of just a single modality (acceleration). However, further sensing modalities such as proximity or other more traditional modalities such as video and audio could also be combined opportunistically to provide richer representations for social behavior understanding.

REFERENCES

[1] X. Alameda-Pineda, J. Staiano, R. Subramanian, L. Batrinca, E. Ricci, B. Lepri, O. Lanz, N. Sebe, Salsa: a novel dataset for multimodal group behavior analysis, IEEE Trans. Pattern Anal. Mach. Intell. 38 (8) (2016) 1707–1720.

[2] X. Alameda-Pineda, Y. Yan, E. Ricci, O. Lanz, N. Sebe, Analyzing free-standing conversational groups: a multimodal approach, in: ACM International Conference on Multimedia, 2015.

[3] N. Ambady, R. Rosenthal, Thin slices of expressive behavior as predictors of interpersonal consequences: a meta-analysis, Psychol. Bull. 111 (2) (1992) 256.

[4] Ling Bao, Stephen Intille, Activity recognition from user-annotated acceleration data, Pervasive Comput. (2004) 1–17.

[5] Laura Cabrera-Quiros, Ekin Gedik, Hayley Hung, Estimating self-assessed personality from body movements and proximity in crowded mingling scenarios, in: Proceedings of the 18th ACM International Conference on Multimodal Interaction, ICMI 2016, New York, NY, USA, ACM, 2016, pp. 238–242.

[6] Laura Cabrera-Quiros, Hayley Hung, Who is where?: Matching people in video to wearable acceleration during crowded mingling events, in: Proceedings of the 2016 ACM on Multimedia Conference, MM '16, New York, NY, USA, ACM, 2016, pp. 267–271.

[7] G. Chittaranjan, H. Hung, Are you a werewolf? Detecting deceptive roles and outcomes in a conversational role-playing game, in: International Conference on Acoustics, Speech and Signal Processing (ICASSP), 2010.

[8] M. Cristani, L. Bazzani, G. Paggetti, A. Fossati, A. Del Bue, D. Tosato, G. Menegaz, V. Murino, Social interaction discovery by statistical analysis of F-formations, in: British Machine Vision Conference, August 2011.

[9] Marco Cristani, Anna Pesarin, Alessandro Vinciarelli, Marco Crocco, Vittorio Murino, Look at who's talking: voice activity detection by automated gesture analysis, in: Constructing Ambient Intelligence, Springer, 2012, pp. 72–80.

[10] T.M.T. Do, D. Gatica-Perez, GroupUs: smartphone proximity data and human interaction type mining, in: 2011 15th Annual International Symposium on Wearable Computers, June 2011, pp. 21–28.

[11] Trinh Minh Do, Daniel Gatica-Perez, Human interaction discovery in smartphone proximity networks, Pers. Ubiquitous Comput. 17 (3) (March 2013) 413–431.

[12] Wen Dong, Bruno Lepri, Fabio Pianesi, Alex Pentland, Modeling functional roles dynamics in small group interactions, IEEE Trans. Multimed. 15 (1) (2013) 83–95.

[13] Nathan Eagle, Alex Pentland, Social serendipity: mobilizing social software, IEEE Pervasive Comput. 4 (2) (April 2005) 28–34.

[14] Nathan Eagle, Alex Sandy Pentland, Reality mining: sensing complex social systems, Pers. Ubiquitous Comput. 10 (4) (2006) 255–268.

[15] Sergio Escalera, Oriol Pujol, Petia Radeva, Jordi Vitrià, María Teresa Anguera, Automatic detection of dominance and expected interest, EURASIP J. Adv. Signal Process. 2010 (2010).

[16] D. Gatica-Perez, I. McCowan, D. Zhang, S. Bengio, Detecting group interest-level in meetings, in: IEEE Int. Conf. on Acoustics, Speech, and Signal Processing (ICASSP), 2005.

[17] Daniel Gatica-Perez, Automatic nonverbal analysis of social interaction in small groups: a review, Image Vis. Comput. 27 (12) (November 2009) 1775–1787.

[18] Ekin Gedik, Hayley Hung, Personalised models for speech detection from body movements using transductive parameter transfer, Pers. Ubiquitous Comput. 21 (4) (Aug. 2017) 723–737.

[19] Juan Lorenzo Hagad, Roberto Legaspi, Masayuki Numao, Merlin Suarez, Predicting levels of rapport in dyadic interactions through automatic detection of posture and posture congruence, in: 2011 IEEE Third International Conference on Privacy, Security, Risk and Trust (PASSAT) and 2011 IEEE Third International Conference on Social Computing (SocialCom), IEEE, 2011, pp. 613–616.

[20] H. Hung, Y. Huang, C. Yeo, D. Gatica-Perez, Associating audio-visual activity cues in a dominance estimation framework, in: Computer Vision and Pattern Recognition Workshop on Human Communicative Behaviour, 2008.

[21] H. Hung, D. Jayagopi, S.O. Ba, J.-M. Odobez, D. Gatica-Perez, Investigating automatic dominance estimation in groups from visual attention and speaking activity, in: International Conference on Multi-modal Interfaces, 2008.

[22] H. Hung, B.J.A. Krose, Detecting F-formations as dominant sets, in: International Conference on Multimodal Interfaces (ICMI), November 2011.

[23] Hayley Hung, Gwenn Englebienne, Jeroen Kools, Classifying social actions with a single accelerometer, in: ACM International Joint Conference on Pervasive and Ubiquitous Computing, ACM, 2013, pp. 207–210 (Oral presentation).

[24] Hayley Hung, Daniel Gatica-Perez, Estimating cohesion in small groups using audio-visual nonverbal behavior, IEEE Trans. Multimed. 12 (6) (October 2010) 563–575.

[25] Hayley Hung, Yan Huang, Gerald Friedland, Daniel Gatica-Perez, Estimating dominance in multi-party meetings using speaker diarization, IEEE Trans. Audio Speech Lang. Process. 19 (4) (May 2011) 847–860.

[26] D. Jayagopi, H. Hung, C. Yeo, D. GaticaPerez, Modeling dominance in group conversations from non-verbal activity cues, IEEE Trans. Audio Speech Lang. Process. (2008).

[27] L. Cabrera-Quiros, A. Demetriou, E. Gedik, L.v.d. Meij, H. Hung, The MatchN-Mingle dataset: a novel multi-sensor resource for the analysis of social interactions and group dynamics in-the-wild during free-standing conversations and speed dates, IEEE Trans. Affect. Comput. (2018), https://doi.org/10.1109/TAFFC.2018.2848914.

[28] Bruno Lepri, Ankur Mani, Alex Pentland, Fabio Pianesi, Honest signals in the recognition of functional relational roles in meetings, in: AAAI Spring Symposium: Human Behavior Modeling, 2009, pp. 31–36.
[29] Anmol Madan, Ron Caneel, Alex Pentland, Voices of Attraction, 2004.
[30] R. Caneel, A. Madan, A. Pentland, Voices of attraction, in: Proceedings of Augmented Cognition (AugCog), HCI, 2005.
[31] Peter Marwedel, Embedded System Design: Embedded Systems Foundations of Cyber-Physical Systems, and the Internet of Things, Springer, 2011.
[32] Aleksandar Matic, Venet Osmani, Alban Maxhuni, Oscar Mayora, Multi-modal mobile sensing of social interactions, in: Pervasive Health, IEEE, 2012, pp. 105–114.
[33] Aleksandar Matic, Venet Osmani, Oscar Mayora-Ibarra, Mobile monitoring of formal and informal social interactions at workplace, in: Proceedings of the 2014 ACM International Joint Conference on Pervasive and Ubiquitous Computing: Adjunct Publication, UbiComp '14 Adjunct, ACM, 2014, pp. 1035–1044.
[34] D. McNeill, Language and Gesture, Cambridge University Press, New York, 2000.
[35] Marjolein C. Nanninga, Yanxia Zhang, Nale Lehmann-Willenbrock, Zoltán Szlávik, Hayley Hung, Estimating verbal expressions of task and social cohesion in meetings by quantifying paralinguistic mimicry, in: Edward Lank, Alessandro Vinciarelli, Eve E. Hoggan, Sriram Subramanian, Stephen A. Brewster (Eds.), Proceedings of the 19th ACM International Conference on Multimodal Interaction, ICMI 2017, Glasgow, United Kingdom, November 13–17, 2017, ACM, 2017, pp. 206–215.
[36] Catharine Oertel, Céline De Looze, Stefan Scherer, Andreas Windmann, Petra Wagner, Nick Campbell, Towards the automatic detection of involvement in conversation, in: Proceedings of the 2010 International Conference on Analysis of Verbal and Nonverbal Communication and Enactment, COST'10, Springer-Verlag, Berlin, Heidelberg, 2011, pp. 163–170.
[37] Daniel Olguin Olguin, Benjamin N. Waber, Taemie Kim, Akshay Mohan, Koji Ara, Alex Pentland, Sensible organizations: technology and methodology for automatically measuring organizational behavior, IEEE Trans. Syst. Man Cybern., Part B 39 (1) (2009) 43–55.
[38] K. Otsuka, J. Yamato, Y. Takemae, H. Murase, Quantifying interpersonal influence in face-to-face conversations based on visual attention patterns, in: Proc. ACM CHI Extended Abstract, Montreal, Apr. 2006.
[39] Emanuele Principi, Rudy Rotili, Martin Wöllmer, Stefano Squartini, Björn Schuller, Dominance detection in a reverberated acoustic scenario, in: Jun Wang, Gary G. Yen, Marios M. Polycarpou (Eds.), International Symposium on Neural Networks, in: Lect. Notes Comput. Sci., vol. 7367, Springer, 2012, pp. 394–402.
[40] N. Raiman, H. Hung, G. Engliebienne, Move, and I will tell you who you are: detecting deceptive roles in low-quality data, in: International Conference on Multimodal Interfaces (ICMI), November 2011.
[41] Rajesh Ranganath, Dan Jurafsky, Dan McFarland, It's not you, it's me: detecting flirting and its misperception in speed-dates, in: Proceedings of the 2009 Conference on Empirical Methods in Natural Language Processing, vol. 1, Association for Computational Linguistics, 2009, pp. 334–342.
[42] Nishkam Ravi, Nikhil Dandekar, Preetham Mysore, Michael L. Littman, Activity recognition from accelerometer data, in: AAAI, vol. 5, 2005, pp. 1541–1546.
[43] Rutger Rienks, Dirk Heylen, Dominance detection in meetings using easily obtainable features, in: Machine Learning for Multimodal Interaction, Springer, 2006, pp. 76–86.
[44] Rutger Rienks, Anton Nijholt, Dirk Heylen, Verbal behavior of the more and the less influential meeting participant, in: Proceedings of the 2007 Workshop on Tagging, Mining and Retrieval of Human Related Activity Information, TMR '07, New York, NY, USA, ACM, 2007, pp. 1–8.

[45] Enver Sangineto, Gloria Zen, Elisa Ricci, Nicu Sebe, We are not all equal: personalizing models for facial expression analysis with transductive parameter transfer, in: Proceedings of the 22nd ACM International Conference on Multimedia, ACM, 2014, pp. 357–366.
[46] Xiaofan Sun, Anton Nijholt, Khiet P. Truong, Maja Pantic, Automatic understanding of affective and social signals by multimodal mimicry recognition, in: Affective Computing and Intelligent Interaction, Springer, 2011, pp. 289–296.
[47] T. Teixeira, D. Jung, A. Savvides, Tasking networked CCTV cameras and mobile phones to identify and localise multiple persons, in: ACM International Joint Conference on Pervasive and Ubiquitous Computing (UbiComp), 2010.
[48] Ryoko Tokuhisa, Ryuta Terashima, Relationship between utterances and "enthusiasm" in non-task-oriented conversational dialogue, in: Proceedings of the 7th SIGdial Workshop on Discourse and Dialogue, SigDIAL '06, Stroudsburg, PA, USA, Association for Computational Linguistics, 2006, pp. 161–167.
[49] A. Veenstra, H. Hung, Do they like me? Using video cues to predict desires during speed-dates, in: International Conference on Computer Vision Workshop on Socially Intelligent Surveillance Monitoring, November 2011.
[50] A. Wilson, H. Benko, CrossMotion: fusing device and image motion for user identification, tracking and device association, in: International Conference on Multimodal Interaction (ICMI), 2014.
[51] Hans-Georg Wolff, Klaus Moser, Effects of networking on career success: a longitudinal study, J. Appl. Psychol. 94 (1) (2009) 196.
[52] Britta Wrede, Elizabeth Shriberg, Spotting "hot spots" in meetings: human judgments and prosodic cues, in: INTERSPEECH, 2003.
[53] Zhou Yu, David Gerritsen, Amy Ogan, Alan Black, Justine Cassell, Automatic prediction of friendship via multi-model dyadic features, in: Proceedings of the SIGDIAL 2013 Conference, Metz, France, Association for Computational Linguistics, August 2013.
[54] Massimo Zancanaro, Bruno Lepri, Fabio Pianesi, Automatic detection of group functional roles in face to face interactions, in: Proceedings of the 8th International Conference on Multimodal Interfaces, ACM, 2006, pp. 28–34.
[55] Lu Zhang, Hayley Hung, Beyond F-formations: determining social involvement in free standing conversing groups from static images, in: 2016 IEEE Conference on Computer Vision and Pattern Recognition (CVPR), IEEE, 2016, pp. 1086–1095.

Chapter 12

Detecting conversational groups in images using clustering games

Sebastiano Vascon*,†, Marcello Pelillo*,†
*European Centre for Living Technology, Venice, Italy †DAIS, Venice, Italy

CONTENTS

12.1 INTRODUCTION

After spending a substantial effort on the automatic analysis of a single person's behavior in images and videos, the attention of the computer vision and multimedia communities has recently shifted to group behavior and social interaction [15,30,12,24,44,18,8,22,43,37]. This is indeed not surprising, as humans are a social species and heavily interact with each other to achieve goals and exchange opinions or states of mind [13,31]. Detecting groups of people and analyzing their behavior is indeed of primary importance in many contexts such as video surveillance [8], social robotics [19], social signal processing [30,18,24,15], and activity recognition [5], etc.

Multimodal Behavior Analysis in the Wild. https://doi.org/10.1016/B978-0-12-814601-9.00024-9
Copyright © 2019 Elsevier Ltd. All rights reserved.

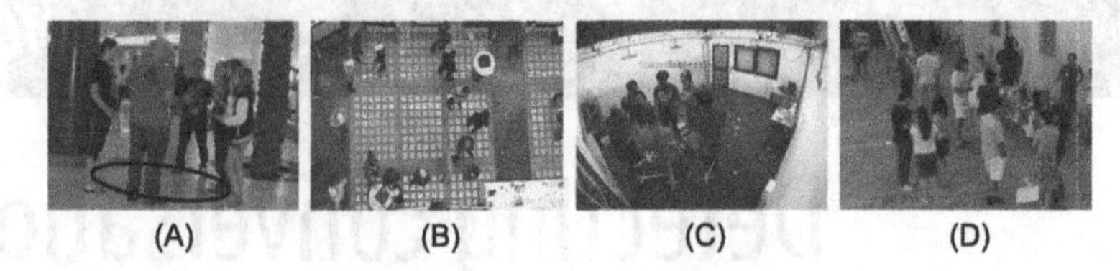

■ **FIGURE 12.1** Examples of conversational groups: A) in black, graphical depiction of overlapping space within an F-formation: the o-space; B) a poster session in a conference, where different groupings are visible; C) a *circular* F-formation; D) a typical surveillance setting where camera is located at 2.5-3 meters from the floor, for which detecting F-formations is a challenging task. Image courtesy of [40].[1]

The way in which people behave in public has been extensively studied by social psychologists, and by exploiting these findings new models to automatically analyze human behavior have recently been proposed. For example, the physical distance between persons has been studied by Hall [16], who found a relation between the space left between individuals and their levels of intimacy. Furthermore, Goffman [14] observed that people interaction can be categorized into two groups: those that are 'unfocused' and those that are 'focused'. The unfocused encounters involve interactions such as avoiding people on the street or greeting a colleague while passing them in the corridor. Focused interactions, instead, concern the involvement of people in a collective activity aimed at a common goal. These include, e.g., playing or watching a football match, conversing, or marching in a band.

A key definition in this field is that of an *F-formation*, or conversational group, namely a type of configuration that arises "whenever two or more individuals in close proximity orient their bodies in such a way that each of them has an easy, direct and equal access to every other participant's transactional segment, and when they maintain such an arrangement" [7, p. 243]. A person's *transactional segment* is defined as the area in front of him/her in which the interactions take place or, more specifically, the area "where hearing and sight are most effective" [6]. In Fig. 12.1A–D some examples of F-formations in real-world settings are shown.

As shown in Fig. 12.2A–E, different classes of F-formations may arise. In the case of two participants, typical F-formation configurations are vis-a-vis, L-shape, and side-by-side, while in the case of more than two participants a natural structure is the circular one which allows to maintain equality in the interaction. Three different social spaces emerge in an F-formation: the *o-space*, the *p-space* and the *r-space*. The most important part is the o-space

[1]Reprinted by permission from Springer Nature Customer Service Centre GmbH: Springer Nature, 12th Asian Conference on Computer Vision: A Game-Theoretic Probabilistic Approach for Detecting Conversational Groups, S. Vascon et al., 2014, https://link.springer.com/chapter/10.1007/978-3-319-16814-2_43.

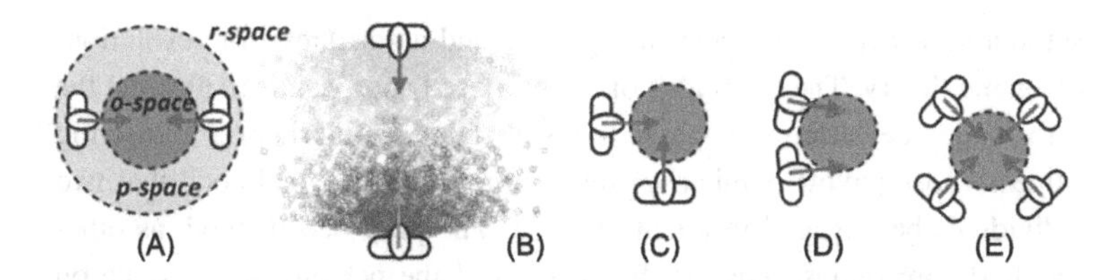

■ **FIGURE 12.2** F-formations; A) components of an F-formation: o-space, p-space, r-space; in this case, a face-to-face F-formation is sketched; B) modeling the frustum of attention by particles: in the intersection stays the o-space; C) L-shape type; D) side-by-side type; E) circular type. Image courtesy of [40].[2]

(see Fig. 12.2A), an empty convex space surrounded by the participants of a social interaction, in which every participant looks inward, and in which no external people are allowed. The p-space is an area that surrounds the o-space, and that contains the bodies of the people actually involved in a conversation. The r-space is the area outward the p-space. Another useful sociological definition, recently studied from a computer vision perspective [45], is that of an *associate*. The associates of an F-formation are people who try to get involved in the group of conversing people, but they do not succeed because either they are not fully accepted by the group, or because they feel unable to converse with the others in it [21].

Therefore detecting groups of conversing people can be seen as the problem of retrieving the o-space of a set of persons. To this end, in the first place it is required to detect the persons' positions and orientations (head or body) and subsequently to group the individuals thus found. These cues, known as low-level features [38], are nowadays readily obtainable, in particular with the advent of the deep-features. We report here some methods in the literature that face this problem [3,20,24,10] while in [38] a valuable review of features to detect conversational groups is reported.

After the low-level features from a scene are extracted, it is necessary to have a model that use them in order to determine whether an interaction between peoples is taking place or not. These models are built on top of the so-called *high-level* features, which are biologically and sociologically inspired models of interaction between people. Typically, like in the case of the specific method proposed in this chapter, these features are based on the notion of transactional segment [38].

Along the lines outlined above, in this chapter we will focus on the sociological notion of an *F-formation*, and we will describe an approach to their

[2]Reprinted by permission from Springer Nature Customer Service Centre GmbH: Springer Nature, 12th Asian Conference on Computer Vision: A Game-Theoretic Probabilistic Approach for Detecting Conversational Groups, S. Vascon et al., 2014, https://link.springer.com/chapter/10.1007/978-3-319-16814-2_43.

automatic detection in images using ideas and algorithms from evolutionary game theory. The main idea of the proposed approach is to abstract the F-formation detection problem in terms of stable equilibria of a "clustering game" whose payoff (similarity) function encodes the likelihood that two individuals be engaged in a conversation. The proposed method, as other standard approaches, requires the position of the persons in the scene on the ground plane as well as their body or head orientation. Once these cues are detected, an interaction model uses them to compute a score between each pair of persons, which in turn determines the payoff function of our grouping game. Conversational groups are then extracted using replicator dynamics from evolutionary game theory. Note that, unlike other clustering techniques, we do not need to know the number of groups in advance, as we extract them sequentially, and our approach works with arbitrary, even asymmetric, similarity functions. Experimental results over a number of benchmark datasets demonstrate the effectiveness of the proposed approach as compared to a number of other approaches in the literature.

12.2 RELATED WORK

In the past few years, several approaches for detecting conversational groups have been developed. Groh et al. [15] proposed a technique that uses the relative shoulder orientation and distance between people. An SVM classifier is then trained on top of this representation to perform the detection. Later, Cristani et al. [8] proposed a Hough voting schema in which the o-space location is found on the basis of the accumulated density estimates. At around the same time, Hung and Kröse [18] developed an algorithm based on the idea of abstracting conversational groups in terms of dominant-set clusters of a weighted graph [27]. These two approaches were compared by Setti et al. [32] who analyzed their strengths and weaknesses. They found that, in the presence of noise, the method of Cristani et al. [8] is more stable if only head orientation information is available, while the approach of Hung and Kröse [18] performs better when only positions (and not orientations) are available. They also found that a substantial improvement in the performances could be obtained combining probabilistic approaches (like [8]) and graph-based clustering methods like [18].

In Setti et al. [33] an improvement over [8] was proposed which aims at robustly handling the physical effect that different sizes of conversational groups would have on the spatial layout of each member of the group. Tran et al. [37] analyzed the temporal patterns of groups to classify collective activities. Choi et al. [4] modeled different forms of group behavior trying to distinguish differing group types.

In Vascon et al. [40,39] the problem of detecting conversational groups was modeled as a combination of a probabilistic interaction model and a deterministic grouping from evolutionary game theory. Following the trend of graph-based methods, Setti et al. [34] proposed a graph-cut based minimization method for detecting conversational groups using proxemic data and Solera et al. [35] developed a method to detect social groups in crowds through a correlation clustering procedure on people trajectories. Ricci et al. [28] proposed a method for jointly estimating head, body orientation and conversing groups while Zhang et al. [45] extended the previous work of [18] creating a model to detect the so-called *associates* of F-formations and taking also into account the geometry (furniture) of the scene.

12.3 CLUSTERING GAMES

Here we provide a brief introduction to some basic concepts of evolutionary game theory which will be instrumental to the development of our group detection framework. For a more comprehensive discussion we refer the reader to [42].

12.3.1 Notations and definitions

According to classical game theory [11], a game of strategy between two players (agents) can be formalized as a triplet $\Gamma = (P, S, \pi)$, where $P = \{1, 2\}$ is the set of players, $S = \{1, \dots, n\}$ is a set of *pure strategies* (or actions) available to each player, and $\pi : S^2 \to \mathbb{R}$ is a *payoff function.* The role of π is to quantify the utility of each strategy profile, which is an (ordered) pair of pure strategies played by the two players. In this context, we shall restrict our discussion to so-called *symmetric games*, where the role of the two players can be exchanged without changing the structure of the game being played. Hence we assume that all the players have the same set of strategies and the same payoff functions, which is in fact not necessarily the case in a more general setting. In this case the payoff function is represented by a matrix $A = (a_{ij}) \in \mathbb{R}^{n \times n}$, where $a_{ij} = \pi(i, j)$ represents the payoff a player receives when strategies i and j are played.

Evolutionary game theory, introduced in the early 1970s by Maynard Smith [25], considers an idealized scenario whereby individuals are repeatedly drawn at random from a large, ideally infinite, population to play a two-player game. In contrast to classical game theory, here players are not supposed to behave rationally or to have complete knowledge of the details of the game. They act instead according to an inherited behavioral pattern, or pure strategy, and it is supposed that some evolutionary selection process operates over time on the distribution of behaviors. Here, and in the sequel, an agent with preassigned strategy $j \in S$ will be called a *j-strategist.*

The state of the population at a given time t can be represented as a n-dimensional vector $\boldsymbol{x}(t)$, where $x_j(t)$ represents the fraction of j-strategists in the population at time t. Hence, the initial distribution of preassigned strategies in the population is given by $\boldsymbol{x}(0)$. The set of all possible states describing a population is given by

$$\Delta = \left\{ \boldsymbol{x} \in \mathbb{R}^n : \sum_{j \in S} x_j = 1 \text{ and } x_j \geq 0 \text{ for all } j \in S \right\},$$

which is called *standard simplex.* Points in the standard simplex are also referred to as *mixed strategies* in game theory. As time passes, the distribution of strategies in the population changes under the effect of a selection mechanism which, by analogy with Darwinian process, aims at spreading the fittest strategies in the population to the detriment of the weakest ones which, in turn, will be driven to extinction (we postpone the formalization of selection mechanisms to Section 12.5). For notational convenience, we drop the time reference t from a population state and we refer to $\boldsymbol{x} \in \Delta$ as a population rather than population state. Moreover, we denote by $\sigma(\boldsymbol{x})$ the *support* of $\boldsymbol{x} \in \Delta$:

$$\sigma(\boldsymbol{x}) = \{j \in S : x_j > 0\},$$

which is the set of strategies that are alive in a given population $\boldsymbol{x}$.

We will find it useful to define the following function $u : \mathbb{R}^n \times \mathbb{R}^n \rightarrow \mathbb{R}$:

$$u\left(\boldsymbol{y}^{(1)}, \boldsymbol{y}^{(2)}\right) = \sum_{(s_1,s_2) \in S^2} \pi(s_1, s_2) \prod_{i \in \{1,2\}} y_{s_i}^{(i)} = \boldsymbol{y}^{(1)\top} A \boldsymbol{y}^{(2)}. \tag{12.1}$$

We will also write $\boldsymbol{e}^j$ to indicate the n-vector with $x_j = 1$ and zero elsewhere. Now, it is easy to see that the expected payoff earned by a j-strategist in a population $\boldsymbol{x} \in \Delta$ is given by

$$u\left(\boldsymbol{e}^j, \boldsymbol{x}\right) = (A\boldsymbol{x})_j = \sum_{s \in S} a_{js}, x_s, \tag{12.2}$$

while the expected payoff over the entire population is given by

$$u(\boldsymbol{x}, \boldsymbol{x}) = \boldsymbol{x}^\top A \boldsymbol{x} = \sum_{j \in S} x_j (A\boldsymbol{x})_j. \tag{12.3}$$

A fundamental notion in game theory is that of an equilibrium [42]. Intuitively, an evolutionary process reaches an equilibrium $\boldsymbol{x} \in \Delta$ when every individual in the population obtains the same expected payoff and no strategy can thus prevail upon the other ones. Formally, $\boldsymbol{x} \in \Delta$ is a *Nash equilibrium* if

$$u(\boldsymbol{y}, \boldsymbol{x}) \leq u(\boldsymbol{x}, \boldsymbol{x}) \tag{12.4}$$

for all $\boldsymbol{y} \in \Delta$. We say that a Nash equilibrium $\boldsymbol{x}$ is *strict* if (12.4) holds strictly for all $\boldsymbol{y} \in \Delta \setminus \{\boldsymbol{x}\}$.

Within a population-based setting the notion of a Nash equilibrium turns out to be too weak as it lacks stability under small perturbations. This motivated J. Maynard Smith [25] to introduce a refinement of the Nash equilibrium concept generally known as an Evolutionary Stable Strategy (ESS). Formally, assume that in a population $\boldsymbol{x} \in \Delta$, a small share ϵ of mutant agents appears, whose distribution of strategies is $\boldsymbol{y} \in \Delta$. The resulting post-entry population is then given by $\boldsymbol{w}_\epsilon = (1-\epsilon)\boldsymbol{x} + \epsilon\boldsymbol{y}$. Biological intuition suggests that evolutionary forces select against mutant individuals if and only if the expected payoff of a mutant agent in the postentry population is lower than that of an individual from the original population, i.e.,

$$u(\boldsymbol{y}, \boldsymbol{w}_\epsilon) < u(\boldsymbol{x}, \boldsymbol{w}_\epsilon)\,. \tag{12.5}$$

Hence, a population $\boldsymbol{x} \in \Delta$ is said to be *evolutionary stable* if inequality (12.5) holds for any distribution of mutant agents $\boldsymbol{y} \in \Delta \setminus \{\boldsymbol{x}\}$, granted the population share of mutants ϵ is sufficiently small. It can be shown [42] that $\boldsymbol{x}$ is an ESS equilibrium if and only if it is a Nash equilibrium and the additional stability property $u(\boldsymbol{x}, \boldsymbol{y}) > u(\boldsymbol{y}, \boldsymbol{y})$ holds for all $\boldsymbol{y} \in \Delta \setminus \{\boldsymbol{x}\}$ such that $u(\boldsymbol{y}, \boldsymbol{x}) = u(\boldsymbol{x}, \boldsymbol{x})$.

12.3.2 Clustering games

An instance of the clustering problem can be described by an edge-weighted graph, which is formally defined as a triplet $G = (V, E, \omega)$, where $V = \{1, \ldots, n\}$ is a finite set of *vertices*, $E \subseteq V \times V$ is the set of edges and $\omega : E \to \mathbb{R}$ is a real-valued function which assigns a weight to each edge. Within our clustering framework the vertices in G correspond to the objects to be clustered, the edges represent neighborhood relationships among objects, and the edge-weights reflect similarity among linked objects.

Given a graph $G = (V, E, \omega)$, representing an instance of a clustering problem, we cast it into a two-player *clustering game* $\Gamma = (P, V, \pi)$ where the players' pure strategies correspond to the objects to be clustered and the payoff function π is proportional to the similarity of the objects/strategies

$(v_1, v_2) \in V^2$ selected by the players:

$$\pi(v_1, v_2) = \begin{cases} \omega(v_1, v_2) & \text{if } (v_1, v_2) \in E\,, \\ 0 & \text{otherwise}\,. \end{cases} \tag{12.6}$$

Our clustering game will be played within an evolutionary setting wherein, as described above, two players are repeatedly drawn at random from a large population. A dynamic evolutionary selection process will then make the population $\boldsymbol{x}$ evolve according to a Darwinian survival-of-the-fittest principle in such a way that, eventually, the better-than-average objects will survive and the others will get extinct. Within this setting, the clusters of a clustering problem instance can be characterized in terms of the ESS's of the corresponding (evolutionary) clustering game, thereby justifying the following definition.

Definition 1 (ESS-cluster). Given an instance of a clustering problem $G = (V, E, \omega)$, an *ESS-cluster* of G is an ESS of the corresponding clustering game.

For the sake of simplicity, when it will be clear from the context, the term ESS-cluster will be used henceforth to refer to either the ESS itself, namely the membership vector $\boldsymbol{x} \in \Delta$, or to its support $\sigma(\boldsymbol{x}) = C \subseteq V$.

The motivation behind the above definition resides in the observation that ESS-clusters do incorporate the two basic properties of a cluster, i.e.,

- *internal coherency*: elements belonging to the cluster should have high mutual similarities;
- *external incoherency*: the overall cluster internal coherency decreases by introducing external elements.

In fact, it has been shown [36] that ESS-clusters are in one-to-one correspondence with *dominant sets* a graph-theoretic notion of cluster which generalizes the notion of a maximal clique to edge-weighted graphs [27].

Interestingly, if the payoff matrix is symmetric, then ESS-clusters can be interpreted in terms of optimization theory. Indeed, consider the following *standard quadratic program* [2]:

$$\begin{aligned} \max \quad & u(\boldsymbol{x}, \boldsymbol{x}) = \boldsymbol{x}^\top A \boldsymbol{x} \\ \text{subject to} \quad & \boldsymbol{x} \in \Delta \subset \mathbb{R}^n\,. \end{aligned} \tag{12.7}$$

The following result holds [17].

Theorem 1. *ESS-clusters are in one-to-one correspondence with strict local maximizers of* (12.7), *provided that the payoff matrix is symmetric.*

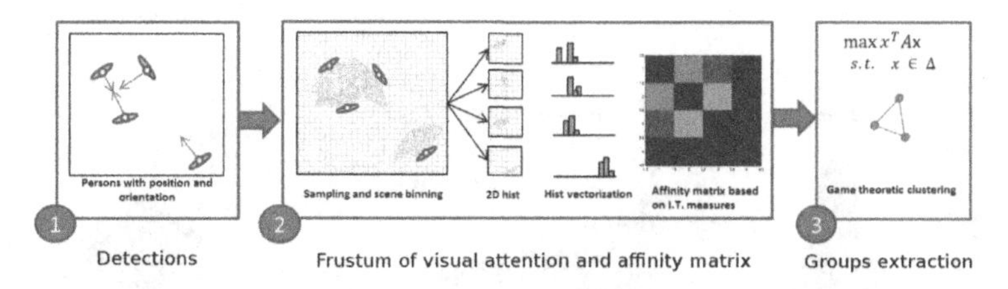

■ **FIGURE 12.3** The pipeline of our detection algorithm. Image courtesy of [38].[3]

Note that program (12.7) is the same as that used by Motzkin and Straus to characterize the maximum cliques of an unweighted (undirected) graph [26]. In fact, if the payoff matrix A is a 0/1 matrix, then the local/global solutions of (12.7) correspond to maximal/maximum cliques in the graph having A as its adjacency matrix.

12.4 CONVERSATIONAL GROUPS AS EQUILIBRIA OF CLUSTERING GAMES

In this section, we shall detail the structure of our detection algorithm which is based on the idea of abstracting conversational groups in terms of ESS-clusters of a clustering game. In our formulation, the pure strategies being played correspond to the persons detected in a given image, and the payoff (similarity) function is obtained using a sociological notion known as the "frustum of attention."

12.4.1 Frustum of attention

The model of the frustum is inspired by the concept of transactional segment introduced by Kendon, which is defined as "the area in front of a person that can be reached easily, and where hearing and sight are most effective" [6]. In the proposed interaction model, both the field of view of each person and the locus of attention of all other senses (for a given body orientation) are taken into account. Body orientation is approximated using the orientation of the head, which is more convenient as occlusion typically makes the inference of the body orientation more difficult.

The frustum is characterized by a direction θ (which could be the person's head or body orientation), an aperture γ (we used $\gamma = 160°$) and a length l. These three components determine the socio-attentional frustum of an individual. The area of the frustum is generated by sampling two distributions,

[3]Reprinted from: Group and Crowd Behavior for Computer Vision, 1st ed., Vascon et al., Chapter 3 – Group Detection and Tracking Using Sociological Features, Copyright (2017), with permission from Elsevier.

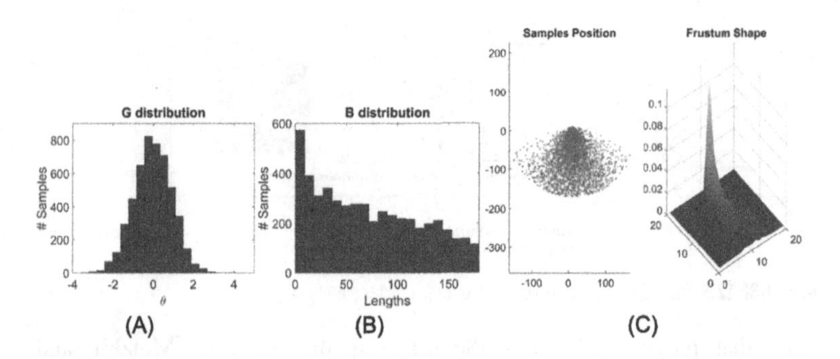

■ **FIGURE 12.4** A) The probability distribution over the orientations; B) the distance from the person; C) the frustum generated using sampling. Image courtesy of [39].[4]

a Gaussian distribution G and a Beta distribution B. The *G distribution* (see Fig. 12.4A) is used to generate samples related to the aperture of the frustum. The Gaussian is thus centered in the provided orientation θ with a variance set to a value such that the full width of the Gaussian distribution corresponds to the desired aperture of the frustum. In a Gaussian distribution the 99% of the samples are located in the range of $[-3\sigma, 3\sigma]$; this range will correspond to the full aperture of the frustum, so setting the variance in a way that the aperture is fully covered becomes an easy task, $\sigma = \frac{1}{3} * \frac{\gamma}{2}$.

The *B distribution* (see Fig. 12.4B) is used to generate samples that are denser nearby the person while decaying far away from him/her. The rationale behind this assumption is that the interactions are more plausible in close proximity rather than at higher distances. To achieve the desired shape of the frustum, we set the distribution parameters $\alpha = 0.8$ and $\beta = 1.1$ (see Fig. 12.4B). The values returned by the B distribution are bounded in [0, 1] and need to be multiplied by the desired length of the frustum l, which depend by the data and by the type of interaction that is required to detect [16].

The samples obtained using these two distributions are thus in polar coordinates (an angle and a distance), to obtain samples in the 2D space is sufficient to apply a simple trigonometric rule to each sample. Given a pair of samples from G and B (G_i, B_i) and the position of a person (p_x, p_y) the 2D position of each sample is:

$$s_x = p_x + \cos(G_i) * B_i * l, \qquad (12.8)$$
$$s_y = p_y + \sin(G_i) * B_i * l.$$

[4]Reprinted from Computer Vision and Image Understanding, Vol. 143, Vascon et al., Detecting conversational groups in images and sequences: A robust game-theoretic approach, pages 11–24, Copyright (2015), with permission from Elsevier.

Having drawn independently n samples from the two distributions and applying the above equations we obtain a set of samples which falls in the human frustum of visual attention (see Fig. 12.4C).

Each person in a scene is thus modeled using his/her frustum. A subsequent step is to bin the frustum to a 2-dimensional histogram h_i of size $N_c \times N_r$ which cover the entire scene. Each h_i is normalized such that it add up to 1. Subsequently, the 2-dimensional histogram h_i is converted to a 1D histogram by concatenating the rows of the matrix. This last operation is performed to make the next task of quantifying pairwise interactions easier.

12.4.2 **Quantifying pairwise interactions**

Two persons are more likely to interact if their frustums overlap and in order to quantify the likelihood of being interactants different measures can be employed. Since the frustum is encoded into a histogram that represents a discrete probability distribution, it is natural to consider information-theoretic measures to model the distance between a pair of frustums.

Given a pair of discrete probability distributions $P = \{p_1, \ldots, p_n\}$ and $Q = \{q_1, \ldots, q_n\}$, a measure of their distance is given by the Kullback–Leibler (KL) divergence, which is defined as

$$D(P\|Q) = \sum_{i=1}^{n} \log p_i \frac{p_i}{q_i} \tag{12.9}$$

and its symmetric version, the Jensen–Shannon (JS) divergence [23], which is defined as:

$$J(P, Q) = \frac{D(P\|M) + D(Q\|M)}{2} \tag{12.10}$$

where $M = \frac{1}{2}(P + Q)$ is the mid-point between P and Q. Hence, given two persons i and j in a scene and their histograms h_i and h_j, the distance between i and j can be calculated either as $D(h_i\|h_j)$ or as $J(h_i, h_j)$.

In order to obtain the payoff matrix $A = (a_{ij})$ of our clustering game (which quantifies similarities as opposed to distances) we used the classical Gaussian kernel:

$$a_{i,j} = \begin{cases} \exp\left\{-\frac{d(h_i, h_j)}{\sigma}\right\} & \text{if } i \neq j\,, \\ 0 & \text{otherwise}\,, \end{cases} \tag{12.11}$$

where the function "d" refers to either Eq. (12.10) or Eq. (12.9). The similarity of a person with him/her-self ($i = j$) is set to zero to avoid groups composed only by singletons. Here, σ acts as a scale parameter which can be determined by cross-validation or by extensive search.

12.4.3 The algorithm

Given an image, with the positions and orientations of all the persons, the pipeline of our detection algorithm (see Fig. 12.3) can finally be summarized in the following steps:

1. For each person $p_i \in P$, generate a frustum f_i based on his/her position and orientation in world coordinates (see Section 12.4.1).
2. Bin the frustum f_i to a 2-dimensional histogram h_i (see Section 12.4.1).
3. Convert the binned frustum h_i to a 1-dimensional histogram through row concatenation.
4. Compute the payoff matrix for each pair of frustums (h_i, h_j) (see Section 12.4.2)
5. Extract the ESS-clusters (conversational groups) of the clustering game (see Section 12.5)

12.5 FINDING ESS-CLUSTERS USING GAME DYNAMICS

Once the problem of detecting conversational groups has been abstracted in terms of finding the evolutionary stable strategies of a clustering game, we need an algorithm for finding them. Evolutionary game theory offers a wide range of principled solutions to accomplish this task, and we now describe a popular class of simple and effective ESS-finding dynamics.

In evolutionary game theory the assumption is made that the game is played over and over, generation after generation, and that the action of natural selection will result in the evolution of the fittest strategies. A general class of evolution equations is given by the following set of ordinary differential equations [42]:

$$\dot{x}_i = x_i(t) g_i(\boldsymbol{x}) \tag{12.12}$$

for $i = 1, \ldots, n$, where a dot signifies the derivative with respect to time and $g = (g_1, \ldots, g_n)$ is a function with open domain containing Δ. Here, the function g_i $(i \in S)$ specifies the rate at which pure strategy i replicates. It is usually required that the growth function g is *regular* [42], which means that it is Lipschitz continuous and that $g(\boldsymbol{x})^\top \boldsymbol{x} = 0$ for all $\boldsymbol{x} \in \Delta$. The former condition guarantees us that the system of the differential equation (12.12) has a unique solution through any initial population state. The latter condition, instead, ensures that the simplex Δ is invariant under (12.12), namely, any trajectory starting in Δ will remain in Δ.

A point x is said to be a *stationary* (or equilibrium) point for our dynamical systems, if $\dot{x}_i = 0$ $(i \in S)$. A stationary point x is (Lyapunov) *stable* if for every neighborhood U of x there exists a neighborhood V of x such that $x(0) \in V$ implies $x(t) \in U$ for all $t \geq 0$. A stationary point is said to be *asymptotically stable* if any trajectory starting in its vicinity will converge to it as $t \to \infty$.

Payoff-monotonic game dynamics represent a wide class of regular selection dynamics for which useful properties hold. Intuitively, for a payoff-monotonic dynamics the strategies associated to higher payoffs will increase at a higher rate. Formally, a regular selection dynamics (12.12) is said to be *payoff-monotonic* if

$$g_i(x) > g_j(x) \Leftrightarrow u\left(e^i, x\right) > u\left(e^j, x\right)$$

for all $x \in \Delta$ and $i, j \in S$.

Although this class contains many different dynamics, it turns out that they share a lot of common properties. To begin, they all have the same set of stationary points. Indeed, $x \in \Delta$ is a stationary point under any payoff monotonic dynamics if and only if $u\left(e^i, x\right) = u(x, x)$ holds for all $i \in \sigma(x)$ [42].

A well-known subclass of payoff-monotonic game dynamics is given by

$$\dot{x}_i = x_i \left(f\left(u\left(e^i, x\right)\right) - \sum_{j \in S} x_j f\left(u\left(e^j, x\right)\right) \right),$$

where $f(u)$ is an increasing function of u. These models arise in modeling the evolution of behavior by way of imitation processes, where players are occasionally given the opportunity to change their own strategies [42].

When f is the identity function, that is, $f(u) = u$, we obtain the standard continuous-time *replicator equations*,

$$\dot{x}_i = x_i \left(u\left(e^i, x\right) - u(x, x)\right), \tag{12.13}$$

whose basic idea is that the average rate of increase $\dot{x}_i / x_i$ equals the difference between the average fitness of strategy i and the mean fitness over the entire population.

Another popular model arises when $f(u) = e^{ku}$, where k is a positive constant. As k tends to 0, the orbits of this dynamics approach those of the standard, first-order replicator model, slowed down by the factor k; moreover, for large values of k, the model approximates the so-called best-reply dynamics [17].

The replicator dynamics, and more in general any payoff monotonic dynamics, have the following properties [42,17]:

Theorem 2. *Under any payoff monotonic dynamics the following holds true:*

- *Nash equilibria are stationary points.*
- *Strict Nash equilibria are asymptotically stable.*
- *Stationary points that are limit of interior trajectories are Nash equilibria.*
- *Stable stationary points are Nash equilibria.*
- *ESSs are asymptotically stable.*

Furthermore, if we restrict our focus to symmetric payoff matrices, i.e. $A = A^\top$, then stronger properties hold, as stated in the following theorem.

Theorem 3. *If $A = A^\top$ then the following holds:*

- $u(\boldsymbol{x}, \boldsymbol{x})$ *is strictly increasing along any non-constant trajectory of* (12.13). *In other words, for all $t \geq 0$ we have $\dot{u}(\boldsymbol{x}, \boldsymbol{x}) > 0$, unless $\boldsymbol{x}$ is a stationary point. Furthermore, any such trajectory converges to a (unique) stationary point;*
- $\boldsymbol{x}$ *is asymptotically stable if and only if $\boldsymbol{x}$ is an ESS.*

In order to implement the continuous-time replicator dynamics one can resort to some iterative method like, e.g., the Runge–Kutta method, to find an approximate solution to the ordinary differential equations. Alternatively, one can adopt the discrete-time counterpart of (12.13), known as discrete-time replicator dynamics, which are given by

$$x_i(t+1) = x_i(t) \frac{u\left(e^i, \boldsymbol{x}\right)}{u\left(\boldsymbol{x}, \boldsymbol{x}\right)} \tag{12.14}$$

for $i \in S$. These are shown to possess many of the convergence properties of their continuous-time counterparts.

In conclusion, the algorithm for extracting conversational groups is as follows:

1. Initialize the vector $\boldsymbol{x}$ (typically at the barycenter of Δ).
2. Iterate (12.14) until convergence.
3. The conversational group C_i corresponds to the support of the converged vector $\sigma(\boldsymbol{x})$.
4. Remove the element of C_i from S and iterate again until all the persons detected in the image are grouped.

Table 12.1 Datasets: multiple #Frames indicate diverse sequences

Dataset	#Sequences	#Frames	Consecutive frames	Automated tracking
CocktailParty	1	320	Y	Y
PosterData	82	1	N	N
CoffeeBreak	2	45,74	Y	Y
GDet	5	132,115,79,17,60	N	Y
Synth	10	10	N	N

Notice that our game-theoretic approach does not require *a priori* knowledge on the number of groups (since it extracts them sequentially) and makes no assumption on the structure of the affinity matrix, being it able to work with asymmetric and even negative similarity functions alike. It also generalizes naturally to hypergraph clustering problems, i.e., in the presence of high-order affinities, in which case the clustering game is played by more than two players [29].

12.6 EXPERIMENTS AND RESULTS

In this section we report some experiments conducted with our game-theoretic detection algorithm. Our approach has been compared with a Hough-based approach [8] and its refined version [32] (HFF), a hierarchical extension thereof [33] (MULTI), a dominant set-based technique [18] (DS). Comparisons with other baselines are not reported here, and we refer the reader to [32,8] for a more comprehensive comparative analysis.

12.6.1 Datasets

The method has been tested on five publicly available datasets (see Table 12.1) which represent the standard benchmark for conversational groups detection. All the aforementioned datasets provide for each person i in the scene his/her x, y position in world coordinate and the head orientation. In three cases the annotations have been done via automatic tracking while on the other two datasets have been performed manually by their respective authors as stated in Table 12.1.

CocktailParty [44]. The CocktailParty dataset is composed of 16 min of video recorded during a cocktail party in a 30 m^2 lab environment involving seven subjects. The dataset is challenging for video analysis due to frequent and persistent occlusions given the highly cluttered scene. The position and horizontal head orientation of the subjects were recorded using a particle filter-based body tracker with head pose estimation. Groups were manually annotated by an expert every 3 s, resulting in a total of 320 distinct frames for evaluation.

PosterData [18]. It consists of 3 h of indoors video in a large atrium of a hotel with over 50 people during a scientific meeting. The meetings involve poster presentations and a coffee break. The cameras were mounted on the ceiling pointing downwards to the hall. The 82 distinct image frames were selected to maximize the differences between images, ambiguity in group membership and varying levels of crowdedness.

CoffeeBreak [8]. The dataset represents a coffee-break scenario of a social event, with maximum 14 individuals organized in groups of 2–3 people. People positions were estimated with a multi-object tracking of the heads, and the detection of the head orientation detection has been performed afterwards. This dataset is particularly challenging because the orientation of the head has solely four possible orientations (front, back, left, right). The tracked positions were projected onto the ground plane. A trained expert annotated the videos indicating the groups in the scenes. The dataset is composed of a total of 45 frames for *Seq1* and 75 frames for *Seq2*.

GDet [8]. The dataset consists of a vending machine area where people take coffee, other drinks and chat. In this case, head orientation considers solely four possible alternatives.

Synth [8]. The Synth dataset is a synthetic dataset in which a trained expert generate 10 different *situations*, with groups of conversing people and singletons. Each situation is repeated 10 times, with slightly varying positions and head orientations of the subjects. Here, noise (in position and orientation) is absent so the dataset is relatively simple.

12.6.2 Evaluation metrics and parameter exploration

The evaluation of the performance follows the standard methodology proposed in [32]. A group is considered as detected correctly if at least $\lceil (T \cdot |G|) \rceil$ of their members are correctly grouped by the algorithm, and if no more than $\lceil (1 - T) \cdot |G| \rceil$ false subjects are identified, where $|G|$ is the cardinality of the labeled group G, and $T = 2/3$. Based on this metrics, we compute *precision*, *recall*, and *F1-score* per frame. Averaging these values over the frames produces the final score.

Different combinations of parameters have been explored and validated on each dataset. In particular, we reported the performance of the proposed method when using the similarity function in Eq. (12.11) with the distance

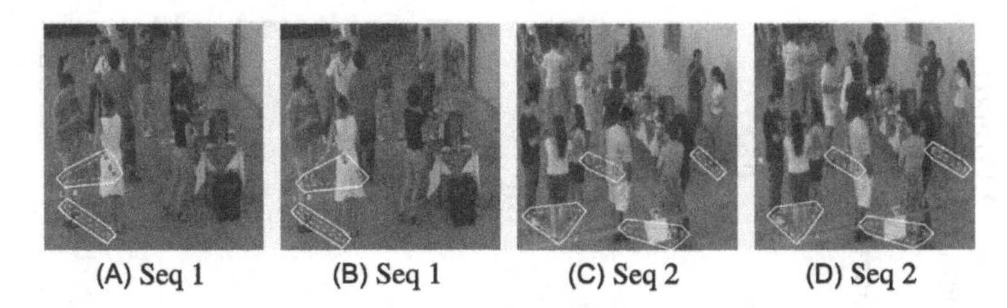

■ **FIGURE 12.5** Qualitative results on the CoffeeBreak dataset compared with the state of the art HFF [8]. In yellow the ground-truth, in green our method and in red HFF. As is evident from (A, B, C, D) HFF often fails in detecting groups of more than two persons while our approach is more stable. Image courtesy of [40].[5]

function in Eq. (12.10) (as suggested by [40]), and changing the value of σ in the range $\{0.1, 0.2, 0.4, 0.5, 0.7, 0.9\}$. To explore the effect of the length of the frustum, we based our analysis on the studies conducted by Hall in [16,9] in which a focused encounter between two persons may occur between 45 cm to 2 m; correspondingly, the parameter l will range in the same interval.

12.6.3 **Experiments**

The proposed method is applied on the above datasets in order to detect all the conversational groups in the proposed scene. The results for the five datasets are reported in Table 12.2 also comprising the used parameters. In Fig. 12.5 the qualitative results of our group detector are shown in comparison with the HFF method [8]. As done in the comparative approaches, we reported here the performances obtained with the best parameter settings using the Jensen–Shannon (Eq. (12.10)) divergence. Results are averaged over 10 runs to assess the stability of the method. For the sake of completeness, we also compared the game-theoretic clustering approach with the well known spectral clustering algorithm. Since the number of groups is unknown, we used the spectral-gap heuristic to determine the number of clusters [41] (see "R-GTCG SC" rows in Table 12.2).

By looking at the results, we can state that in just one case the proposed method performs comparably with the state of the art which is the precision in the PosterData dataset where the methods [18,32] obtain a 1% improvement over the game-theoretic approach. In all the other datasets the performances are definitely superior, reaching 100% of precision/recall/f1

[5]Reprinted by permission from Springer Nature Customer Service Centre GmbH: Springer Nature, 12th Asian Conference on Computer Vision: A Game-Theoretic Probabilistic Approach for Detecting Conversational Groups, S. Vascon et al., 2014, https://link.springer.com/chapter/10.1007/978-3-319-16814-2_43.

Table 12.2 Results on frame: only the best results are shown while the parameters are discussed in the paper (σ in Eq. (12.11) and l in Eq. (12.8)). The comparative methods are IRPM [1], HFF [8], DS [32], MULTISCALE [33], GTCG [40], "R−GTCG" our method and "R-GTCG SC" the results of our method using the Spectral Clustering technique instead of the game-theoretic-clustering

	CoffeeBreak (S1+S2)			PosterData			Gdet		
Method*	***Prec***	***Rec***	***F1***	***Prec***	***Rec***	***F1***	***Prec***	***Rec***	***F1***
IRPM [1,32]	0.60	0.41	0.49	–	–	–	–	–	–
HFF [32]	0.82	0.83	0.82	**0.93**	**0.96**	**0.94**	0.67	0.57	0.62
DS ([18,32])*	0.68	0.65	0.66	**0.93**	0.92	0.92	–	–	–
MULTISCALE [33]	0.82	0.77	0.80	–	–	–	–	–	–
GTCG [40] KL	0.80	0.84	0.82	0.90	0.94	0.92	**0.76**	0.75	0.75
R-GTCG SC	0.52	0.59	0.55	0.26	0.27	0.26	0.75	0.75	0.75
R-GTCG	**0.86**	**0.88**	**0.87**	0.92	**0.96**	**0.94**	**0.76**	**0.76**	**0.76**
		$\sigma = 0.2, l = 145$			$\sigma = 0.25, l = 115$			$\sigma = 0.7, l = 180$	

	CocktailParty			Synth		
Method*	***Prec***	***Rec***	***F1***	***Prec***	***Rec***	***F1***
IRPM [1,32]	–	–	–	0.71	0.54	0.61
HFF ([8,33])	0.59	0.74	0.66	0.73	0.83	0.78
MULTISCALE [33]	0.69	0.74	0.71	0.86	0.94	0.90
GTCG [40] KL	0.85	0.81	0.83	**1.00**	**1.00**	**1.00**
R-GTCG SC	0.77	0.72	0.74	0.40	0.90	0.56
R-GTCG	**0.87**	**0.82**	**0.84**	**1.00**	**1.00**	**1.00**
		$\sigma = 0.6, l = 170$			$\sigma = 0.1, l = 75$	

* Note that in [32] the parameters for the DS method were not fully optimised.

* In case of double citations, the first one refers to the original method while the latter refers to the more recent paper from which the results have been taken.

in the Synthetic dataset while in the GDet and CocktailParty an almost 10% improvement is achieved.

In particular, our findings suggest that the proposed approach is to be preferred over the others under a wide variety of different scenarios and, in general, the performance is very stable under both noisy (real) and ideal (synthetic) dataset. For example, we have the highest performance in the CoffeeBreak even if it is a very noisy dataset in terms of head orientation since only four orientations are possible, while in the Synthetic dataset which represents an ideal case we reach 100% in precision and recall. From the experiments and findings in [40,39], it is clear that the Jensen–Shannon measure produces the highest and more stable performance. This suggests that, while modeling pairwise social interactions, it is reasonable to assume mutuality in the relations implying a symmetric affinity in the model (like the JS measure).

12.7 CONCLUSIONS

In this chapter, we have described an approach for detecting conversational groups (a.k.a. F-formations) in images which is based on the idea of finding evolutionary stable strategies of a non-cooperative "clustering game" (ESS-clusters). The approach improves upon existing methods by building a stochastic model of social attention which captures the likelihood that two individuals are engaged in a conversation. This is used to derive a payoff (similarity) function between detected individuals which defines the underlying clustering game. As it turns out, the stable equilibria of this game represent maximally coherent groups, and we used simple and effective evolutionary game dynamics to extract them. Extensive experimental results provide evidence of the effectiveness of our approach as compared to other methods in the literature.

REFERENCES

[1] L. Bazzani, M. Cristani, D. Tosato, M. Farenzena, G. Paggetti, G. Menegaz, V. Murino, Social interactions by visual focus of attention in a three-dimensional environment, Expert Syst. 30 (2) (2013) 115–127.

[2] I.M. Bomze, On standard quadratic optimization problems, J. Glob. Optim. 13 (4) (Dec. 1998) 369–387.

[3] C. Chen, J. Odobez, We are not contortionists: coupled adaptive learning for head and body orientation estimation in surveillance video, in: The IEEE Conference on Computer Vision and Pattern Recognition (CVPR), IEEE, 2012, pp. 1544–1551.

[4] W. Choi, Y.W. Chao, C. Pantofaru, S. Savarese, Discovering groups of people in images, in: European Conference on Computer Vision (ECCV), 2014.

[5] W. Choi, S. Savarese, A unified framework for multi-target tracking and collective activity recognition, in: European Conference on Computer Vision (ECCV), Springer, 2012, pp. 215–230.

[6] T. Matthew Ciolek, The proxemics lexicon: a first approximation, J. Nonverbal Behav. 8 (1) (Sep. 1983) 55–79.

[7] T. Matthew Ciolek, A. Kendon, Environment and the spatial arrangement of conversational encounters, Sociol. Inq. 50 (3–4) (1980) 237–271.

[8] M. Cristani, L. Bazzani, G. Paggetti, A. Fossati, D. Tosato, A. Del Bue, G. Menegaz, V. Murino, Social interaction discovery by statistical analysis of F-formations, in: Proc. of BMVC, BMVA Press, 2011, pp. 23.1–23.12.

[9] M. Cristani, G. Paggetti, A. Vinciarelli, L. Bazzani, G. Menegaz, V. Murino, Towards computational proxemics: inferring social relations from interpersonal distances, in: SocialCom/PASSAT 2011, 2011, pp. 290–297.

[10] L. Donghoon, Y. Ming-Hsuan, O. Songhwai, Fast and accurate head pose estimation via random projection forests, in: The IEEE International Conference on Computer Vision (ICCV), December 2015.

[11] D. Fudenberg, J. Tirole, Game Theory, MIT Press, Cambridge, MA, 1991.

[12] T. Gan, Y. Wong, D. Zhang, M.S. Kankanhalli, Temporal encoded F-formation system for social interaction detection, in: Proceedings of the 21st ACM International Conference on Multimedia, MM '13, ACM, New York, NY, USA, 2013, pp. 937–946.

[13] H. Garfinkel, Studies in Ethnomethodology, Prentice-Hall, 1967.

[14] E. Goffman, Behavior in Public Places: Notes on the Social Organization of Gatherings, Free Press, 1966.

[15] G. Groh, A. Lehmann, J. Reimers, M.R. Frieß, L. Schwarz, Detecting social situations from interaction geometry, in: 2010 IEEE Second International Conference on Social Computing (SocialCom), IEEE, 2010, pp. 1–8.

[16] E.T. Hall, The Hidden Dimension, Doubleday, Garden City, NY, USA, October 1966.

[17] J. Hofbauer, K. Sigmund, Evolutionary Games and Population Dynamics, Cambridge University Press, Cambridge, UK, 1998.

[18] H. Hung, B. Kröse, Detecting F-formations as dominant sets, in: Int. Conf. on Multimodal Interaction (ICMI), 2011.

[19] H. Hüttenrauch, K.S. Eklundh, A. Green, E.A. Topp, Investigating spatial relationships in human–robot interaction, in: 2006 IEEE/RSJ International Conference on Intelligent Robots and Systems, IEEE, 2006, pp. 5052–5059.

[20] V. Jain, J.L. Crowley, Head pose estimation using multi-scale gaussian derivatives, in: Image Analysis, Springer, 2013, pp. 319–328.

[21] A. Kendon, Conducting Interaction: Patterns of Behavior in Focused Encounters, vol. 7, CUP Archive, 1990.

[22] T. Lan, Y. Wang, W. Yang, S.N. Robinovitch, G. Mori, Discriminative latent models for recognizing contextual group activities, IEEE Trans. Pattern Anal. Mach. Intell. 34 (8) (August 2012) 1549–1562.

[23] J. Lin, Divergence measures based on the Shannon entropy, IEEE Trans. Inf. Theory 37 (1991) 145–151.

[24] M. Marin-Jimenez, A. Zisserman, V. Ferrari, Here's looking at you, kid. Detecting people looking at each other in videos, in: British Machine Vision Conference, 2011.

[25] J. Maynard Smith, Evolution and the Theory of Games, Cambridge University Press, Cambridge, UK, 1982.

[26] T.S. Motzkin, E.G. Straus, Maxima for graphs and a new proof of a theorem of Turán, Can. J. Math. 17 (1965) 533–540.

[27] M. Pavan, M. Pelillo, Dominant sets and pairwise clustering, IEEE Trans. Pattern Anal. Mach. Intell. 29 (1) (2007) 167–172.

[28] E. Ricci, J. Varadarajan, R. Subramanian, S. Rota Bulo, N. Ahuja, O. Lanz, Uncovering interactions and interactors: joint estimation of head, body orientation and F-formations from surveillance videos, in: The IEEE International Conference on Computer Vision (ICCV), December 2015.

[29] S. Rota Bulò, M. Pelillo, A game-theoretic approach to hypergraph clustering, IEEE Trans. Pattern Anal. Mach. Intell. 35 (6) (June 2013) 1312–1327.
[30] L. Ruonan, P. Parker, Z. Todd, Finding group interactions in social clutter, in: The IEEE Conference on Computer Vision and Pattern Recognition (CVPR), June 2013.
[31] D. Schweingruber, C. McPhail, A method for systematically observing and recording collective action, Sociol. Methods Res. 27 (4) (1999) 451–498.
[32] F. Setti, H. Hung, M. Cristani, Group detection in still images by F-formation modeling: a comparative study, in: WIAMIS, 2013.
[33] F. Setti, O. Lanz, R. Ferrario, V. Murino, M. Cristani, Multi-scale F-formation discovery for group detection, in: International Conference on Image Processing (ICIP), 2013.
[34] F. Setti, C. Russell, C. Bassetti, M. Cristani, F-formation detection: individuating free-standing conversational groups in images, PLoS ONE 10 (5) (2015).
[35] F. Solera, S. Calderara, R. Cucchiara, Socially constrained structural learning for groups detection in crowd, IEEE Trans. Pattern Anal. Mach. Intell. 38 (5) (May 2016) 995–1008.
[36] A. Torsello, S. Rota Bulò, M. Pelillo, Grouping with asymmetric affinities: a game-theoretic perspective, in: The IEEE Conference on Computer Vision and Pattern Recognition (CVPR), 2006, pp. 292–299.
[37] K.N. Tran, A. Gala, I.A. Kakadiaris, S.K. Shah, Activity analysis in crowded environments using social cues for group discovery and human interaction modeling, Pattern Recognit. Lett. 44 (2013) 49–57.
[38] S. Vascon, L. Bazzani, Group detection and tracking using sociological features, in: Vittorio Murino, Marco Cristani, Shishir Shah, Silvio Savarese (Eds.), Group and Crowd Behavior for Computer Vision, Academic Press, 2017, pp. 29–66 (Chapter 3).
[39] S. Vascon, E.Z. Mequanint, M. Cristani, H. Hung, M. Pelillo, V. Murino, Detecting conversational groups in images and sequences: a robust game-theoretic approach, in: Inference and Learning of Graphical Models Theory and Applications in Computer Vision and Image Analysis, Comput. Vis. Image Underst. 143 (2016) 11–24.
[40] S. Vascon, Z.E. Mequanint, M. Cristani, H. Hung, M. Pelillo, V. Murino, A game-theoretic probabilistic approach for detecting conversational groups, in: Proceedings, Asian Conference on Computer Vision (ACCV), Heidelberg, Germany, in: Lect. Notes Comput. Sci., Springer, November 2014.
[41] U. von Luxburg, A tutorial on spectral clustering, Stat. Comput. 17 (4) (2007) 395–416.
[42] J.W. Weibull, Evolutionary Game Theory, Cambridge University Press, Cambridge, UK, 1995.
[43] T. Yu, S. Lim, K.A. Patwardhan, N. Krahnstoever, Monitoring, recognizing and discovering social networks, in: The IEEE Conference on Computer Vision and Pattern Recognition (CVPR), 2009.
[44] G. Zen, B. Lepri, E. Ricci, O. Lanz, Space speaks: towards socially and personality aware visual surveillance, in: 1st ACM International Workshop on Multimodal Pervasive Video Analysis, 2010, pp. 37–42.
[45] L. Zhang, H. Hung, Beyond f-formations: determining social involvement in free standing conversing groups from static images, in: The IEEE Conference on Computer Vision and Pattern Recognition (CVPR), IEEE, 2016, pp. 1086–1095.

Chapter 13

We are less free than how we think: Regular patterns in nonverbal communication☆

Alessandro Vinciarelli*, Anna Esposito†, Mohammad Tayarani*, Giorgio Roffo*, Filomena Scibelli†, Francesco Perrone*, Dong-Bach Vo*

*University of Glasgow, School of Computing Science, Glasgow, UK †Università degli Studi della Campania L. Vanvitelli, Dipartimento di Psicologia, Caserta, Italy

CONTENTS

13.1 INTRODUCTION

Everyday life behavior in naturalistic settings is typically defined as "*spontaneous*", an adjective that, according to the most common English dictionaries, accounts for the lack of planning, rules or any other inhibitions or

☆This work was supported by the Engineering and Physical Sciences Research Council (EPSRC) through the projects "*School Attachment Monitor*" (EP/M025055/1) and "*Socially Competent Robots*" (EP/N035305/1).

Multimodal Behavior Analysis in the Wild. https://doi.org/10.1016/B978-0-12-814601-9.00030-4
Copyright © 2019 Elsevier Ltd. All rights reserved.

constraints: "*done in a natural, often sudden way, without any planning or without being forced*" (Cambridge Dictionary), "*Spontaneous acts are not planned or arranged, but are done because someone suddenly wants to do them*" (Collins English Dictionary), "*proceeding from natural feeling or native tendency without external constraint*" (Merriam Webster), "*Performed or occurring as a result of a sudden impulse or inclination and without premeditation or external stimulus*" (Oxford Dictionary), etc. On the other hand, social psychology has shown that everyday behavior tends to follow principles and laws that result into stable behavioral patterns that can be not only observed, but also analyzed and measured in statistical terms. Thus, at least in the case of everyday behavior, the actual meaning of the adjective "*spontaneous*" seems to be the tendency of people to effortlessly adopt and display, typically outside conscious awareness, recognizable behavioral patterns.

There are at least two reasons behind the presence of stable behavioral patterns. The first is that behavior—in particular its nonverbal component (facial expressions, gestures, vocalizations, etc.)—contributes to communication by adding layers of meaning to the words being exchanged. In other words, people that accompany the same sentence (e.g., "*I am happy*") with different behaviors (e.g., frowning and smiling) are likely to mean something different. This requires the adoption of stable behavioral patterns because communication can take place effectively only when there is coherence between the meaning that someone tries to convey and the signals adopted to convey it [37]. The second important reason behind the existence of stable behavioral patterns is that no smooth interaction is possible if the behavior of its participants is not, at least to a certain extent, predictable [30]. For example, during a conversation people that talk expect others to listen and, when this does not happen, people tend to adopt repairing mechanisms that bring back the interactants to a situation where one person speaks and the others listen (unless there is an ongoing conflict and then people compete for the floor or there is no intention to be involved in the conversation for one or more interactants).

The considerations above suggest that human behavior, while being a complex and rich phenomenon, is not random and it is possible to detect order in it. Such an observation plays a crucial role in any discipline that tries to make sense of human behavior, including social psychology, anthropology, sociology and, more recently, technological domains aimed at seamless interaction between people and machines like, e.g., Human–Robot Interaction [13], Affective Computing [21], Computational Paralinguistics [28] or Social Signal Processing [32]. In particular, the possibility to use human behavior as a physical, machine detectable evidence of social and psychological phenomenon—if behavior can be perceived through the senses and interpreted unconsciously, it can be detected via sensors—can help to make machines socially intelligent, i.e., capable to understand the social landscape in the same way as people do [33].

The rest of this chapter aims at showing what the expression "*stable behavioral pattern*" means exactly through several experimental examples. In particular, the chapter aims at providing the reader with a few basic techniques that can help the detection of behavioral patterns in data, with the focus being on conversations as these are the primary site of human sociality. Overall, the chapter shows that human behavior is far less free than how it might look like, but this, far from being a constraint or an inhibition, is the very basis for social interaction. Section 13.2 shows how the occurrence of certain nonverbal behavioral cues accounts for important social dimensions such as gender and role; Section 13.3 shows that the sequence of speakers in a conflictual conversation provides information on who agrees with whom; Section 13.4 shows how people tend to imitate the behavioral patterns of others and the final Section 13.5 draws some conclusions.

13.2 **ON SPOTTING CUES: HOW MANY AND WHEN**

One of the key-principles of all disciplines based on behavior observation (ethology, social psychology, etc.) can be formulated as follows:

> *"[...] the circumstances in which an activity is performed and those in which it never occurs [provide] clues as to what the behavior pattern might be for (its function)" [19].*

On such a basis, this section shows that one of the simplest possible observations about behavior—how many times a given cue takes place and when—can provide information about stable behavioral patterns and, most importantly, about the social and psychological phenomena underlying the patterns. The observations are made over the *SSPNet Mobile Corpus* [22], a collection of 60 phone calls between 120 unacquainted individuals (708 min and 24 s in total). The conversations revolve around the *Winter Survival Task* [15], a scenario that requires two people to identify, in a list of 12 objects, the items most likely to increase the chances of survival for people that crash with a plane in a polar area. The main reason to use such a scenario is that it allows unacquainted individuals to start and sustain a conversation without the need to find a topic. Furthermore, since it happens rarely that the people involved in the experiments are expert in survival techniques—only 1 out of 120 subjects in the case of the SSPNet Mobile Corpus—the interaction tends to be driven by actual social and psychological factors rather than by competence and knowledge differences about the topic of conversation.

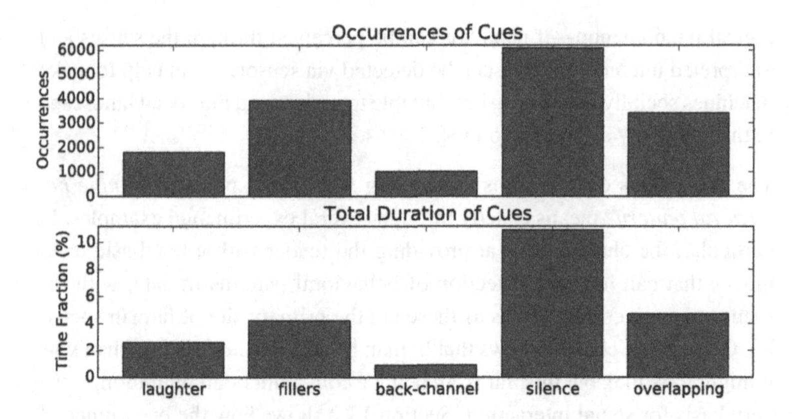

■ **FIGURE 13.1** The upper chart shows the number of occurrences for each cue, while the lower chart shows the fraction of the total corpus time every cue accounts for.

In the case of the SSPNet Mobile Corpus, the conversations take place via phone and, hence, the participants can use only paralinguistics, i.e., vocal nonverbal behavioral cues. Given that many cues (e.g., laughter or pauses) require one to stop speaking, the amount of time spent in nonverbal communication can give a measure of how much such a phenomenon is important in communication (see below).

13.2.1 **The cues**

The Corpus has been annotated in terms of five major nonverbal cues, namely laughter, fillers, back-channel, silence and overlapping speech (or interruptions). The motivation behind such a choice is that these cues are among the most common (see below for the statistics) and widely investigated in the literature [17,25]. The left chart of Fig. 13.1 provides the number of times each of the cues above has been observed in the corpus, while the right chart shows how much time each of the cues accounts for in total. Overall, the total number of occurrences is 16,235, corresponding to 23.5% of the total time in the corpus. While being under time pressure—the Winter Survival Task must be completed as quickly as possible—the participants invest roughly one quarter of their time in nonverbal communication rather than in speaking. This is a clear indication of how important the cues mentioned above are.

Laughter is one of the first nonverbal behavioral cues that have been addressed in scientific terms. The first studies date back to the seminal work by Darwin about the expression of emotions [9] and the behavior of children [10]. In more recent times, laughter has been investigated in social psychology and cognitive sciences and one of its most common definitions is as follows: "*a common, species-typical human vocal act and auditory signal that is important in social discourse*" [24]. According to the literature (see [23] for a survey), the most common patterns associated to laughter are that women laugh more than men, that people tend to laugh more when they listen than when they talk and that, in general, people laugh at the end of sentences. Furthermore, recent work has shown that laughter tends to take place in correspondence of topic changes [5]. Fig. 13.1 shows that the SSPNet Mobile corpus includes 1805 laughter occurrences for a total duration of 1114.8 s (2.6% of the total Corpus length), corresponding to one occurrence every 23.5 s, on average.

When people involved in a conversation want to hold the floor, but do not know what to say next, they tend to use expressions like "*ehm*" or "*uhm*," called *fillers*, to signal the intention to keep speaking. The reason behind the name is that they "*are characteristically associated with planning problems [...] planned for, formulated, and produced as parts of utterances just as any word is*" [8]. Fig. 13.1 shows that the 120 subjects of the SSPNet Mobile Corpus utter 3912 fillers that correspond to a total length of 1815.9 s (4.2% of the Corpus time). The average interval between one filler and the other is 10.9 s. The symmetric cue is *back-channel*, i.e., a short utterance like "*ah-ah*" or "*yeah*" that is "*produced by one participant in a conversation while the other is talking*" [36]. In general the goal of back-channel is to signal to others attention and agreement while implicitly saying that they can continue to speak. The total number of back-channel episodes in the Corpus is 1015 (one episode every 41.9 s, on average), for a total of 407.1 s (0.9% of the Corpus time).

The time intervals during which both participants of a conversation were not speaking have been labeled as *silence*. Such a definition does not take into account the multiple functions of such a cue—6091 occurrences for a total of 4670.6 s (10.9% of the corpus length)—but it still allows one to test whether there is any relationship between the occurrences of silence and social or psychological phenomena of interest. The same applies to *overlapping speech*, i.e., the time intervals during which both people involved in the same conversation speak at the same time. Such a phenomenon can take place for different reasons and with different functions, but it tends to be an exception rather than the norm [27]. For this reason, it is important to count the occurrences of such a cue that are 3412 for a total of 2000.5 s (4.7% of the corpus time).

13.2.2 **Methodology**

Each of the cues annotated in a corpus can be thought of as a triple (c_i, t_i, d_i), where $i = 1, \ldots, N$ (N is the total number of cues that have been counted), $c_i \in \mathcal{C} = \{C_1, \ldots, C_L\}$ is one of the L cues that have been counted (in the case of the SSPNet Mobile Corpus, the cues are laughter, filler, back-channel, silence and overlapping speech), t_i is the time at which the cue i begins, and d_i is its duration. Such a notation allows one to define the number N_c of occurrences of a given cue as follows:

$$N_c = \left|\{(c_i, t_i, d_i) : c_i = c\}\right|, \tag{13.1}$$

where $|.|$ denotes the cardinality of a set.

The corpus can be segmented into time intervals according to the value of a variable V that corresponds to a factor of interest for the analysis of the data. For example, in the case of the gender, the variable can take two values—v_1 and v_2—that correspond to *female* and *male*, respectively. In this way, it is possible to count the number of times a cue c takes place when $v = v_k$ as follows:

$$N_c^{(k)} = \left|\{(c_i, t_i, d_i) : c_i = c, [t_i, t_i + d_i] \in V_k\}\right|, \tag{13.2}$$

where V_k is the set of all corpus segments in which $v = v_k$.

The notations above allow one to define the following χ^2 variable:

$$\chi^2 = \sum_{k=1}^{K} \frac{N_c^{(k)} - N_c E_k}{E_k}, \tag{13.3}$$

where E_k is the fraction of total corpus length in which the value of V is v_k. The particularity of the χ^2 variable is that its probability density function is known in the case there is no relationship between the cue c and the variable v. In other words, the probability density function is known when the differences in $N_c^{(k)}$ do not depend on the interplay between the cue and the factor underlying the variable V, but on simple statistical fluctuations. This makes it possible to estimate the probability to obtain a value of χ^2 at least as high as the one estimated with Eq. (13.3) when the occurrence of a cue does not depend on the factor underlying V. When such a probability is lower than a value α, then in it possible to say that there is an *effect* with confidence level α, i.e., that there is a relationship between the cue and the factor underlying V and the probability that such an observation is a false positive—meaning that it is the result of chance—is lower than α. In general, the literature accepts an effect when the probability α, called the *p-value*, is lower than 0.05 or 0.01.

When the approach above is used several times with confidence level α, then the probability that none of the observed effects is the result of chance is $p = (1 - \alpha)^M$, where M is the number of statistical inferences that someone

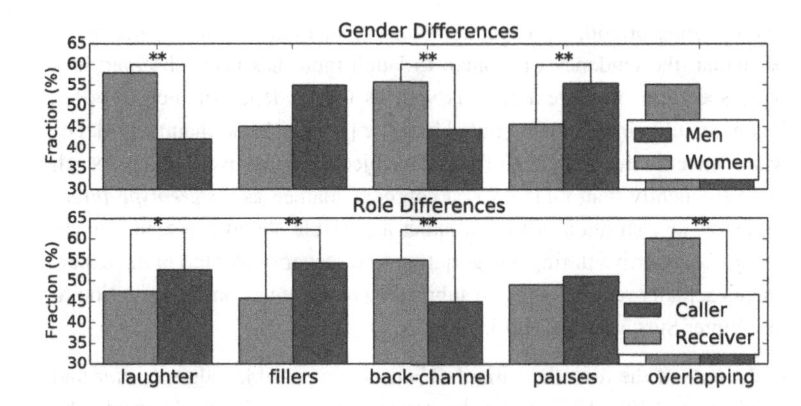

■ **FIGURE 13.2** The upper chart shows the difference between male and female subjects, while the lower one shows the difference between callers and receivers. The single and double stars show effects significant at 5% and 1%, respectively (after Bonferroni correction).

makes out of the data (or the total number of p-values that are estimated). For this reason, it is common to apply a correction, i.e., to modify the value of α to ensure that the risk of false positives is eliminated or, at least, limited to a point that possible false positives do not change the conclusions that can be made out of the observed effects. The correction most commonly applied is the *Bonferroni* one. Following such an approach, an effect is considered to be significant at level α only when the value is lower than α/M. Such a correction aims at eliminating all false positives, but this happens at the cost of eliminating many true positives as well. For this reason, most recent approaches apply the *False Discovery Rate* (FDR) correction [3]. This correction aims at ensuring that false positives, if any, are sufficiently few not to change the conclusions that can be made out of the observed effects. In the case of the FDR, the values are ordered from the smallest to the largest and the one that ranks kth is accepted if it is lower than $\alpha M/k$.

13.2.3 **Results**

The methodology described earlier has been applied to the SSPNet Mobile Corpus using gender and role. In the case of gender, the variable V can take two values, *female* and *male*, and the results of the analysis are illustrated in Fig. 13.2. The chart shows that female subjects tend to display more frequently than male ones laughter, back-channel and overlapping speech (meaning that they start speaking when someone else does more frequently). In other words, the distribution of such cues is not random, but it changes according to the gender of speaker, thus showing that there are different behavioral patterns for female and male subjects (at least in the corpus under analysis).

The literature provides confirmation and explanations for these patterns. In particular, the tendency of woman to laugh more has been observed in a wide spectrum of context [23] as well as the tendency of men to adopt higher-status behaviors like limited laughter [18] and back-channel [14]. For what concerns the tendency of female subjects to start overlapping speech more frequently than male ones, it can be explained as a *stereotype-threat* behavior, i.e., an attempt to contradict a possible stereotype abut women being less assertive during a negotiation (the scenario adopted in the experiments actually requires the two subjects to negotiate a common solution to the Winter Survival Task) [29].

In the case of the role, the variable V takes two possible values (*caller* and *receiver*) and Fig. 13.2 shows the patterns that have been observed. The callers tend to display fillers more frequently than expected and, viceversa, they ten to initiate overlapping less frequently than how it would happen in absence of effects. Such a pattern is compatible with a status difference, i.e., with the tendency of the receivers to behave like if they were above the callers in terms of social verticality and the tendency of these latter to behave like if they are hierarchically inferior [20,26]. As a confirmation of these behavioral patterns, the receivers tend to win the negotiations involved in the Winter Survival Task significantly more frequently than the callers [34].

Overall, the observations above show that people actually display detectable behavioral patterns that can be identified with a methodology as simple as counting the cues and showing, through a statistical significance test, whether they tend to occur more or less frequently according to the value of an independent variable V expected to account for an underlying social or psychological phenomenon (gender and role in the case of this section). In other words, the observations of this section show that nonverbal behavior, especially when it comes to cues that are displayed spontaneously and result from the interaction with others, is not random but it tends to follow behavioral patterns.

13.3 ON FOLLOWING TURNS: WHO TALKS WITH WHOM

One of the main tenets of conversation analysis is that people tend to speak one at a time and, in general, overlapping speech and silence tend to result from errors in turn-taking, the body of practices underlying the switch from one speaker to the other, more than from the actual intentions of interacting speakers:

> *"Talk by more than one person at a time in the same conversation is one of the two major departures that occur from what appears to be*

a basic design feature of conversation, [...] namely 'one at a time' (the other departure is silence, i.e. fewer than one at a time)" [27].

This seems to contradict the observation of Section 13.2 that overlapping speech and silence account, jointly, for roughly 15% of the total time in a corpus of conversations. However, the data of Section 13.2 shows that the average length of a silence or an overlapping speech segment are lower than a second. This shows that, however frequent, such episodes tend to be short and people actually tend to stay in a situation where one person speaks and the others listen.

Following up on the above, it possible to think of a conversation as a sequence of triples $(s_i, t_i, \Delta t_i)$, where $i = 1, \ldots, N$ (N is the total number of turns in a conversation), $s_i \in A = \{a_1, \ldots, a_G\}$ is one of the G speakers involved in a conversation, t_i is the time at which turn i starts and Δt_i is the duration of the turn. The rest of this section shows how such basic information can be used to detect the structure of a conflict.

13.3.1 Conflict

The literature proposes many definitions of conflict and each of them captures different aspects of the phenomenon (see the contributions in [12] for a wide array of theoretic perspectives and approaches). However, there is a point that all researchers investigating conflict appear to agree upon, namely that the phenomenon takes place whenever multiple parties try to achieve incompatible goals, meaning that one of the parties can achieve its goals only if the others do not (or at least such is the perception of the situation): "*conflict is a process in which one party perceives that its interests are being opposed or negatively affected by another party*" [35], "*[conflict takes place] to the extent that the attainment of the goal by one party precludes its attainment by the other*" [16], "*Conflict is perceived [...] as the perceived incompatibilities by parties of the views, wishes, and desires that each holds*" [2], etc.

Goals cannot be accessed directly, but can only be inferred from the observable behavior, both verbal and nonverbal, of an individual. For this reason, the automatic analysis of conflict consists mainly of detecting the physical traces that conflict leaves in behavior (what people do different or peculiar when they are involved in conflict) and using them to predict whether there is conflict or what are the characteristics of an ongoing conflict. In the particular case of turns, the main physical trace of conflict is the tendency of people to react immediately to interlocutors they disagree with [4]. This means that speaker adjacency statistics should provide indications on who agrees or disagrees with whom and, hence, on the presence of possible groups that oppose one another.

13.3.2 Methodology

The beginning of this section shows that a conversation can be thought of, with good approximation, as a sequence of triples $(s_i, t_i, \Delta t_i)$, with each of these corresponding to a turn. This section shows that the sequence of the speakers—meaning who speaks before and after whom—far from being random provides information about the composition of groups, if any, that oppose one another during a discussion. The main reason why this is possible is that people involved in conversations—and more in general in social interactions—tend to adopt *preference structures*, i.e., they tend to behave in a certain way because any alternative way of behaving would be interpreted in a wrong way. In the case of conflict, the preference structure is that people that do not react immediately to someone they disagree with will be considered to agree or to have no arguments at disposition.

In formal terms, this means that we can attribute each speaker involved in a conversation three possible labels, namely g_1 (the speaker belongs to group 1), g_2 (the speaker belongs to group 2) or g_0 (the speaker is neutral). Since every speaker is assigned always the same label, the sequence of the speakers can be mapped into a sequence of symbols $X = (x_1, \ldots, x_N)$, where $x_i \in \{g_0, g_1, g_2\}$ and $x_i = f(s_i)$—$f(s_i)$ is a function that maps a speaker into one of the three possible labels.

The simplest probabilistic model of the sequence $X = (x_1, \ldots, x_N)$ is the *Markov chain*:

$$p(X) = p_{x_1} \prod_{k=2}^{T} p(x_{k-1}|x_k), \tag{13.4}$$

where $p(x_1)$ is the probability of starting with label x_1 and $p(x_{k-1}|x_k)$ is the probability of a transition from x_{k-1} to x_k.

13.3.3 Results

Such a model has been adopted to analyze the conflict in the political debates of the Canal 9 Corpus, a collection of 45 political debates involving a total of 190 persons speaking for 27 h and 56 min [31]. The main reason for using political debates is that these are built around conflict. In other words, following the definition at the beginning of this section, if one of the participants achieves its goals, the others do not. Furthermore, the political debates have a structure that allows one to make simplifying assumptions, namely that the number of participants is always the same (5 in the debates of the Corpus), that the number of members per group is always the same (2 in the debates of the Corpus) and that there is always a neutral moderator (1 person in the debates of the Corpus). In this way, the number of functions $f(s)$ mapping the speakers to the labels is limited to $L = 15$, i.e., there are only 15 ways to assign the labels to the speakers when taking into account the assumptions above.

Given that each of the 15 functions gives rise to a different sequence X_j of labels, it is possible to identify the function that produces the most probable sequence:

$$X^* = \arg\max_{j=1,\ldots,L} p(X_j). \tag{13.5}$$

The function that generates X^* is the one most likely to group the debate participants in the correct way, i.e., to correctly identify the composition of the groups and the moderator. The experiments performed over the Canal 9 Corpus with the approach above show that the grouping is correct in 64.5% of the cases, thus confirming that the sequence of the speakers is not random, but it follows detectable behavioral patterns that are stable enough to allow the development of automatic analysis approaches.

13.4 ON SPEECH DANCING: WHO IMITATES WHOM

Human sciences show that people change their way of communicating according to the way others communicate to them. The phenomenon takes place in different ways and, correspondingly, it takes different names in the literature. The unconscious tendency to imitate others is typically referred to as *mimicry* [7], the adoption of similar timing patterns in behavior is typically called *synchrony* [11] and, finally, the tendency to enhance similarity or dissimilarity in observable behavior typically goes under the name of *adaptation* [6].

This section focuses on one of the many facets of adaptation, i.e., the tendency to change acoustic properties of speech in relationship to the way others speak in a given conversation. In particular, this work investigates the interplay between the acoustic adaptation patterns that people display in a conversation and their performance in addressing a collaborative task. The problem is important because it can provide insights about the role of adaptation in collaborative problem solving. Furthermore, it can give indications on how to improve the efficiency of people that address a task together.

13.4.1 Methodology

The words of a conversation can be mapped into sequences $X = (\vec{x}_1, \ldots, \vec{x}_T)$ of *observation vectors*, i.e., vectors where every component is a physical measurement extracted from the speech signal. The reason why X includes multiple vectors is that these are extracted at regular time steps from short signal segments. The motivation behind such a representation is that it allows the application of statistical modeling approaches.

The goal of the methodology presented in this section is to measure the tendency of A and B, the two speakers involved in a conversation, to become more or less similar over time. If $p(X|\theta_A)$ and $p(X|\theta_B)$ are probability distributions trained over spoken material of A and B, respectively, then it is possible to estimate the following likelihood ratio for every word $w_k^{(A)}$ in a conversation, where the A index means that the word has been uttered by A:

$$d_k(A, B) = \log \frac{p(X_k^{(A)}|\theta_B)}{p(X_k^{(A)}|\theta_A)}. \tag{13.6}$$

The expression of $d_k(B, A)$ can be obtained by switching A and B in the equation above and, in general, $d_k(A, B) \neq d_k(B, A)$. If $d_k(A, B) > 0$, it means that $p(X|\theta_B)$ explains $w_k^{(A)}$ better than $p(X|\theta_B)$. Vice versa, if $d_k(A, B) < 0$, it means that it is the model of speaker A that explains the data better than the model of speaker B. Similar considerations apply to $d_k(B, A)$.

In the experiments of this chapter, $p(X|\theta)$ is estimated with hidden Markov models (the speaker index is omitted for making the notation lighter), a probability density function defined over joint sequences of states and feature vectors (X, S), where $X = (\vec{x}_1, \dots, \vec{x}_T)$ and $S = (s_1, \dots, s_T)$ with every s_i belonging to a finite set $V = \{v_1, \dots, v_L\}$:

$$p(X, S|\theta) = \pi_{s_1} b_{s_1}(\vec{x}_1) \cdot \prod_{k=2}^{T} a_{s_k s_{k-1}} b_{s_k}(\vec{x}_k), \tag{13.7}$$

where π_{s_1} is the probability of s_1 being the first state of the sequence, $a_{s_k s_{k-1}}$ is the probability of a transition between s_{k-1} and s_k, $b_{s_k}(\vec{x})$ is the probability of observing a vector $\vec{x}$ when the state is s_k, and T is the total number of vectors in sequence X. In the experiments of this work, the value of $p(X|\theta)$ is approximated as follows:

$$p(X|\theta) = \arg\max_{S \in \mathcal{S}_T} p(X, S|\theta), \tag{13.8}$$

where $\mathcal{S}_T$ is the set of sequences of all possible states of length T.

Given that the features adopted in the experiments are continuous, the emission probability functions $b_{v_k}(\vec{x})$ are mixtures of Gaussians, i.e., weighted sums of multivariate normal distributions:

$$b_{v_k}(\vec{x}) = \sum_{l=1}^{G} \alpha_l^{(k)} \mathcal{N}(\vec{x}|\vec{\mu}_l^{(k)}, \Sigma_l^{(k)}), \tag{13.9}$$

where $\sum_{l=1}^{G} \alpha_l^{(k)} = 1$, $\vec{\mu}_l^{(k)}$ is the mean of Gaussian l in the mixture of state v_k and $\Sigma_l^{(k)}$ is the covariance matrix in the same Gaussian.

One of the main physical evidences of adaptation in human–human communication is that the behavior of people changes over time: "*A necessary although not sufficient condition for establishing adaptation is showing that at least one person's behaviors change over time*" [6]. In the experiments of this work, the behaviors of interest are the acoustic properties of speech, represented in terms of feature vectors extracted from speech samples of A and B.

If $t_k^{(A)}$ is the time at which word $w_k^{(A)}$ starts, then it is possible to estimate the *Spearman* correlation coefficient $\rho(A, B)$ between $d_k(A, B)$ and $t_k^{(A)}$:

$$\rho(A, B) = 1 - \frac{6 \cdot \sum_{j=1}^{N_A} (\Delta r_j)^2}{N_A \cdot (N_A^2 - 1)}, \tag{13.10}$$

where Δr_j is the difference between, on the one hand, the rank of $d_j(A, B)$ across all observed values of $d_k(A, B)$ and, on the other hand, the rank of $t_j^{(A)}$ across all observed values of $t_k^{(A)}$. The value of $\rho(B, A)$ can be obtained by simply switching A and B in the equations above. The motivation behind the choice of the Spearman coefficient is that it is more robust to the presence of outliers with respect to other measurements of the correlation like, e.g., the *Pearson* correlation coefficient. In fact, the Spearman coefficient does not take into account the value of the variables, but their ranking among the observed values. In this way, it is not possible for one outlier to change the value of the coefficient to a significant extent.

When $\rho(A, B)$ is positive and statistically significant, it means that A tends to become more similar to B in terms of distribution of the features that are extracted from speech. This seems to correspond to the "*exhibition of similarity between two interactants, regardless of etiology, intentionality or partner influence*" [6], a phenomenon called *matching*. Conversely, when $\rho(A, B)$ is negative and statistically significant, it means that A tends to become increasingly more different—in terms of features' distribution—from B over time. This appears to correspond to the "*exhibition of dissimilar behavior [...] regardless of etiology, intent or partner behavior*" [6], a phenomenon called *complementarity*. If $\rho(A, B)$ is not statistically significant, it means that no matching and complementarity take place or, if they do,

Table 13.1 This table shows the adaptation patterns that a pair of speakers A and B can display. Symbols $+$, $-$ and $=$ stand for matching, complementarity and lack of observable effects, respectively. Acronyms CO, DI, CM and IN stand for convergence, divergence, complementarity and independence, respectively

$A+B+$ (CO)	$A+B=$ (CO)	$A+B-$ (CM)
$A=B+$ (CO)	$A=B=$ (IN)	$A=B-$ (DI)
$A-B+$ (CM)	$A-B=$ (DI)	$A-B-$ (DI)

they are too weak to be observed. All the considerations made above about $\rho(A,B)$ apply to $\rho(B,A)$ as well.

Following the definitions above, the combination of $\rho(A,B)$ and $\rho(B,A)$ provides a representation of the adaptation pattern, if any, that a dyad of interacting speakers displays. Table 13.1 shows all possible combinations, where the symbols "+", "−" and "=" account for *matching* (the correlation is positive and statistically significant), *complementarity* (the correlation is negative and statistically significant), or *lack of observable effects* (the correlation is not statistically significant), respectively. In every cell, the symbol on the left is $\rho(A,B)$ while the other one is $\rho(B,A)$. The combinations result into the following adaptation patterns:

- *Convergence* (CO): the correlation is positive and statistically significant for at least one of the two speakers and no negative and statistically significant correlations are observed (cells "++", "+=" and "=+").
- *Divergence* (DI): the correlation is negative and statistically significant for at least one of the two speakers and no positive and statistically significant correlations are observed (cells "−−", "−=" and "=−").
- *Compensation* (CM): the correlation is statistically significant for both speakers, but with opposite sign (cells "+−", "−+").
- *Independence* (IN): both correlations are not statistically significant and adaptation patterns, if any, are too weak to be detected (cell "==").

The experiments of this chapter aim not only at detecting the adaptation patterns above, but also at showing whether they have any relationship with the performance of the subjects, i.e., whether certain patterns tend to appear more or less frequently when the subjects need more time to address a task.

13.4.2 **Results**

The experiments of this work have been performed over a collection of 12 dyadic conversations between fully unacquainted individuals (24 in total). All subjects are female that were born and raised in Glasgow (United Kingdom). The conversations revolve around the *Diapix UK Task* [1], a scenario commonly adopted in human–human communication studies. The task is based on 12 pairs of almost identical pictures—referred to as the *sub-tasks* hereafter—that differ only for a few, minor details (e.g., the t-shirt of a given person has two different colors in the two versions of the same picture). The 12 sub-tasks can be split into three groups, each including *four* pairs of pictures, corresponding to the different scenes being portrayed, namely the *Beach* (B), the *Farm* (F) and the *Street* (S). The number of differences to be spotted is the same for all sub-tasks and the subjects involved in the experiments are asked to complete the task as quickly as possible.

As a result of the scenario, each of the 12 conversations of the corpus can be split into 12 time intervals corresponding to the sub-tasks. Thus, the data includes $12 \times 12 = 144$ conversation segments during which a dyad addresses a particular sub-task, i.e., it spots the differences for a particular pair of pictures. The segments serve as analysis units for the experiments and, on average, they contain 1502.5 words for a total of 216,368 words in the whole corpus.

During the conversations, the two subjects addressing the task together sit back-to-back and are separated by a non-transparent curtain. This ensures that the two members of the same dyad can communicate via speech, but not through other channels like, e.g., facial expressions or gestures. Furthermore, the setup ensures that each member of a dyad can see one of the pictures belonging to a Diapix sub-task, but not the other. Thus, the subjects can complete the sub-tasks only by actively engaging in conversation with their partners. As a result, all subjects have uttered at least 40.1% of the words in the conversations they were involved in. Moreover, the deviation with respect to a uniform distribution (meaning that the speakers utter the same number of words) is statistically significant only in one case.

Since the subjects are asked to spot the differences between two pictures as quickly as possible, the amount of time needed to address a sub-task can be used as a measure of performance: The shorter it takes to complete a sub-task, the higher the performance of a dyad. The lower plot of Fig. 13.3 shows the average amount of time required to address each of the sub-tasks (the error bars correspond to the standard deviations). Some of these require, on average, more time to be completed, but the differences are not statistically significant according to a *t*-test. Thus, the performance of the dyads is, on average, the same over all sub-tasks and none of these appears to be more challenging than the others to a statistically significant extent.

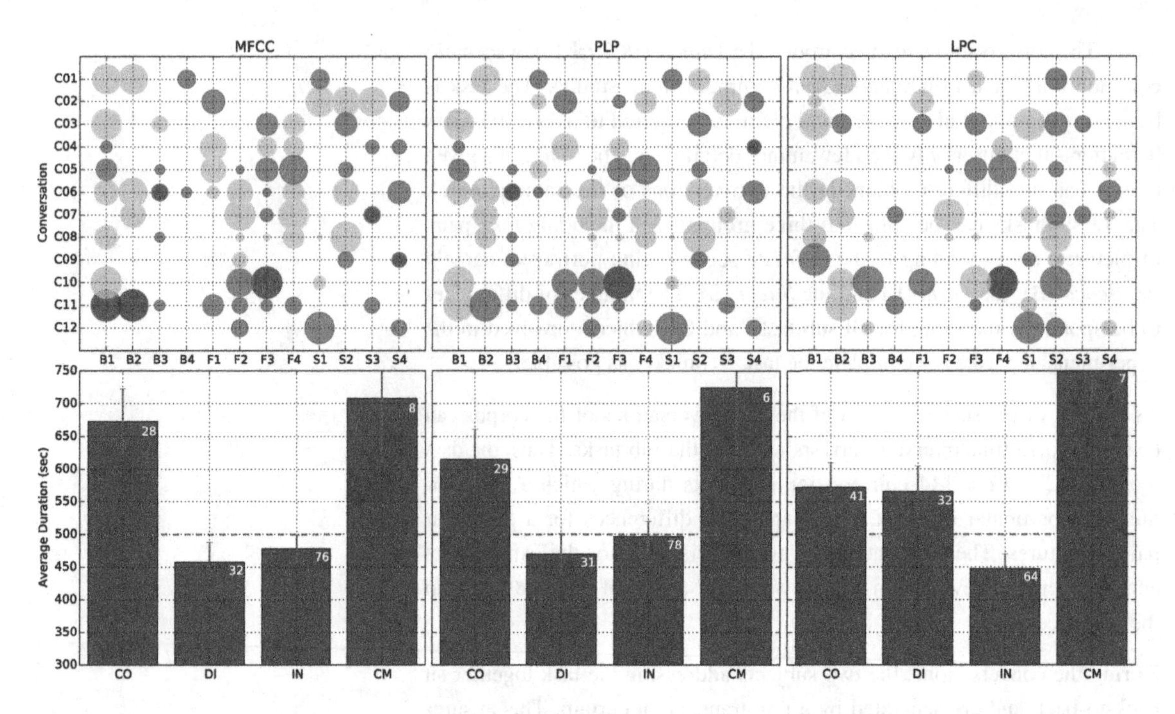

■ **FIGURE 13.3** The scatter plots show the correlations that are statistically significant (the size of the bubble is proportional to the time needed to complete the corresponding sub-task), blue bubbles correspond to complementarity patterns, red bubbles to divergence patterns, and yellow bubbles to convergence patterns. The charts show the average duration of the task where a certain pattern is observed. The error bar is the value of σ/n, where σ is the standard deviation and n is the number of sub-tasks in which the pattern is observed.

The experiments have been performed with HMMs that have only one state and adopt *Mixtures of Gaussians* (MoGs) as emission probability functions:

$$p(X|\theta) = \pi_{s_1} b_{s_1}(\vec{x}_1) \cdot \prod_{k=2}^{T} a_{s_k s_{k-1}} b_{s_k}(\vec{x}_k) = \prod_{k=1}^{T} b(\vec{x}_k), \tag{13.11}$$

where $b(\vec{x}_k)$ is the MoG (used as emission probability function). The equation above holds because $\pi_{s_1} = 1$ (there is only one state and its probability of being the first state is consequently 1) and $a_{s_k s_{k-1}} = a = 1$ (the only possible transition is between the only available state and itself). MoGs are word independent because the same model is used for all words uttered by a given speaker. Furthermore, MoGs are time independent because the value of $p(X|\theta)$ depends on the vectors included in X, but not on their order.

The main goal behind the use of MoGs is to verify whether adaptation takes place irrespectively of the words being uttered and of possible temporal patterns in acoustic properties. This is important because it can show whether the patterns detected by the approach are just a side effect of lexical alignment—the tendency to use the same words across participants of the same conversation—or the result of actual accommodation taking place at the acoustic level.

In the experiments of this work, the number G of Gaussians in the mixtures is 10. This parameter has been set *a priori* and no other values have been tested. The choice of G results from two conflicting requirements. On the one hand, G needs to be large to approximate as closely as possible the actual distribution of the data. On the other hand, the larger is G, the larger is the number of parameters and, hence, the larger is the amount of training material needed. A different MoG has been trained for each speaker of the corpus. The training has been performed using the *Expectation–Maximization* algorithm and all the words uttered by a given speaker have been used as training material. The motivation behind this choice is that the goal of the experiments is not to detect adaptation in unseen data, but to analyze the use of adaptation in a corpus of interactions. In other words, the assumption behind the approach proposed in this work is that the whole material to be analyzed is at disposition at the moment of the training.

Fig. 13.3 shows the results for different ways of mapping speech into vectors, called MFCC, PLP and LPC, respectively (the dimensionality of the feature vectors is 12 in the three cases). The bubble plots show the accommodation patterns detected in individual Diapix UK sub-tasks. The size of the bubbles is proportional to the amount of time needed to complete the sub-task. Whenever a bubble is missing, it means that no statistically significant correlations have been observed (the detected pattern is independence). Statistically significant values of $\rho(A, B)$ and $\rho(B, A)$ can be observed over the whole range of conversation lengths observed. This shows that the detected patterns are not the mere effect of conversation length, but the result of the way the subjects speak. Overall, the number of $\rho(A, B)$ and $\rho(B, A)$ values that are statistically significant with $p < 0.05$ (according to a t-test) is 116 in the case of MFCC, 114 in the case of PLP and 127 in the case of LPC. According to a binomial test, the probability to get such a result by chance is lower than 10^{-6} in all cases. This seems to further confirm that the observations result from patterns actually detected in the data.

The use of different feature extraction processes leads to the detection of adaptation patterns in different sub-tasks. The agreement between two feature extraction processes can be measured with the percentage of cases in which the detected pattern is the same for a given sub-task. In the case of MFCC and PLP, the percentage is 83.3%, while it is 51.4% and 52.8% for the comparison MFCC vs LPC and PLP vs LPC, respectively. In other words, while MFCC and PLP tend to agree with each other, LPC tends to disagree with both the other feature extraction approaches. One possible explanation is that MFCC and PLP account for perceptual information while LPC does not. Hence, LPC allows one to detect adaptation patterns in correspondence of acoustic properties that the other feature sets do not capture and vice versa.

Following the indications of communication studies [6], the investigation of the interplay between detected adaptation patterns and other observable aspects of the interaction can provide indirect confirmation that the approach captures actual communicative phenomena. For this reason, the charts in the lower part of Fig. 13.3 show the average amount of time required to complete the sub-tasks in correspondence of different detected adaptation patterns. In the case of MFCC, there is a statistically significant difference between the length of the sub-tasks in which the detected pattern is convergence and those in which it is divergence or independence ($p < 0.001$ in both cases according to a t-test). The same can be observed for PLP ($p < 0.001$ according to a t-test), but, in this case, there are other statistically significant differences as well. In particular, between divergence and complementarity ($p < 0.05$ according to a t-test) and between independence and complementarity ($p < 0.05$ according to a t-test).

Overall, the findings above show that the outcomes of the approach do not distribute randomly but tend to be associated to the difficulty that the subjects experience when they address a given sub-task, at least in the case of MFCC and PLP. In fact, the subjects are asked to complete the tasks as quickly as they can and, hence, they take more time when they encounter more difficulties.

The relationship between the occurrence of certain adaptation patterns and the amount of time needed to complete a sub-task provides indirect confirmation that the approach captures actual adaptation patterns. The lack of effects for LPC further confirms such a hypothesis. In fact, MFCC and PLP account for speech properties that the subjects can perceive and, hence, they can react to by displaying adaptation. This is not the case for LPC that tends to disagree with the other feature extraction processes (see above) and probably accounts for information that the subjects do not manage to use when they display adaptation patterns.

13.5 CONCLUSIONS

This chapter has shown that human behavior is not random, even when it is spontaneous and not scripted, but it follows principles and laws that result into regular and detectable behavioral patterns. The exact meaning of such an expression changes according to the cases, but it typically corresponds to behavioral cues that change their observable and statistical properties according to underlying social and psychological phenomena. The example of Section 13.2 shows that the very number of times people display a given cue—e.g., laughter or fillers—depends on two major social characteristics, namely gender and role. In the case of Section 13.3, the focus is on turns and their sequence. The example shows that the very sequence of speakers—who talks to whom—provides information about conflict in political debates. Finally, the experiments of Section 13.4 show that people collaborating on a task make their speaking style more or less similar to the speaking style of their interlocutors and, furthermore, such a phenomenon interacts with the amount of time required to complete a task. In all cases, the existence of the patterns is the basis for the development of automatic analysis approaches.

REFERENCES

[1] R. Baker, V. Hazan, DiapixUK: task materials for the elicitation of multiple spontaneous speech dialogs, Behav. Res. Methods 43 (3) (2011) 761–770.

[2] C. Bell, F. Song, Emotions in the conflict process: an application of the cognitive appraisal model of emotions to conflict management, Int. J. Confl. Manage. 16 (1) (2005) 30–54.

[3] Y. Benjamini, Y. Hochberg, Controlling the False Discovery Rate: a practical and powerful approach to multiple testing, J. R. Stat. Soc. B (1995) 289–300.

[4] J. Bilmes, The concept of preference in conversation analysis, Lang. Soc. 17 (02) (1988) 161–181.

[5] F. Bonin, N. Campbell, C. Vogel, Time for laughter, Knowl.-Based Syst. 71 (2014) 15–24.

[6] J.K. Burgoon, L.A. Stern, L. Dillman, Interpersonal Adaptation, Cambridge University Press, 1995.

[7] Tanya L. Chartrand, Amy N. Dalton, Mimicry: its ubiquity, importance and functionality, in: Oxford Handbook of Human Action, vol. 2, 2009, pp. 458–483.

[8] H.H. Clark, J.E. Fox Tree, Using "uh" and "um" in spontaneous speaking, Cognition 84 (1) (2002) 73–111.

[9] C. Darwin, The Expression of Emotion in Man and Animals, John Murray, 1872.

[10] C. Darwin, A biographical sketch of an infant, Mind 7 (1) (1877) 285–294.

[11] E. Delaherche, M. Chetouani, M. Mahdhaoui, C. Saint-Georges, S. Viaux, D. Cohen, Interpersonal synchrony: a survey of evaluation methods across disciplines, IEEE Trans. Affect. Comput. 3 (3) (2012) 349–365.

[12] F. D'Errico, I. Poggi, A. Vinciarelli, L. Vincze (Eds.), Conflict and Multimodal Communication, Springer Verlag, 2015.

[13] T. Fong, I. Nourbakhsh, K. Dautenhahn, A survey of socially interactive robots, Robot. Auton. Syst. 42 (3) (2003) 143–166.

[14] J.A.A. Hall, E.J. Coats, L. Smith LeBeau, Nonverbal behavior and the vertical dimension of social relations: a meta-analysis, Psychol. Bull. 131 (6) (2005) 898–924.

[15] M. Joshi, E.B. Davis, R. Kathuria, C.K. Weidner, Experiential learning process: exploring teaching and learning of strategic management framework through the winter survival exercise, J. Manag. Educ. 29 (5) (2005) 672–695.

[16] C.M. Judd, Cognitive effects of attitude conflict resolution, J. Confl. Resolut. 22 (3) (1978) 483–498.

[17] M.L. Knapp, J.A. Hall, Nonverbal Communication in Human Interaction, Harcourt Brace College Publishers, 1972.

[18] A. Leffler, D.L. Gillespie, J.C. Conaty, The effects of status differentiation on nonverbal behavior, Soc. Psychol. Q. 45 (3) (1982) 153–161.

[19] P. Martin, P. Bateson, Measuring Behaviour, Cambridge University Press, 2007.
[20] Julian A. Oldmeadow, Michael J. Platow, Margaret Foddy, Donna Anderson, Self-categorization, status, and social influence, Soc. Psychol. Q. 66 (2) (2003) 138–152.
[21] R.W. Picard, Affective Computing, MIT Press, 2000.
[22] A. Polychroniou, H. Salamin, A. Vinciarelli, The SSPNet Mobile Corpus: social signal processing over mobile phones, in: Proceedings Language Resources and Evaluation Conference, 2014, pp. 1492–1498.
[23] R.P. Provine, Laughter punctuates speech: linguistic, social and gender context of laughter, Ethology 95 (4) (1993) 291–298.
[24] R.R. Provine, Y.L. Yong, Laughter: a stereotyped human vocalization, Ethology 89 (2) (1991) 115–124.
[25] V.P. Richmond, J.C. McCroskey, Nonverbal Behaviors in Interpersonal Relations, Allyn and Bacon, 1995.
[26] V.P. Richmond, J.C. McCroskey, S.K. Payne, Nonverbal Behavior in Interpersonal Relations, Prentice Hall, 1991.
[27] E.A. Schegloff, Overlapping talk and the organization of turn-taking for conversation, Lang. Soc. 29 (01) (2000) 1–63.
[28] B. Schuller, A. Batliner, Computational Paralinguistics: Emotion, Affect and Personality in Speech and Language Processing, John Wiley & Sons, 2013.
[29] L.L. Thompson, J. Wang, B.C. Gunia, Negotiation, Annu. Rev. Psychol. 61 (2010) 491–515.
[30] H.L. Tischler, Introduction to Sociology, Harcourt Brace College Publishers, 1990.
[31] A. Vinciarelli, A. Dielmann, S. Favre, H. Salamin, Canal9: a database of political debates for analysis of social interactions, in: Proceedings of International Conference on Affective Computing and Intelligent Interaction and Workshops, 2009, pp. 1–4.
[32] A. Vinciarelli, M. Pantic, H. Bourlard, Social signal processing: survey of an emerging domain, Image Vis. Comput. 27 (12) (2009) 1743–1759.
[33] A. Vinciarelli, M. Pantic, D. Heylen, C. Pelachaud, I. Poggi, F. D'Errico, M. Schroeder, Bridging the gap between social animal and unsocial machine: a survey of social signal processing, IEEE Trans. Affect. Comput. 3 (1) (2012) 69–87.
[34] A. Vinciarelli, H. Salamin, A. Polychroniou, Negotiating over mobile phones: calling or being called can make the difference, Cogn. Comput. 6 (4) (2014) 677–688.
[35] J.A. Wall, R. Roberts Callister, Conflict and its management, J. Manag. 21 (3) (1995) 515–558.
[36] N. Ward, W. Tsukahara, Prosodic features which cue back-channel responses in English and Japanese, J. Pragmat. 32 (8) (2000) 1177–1207.
[37] T. Wharton, Pragmatics and Nonverbal Communication, Cambridge University Press, 2009.

Chapter 14

Crowd behavior analysis from fixed and moving cameras

Pierre Bour, Emile Cribelier, Vasileios Argyriou
Kingston University, Faculty of Science, Engineering and Computing, Surrey, UK

CONTENTS

14.1 INTRODUCTION

Crowd behavior analysis and recognition is becoming an important research topic in video surveillance for public places, robotics interacting with groups of people (e.g. tourist guides), market research and management, architecture and design of public areas, and entertainment providing models for accurate simulation with applications in films, games and CGI. With the advent of machine learning and computer vision, and due to the increase of population and diversity of human activities resulting in more frequent crowded scenes in the real world, automatic crowd behavior recognition and analysis has become essential these days. Certain events that involve a large number of people gathering together (crowding) such as public assemblies, concerts, sports, demonstrations, protests, bring enormous challenges

Multimodal Behavior Analysis in the Wild. https://doi.org/10.1016/B978-0-12-814601-9.00023-7
Copyright © 2019 Elsevier Ltd. All rights reserved.

to public management, security or safety. Therefore, recognizing behaviors and predicting people's activities from a video are major concerns in the field of computer vision.

The security of public events involving groups of people or crowds has always been of high concern to public authorities, due to the high level of risk. Automated crowd behavior understanding or analysis has already attracted much research attention in the computer vision community [73]. Although many algorithms have been developed to track, recognize and understand the behaviors of humans and objects in video sequences [42], they were mainly applicable in low density scenarios and controlled conditions [102, 141,142]. Considering crowded, scenes in the wild, the problem of behavior analysis can not be handled well, due to the visual occlusions and ambiguities, as well as the complex behaviors and scene semantics, the large number of individuals present that decrease the accuracy of tracking and detection algorithms, make crowd analysis a challenging task. Furthermore, analysis of human activity in crowded scenes is one of the most challenging topics in computer vision and machine learning. In comparison to solutions focusing on understanding of actions performed by individuals [25,26], which is a problem yet to be fully solved, crowd behavior analysis faces even more challenges like self-organizing activities and emergent behaviors [75]. Under such circumstances and due to the increased practical demand it is becoming an important research direction attracting lots of novel contributions [11,19,42,94,103,115,124].

Recognizing behaviors and predicting people's activities from a video are major concerns in the field of computer vision. Automatic video surveillance has become necessary these days. The goal of most video surveillance systems is to provide security, but they can be used for other purposes such as estimating flows or improving the layout of public spaces. Indeed, in addition to increasing the effectiveness of surveillance in public places, they allow observation of crowd flows, resource management, and real-time detection of critical situations. The analysis of crowd behavior is very useful in a large number of applications [2], among which are the following.

- **The design and layout of spaces:** the analysis of human and crowd behavior indoors and outdoors makes it possible to estimate the attendance rate of each space and the corresponding paths that are followed. It provides guidelines to adapt the design and layout of a space according to the expected crowd behavior and habits.
- **Human or crowd-machine interaction:** systems for recognizing the actions of a person or groups allow to develop interfaces that interact with multiple people such as tourist or airport guide robots.

- **VR/AR in games and movies:** virtual environments and augmented reality systems in games or movies could improve immersion by taking into account people or crowd behaviors.
- **Indexing videos:** automatic extraction of information from videos improves their archiving and consultation. With the advent of sharing sites and the increase in storage capacity, it has become necessary to develop applications that can access videos quickly and efficiently. Knowledge of the events contained in the videos makes it possible to improve the relevance of the answers proposed to users wishing to find content involving certain human or crowd actions.
- **Crowd management:** the automatic study of crowd behavior makes it possible to develop crowd management strategies, particularly for popular and high attendance events (e.g. sports events, concerts, public events, etc.). The study of crowds makes it possible to anticipate risky behaviors and to avoid accidents caused by the disorganized movement of crowds.
- **Automatic video surveillance:** it automatically detects and provides a variety of information regarding events or abnormalities, assisting the security personnel. For example, automatic control of the entrances and exits of certain area zones, identification and recognition of persons, detection of unusual activities, etc.

Traditionally, crowd behaviors can be modeled in a short-term manner using video sequences covering a short temporal period in one instance. These models are able to describe simple actions using pattern recognition algorithms with reasonable precisions. An intrinsic challenge still exists in deducing the underlying consequences of a chain of short-term crowd actions. It is evident that motion information and actions are critical to crowd dynamics. However, modeling the long-term crowd behaviors is a very challenging task, due to the difficulties in inferring the overall picture of crowd behavior development based on the underlying long-term behavior context by studying the apparently individual and disconnected short-term actions in both the spatial and the temporal domains. The computer vision and analytic models will need to incorporate corresponding socio-psychological knowledge, to help understanding the links between short-term and long-term crowd behaviors, and the underlying social factors of such behaviors. Following this rationale, many researchers approach the overall problem in two interlinked ways—short-term and long-term crowd behavior analysis and modeling, advancing the state-of-the-art in a number of key problem domains in the field.

Crowd analysis is still a relatively new field in surveillance systems [97]. Tracking the behavior of masses of people and identifying certain behaviors among them, is a challenging task as there are various context-driven factors that may characterize an event. Moreover, single individual tracking cannot be readily applied, since input channels are exposed to heavily occluded environments and holistic techniques need to be employed [42], especially when capturing takes place in unconstrained environments. To this end, appearance learning techniques are common for crowd feature extraction and foreground segmentation, in the form of density estimation through texture and flow related features or low-level analysis techniques utilizing region of interest detection approaches and contours [43]. For short-term crowd behavior learning and event or abnormality detection, crowd motion pattern extraction is also utilized. A common approach is the use of particle flows [124], where a grid of particles is placed over video frames and variants of social force modeling are applied for retrieving spatio-temporal interactions. Classification or regression techniques are also common, such as hidden Markov models [91], and particle swarm optimization [116] applied to velocity, directions and unusual motion patterns.

The detection of unexpected behaviors is difficult in crowds due to occlusions, and to be efficient it depends also on crowd density and available computing time. There are two different approaches, the object-based approaches which take each person as an object in the crowd, and holistic approaches which take the crowd as a single object. Holistic methods usually still have some occlusion problems and are sensitive to perspective variations. The main methodologies used for crowd behavior analysis are histograms of motion directions, optical flow, crowd density estimation and person tracking. These approaches will be discussed in more details bellow.

The remainder of this chapter is organized as follows. In the following section concepts related to macro and micro crowd modeling are introduced. In Section 14.3 approaches for crowd behavior analysis considering motion information from fixed cameras are presented. Section 14.4 provides a detailed review on the topic of density estimation considering holistic and object-based approaches for crowd behavior recognition and Section 14.5 elaborates convolutional neural network (CNN)-based architectures for crowd density estimation and the detection of abnormal events. In Section 14.6 methods for events and pedestrian detection in crowds from moving cameras are detailed. Section 14.7 provides information about metrics and datasets used for crowd behavior analysis in comparative studies. Finally, conclusions are drawn and current trends are discussed.

14.2 MICROSCOPIC AND MACROSCOPIC CROWD MODELING

Crowd modeling and dynamics aim to analyze and simulate how and where crowds form and move above the critical density level [133], and how the environment and individuals in the crowd interact. These studies were mainly conducted to understand and predict crowd behaviors, support planning of urban infrastructure considering also risk evaluation, and market analysis [50,80,113]. Simulating behavior has been an area of intense interest in recent years and the ability to simulate how someone in a given environment is likely to interact with the world around them has a huge range of applications including pedestrian facility suitability and capacity analysis [18], computer graphics and gaming [87], the social sciences [89] and engineering [148].

Crowd behavior modeling can be performed considering either microscopic (agent-based) or macroscopic (continuum-based) approaches. Macroscopic solutions work better for medium and high density crowds, while the agent-based ones are more suitable for low density scenes, considering the movement and behavior of each pedestrian in the crowd.

Macroscopic approaches applied to large crowds, involve treating the agents as a whole, usually as a fluid or continuum, which responds to local influences [82,138]. Macroscopic methods present a number of advantages, the computational requirements to process large scale crowds is lower and the effects of unforseen behavior of individual agents has less impact on the crowd overall. The fundamental issue with this approach is the tendency to model the direction of movement and speed of a pedestrian as a single flow-density relationship with those around them. This does not easily upscale to more complex problems, including multidirectional simulations, nor does it take into account the motivations of the individual pedestrians as they are all represented identically.

The second method, microscopic, treats pedestrians as discrete individuals in a simulation. This allows a better level of granularity on the motivations behind an agents direction, speed and position. This provides the functionality to allow decision making properties and risk related variables to be considered for agents individually. Due to the individual nature of each agent, these methods tend to have a higher computational cost. However the microscopic models are tending to be favored for their ability to better emulate human behavior.

Simulating behavior is often considered to have started with Reynolds' work in 1987 with the emulation of animal behavior [120]. Within this work

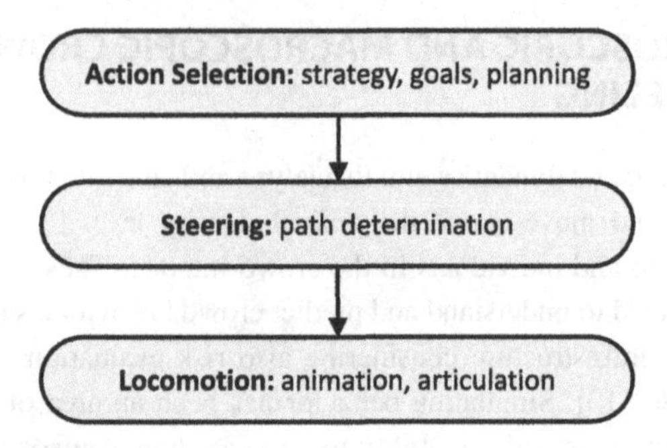

■ **FIGURE 14.1** The three layer simulation hierarchy.

the concept of steering simulation was introduced, whereby each individual agent in a scene has its movement governed by a set of rules. This approach was further developed by Reynolds [119], and here the steering behaviors model was better defined for gaming applications as a three layer hierarchy (see Fig. 14.1).

Helbing et al. [69] introduced the Social Force Model (SFM) which uses potential fields defined by neighboring agents to impart an acceleration to each agent. SFM's compute the trajectory of each agent by applying a series of forces to each agent that depend on the relative positions and velocities of nearby agents and the goal of the agent. One observed effect of the social force model is the emergence of behaviors within the agents present in crowds, such as line forming in tight areas. Another key advantage of this model was the use of variables that related to physical principles in our world.

Another example of this type of approach was proposed by Xi et al. [148], in which a dense model is proposed based on the inclusion of a number of decision making factors to each agent. In more details, a model was suggested integrating extended decision field theory for tactical level human decision making, the social force model to represent physical interactions and congestions among people and the environment considering a dynamic planning algorithm involving AND/OR graphs.

Karamouzas et al. [85] used a predictive collision avoidance model which focuses on modeling a human's ability to predict future collisions and take a minimal avoiding action. Zheng et al. [153,154] used a rudimentary form of behavior modeling for their human interaction contribution to risk. Using an evaluation of the average movement of a human captured using kinect skeletal data, a local disturbance field for a human in an environment was created. This was extended with a second disturbance field constructed of all the possible paths a human might make through an environment creating a global disturbance field for the scene.

Stroeve et al. [134] attempted to model the situational awareness (SA) of airport workers in the context of human performance modeling in accident risk assessment. Endsley [53] defines SA as follows: situation awareness is the perception of elements in the environment within a volume of time and space, the comprehension of their meaning, and the projection of their status in the near future. Kim et al. [87] added a stressor component into their simulation algorithm. The addition of this element allows for agents in a scenario to disregard common psychological behavior effects that maintain coherent movement such as separation from those around. Instead the agent focuses on reaching their perceived goal as fast as possible. Through this addition the ability to model instances such as building evacuations under stress become more accurate. Klugl et al. [89] created a model that aims to simulate an agents response to the evacuation of a train with an engine on fire in a tunnel. Human behavior is simulated such that an agent is able to take into account information related to the perceived heat and smoke levels, as well as input from other agents pertaining to the source of the fire. This aims to provide insight into human behavior in emergency situations. Also it is worth mentioning the work in [20,21,79,114]: SFM principles were utilized to detect abnormal events in the crowd or to learn and detect collective behavior patterns.

Recently, several crowd models have been utilized for the purposes of crowd behavior analysis [1,10,15,114,115,124]. For these applications usually holistic properties of the scene were considered and many dynamic crowd models have been proposed [11,108,115,124]. Fluid dynamics was introduced for crowd simulation in [63] and a combination of physics-based models with real data was suggested in [133].

14.3 MOTION INFORMATION FOR CROWD REPRESENTATION FROM FIXED CAMERAS

14.3.1 Pre-processing and selection of areas of interest

The analysis of the behavior of people and groups relies mainly on the analysis of motion in the video. Before proceeding to motion based solutions, it is common practice to define or estimate first regions where motion is supposed to occur. These areas are also called regions of interest. A region of interest may be a point, a line, a random shape, or an image set. It can be manually defined or estimated automatically using for example background extraction methods or by detecting points of interest. The simplest solution to delimit the area of interest is a manual definition of a movement zone. This is a practical solution in situations where the movement zones are known a priori and do not cover the entire scene. Also a motion area can be defined as a line (in the case of flow estimation) or as a mask defining motion zones. A motion mask can be used in conjunction with other motion zone detection methods to restrict further processing areas to specific locations.

Background modeling is used in different applications such as video surveillance or multimedia. The easiest way to separate the front of the background is to acquire an image of the background that contains no moving objects. A subtraction of images using an estimated threshold is then carried out between each new image acquired and the initial background image. However, a single background image is often not enough or unavailable because it is constantly modified by various events (e.g. lighting changes, objects entering or exiting the scene, etc.). In order to overcome the problems of robustness and adaptation, numerous works on background subtraction have been proposed based on Gaussian Mixture Models (GMMs) [104], CNNS [22], etc. In order to improve further the preprocessing processes, points, objects or areas of interest (e.g. cars, lines, etc.) that represent and provide useful information may be considered too. In general, points of interest are pixels in an image that are likely to be tracked effectively in the video sequence (e.g. Harris corner detector).

14.3.2 Motion-based crowd behavior analysis

Optical flow approaches are based on the first degree Taylor series expansion of the change at each pixel. Given two images $I(x, y, t_n)$ and $I(x, y, t_{n+1})$, the optical flow is defined under the assumption of brightness constancy as

$$I(x, y, t) = I(x + dx, y + dy, t + dt) \tag{14.1}$$

where dx and dy correspond to the motion vector over dt. Considering now the Taylor series expansion we obtain

$$\begin{aligned} I(x + dx, y + dy, t + dt) = I(x, y, t) + \\ \frac{\partial I}{\partial x}dx + \frac{\partial I}{\partial y}dy + \frac{\partial I}{\partial t}dt + ..., \end{aligned} \tag{14.2}$$

where the higher order terms can be ignored and by taking into account Eq. (14.1) and dividing throughout by dt we obtain

$$\frac{\partial I}{\partial x}\frac{dx}{dt} + \frac{\partial I}{\partial y}\frac{dy}{dt} + \frac{\partial I}{\partial t} = 0 \tag{14.3}$$

where $\partial I/\partial x, \partial I/\partial y$ and $\partial I/\partial t$ are the image derivatives. Since we have one equation with two unknowns, dx and dy, for each pixel, additional constraints are required, and they are usually provided by neighboring pixels.

Thus aiming to have a more compact representation, the equation below is derived where V_x, V_y, are the x and y components of the velocity or optical flow of $I(x, y, t)$ and I_x, I_y and I_t are the derivatives of the image at (x, y, t) in the corresponding directions.

$$I_x V_x + I_y V_y = -I_t \tag{14.4}$$

The solution as given by Lucas and Kanade [12,16,96] is a non-iterative method, which assumes a locally constant flow. Based on that the flow (V_x, V_y) is constant in a small window, of size $m \times m$ with $m > 1$, centered at pixel x, y and numbering the pixels as $1, ..., n$, a set of equations is obtained:

$$\begin{bmatrix} I_{x1} & I_{y1} \\ \vdots & \vdots \\ I_{xn} & I_{yn} \end{bmatrix} \begin{bmatrix} V_x \\ V_y \end{bmatrix} = \begin{bmatrix} -I_{t1} \\ \vdots \\ -I_{tn} \end{bmatrix} \Rightarrow A\vec{M} = -b \Rightarrow$$
$$\vec{M} = (A^T A)^{-1} A^T (-b) \tag{14.5}$$

This means that the optical flow can be estimated by calculating the derivatives of an image in all three dimensions. In order to give more prominence to the center pixel of the window, a weighting function $W(i, j)$, with $i, j \in [1, ..., m]$ could be added. Gaussian functions are preferred for this purpose, but other functions are also suitable. Horn and Schunck [71] added a regularization term, in order to enforce global smoothness, aiming to minimize the equation

$$\int_{\Omega} \left(\frac{\partial I}{\partial x} dx + \frac{\partial I}{\partial y} dy + \frac{\partial I}{\partial t} + \alpha(|\nabla dx|^2 + |\nabla dy|^2) \right) dxdy \tag{14.6}$$

The additional term helps to reduce the outliers due to incorrect matches. The current state of the art methods are based either on energy minimization [31] or on deep CNN architectures [49,76], providing significant accuracy with acceptable computational complexity.

Optical flow is used extensively in crowd behavior analysis and the proposed solutions are considered holistic approaches. Therefore, it is not expected that one may to detect or segment each person of the crowd but the crowd is considered as a single entity. The advantage of holistic approaches based on optical flow is that computational complexity remains relatively low. Andrade et al. in [13] presented a framework for crowd behavior analysis that is based on optical flow information from video sequences aiming to represent the crowd behavior as flow variations over time. The flow features are encoded with Hidden Markov Models to allow for the detection of emergency or abnormal events in the crowd (see Fig. 14.2). In order to increase the detection sensitivity, a local modeling approach is used. The results with simulated crowds show the effectiveness on detecting abnormalities in dense crowds. Similar approaches considering flow information and holistic representations were proposed in [14].

FIGURE 14.2 Left) Image from the original crowd. Right) Color coded representation of the optical flow.

Dee et al. in [45] proposed an approach based on motion information and tracklets considering Kanade–Lucas–Tomasi (KTL) features tracked over time generating a distribution along eight equally spaced directions (up, up-left, left, etc). An example of the concatenated distributed features is shown in Fig. 14.3(left). The scene is divided in $M \times N$ blocks and the histogram of motion direction (HMD) is calculated for each one of them. In order to estimate the location of an event of interest, the $L2$ distance is calculated. Fig. 14.3(right) shows a visualization of an event localization.

FIGURE 14.3 Left) Concatenation of KTL features. Right) Locations of event of interest.

14.4 CROWD BEHAVIOR AND DENSITY ANALYSIS

In crowd analysis people counting and density estimation are important problems and since crowd density is one of the basic descriptions of the crowd status, it can provide significant information about behaviors and certain events. It would be a great help to make appropriate decisions for emergency and safety controls. Furthermore, it can be used for market analysis and, specifically, it could be used for developing service providers in public places, or supplying the current state and queues of waiting customers.

There are two main approaches to calculate the crowd density. The direct approach is to segment and detect each person in the crowd, but this method is sensitive to occlusions. In the second approach people counting is carried out normally using the measurements of features with learning mechanisms or based on a statistical analysis of the whole crowd, aiming to achieve accurate counting estimates.

14.4.1 Person detection and tracking in crowded scenes

Another way to analyze crowd behavior is to track every person in the crowd. These methods have two main issues to overcome. First, the computation time increases significantly with the number of people that it is expected to track. Secondly, tracking is very sensitive to occlusions and therefore it operates well only for a low density crowd.

■ **FIGURE 14.4** Bottom-up approaches for human detection and pose estimation in crowded scenes.

In order to deal with occlusions in high density crowds, bottom–up approaches can be utilized aiming to detect and track separately human body parts (e.g. head, arms, etc.) [32,35,51,146] as is shown on Fig. 14.4. To improve the performance of people detection crowd density constraints can be imposed as it was proposed in [121], where person detection was formulated as the optimization of a joint energy function combining crowd density estimation and the localization of individual people (see Fig. 14.5).

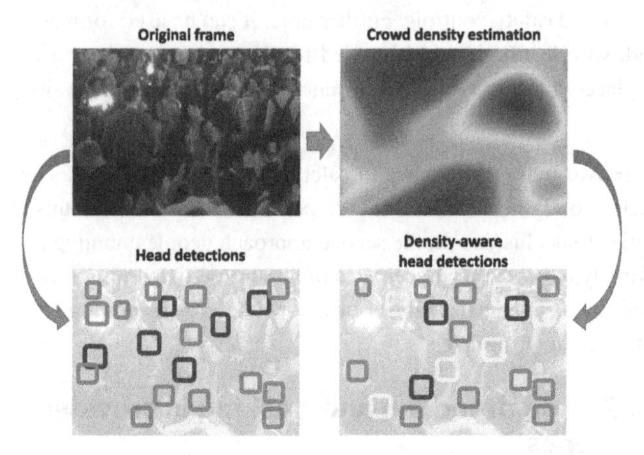

■ **FIGURE 14.5** Enhancement of head detection by crowd density.

14.4.2 Low level features for crowd density estimation

Pixel-based methods rely on local features (such as individual pixel analysis obtained through background subtraction models or edge detection) to estimate the number of people in a scene. Since very low-level features are used, this class is mostly focused on density estimation rather than precise people counting. Regazzoni et al. in [118] combined a background subtraction technique and detected edges to estimate crowd density. Their method also includes geometrical correction due to the camera perspective transformation, to account for the fact that the size of the same person may change at different locations. Similar methodologies based on pixel level features were proposed in [27,38,43,45,98].

Texture is one of the important features used in identifying objects or regions of interest in an image. And it is used to estimate the number of people rather than identifying individuals in a scene. In [147] Xinyu et al. analyze the texture information based on the contour of human blobs in order to learn different levels of group scales. In more detail a perspective projection model is used to generate multi-resolution image cells aiming to obtain better density estimation in crowded scenes. In order to achieve a uniform representation of texture features, the cell size is normalized and a technique of searching the extrema in the Harris–Laplacian space is also applied. The texture feature vectors that now have reduced instability of texture feature measurements, are extracted from each input image cell and a support vector machine (SVM) method is applied to estimate the crowd density.

Brostow et al. in [30] proposed a texture-based approach for abnormal behavior detection in crowds by learning a shape model to estimate the accurate number of people in the scene. The authors in [4,52] proposed that images of dense crowds tend to present fine textures, while images of low density crowds tend to present coarse texture. In their work, a statistics based method of Gray Level Dependence Matrices (GLDMs) was applied on the acquired images for extracting crowd density features. Four GLDM measures were used: contrast, homogeneity, entropy, and energy. Then, Self-Organizing Mapping was applied on these features to classify the crowd images according to five density levels (e.g. very low, low, moderate, high, and very high) as depicted in Fig. 14.6. Also similar approaches for crowd counting that are worth mentioning were proposed in [3,36,68,90,147].

FIGURE 14.6 Different crowd density, very high, high, medium, low, very low.

Methods that rely on object-level analysis try to identify individual objects and humans in a scene. They tend to produce more accurate results when compared to pixel-level analysis or texture based approaches, but identifying individuals in a single image or a video sequence is mostly feasible in lower density crowds.

For example Brostow and Cipolla in [30] presented an unsupervised Bayesian clustering method to detect independent movements in a crowd. Their hypothesis is that a pair of points that move together should be part of the same entity. An optical flow algorithm combined with an exhaustive search (the search region is defined by ground-plane camera calibration) based on the normalized cross correlation is used to track certain image features. An unsupervised Bayesian clustering algorithm then is applied to group such features, aiming to identify each individual moving in the crowd. Similar object based approaches using Bayesian human segmentation and independent motion estimation were introduced in [30,36,52,77,78,84,91, 117,123,137,152].

14.5 CNN-BASED CROWD ANALYSIS METHODS FOR SURVEILLANCE AND ANOMALY DETECTION

Crowd behavior analysis in extremely dense scenes is important for video surveillance and anomaly detection. People counting and event detection are essential for crowd analysis but they become especially challenging

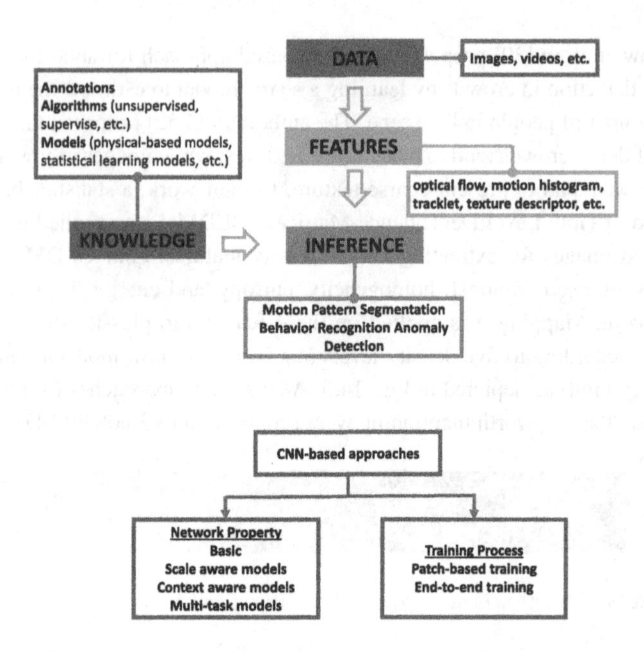

FIGURE 14.7 Generic framework for crowd behavior analysis and a categorization of CNN-based solutions.

tasks due to severe occlusions, cluttered scenes and perspective distortions. Existing methods as discussed above are based on human detectors or estimate the density of crowds using texture or other features combined with learning mechanisms and statistical models. The selected features in these cases are hand-crafted, such as SIFT, HOG etc, and thus are prone to failing in extreme situations and scenes. Over the last few years end-to-end deep CNN solutions were proposed for crowd analysis in extremely dense scenes. These methods aim to design deep CNN models to automatically learn effective features for event detection, people counting and density estimation. The general frameworks for crowd analysis using features and CNN-based approaches are shown in Fig. 14.7. Furthermore, background subtraction is not essential, since its influence is reduced by increasing the negative samples during the training stage.

Therefore, CNN-based approaches demonstrated significant improvements over previous feature-based methods, thus motivating more researchers to explore further similar approaches for related crowd analysis problems. Wang et al. in [145] proposed an end-to-end deep CNN regression model for counting people from images in extremely dense crowds. In [64,88,145] AlexNet style architectures were adopted, where the final fully connected layer was replaced with a single neuron layer for predicting the total amount of people. Besides, in order to reduce false responses, due to background like buildings and trees in the images, training data was augmented with additional negative samples whose ground truth count was set as zero.

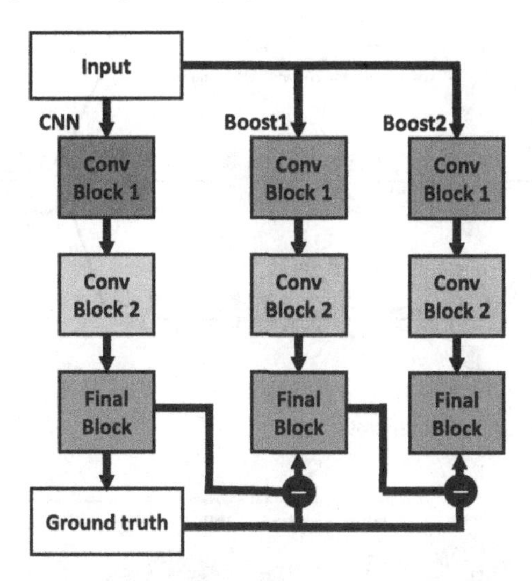

■ **FIGURE 14.8** The CNN architecture proposed in [143].

More recently Zhao et al. in [151] addressed a higher level cognitive task of counting people that cross a line. Though the task is a video-based application, it comprises of a CNN-based model that is trained with pixel-level supervision maps similar to single image crowd density estimation methods. Their approach consists of a two-phase training scheme that decomposes the original counting problem into two sub-problems: estimating a crowd density map and a crowd velocity map where the two tasks share the initial set of layers enabling them to learn more effectively. The estimated crowd density and crowd velocity maps are then multiplied element-wise to generate the crowd counting maps. A significant number of solutions based on deep CNN architectures were proposed recently and it is worth mentioning the work in [28,39,92,94,101,107,128,130,136,143,150]. In more detail in [143] a CNN ensemble model is used with a sample selection algorithm and each layer is trained to estimate the residual error of the previous layer (see Fig. 14.8). It improves the training process by reducing the effect of low quality samples, things like trivial samples or outliers. So when samples are correctly classified on early training stages, those are considered trivial samples. Therefore, it is less presented on the later training stages to improve the generalization performance of the model. In [150] the authors proposed a multi-column network, comprised by three columns corresponding to filters with receptive field of different sizes (see Fig. 14.9). Regarding the work in [107] the Hydra-CNN architecture is proposed using a pyramid of input patches providing a scale aware counting model and a density map with size equal to 1/4 of the input image (see Fig. 14.10).

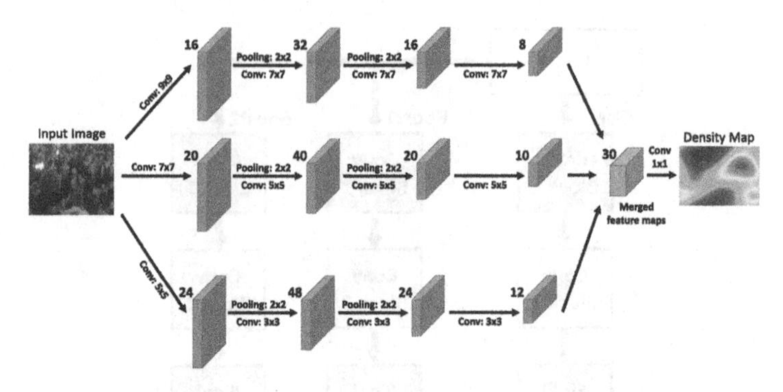

■ **FIGURE 14.9** The CNN architecture proposed in [150].

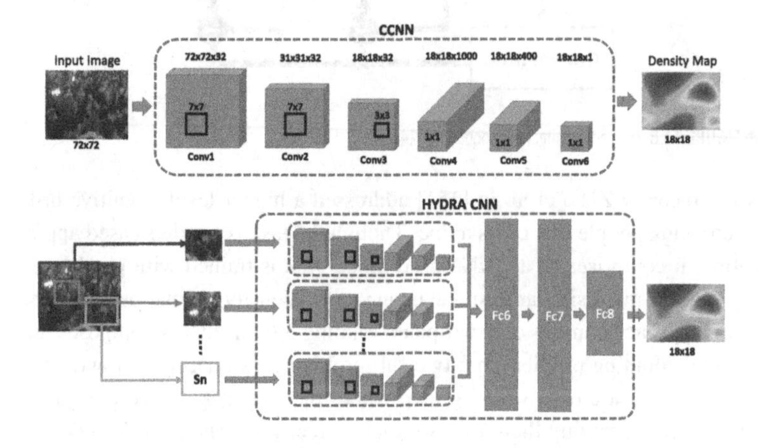

■ **FIGURE 14.10** The CNN architecture proposed in [107].

Crowd analysis is affected by many factors such as occlusion and similarity of appearance between people and background objects, and large variability of camera view-points. Current state-of-the art approaches tackle these factors by using advanced CNN architectures and recurrent networks. In [125] the authors proposed a switching convolutional neural network that leverages variation of crowd density within an image to improve the accuracy and localization of the predicted crowd count. This work supports independent CNN regressors designed to have different receptive fields and a switch classifier was trained to relay the crowd scene patch to the best CNN regressor. In more detail, a VGG-based switch classifier and regressors of a multi-column network are used (see Fig. 14.11). The switch classifier decides the optimal regressor for accurate counting on an input patch, while

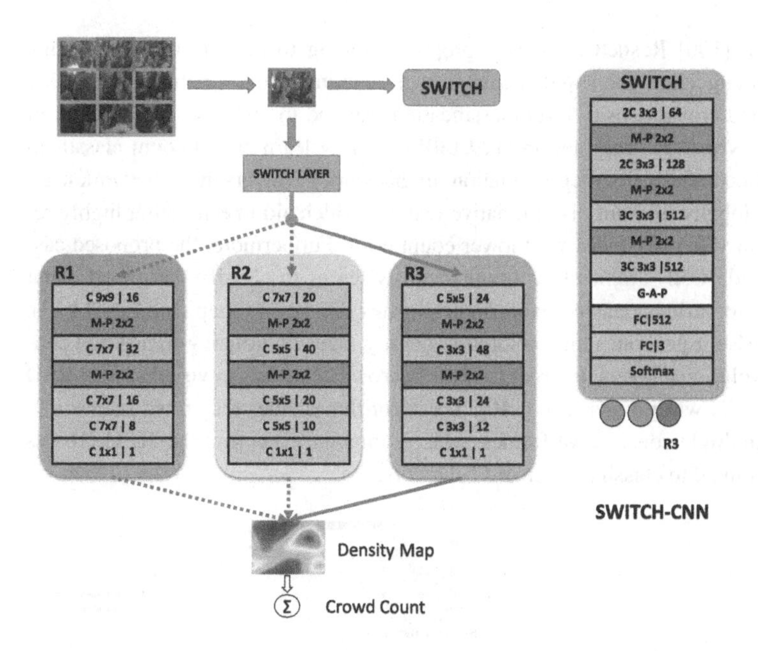

■ **FIGURE 14.11** The CNN architecture proposed in [125].

the regressors are trained to estimate density maps for different crowd density variations.

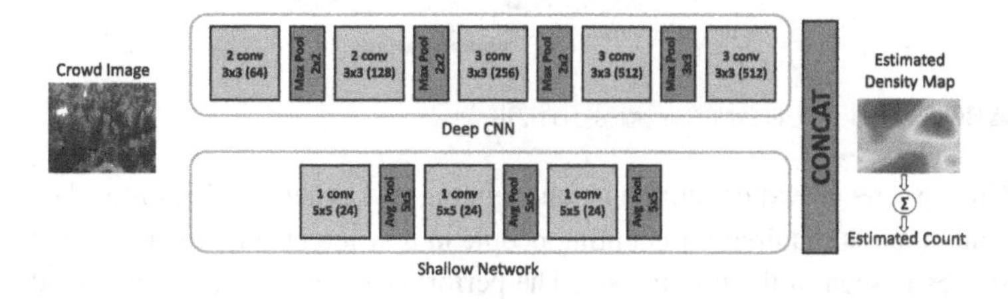

■ **FIGURE 14.12** The CNN architecture proposed in [28].

The authors in [28] introduced a deep learning framework for estimating crowd density from static images of highly dense crowds (see Fig. 14.12). The combination of deep and shallow fully convolutional models is considered and an extensive data augmentation method is applied. Also, patches from multi-scale image representations are used to make the system robust to scale variations. Furthermore, since they combined deep and shallow networks to predict the density map for a given crowd image, this approach helps to estimate both the high-level semantic information (face/body detectors) and the low-level features (blob detectors), that are necessary for crowd counting under large scale variations.

In [100] ResnetCrowd was proposed aiming to provide a deep residual architecture for simultaneous crowd counting, violent behavior detection and crowd density level classification. An end-to-end cascaded network of CNNs was suggested in [129,130] to jointly learn crowd count classification and density map estimation. In this work the layers in the network learn globally relevant discriminative features which aid in estimating highly refined density maps with lower count error. Furthermore, the proposed cascade CNN architecture simultaneously learns to classify the crowd count into various density levels and estimates the density map (see Fig. 14.13). Also it generates high resolution density maps by utilizing transposed convolutional layers. In order to identify crowd behaviors in visual scenes, a 3D CNN was proposed in [110]. The algorithm divides the crowd scenes into multiple consecutive frames, which forms the input and the 3D CNN was trained to classify the crowd behaviors.

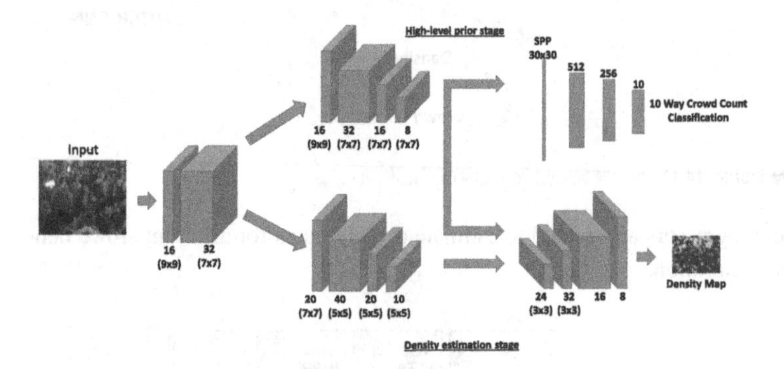

■ **FIGURE 14.13** The CNN architecture proposed in [129].

Cross-scene crowd counting is a challenging task where no laborious data annotation is required for counting people in new target surveillance crowd scenes unseen in the training set. The performance of most existing crowd counting methods drops significantly when they are applied to an unseen scene. To address this problem of crowd counting in unseen scenes, a CNN was proposed in [149], which was trained alternatively with two related learning objectives, crowd density and crowd count. This switchable learning approach is able to obtain a better local optimum for both objectives. Also, a data-driven method to fine-tune the trained CNN model for a given target scene was suggested aiming to handle unseen crowd scenes. In general, the important aspect of this work is that it aims to fine-tune the model using training samples that are similar to the target scenes. This makes the network more robust to cross-scene crowd counting (see Fig. 14.14).

In [34] a CNN-based method to monitor the number of crowd flow, such as the number of entering or leaving people in high density crowds was in-

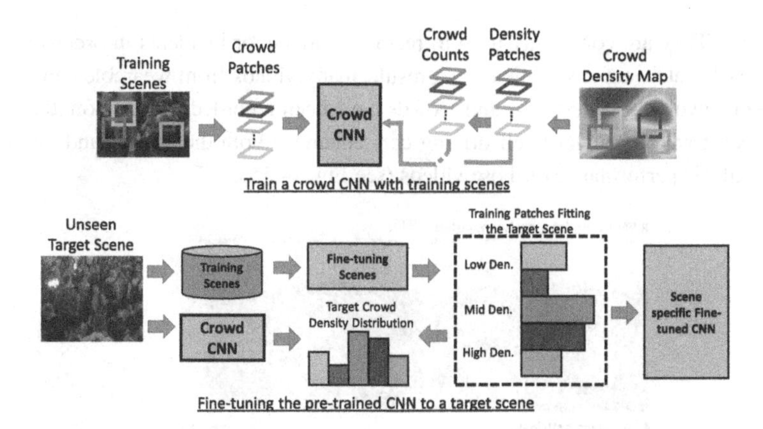

■ **FIGURE 14.14** The CNN architecture proposed in [149].

troduced. This work combines classification CNNs with regression CNNs, aiming to increase the overall robustness. The authors in [127] focused on learning dynamic representations, and how they can be combined with appearance features for video analysis, and therefore a spatio-temporal CNN was proposed. This architecture is based on the decomposition of 3D feature maps into 2D spatio- and 2D temporal-slices representations. In [112], a crowd density estimation method using ConvNet, a deep convolutional neural network, was introduced. Similar solutions were proposed in [109] for crowd density estimation and an improved convolutional neural network was combined with traditional texture features calculated by the convolutional layer. In general, crowd density estimation is an important element for crowd monitoring, control and behavior understanding.

14.6 CROWD ANALYSIS USING MOVING SENSORS

The use of moving cameras for crowd behavior analysis could be a very interesting approach in several situations and applications. One of the main issues with fixed cameras for the analysis of crowd behavior is the density and more specifically the resulting number of occlusions. Moving cameras allow us to change the perspective view at any time. In the case of crowd analysis, the viewpoint can be changed moving from aside to a top–down view aiming to minimize the occlusions. On the other hand, the methods used with fixed cameras will not provide similar results to the moving ones, since tracking and motion are more difficult to estimate due to the moving background.

Examples of moving cameras are the car dash-cams and the body cameras or the camera glasses that are worn by security personnel and police officers. They are commonly used to record accidents or incidents in order to clarify later responsibilities. As a result, many videos from wearable cameras involving pedestrians and crowds have been recorded. In addition, the ever greater growth of self-driving cars enhanced both the quality and the analysis performance of those videos (see Fig. 14.15).

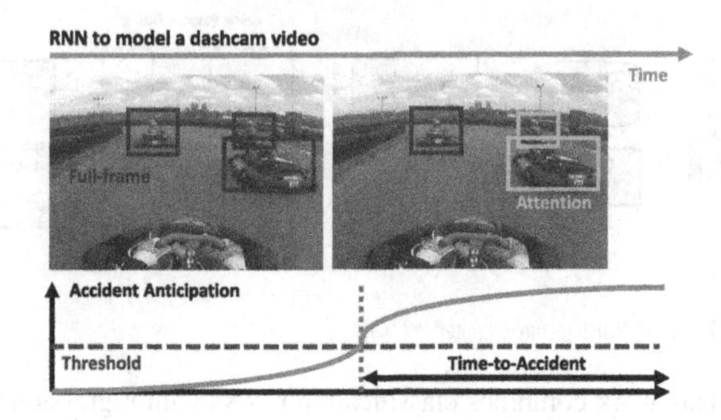

■ **FIGURE 14.15** Example methods for dash-cam video analysis [65].

Some of the proposed approaches include multiple model designs. In [65, 81] a dynamic-spacial-attention (DSA) mechanism initially learns to distribute soft attention to candidate objects in each frame dynamically and then RNNs for sequence modeling are used with long–short-term memory (LSTM) [70] cells aiming to model the long-term dependencies of all cues.

The analysis of First Person Videos aiming to understand the environment surrounding the wearer is a difficult task and the authors in [106] proposed a method for detecting and mapping the presence of people and crowds. In order to build a robust representation, features extracted at the crowd level were used aiming to handle the variations and occlusions inside a crowd. To this aim, convolutional neural networks have been exploited obtaining high accuracy on the recognition of crowds and behaviors.

Another case of moving cameras that has significant applications is the Unmanned Aerial Vehicles (UAVs) or drones. In these cases different types of sensors may be available, including depth or thermal cameras.

Unmanned aerial vehicles (UAVs) and drones are commonly used for different kind of tasks due to their price drop over the last years. As a consequence, more researchers are exploring the potentials of using UAVs for pedestrian or crowd detection and tracking. Pedestrian detection from the images captured by drones is a difficult task due to platform motion, image instability and the relatively small size of the captured objects, but also due to variation of camera orientation and illumination. The approach proposed in [99] utilizes thermal imagery. However, even using thermal cameras, the detection of pedestrians and groups is still a challenging problem because of the human thermal signature variation due to weather changes. Fig. 14.16 shows thermal signatures of pedestrians on a street recorded by an UAV. On these images pedestrians appear to be white blobs and the background or other objects can also produce similar shape and intensity blobs. The authors in [99] used Regions of Interest (ROI) extraction and ROI classification. They proposed a blob extraction method based on regional gradient features. The regions are then fully analyzed by a linear Support Vector Machine (SVM) [41] classifier based on a weighted fusion of Histogram of Oriented Gradients (HOG) [44] and a Discrete Cosine Transform (DCT) descriptor aiming to detect crowds and groups of pedestrians with higher accuracy. Additionally, the optical flow is estimated using an extension of the Lucas-Kanade method [29], in order to eliminate the motion induced by the UAVs.

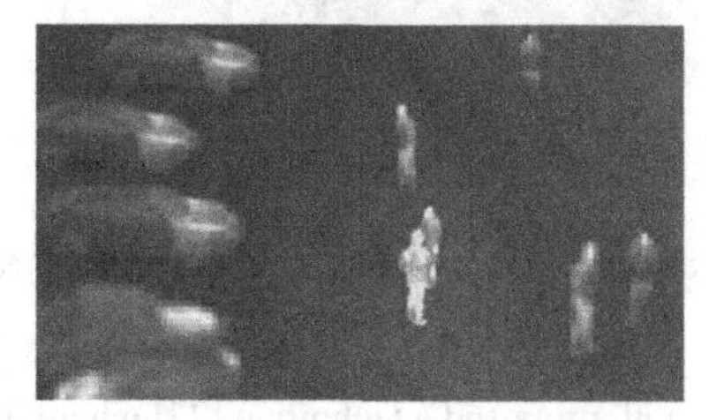

FIGURE 14.16 Thermal signatures of pedestrians on the street.

The authors in [66] applied cascaded Haar classifiers with additional multivariate Gaussian shape matching for secondary confirmation to achieve people detection from UAV thermal images. In [111] a cascaded Haar feature-based body part detector was suggested considering the head, upper body and legs as the characteristic parts to generate individual body part classifiers. Furthermore, in order to increase the identification and location accuracy of pedestrians and groups of people, a particle filter-based approach was also introduced.

The approach in [5–8] combines two algorithms for feature extraction, Local Binary Patterns (LBPs) and Haar-like features. Local Binary Patterns (LBPs) was presented in [105] as a texture descriptor for object detection, and compares a central pixel with its neighbors. They used Adaptive Boosting (AdaBoost) as a training algorithm and a combination of Haar-LBP features due to their low computational complexity. Based on their methodology 70% of the data was used for training and the remaining 30% for testing, and the authors reported results with UAV data in different scenarios.

In [132] a ground plane extractor was proposed to approximate its orientation in 3D according to the position of the camera. With this method, it is possible to estimate the height of an object for each region in an image, which allows one to search pedestrians in certain areas. Another approach to estimate objects' height was proposed in [135], based on the homography of the ground plane and the approximate known real size of the objects (see Fig. 14.17). In this case the objects to detect are pedestrians and their heights were selected to be in between 160 cm and 185 cm. Rectangular regions were cut from the image and transformed into a standard size, where each region was evaluated at the same scale.

■ **FIGURE 14.17** Example of approaches based on ground plane extraction to detect pedestrians and their heights.

In 2012, Benenson et al. proposed a technique [23] where, instead of evaluating a single model on multiple feature layers, they created a set of models and used this set on the same features. This multi-model approach was combined with possible positions of pedestrians in the images based on ground plane estimation.

Appearance-based methods have proved to be accurate enough even in the presence of complex lighting variations or cluttered background. They are typically based on Deformable Part Models (DPMs), Convolutional Neural Networks (CNNs) and Random Forests. Recently, in [47] the authors proposed the Aggregate Channel Feature (ACF), which was a combination of a generalized feature approximation technique based on [48] and using all possible rectangles in a grid. The reduction of the amount of features, combined with the improved AdaBoost training method [17] resulted in a detector which was both accurate and fast.

14.7 METRICS AND DATASETS

14.7.1 Metrics for performance evaluation

Metrics for human detection have been applied in crowd analyses, which aim to estimate and evaluate all possible error types given the ground truth in different modalities.

True Positive TP**:** number of detected real human locations that are correctly detected.
False Positive FP**:** number of detected human locations that are not real.
True Negative TN**:** number of non-existing human locations that are not detected, and therefore correctly discarded.
False Negative FN**:** number of real human locations that are not detected.

The standard PASCAL 50% overlap criterion [24] is used for the association between a detected location and the ground truth. These metrics are usually expressed as precision and recall that are broadly used for detection accuracy evaluation, with exactness or quality being expressed as precision, and completeness or quantity as recall. We have

$$Precision = \frac{TP}{TP + FP} \tag{14.7}$$

$$Recall = \frac{TP}{P} = \frac{TP}{TP + FN} \tag{14.8}$$

Similar metrics are the sensibility and specificity,

$$Sensitivity = \frac{TP}{P} = \frac{TP}{TP + FN} \tag{14.9}$$

$$Specificity = \frac{TN}{N} = \frac{TN}{FP + TN} \tag{14.10}$$

Also, Accuracy is widely used and it is defined as

$$Accuracy = \frac{TP + TN}{P + N} = \frac{TP + TN}{TP + FN + FP + TN} \tag{14.11}$$

Furthermore, the F1 score is commonly considered to detection performance evaluation.

$$F1Score = 2 * (Recall * Precision)/(Recall + Precision) \tag{14.12}$$

but since it does not reflect properly the false positives, it is complemented by the FP metric or the $Precision$ to allow conclusions to be drawn. These metrics are applicable to tasks related to the recognition of discrete crowd attributes (e.g. ‘who’, ‘where’, ‘what’) or of human activities.

The main issue with these metrics is that temporal consistence is not considered and they cannot fully exploit cases such as tracking over time. An initial approach could be to consider the same metrics on a frame by frame basis. However, these metrics do not reflect all the complexity and types of error in multi-target scenarios, common in crowds. Therefore, metrics that take into consideration consistence over long periods of frames and identity exchanges between humans in a crowd are utilized in these cases. Two measures of tracking performance are proposed by the CLEAR MOT performance metrics [86]. First we have the multiple person tracker accuracy (MOTA), and second the multiple person tracker precision (MOTP), defined as follows:

$$MOTA = 1 - \frac{\sum FN + FP + Id_{switch}}{TP + FN} \tag{14.13}$$

$$MOTP = \frac{\sum e_{TP}}{\sum TP} \tag{14.14}$$

where e_{TP} is the distance between the ground-truth persons' locations and the associated detected locations.

Finally, efficiency or computational cost should be part of crowd behavior analysis performance metrics. Since detection and classification tasks are expected to be performed in real time or near real time, complexity (i.e. the average time in seconds per frame for specific hardware and software) is an important evaluation criterion. Also if it is possible the theoretical complexity, described by means of the O-notation, could be utilized in a comparative study.

14.7.2 Datasets for crowd behavior analysis

Over the last few years several datasets for crowd behavior analysis were published and in this section some of them are listed providing short descriptions, information about their size, labeling and current state of the art results.

The **UCSD Anomaly Detection Dataset** [46] includes videos of pedestrians captured from a fixed elevated camera of varying density, ranging from sparse to very crowded. In this dataset abnormal events are considered either the presence of non-pedestrian objects or anomalous pedestrian motion patterns. For each clip, the ground truth annotation includes a binary flag per frame, indicating whether an anomaly is present at that frame. In addition, a subset of 10 clips for PEDs1 and 12 clips for PEDs2 are provided with manually generated pixel-level binary masks, which identify the regions containing anomalies. This is intended to enable the evaluation of performance with respect to ability of algorithms to localize anomalies.

The **Violent-Flows—Crowd Violence Non-violence Database and benchmark** [140] is focused on crowd violence. Real-world videos are part of this database and standard evaluation methodologies are available aiming to provide violent/non-violent classifications and violence outbreak detections. The average length of a video clip is 3.60 s.

The **CUHK Crowd Dataset** [144] includes crowd videos with various densities and perspective scales, collected from many different environments, e.g. streets, shopping malls, airports, and parks. There is annotation at a sample of frames and the total length of the database is 120 min. Regarding the available behavior classes, some of them are listed below: highly mixed pedestrian walking; Crowd walking following a mainstream and well organized; Crowd walking following a mainstream but poorly organized; Crowd merge; Crowd split; Crowd crossing in opposite directions; Intervened escalator traffic; and Smooth escalator traffic. The current state of the art using holistic features is above 70% accuracy.

The **UCF50** [11,19] is an action recognition dataset with 50 action categories, consisting of realistic videos taken from YouTube. This data set is an extension of YouTube Action data set (UCF11) which has 11 action categories. Most of the available action recognition datasets are not realistic and are staged by actors. For all the 50 categories, the videos are grouped into 25 groups, where each group consists of more than four action clips. The video clips in the same group may share some common features, such as the same person, similar background, similar viewpoint, and so on. Some of the action categories of this database are baseball pitch, basketball shooting, bench press, biking, billiards shot, breaststroke, clean and jerk, diving, drumming, fencing, golf swing, playing guitar, high jump, horse race, horse riding, hula hoop, javelin throw, jump rope, jumping jack, kayaking, lunges, military parade, playing piano, pizza tossing, pole vault, pommel horse, pull ups, punch, push ups, rope climbing, rowing, skate boarding, skiing, swing, playing violin, etc. Another extension of UCF includes videos of high density moving objects including vehicles and crowds. The video sequences are mainly from the web and the PETS2009 [62].

Rodriguez's Dataset [121] is a collection of videos from websites such as YouTube including different types of cameras and providing ground-truth trajectories for some randomly selected individuals.

The **Mall Dataset** [37] was collected under diverse illumination conditions or crowd densities and a surveillance camera located in a shopping mall was utilized for the acquisition process. The videos contain pedestrians, while occlusions are present due to scene objects. The dataset is annotated, the resolution is low (320×240) and there is no variation in the scene perspective since the camera is fixed.

The **WorldExpo '10 Dataset** [74] is focused primarily on crowd density and counting. The video sequences are annotated and are captured by multiple surveillance cameras during the Shanghai 2010 WorldExpo event. There is a large diversity in the scene types including videos with disjoint bird views.

Datasets and methodologies focusing on crowd behavior recognition for single person behavior [56–58] or dyadic interactions [126] are also available and are commonly used in most work. In these cases, the classification tasks on crowd analysis consider simple attributes (e.g. "where", "who", "why") [59,60,83,131,139] but not localization or other fine-grained tasks. The work in [61] uses a small dataset of 31 sequences for 6 basic reactions with expensive frame-level annotations, and Zhang in [40,149,150] considers 4 reactions only and uses video-level annotations with about 100 videos being available, reporting moderate degree of success. Finally, most of these databases treat the crowd analysis problem holistically reporting average levels of accuracy. Furthermore, they do not support different acquisition conditions or viewpoints (e.g. static/moving cameras, full, medium and close-up shots, aerial shots).

Another dataset recording from moving cameras was presented in [54,55] and the data was recorded using a pair of AVT Marlins F033C mounted on moving vehicles, with a resolution of 640 × 480 (bayered), and a frame rate of 13–14 FPS. The unbayered images for both cameras are provided, the camera calibration, and if available, the set of bounding box annotations. Depth maps were created based on this data using the publicly available belief-propagation-based stereo algorithm of Huttenlocher and Felzenszwalb. In [9] a dataset with low-resolution RGB videos and ground truth trajectories from multiple fixed and moving cameras monitoring the same scenes (indoor and outdoor) is provided focusing on human or object detection and tracking.

The dataset presented in [67,93] contains video sequences in unconstrained environments filmed with both static and moving cameras. Also, the tracking and evaluation are preformed in image coordinates. Finally, it is worth to mention the datasets presented in [33,72,95,122,127] providing videos for crowd and pedestrian behavior analysis.

14.8 CONCLUSIONS

We have presented an overview of recent and state-of-the-art techniques for crowd behavior analysis considering both fixed and moving cameras. The main research topics that were presented are crowd modeling using microscopic and macroscopic approaches, crowd analysis methods based on motion information, density estimation considering person detection and tracking methodologies, and low level pattern based solutions. Due to its importance in applications related to security and market analysis, anomaly detection in crowded scenes was discussed and state of the art approaches were presented. Furthermore, advances in CNN-based methods for crowd counting and density estimation are reviewed focusing on density estimation and the detection of abnormal events. We also reviewed methodologies for crowd and pedestrian behavior analysis from moving cameras such as UAVs, drones, dash-cams, body cameras and camera glasses discussing also the related most compelling challenges and issues. Finally, evaluation metrics for crowd behavior analysis are presented and datasets captured from fixed and moving cameras for crowd analysis are summarized.

REFERENCES

[1] M.P. Ross, A.A. Sodemann, B.J. Borghetti, A review of anomaly detection in automated surveillance, IEEE Trans. Syst. Man Cybern., Part C, Appl. Rev. 42 (6) (2012) 1257–1272.

[2] J.H. Yin, A.C. Davies, S.A. Velasti, Crowd monitoring using image processing, IEE Electron. Commun. Eng. J. 7 (1) (1995) 37–47.

[3] Z. Liang, A. Chan, N. Vasconcelos, Privacy preserving crowd monitoring: counting people without people models or tracking, in: Proc. IEEE Conf. Computer Vision and Pattern Recognition, 2008, pp. 1–7.

[4] R. Lotufo, A. Marana, L. da Costa, S. Velastin, On the efficacy of texture analysis for crowd monitoring, in: Proc. Int. Symp. Computer Graphics, Image Processing, and Vision (SIBGRAPI'98), Washington, DC, 1998, p. 354.

[5] W.G. Aguilar, M.A. Luna, J.F. Moya, V. Abad, H. Parra, H. Ruiz, Pedestrian detection for UAVs using cascade classifiers with meanshift, in: 2017 IEEE 11th International Conference on Semantic Computing (ICSC), Jan. 2017, pp. 509–514.

[6] Wilbert G. Aguilar, Marco A. Luna, Julio F. Moya, Vanessa Abad, Hugo Ruiz, Humberto Parra, Cecilio Angulo, Pedestrian Detection for UAVs Using Cascade Classifiers and Saliency Maps, Springer International Publishing, 2017, pp. 563–574.

[7] Wilbert G. Aguilar, Marco A. Luna, Julio F. Moya, Marco P. Luna, Vanessa Abad, Hugo Ruiz, Humberto Parra, Real-Time Detection and Simulation of Abnormal Crowd Behavior, Springer International Publishing, 2017, pp. 420–428.

[8] Wilbert G. Aguilar, Marco A. Luna, Hugo Ruiz, Julio F. Moya, Marco P. Luna, Vanessa Abad, Humberto Parra, Statistical Abnormal Crowd Behavior Detection and Simulation for Real-Time Applications, Springer International Publishing, 2017, pp. 671–682.

[9] Alexandre Alahi, Pierre Vandergheynst, Michel Bierlaire, Murat Kunt, Cascade of descriptors to detect and track objects across any network of cameras, Comput. Vis. Image Underst. 114 (6) (June 2010) 624–640.

[10] S. Ali, Taming Crowded Visual Scenes, Ph.D. dissertation, University of Central Florida, 2008.

[11] S. Ali, M. Shah, A Lagrangian particle dynamics approach for crowd flow segmentation and stability analysis, in: IEEE Conference on Computer Vision and Pattern Recognition, 2007, pp. 1–6.

[12] Tomer Amiaz, Eyal Lubetzky, Nahum Kiryati, Coarse to over-fine optical flow estimation, Pattern Recognit. 40 (9) (2007) 2496–2503.

[13] E.L. Andrade, S. Blunsden, R.B. Fisher, Hidden Markov models for optical flow analysis in crowds, in: 18th International Conference on Pattern Recognition (ICPR'06), vol. 1, 2006, pp. 460–463.

[14] E.L. Andrade, S. Blunsden, R.B. Fisher, Modelling crowd scenes for event detection, in: 18th International Conference on Pattern Recognition (ICPR'06), vol. 1, 2006, pp. 175–178.

[15] E.L. Andrade, R.B. Fisher, Simulation of crowd problems for computer vision, in: First International Workshop on Crowd Simulation, vol. 3, 2005, pp. 71–80.

[16] Christoph Hoog Antink, Tarunraj Singh, Puneet Singla, Matthew Podgorsak, Advanced Lucas Kanade optical flow for deformable image registration, J. Crit. Care 27 (3) (2012) e14.

[17] Ron Appel, Thomas Fuchs, Piotr Dollar, Pietro Perona, Quickly boosting decision trees – pruning underachieving features early, June 2013.

[18] Miho Asano, Takamasa Iryo, Masao Kuwahara, A pedestrian model considering anticipatory behaviour for capacity evaluation, in: Transportation and Traffic Theory, vol. 18, 2009, p. 28.

[19] B.E. Moore, B. Solmaz, M. Shah, Identifying behaviors in crowd scenes using stability analysis for dynamical systems, IEEE Trans. Pattern Anal. Mach. Intell. 34 (10) (2012) 2064–2070.

[20] X. Wang, B. Zhou, X. Tang, Random field topic model for semantic region analysis in crowded scenes from tracklets, in: IEEE Conference on Computer Vision and Pattern Recognition, 2011, pp. 3441–3448.

[21] X. Wang, B. Zhou, X. Tang, Understanding collective crowd behaviors: learning a mixture model of dynamic pedestrian-agents, in: IEEE Conference on Computer Vision and Pattern Recognition, 2012, pp. 2871–2878.

[22] Mohammadreza Babaee, Duc Tung Dinh, Gerhard Rigoll, A deep convolutional neural network for background subtraction, CoRR arXiv:1702.01731, 2017.

[23] R. Benenson, M. Mathias, R. Timofte, L. Van Gool, Pedestrian detection at 100 frames per second, in: 2012 IEEE Conference on Computer Vision and Pattern Recognition, June 2012, pp. 2903–2910.

[24] B. Benfold, I. Reid, Stable multi-target tracking in real-time surveillance video, in: Proceedings of the IEEE Computer Society Conference on Computer Vision and Pattern Recognition, 2011, pp. 3457–3464.

[25] V. Bloom, V. Argyriou, D. Makris, G3Di: a gaming interaction dataset with a real time detection and evaluation framework, in: ECCV Workshops, vol. 1, 2014, pp. 698–712.

[26] V. Bloom, D. Makris, V. Argyriou, Clustered spatio-temporal manifolds for online action recognition, in: 2014 22nd International Conference on Pattern Recognition (ICPR), 2014, pp. 3963–3968.

[27] B.A. Boghossian, S.A. Velastin, Motion-based machine vision techniques for the management of large crowds, in: The 6th IEEE International Conference on Electronics, Circuits and Systems, 1999, Proceedings of ICECS '99, vol. 2, Sep. 1999, pp. 961–964.

[28] Lokesh Boominathan, Srinivas S.S. Kruthiventi, R. Venkatesh Babu, CrowdNet: a deep convolutional network for dense crowd counting, in: Proceedings of the 2016 ACM on Multimedia Conference, MM '16, 2016, pp. 640–644.

[29] J. Bouguet, Pyramidal Implementation of the Lucas Kanade Feature Tracker, Intel Corporation, Microprocessor Research Labs, 2000.

[30] G.J. Brostow, R. Cipolla, Unsupervised Bayesian detection of independent motion in crowds, in: 2006 IEEE Computer Society Conference on Computer Vision and Pattern Recognition (CVPR'06), vol. 1, June 2006, pp. 594–601.

[31] T. Brox, A. Bruhn, N. Papenberg, J. Weickert, High accuracy optical flow estimation based on a theory for warping, in: ECCV, 2004, pp. 25–36.

[32] A. Bulat, G. Tzimiropoulos, Human pose estimation via convolutional part heatmap regression, in: ECCV, 2016.

[33] S. Gong, C.C. Loy, T. Xiang, From semi-supervised to transfer counting of crowds, in: Proceedings of IEEE International Conference on Computer Vision, 2013, pp. 2256–2263.
[34] Lijun Cao, Xu Zhang, Weiqiang Ren, Kaiqi Huang, Large scale crowd analysis based on convolutional neural network, Pattern Recognit. 48 (2015) 04.
[35] Zhe Cao, Tomas Simon, Shih-En Wei, Yaser Sheikh, Realtime multi-person 2D pose estimation using part affinity fields, in: CVPR, 2017.
[36] A.B. Chan, N. Vasconcelos, Bayesian Poisson regression for crowd counting, in: Proc. IEEE Int. Conf. Computer Vision, 2009, pp. 1–7.
[37] C.C. Loy, S. Gong, T. Xiang, K. Chen, Feature mining for localised crowd counting, in: European Conference on Computer Vision, 2012.
[38] Siu-Yeung Cho, T.W.S. Chow, Chi-Tat Leung, A neural-based crowd estimation by hybrid global learning algorithm, IEEE Trans. Syst. Man Cybern., Part B, Cybern. 29 (4) (Aug. 1999) 535–541.
[39] U. Meier, J. Schmidhuber, D. Ciregan, Multi-column deep neural networks for image classification, in: 2012 IEEE Conference on Computer Vision and Pattern Recognition (CVPR), IEEE, 2012, pp. 3642–3649.
[40] Ming Ma, Lifeng Sun, Bo Li, Cong Zhang, Jiangchuan Liu, Seeker: topic-aware viewing pattern prediction in crowdsourced interactive live streaming, in: NOSSDAV, 2017, pp. 25–30.
[41] Corinna Cortes, Vladimir Vapnik, Support-vector networks, Mach. Learn. 20 (3) (Sep. 1995) 273–297.
[42] S.R. Musse, C.S. Jacques Jr., C.R. Jung, Crowd analysis using computer vision techniques, IEEE Signal Process. Mag. 27 (5) (2010) 66–77.
[43] H.H. González-Baños, D.B. Yang, L.J. Guibas, Counting people in crowds with a real-time network of simple image sensors, in: ICCV, 2003, pp. 122–129.
[44] N. Dalal, B. Triggs, Histogram of oriented gradients for human detection, in: Proceeding of the IEEE Conference on Computer Vision and Pattern Recognition (CVPR), San Diego, June 2005.
[45] H.M. Dee, A. Caplier, Crowd behaviour analysis using histograms of motion direction, in: 2010 IEEE International Conference on Image Processing, Sept. 2010, pp. 1545–1548.
[46] UCSD Anomaly Detection, http://www.svcl.ucsd.edu/projects/anomaly/dataset.htm, 2013.
[47] Piotr Dollar, Ron Appel, Serge Belongie, Pietro Perona, Fast feature pyramids for object detection, IEEE Trans. Pattern Anal. Mach. Intell. 36 (8) (2014) 1532–1545.
[48] Piotr Dollar, Serge Belongie, Pietro Perona, The fastest pedestrian detector in the west, in: Proc. BMVC, 2010, pp. 68.1–11.
[49] A. Dosovitskiy, P. Fischerand, E. Ilg, P. Hausser, C. Hazirbas, V. Golkov, T. Brox, FlowNet: learning optical flow with convolutional networks, in: IEEE ICCV, 2015, pp. 2758–2766.
[50] R. Dupre, V. Argyriou, A human and group behaviour simulation evaluation framework utilising composition and video analysis, arXiv preprint, arXiv:1707.02655, 2017.
[51] B. Andres, M. Andriluka, E. Insafutdinov, L. Pishchulin, B. Schiele, DeeperCut: a deeper, stronger, and faster multiperson pose estimation model, in: ECCV, 2016.
[52] B. Seemann, E. Leibe, B. Schiele, Pedestrian detection in crowded scenes, in: Proc. IEEE Conf. Computer Vision and Pattern Recognition, Washington, DC, 2005, pp. 878–885.
[53] Mica R. Endsley, Toward a theory of situation awareness in dynamic systems, Hum. Factors 37 (1) (1995) 32–64.
[54] A. Ess, B. Leibe, L. Van Gool, Depth and appearance for mobile scene analysis, in: International Conference on Computer Vision (ICCV'07), October 2007.

[55] A. Ess, B. Leibe, K. Schindler, L. van Gool, A mobile vision system for robust multi-person tracking, in: IEEE Conference on Computer Vision and Pattern Recognition (CVPR'08), IEEE Press, June 2008.
[56] A. Dhall, R. Goecke, S. Lucey, T. Gedeon, Collecting large, richly annotated facial-expression databases from movies, IEEE Multimed. 19 (3) (2012) 34–41.
[57] D. McDuff, et al., Affectiva-MIT facial expression dataset (am-fed): naturalistic and spontaneous facial expressions collected in-the-wild, in: CVPR-W, 2013.
[58] D. McDuff, et al., Predicting ad liking and purchase intent: large-scale analysis of facial responses to ADS, in: IEEE TAC, 2015.
[59] F. Heilbron, et al., ActivityNet: a large-scale video benchmark for human activity understanding, in: CVPR, 2015.
[60] J. Shao, et al., Slicing convolutional neural network for crowd video understanding, in: CVPR, 2016.
[61] R. Hamidreza, et al., Emotion-based crowd representation for abnormality detection, in: BMVC, 2016.
[62] J. Ferryman, A. Shahrokni, Pets2009: dataset and challenge, in: 2009 Twelfth IEEE International Workshop on Performance Evaluation of Tracking and Surveillance, Dec. 2009, pp. 1–6.
[63] J.H. Ferziger, M. Peric, Computational Methods for Fluid Dynamics, vol. 3, Springer, Berlin, 1996.
[64] P. Xu, X. Li, Q. Liu, M. Ye, C. Zhu, M. Fu, Fast crowd density estimation with convolutional neural networks, Eng. Appl. Artif. Intell. 43 (2015) 81–88.
[65] Yu Xiang, Min Sun, Fu-Hsiang Chan, Yu-Ting Chen, Anticipating accidents in dashcam videos, in: ACCV, 2016.
[66] J. Han, A. Gaszczak, T.P. Breack, Real-time people and vehicle detection from UAV imagery, in: Proceedings of the Intelligent Robots and Computer Vision XXVIII: Algorithms and Techniques, San Francisco, CA, USA, January 2011.
[67] A. Geiger, P. Lenz, R. Urtasun, Are we ready for autonomous driving? The KITTI vision benchmark suite, in: Proceedings of the IEEE Computer Society Conference on Computer Vision and Pattern Recognition, 2012.
[68] M. Nixon, H. Rahmalan, J. Carter, On crowd density estimation for surveillance, in: Proc. Institution of Engineering and Technology Conf. Crime and Security, 2006, pp. 540–545.
[69] Dirk Helbing, Peter Molnar, Self-organization phenomena in pedestrian crowds, Condens. Matter (1998) 569–577.
[70] J. Schmidhuber, S. Hochreiter, Long short-term memory, Neural Comput. 9 (1997) 1735–1780.
[71] B. Horn, B. Schunck, Determining optical flow, Artif. Intell. 17 (1–3) (1981) 185–203.
[72] Sabine Sternig, Thomas Mauthner, Manfred Klopschitz, Peter M. Roth, Horst Possegger, Matthias Rüther, Horst Bischof, Unsupervised calibration of camera networks and virtual PTZ camera, in: Proc. Computer Vision Winter Workshop (CVWW), 2014.
[73] L. Hartung, I. Saleemi, M. Shah, Scene understanding by statistical modeling of motion patterns, in: IEEE Conference on Computer Vision and Pattern Recognition, 2010, pp. 2069–2076.
[74] I. Saleemi, C. Seibert, M. Shah, H. Idrees, Multi-source multiscale counting in extremely dense crowd images, in: Proceedings of the IEEE Conference on Computer Vision and Pattern Recognition, 2013, pp. 2547–2554.
[75] D. Helbing, P. Molnár, I.J. Farkas, K. Bolay, Self-organizing pedestrian movement, Environ. Plan. B, Plan. Des. 28 (3) (2001) 361–383.
[76] E. Ilg, N. Mayer, T. Saikia, M. Keuper, A. Dosovitskiy, T. Brox, FlowNet 2.0: evolution of optical flow estimation with deep networks, in: IEEE Conference on Computer Vision and Pattern Recognition (CVPR), Jul. 2017.

[77] J. Rittsche, P.H. Tu, N. Krahnstoeve, Simultaneous estimation of segmentation and shape, in: Proc. Computer Vision and Pattern Recognition, Washington, DC, vol. 2, 2005, pp. 486–493.
[78] C.C. Loy, J. Shao, X. Wang, Scene-independent group profiling in crowd, in: Proceedings of IEEE Conference on Computer Vision and Pattern Recognition (CVPR), 2014.
[79] X. Yang, J. Zhao, Y. Xu, Q. Yan, Crowd instability analysis using velocity-field based social force model, in: IEEE Conference on Visual Communications and Image Processing, 2011, pp. 1–4.
[80] Konrad Jablonski, Vasileios Argyriou, Darrel Greenhill, Crowd Simulation for dynamic environments based on information spreading and agents' personal interests, Transp. Res. Proc. 2 (2014) 412–417.
[81] A. Singh, H.S. Koppula, S. Soh, A. Saxena, A. Jain, Recurrent neural networks for driver activity anticipation via sensory-fusion architecture, in: ICRA, 2016.
[82] Hao Jiang, Wenbin Xu, Tianlu Mao, Chunpeng Li, Shihong Xia, Zhaoqi Wang, Continuum crowd simulation in complex environments, Comput. Graph. 34 (5) (2010) 537–544.
[83] Chen Change Loy, Jing Shao, Kai Kang, Xiaogang Wang, Deeply learned attributes for crowded scene understanding, in: CVPR, 2015.
[84] M.J. Jones, D. Snow, Pedestrian detection using boosted features over many frames, in: Proc. Int. Conf. Pattern Recognition, 2008, pp. 1–4.
[85] Ioannis Karamouzas, Peter Heil, Pascal Van Beek, Mark H. Overmars, A predictive collision avoidance model for pedestrian simulation, Lect. Notes Comput. Sci. 5884 (2009) 41–52.
[86] Keni Bernardin, Rainer Stiefelhagen, Evaluating multiple object tracking performance: the clear mot metrics, Int. J. Image Video Process. 2008 (1) (2008) 246–309, https://doi.org/10.1155/2008/246309.
[87] Sujeong Kim, Stephen J. Guy, Dinesh Manocha, Ming C. Lin, Interactive simulation of dynamic crowd behaviors using general adaptation syndrome theory, in: ACM SIGGRAPH Symposium on Interactive 3D Graphics and Games – I3D'12, 2012, p. 55.
[88] A.K. Jain, J.C. Klontz, A Case Study on Unconstrained Facial Recognition Using the Boston Marathon Bombings Suspects, Tech. Michigan State University, 2013, p. 119.
[89] Franziska Klugl, Georg Klubertanz, Guido Rindsfuser, Agent-based pedestrian simulation of train evacuation integrating environmental data, in: Lect. Notes Comput. Sci., vol. 5803, 2009, pp. 631–638.
[90] D. Kong, D. Gray, Hai Tao, A viewpoint invariant approach for crowd counting, in: 18th International Conference on Pattern Recognition (ICPR'06), vol. 3, 2006, pp. 1187–1190.
[91] L. Kratz, K. Nishino, Anomaly detection in extremely crowded scenes using spatio-temporal motion pattern models, in: CVPR, 2009, pp. 1446–1453.
[92] K. Hotta, T. Kurita, S. Kumagai, Mixture of counting CNNs: adaptive integration of CNNs specialized to specific appearance for crowd counting, arXiv preprint, arXiv: 1703.09393, 2017.
[93] A. Milan, I. Reid, S. Roth, L. Leal-Taixe, K. Schindler, MOTChallenge 2015: towards a benchmark for multi-target tracking, arXiv:1504.01942 [cs], 2015.
[94] T. Li, H. Chang, M. Wang, B. Ni, R. Hong, S. Yan, Crowded scene analysis: a survey, IEEE Trans. Circuits Syst. Video Technol. 25 (3) (March 2015) 367–386.
[95] Mei Kuan Lim, Ven Jyn Kok, Chen Change Loy, Chee Seng Chan, Crowd saliency detection via global similarity structure, in: 22nd International Conference on Pattern Recognition, ICPR 2014, Stockholm, Sweden, August 24–28, 2014, 2014, pp. 3957–3962.

[96] B. Lucas, T. Kanade, An iterative image registration technique with an application to stereo vision, in: Proc. IJCAI, 1981, pp. 674–679.
[97] V. Murino, M. Cristani, A. Vinciarelli, Socially intelligent surveillance and monitoring: analysing social dimensions of physical space, in: CVPRW, 2010, pp. 51–58.
[98] R. Ma, L. Li, W. Huang, Q. Tian, On pixel count based crowd density estimation for visual surveillance, in: IEEE Conference on Cybernetics and Intelligent Systems, Singapore, 2004, vol. 1, 2004, pp. 170–173.
[99] Yalong Ma, Xinkai Wu, Guizhen Yu, Yongzheng Xu, Yunpeng Wang, Pedestrian detection and tracking from low-resolution unmanned aerial vehicle thermal imagery, Sensors 16 (4) (2016).
[100] M. Marsden, K. McGuinness, S. Little, N.E. O'Connor, ResnetCrowd: a residual deep learning architecture for crowd counting, violent behaviour detection and crowd density level classification, in: AVSS, 2017.
[101] K. McGuiness, S. Little, N.E. O'Connor, M. Marsden, Fully convolutional crowd counting on highly congested scenes, arXiv preprint, arXiv:1612.00220, 2016.
[102] B.T. Morris, M.M. Trivedi, A survey of vision-based trajectory learning and analysis for surveillance, IEEE Trans. Circuits Syst. Video Technol. 18 (8) (2008) 1114–1127.
[103] S. Mohd Hashim, N. Amir Sjarif, S. Shamsuddin, S. Yuhaniz, Crowd analysis and its applications, in: Software Engineering and Computer Systems, vol. 179, 2011, pp. 687–697.
[104] A. Nurhadiyatna, W. Jatmiko, B. Hardjono, A. Wibisono, I. Sina, P. Mursanto, Background subtraction using Gaussian mixture model enhanced by hole filling algorithm (GMMHF), in: 2013 IEEE International Conference on Systems, Man, and Cybernetics, Oct. 2013, pp. 4006–4011.
[105] Timo Ojala, Matti Pietikäinen, Topi Mäenpää, A Generalized Local Binary Pattern Operator for Multiresolution Gray Scale and Rotation Invariant Texture Classification, Springer Berlin Heidelberg, Berlin, Heidelberg, 2001, pp. 399–408.
[106] Juan Sebastian Olier, Carlo Regazzoni, Lucio Marcenaro, Matthias Rauterberg, Convolutional neural networks for detecting and mapping crowds in first person vision applications, in: Advances in Computational Intelligence, Springer International Publishing, ISBN 978-3-319-19258-1, 2015, pp. 475–485.
[107] R.J. Lopez-Sastre, D. Onoro-Rubio, Towards perspective-free object counting with deep learning, in: European Conference on Computer Vision, Springer, 2016, pp. 615–629.
[108] N. Courty, P. Allain, T. Corpetti, Crowd flow characterization with optimal control theory, in: Asian Conference of Computer Vision, 2009, pp. 279–290.
[109] Pan Shao-Yun, Guo Jie, Huang Zheng, An improved convolutional neural network on crowd density estimation, ITM Web Conf. 7 (2016) 05009, https://doi.org/10.1051/itmconf/20160705009.
[110] Divya R. Pillai, Crowd behavior analysis using 3D convolutional neural network, Int. J. Sci. Eng. Res. 5 (6) (2014) 1102, ISSN 2229-5518.
[111] M. Chli, R. Siegwart, J. Portmann, S. Lynen, People detection and tracking from thermal views, in: Proceedings of the IEEE International Conference on Robotics and Automatisation (ICRA), Hong Kong, China, May 2014.
[112] Shiliang Pu, Tao Song, Yuan Zhang, Di Xie, Estimation of crowd density in surveillance scenes based on deep convolutional neural network, Proc. Comput. Sci. 111 (2017) 154–159.
[113] G. Longo, R. Brambilla, B. Rudofsky, For Pedestrians Only: Planning, Design, and Management of Traffic-Free Zones, Whitney Library of Design, New York, 1977.
[114] A. Oyama, R. Mehran, M. Shah, Abnormal crowd behavior detection using social force model, in: IEEE Conference on Computer Vision and Pattern Recognition, 2009, pp. 935–942.

[115] B.E. Moore, R. Mehran, M. Shah, A streakline representation of flow in crowded scenes, in: European Conference on Computer Vision, 2010, pp. 439–452.
[116] M. Cristani, R. Raghavendra, A. Del Bue, V. Murino, Optimizing interaction force for global anomaly detection in crowded scenes, in: ICCV Workshops, 2011, pp. 136–143.
[117] V. Rabaud, S. Belongie, Counting crowded moving objects, in: Proc. IEEE Conf. Computer Vision and Pattern Recognition, 2006, pp. 705–711.
[118] C.S. Regazzoni, A. Tesei, V. Murino, A real-time vision system for crowding monitoring, in: International Conference on Industrial Electronics, Control, and Instrumentation, 1993. Proceedings of the IECON '93, vol. 3, Nov. 1993, pp. 1860–1864.
[119] C.W. Reynolds, Steering Behaviors for Autonomous Characters, 1999, pp. 763–782.
[120] Craig W. Reynolds, Flocks, herds and schools: a distributed behavioral model, in: ACM SIGGRAPH Comput. Graph., vol. 21, ACM, 1987, pp. 25–34.
[121] M. Rodriguez, J. Sivic, I. Laptev, J.Y. Audibert, Density-Aware Person Detection and Tracking in Crowds, 2011.
[122] D. Russell, S. Gong, Multi-layered decomposition of recurrent scenes, in: European Conference on Computer Vision (ECCV), Marseille, France, October 2008.
[123] J.-Y. Chen, S.-F. Lin, H.-X. Chao, Estimation of number of people in crowded scenes using perspective transformation, IEEE Trans. Syst. Man Cybern. A 31 (2001) 645–654.
[124] B.E. Moore, S. Wu, M. Shah, Chaotic invariants of Lagrangian particle trajectories for anomaly detection in crowded scenes, in: IEEE Conference on Computer Vision and Pattern Recognition, 2010, pp. 2054–2060.
[125] Deepak Babu Sam, Shiv Surya, R. Venkatesh Babu, Switching convolutional neural network for crowd counting, CoRR-CVPR, arXiv:1708.00199, 2017.
[126] SEWA, http://db.sewaproject.eu, 2016.
[127] J. Shao, C.C. Loy, K. Kang, X. Wang, Slicing convolutional neural network for crowd video understanding, in: 2016 IEEE Conference on Computer Vision and Pattern Recognition (CVPR), June 2016, pp. 5620–5628.
[128] B. Sheng, C. Shen, G. Lin, J. Li, W. Yang, C. Sun, Crowd counting via weighted VLAD on dense attribute feature maps, IEEE Trans. Circuits Syst. Video Technol. 28 (8) (2018) 1788–1797, https://doi.org/10.1109/TCSVT.2016.2637379.
[129] V. Sindagi, V. Patel, CNN-based cascaded multi-task learning of high-level prior and density estimation for crowd counting, in: 2017 IEEE International Conference on Advanced Video and Signal Based Surveillance (AVSS), IEEE, 2017.
[130] Vishwanath Sindagi, Vishal M. Patel, CNN-based cascaded multi-task learning of high-level prior and density estimation for crowd counting, CoRR arXiv:1707.09605, 2017.
[131] Vishwanath A. Sindagi, Vishal M. Patel, Generating high-quality crowd density maps using contextual pyramid CNNs, CoRR arXiv:1708.00953, 2017.
[132] F. De Smedt, D. Hulens, T. Goedemé, On-board real-time tracking of pedestrians on a UAV, in: 2015 IEEE Conference on Computer Vision and Pattern Recognition Workshops (CVPRW), June 2015, pp. 1–8.
[133] G.K. Still, Crowd Dynamics, Ph.D. dissertation, University of Warwick, 2000.
[134] Sybert Stroeve, Henk Blom, Marco van der Park, Multi-agent situation awareness error evolution in accident risk modelling, in: 5th USA/Europe Air Traffic Management R&D Seminar (June), 2003, pp. 23–27.
[135] Patrick Sudowe, Bastian Leibe, Efficient use of geometric constraints for sliding-window object detection in video, in: Computer Vision Systems, Springer, Berlin, Heidelberg, ISBN 978-3-642-23968-7, 2011, pp. 11–20.
[136] W. Liu, Y. Jia, P. Sermanet, S. Reed, D. Anguelov, D. Erhan, V. Vanhoucke, A. Rabinovich, C. Szegedy, Going deeper with convolutions, in: Proceedings of the IEEE Conference on Computer Vision and Pattern Recognition, 2015, pp. 1–9.

[137] O. Kliper-Gross, T. Hassner, Y. Itcher, Violent flows: real-time detection of violent crowd behavior, in: CVPRW, IEEE, 2012, pp. 1–6.
[138] Adrien Treuille, Seth Cooper, Zoran Popović, Continuum crowds, ACM Trans. Graph. 25 (3) (2006) 1160.
[139] V.A. Sindagi, V.M. Patel, A survey of recent advances in CNN-based single image crowd counting and density estimation, Pattern Recognit. Lett. 107 (2018) 3–16.
[140] VIOLENCE-FLOWS, http://www.openu.ac.il/home/hassner/data/violentflows/index.html, 2012.
[141] L. Wang, W. Hu, T. Tan, S. Maybank, A survey on visual surveillance of object motion and behaviors, IEEE Trans. Syst. Man Cybern., Part C, Appl. Rev. 34 (3) (2004) 334–352.
[142] Z. Fu, D. Xie, T. Tan, W. Hu, X. Xiao, S. Maybank, A system for learning statistical motion patterns, IEEE Trans. Pattern Anal. Mach. Intell. 28 (9) (2006) 1450–1464.
[143] L. Wolf, E. Walach, Learning to count with CNN boosting, in: European Conference on Computer Vision, Springer, 2016, pp. 660–676.
[144] X. Wang, X. Ma, W.E.L. Grimson, Unsupervised activity perception in crowded and complicated scenes using hierarchical Bayesian models, IEEE Trans. Pattern Anal. Mach. Intell. 31 (3) (2009) 539–555.
[145] H. Zhang, L. Yang, S. Liu, X. Cao, C. Wang, Deep people counting in extremely dense crowds, in: Proceedings of the 23rd ACM International Conference on Multimedia, ACM, 2015, pp. 1299–1302.
[146] Shih-En Wei, Varun Ramakrishna, Takeo Kanade, Yaser Sheikh, Convolutional pose machines, in: CVPR, 2016.
[147] X. Wu, G. Liang, K.K. Lee, Y. Xu, Crowd density estimation using texture analysis and learning, in: 2006 IEEE International Conference on Robotics and Biomimetics, Dec. 2006, pp. 214–219.
[148] Hui Xi, Young-Jun Son, Seungho Lee, An integrated pedestrian behavior model based on extended decision field theory and social force model, in: Proceedings of the Winter Simulation Conference (WSC '10), 2010, pp. 824–836.
[149] Cong Zhang, Hongsheng Li, X. Wang, Xiaokang Yang, Cross-scene crowd counting via deep convolutional neural networks, in: 2015 IEEE Conference on Computer Vision and Pattern Recognition (CVPR), June 2015, pp. 833–841.
[150] D. Zhou, S. Chen, S. Gao, Y. Ma, Y. Zhang, Single image crowd counting via multicolumn convolutional neural network, in: Proceedings of the IEEE Conference on Computer Vision and Pattern Recognition, 2016, pp. 589–597.
[151] H. Li, R. Zhao, X. Wang, Z. Zhao, Crossing-line crowd counting with two-phase deep neural networks, in: European Conference on Computer Vision, Springer, 2016, pp. 712–726.
[152] T. Zhao, R. Nevatia, Bayesian human segmentation in crowded situations, in: Proc. IEEE Conf. Computer Vision and Pattern Recognition, vol. 2, 2003, pp. 459–466.
[153] Bo Zheng, Yibiao Zhao, Joey Yu, Katsushi Ikeuchi, Song Chun Zhu, Scene understanding by reasoning stability and safety, Int. J. Comput. Vis. 112 (2) (2015) 221–238.
[154] B. Zheng, Y. Zhao, J.C. Yu, K. Ikeuchi, S. Zhu, Detecting potential falling objects by inferring human action and natural disturbance, in: 2014 IEEE International Conference on Robotics and Automation (ICRA), Hong Kong, 2014, pp. 3417–3424.

Chapter 15

Towards multi-modality invariance: A study in visual representation

Lingxi Xie*, Qi Tian†

*The Johns Hopkins University, Department of Computer Science, Baltimore, MD, USA †The University of Texas at San Antonio, Department of Computer Science, San Antonio, TX, USA

CONTENTS

15.1 INTRODUCTION AND RELATED WORK

Recent years have witnessed the rapid progress on artificial intelligence, in which a large number of intelligent systems based on multi-modal content analysis are applied in real-world scenarios. In this chapter, we consider an application built on fine-grained object recognition which integrates audio and video signals. The system is illustrated in Fig. 15.1. The user samples a video (audio included) and sends it to the server, which is asked to recognize the contents in this clip. Of course, we are pursuing two submodules for processing signals in two different modalities, in which visual understanding takes the main role and audio analysis serves as an auxiliary. The

Multimodal Behavior Analysis in the Wild. https://doi.org/10.1016/B978-0-12-814601-9.00029-8
Copyright © 2019 Elsevier Ltd. All rights reserved.

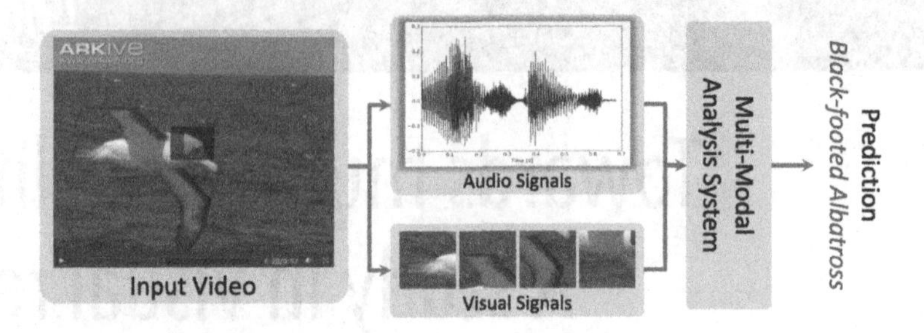

■ **FIGURE 15.1** Given an input video, our task is to extract the audio and visual signals from it and predict the class of the object. A multi-modal content analysis system must be built. This section provides a study towards invariance in visual representation, which is essential in this system.

main goal of this chapter is to provide a comprehensive study in the former module, *i.e.*, visual content analysis.

Visual representation is a long-lasting battle in computer vision. It lays the foundation of a series of problems towards understanding visual contents, including image classification, instance retrieval, object detection, semantic segmentation, *etc.* A basic idea is to transfer the raw image data, *e.g.*, in pixel-wise representation, from image space into an element in feature space, *i.e.*, a high-dimensional feature vector. This vector can then be used by various algorithms, such as nearest neighbor classifier and support vector machines.

This chapter focuses on two typical models. One of them is the Bag-of-Visual-Words (BoVW) model [12], a statistics-based algorithm in which local features are extracted, encoded and summarized into global image representation. It starts with describing local patches using handcrafted descriptors, such as SIFT [41], HOG [13] and LCS [11]. Although these descriptors can be automatically detected using operators such as DoG [41] and MSER [44], the dense sampling strategy [6,61] often works better on classification tasks. Next, a visual vocabulary (codebook) is trained to estimate the feature space distribution. The codebook is often computed with iterative algorithms such as K-Means or GMM. Descriptors are then projected to the codebook, using encoding techniques, such as hard quantization, sparse coding [74], LLC encoding [65], super-vector encoding [83], Fisher vector encoding [56], *etc.* On the final stage, quantized feature vectors are aggregated as compact image representation. Sum pooling, max-pooling and ℓ_p-norm pooling [17] can be different choices, and visual phrases [79,70] and/or spatial pyramids [21,32] are constructed for richer spatial context modeling. The representation vectors are then summarized [71] and fed into machine learning algorithms such as the SVM.

The second one is the Convolutional Neural Network (CNN), one of the most important models in deep learning. It serves as a hierarchical model for large-scale visual recognition. It is based on a network with enough neurons is able to fit any complicated data distribution. In the early years, neural networks were shown effective for simple recognition tasks such as digit recognition [33]. More recently, the availability of large-scale training data (*e.g.*, ImageNet [14]) and powerful GPUs makes it possible to train deep CNNs [30] which significantly outperform the BoVW-based models. A CNN is composed of several stacked layers. In each of them, responses from the previous layer are convolved and activated by a differentiable function. Hence, a CNN can be considered as a composite function, and it is trained by back-propagating error signals defined by the difference between supervised and predicted labels at the top level. Recently, efficient methods were proposed to help CNNs converge faster and prevent over-fitting, such as ReLU activation [30], dropout and batch normalization [25]. It is believed that deeper networks produce better recognition results [57,58]. The intermediate responses of CNNs, or the so-called deep features, serve as an efficient image description or a set of latent visual attributes [15]. They can be used for various types of vision applications, including image classification [26], image retrieval [53,68] and object detection [19]. A discussion of how different CNN configurations impact deep feature performance is available in [10]. Visualization also helps understanding the behavior of CNN models [76].

One of the major shortcomings of the BoVW and CNN models is the unsatisfied stability of image representation. An important way of improvement is to study the *invariance* of local descriptors or patch operators. SIFT [41] achieves scale and rotation invariance by selecting the maxima in the scale space, and picking up a dominant orientation, via gradient computation, and rotating the local patch accordingly. Other examples include Shape Context [5], SURF [4], BRIEF [7], BRISK [36], ORB [54], FREAK [1], *etc.* Radial transform [59] and polar analysis [40] play important roles in generating rotation-invariant local features. In some vision tasks such as fine-grained recognition, objects might have different left/right orientations. Since handcrafted descriptors (such as SIFT) and convolution operations are not reversal-invariant, feature representation of an image and its reversed version might be totally different. To this point, researchers propose to augment the image datasets by adding a reversed copy for each original image, and perform classification on the enlarged training and testing sets [9,8]. In [49], it is even suggested to learn a larger image transformation set for data augmentation. Similar strategies are also adopted in the CNN training process, including a popular method which adds reversal on each training sample with a probability of 50%, which, as a part of data augmentation, is often cooperated with other techniques such as image cropping [30]. Although data augmentation improves the recognition accuracy consistently, it brings heavier computational overheads, *e.g.*, almost doubled time and memory consumptions on the online testing stage of the BoVW model, or the requirement of more training epochs to make the CNN training process converge.

This chapter presents an alternative idea, *i.e.*, designing reversal-invariant representation of local patterns for both the BoVW and CNN models. On the BoVW model, we start with observing the difference between the original and reversed descriptors, and then suggest computing the orientation of each descriptor so that we can cancel out the impact of image reversal. For orientation estimation, we adopt an approximated summation on the gradient-based histograms of SIFT. Based on this theory, we propose **Max-SIFT** and **RIDE** (Reversal-Invariant Descriptor Enhancement), two simple, fast yet generalized algorithms which bring reversal invariance to local descriptors. Both Max-SIFT and RIDE guarantee to generate identical representation for an image and its left–right reversed copy. Experiments reveal that Max-SIFT and RIDE produce consistent accuracy improvement to image classification. RIDE outperforms data augmentation with higher recognition rates and lower time/memory consumptions. Also, we generalize this idea to the state-of-the-art CNN architectures. We design a reversal-invariant convolution operation (**RI-Conv**) and plug it into conventional CNNs, so that we can train reversal-invariant deep networks, which generate reversal-invariant deep features directly (without requiring post-processing). RI-Conv enjoys the advantage of enlarging the network capacity without increasing the model complexity. Experiments verify the effectiveness of our algorithms, demonstrating the importance of reversal invariance in training efficient CNN models and transferring deep features.

15.2 VARIANCES IN VISUAL REPRESENTATION

People often take pictures without caring about the left/right orientation, since an image and its left–right reversed copy often have the same semantic meaning. Consequently, there exist both left-oriented and right-oriented objects in almost every popular image dataset, especially in the case of fine-grained object recognition on *animals*, *man-made tools*, *etc*. For example, among 11,788 images of the Bird-200 dataset [64], at least 5000 *birds* are oriented to the left and other 5000 oriented to the right. In the Aircraft-100 dataset [43] with 10,000 images, we can also find more than 4800 left-oriented and more than 4500 right-oriented *aircrafts*, respectively.

However, we argue that most image representation models are sensitive to image reversal, *i.e.*, the features extracted from an image and its reversed version may be completely different. Let us take a simple case study using the BoVW model which encodes SIFT with the Fisher vectors [50]. Detailed settings are shown in Section 15.3.4. We perform image classification and retrieval tasks on the Aircraft-100 dataset [43]. We choose this dataset mainly because that the orientation of an *aircraft* is more easily determined than, say, of a *bird*. Based on the original dataset, we manually reverse all the left-oriented images, generating a right-aligned dataset.

With the standard training/testing split (around 2/3 images are used for training and others for testing), the recognition rate is 53.13% on the *original* dataset and rises up to 63.94% on the *right-aligned* dataset, with a more-than-10% absolute accuracy gain (a more-than-20% relative gain). This implies that orientation alignment brings a large benefit to fine-grained object recognition.

Still on the Aircraft-100 dataset. To diagnose, we use all (10,000) images in the right-aligned dataset for training, and evaluate the model with the entire right-aligned and left-aligned datasets, respectively. When we test on the right-aligned dataset, *i.e.*, the training images are identical to the testing image, the classification accuracy is 99.73% (not surprising since we are just performing self-validation). However, when we test on the left-aligned dataset, *i.e.*, each testing image is the reversed version of a training image, the accuracy drops dramatically to 46.84%. This experiment reveals that a model learned from right-oriented objects may not recognize left-oriented objects very well.

Lastly, we perform image retrieval on the right-aligned dataset to directly evaluate the feature quality. Given a query image, we sort the candidates according to the ℓ_2-distance between the representation vectors. Some typical results are shown in Fig. 15.2. When the query is of the same orientation (right) with the database, the search result is satisfying (mAP is 0.4143, the first false-positive is ranked at #18). However, if the query image is reversed, its feature representation changes thoroughly, and the retrieval accuracy drops dramatically (mAP is 0.0025, the first true-positive is ranked at #388). It is worth noting, in the latter case, that the reversed version of the query image is ranked at #514. This means that more than 500 images, most of them coming from different categories, are more similar to the query than its reversed copy, because the image feature is not reversal-invariant.

Although all the above experiments are based on the BoVW model with SIFT and Fisher vectors, we emphasize that similar trouble also arises in the case of extracting deep features from a pre-trained neural network. Since

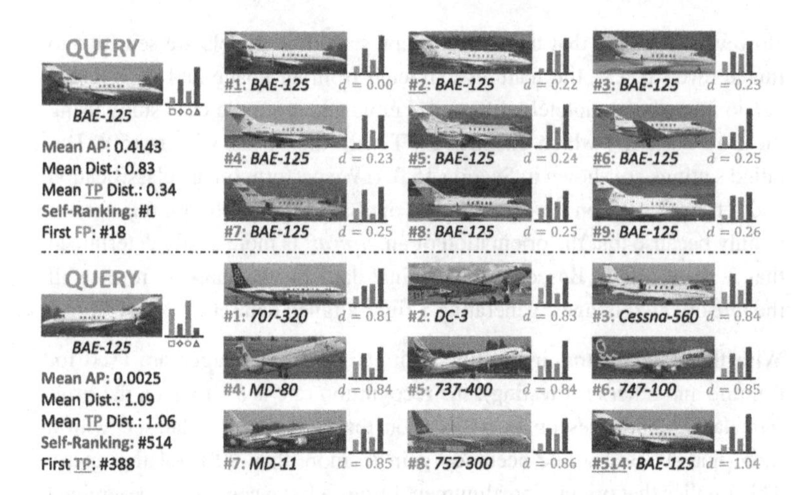

FIGURE 15.2 Content-based image retrieval on the right-oriented Aircraft-100 dataset (best viewed in color). We use the same query image with different orientations. Abbreviations: AP for average precision, TP for true-positive, FP for false-positive.

convolution is not reversal invariance, the features extracted on an image and its reversed version are often different, even when the network is trained with data augmentation (each training image is reversed with a 50% probability). We will present a detailed analysis of this point in Section 15.4.

Since an image and its reversed copy might have totally different feature representation, in a fine-grained dataset containing both left-oriented and right-oriented objects, we are implicitly partitioning the images of each class into two (or even more) prototypes. Consequently, the number of training images of each prototype is reduced and the risk of over-fitting increased. With this observation, some algorithms [9,8] augment the dataset by generating a reversed copy for each image to increase the number of training cases of each prototype, meanwhile the testing stage of deep networks often involves image reversal followed by score average [30,57]. We propose a different idea that generates reversal-invariant image representation in a bottom-up manner.

15.3 REVERSAL INVARIANCE IN BOVW

This section introduces reversal invariance to the BoVW model by designing reversal-invariant local descriptors. We first discuss the basic principle of designing reversal-invariant descriptors, and then provide a simple solution named Max-SIFT. After that, we generalize Max-SIFT as RIDE, and we show that it can be applied to more types of local descriptors. Experiments

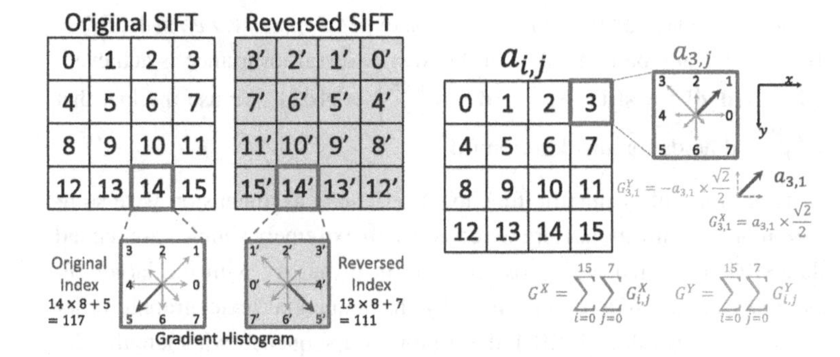

■ **FIGURE 15.3** Left: the structure of a SIFT descriptor and its reversed copy. Right: estimating the global gradient of a SIFT descriptor.

on the BoVW model and Fisher vector encoding verify the effectiveness of our algorithms.

15.3.1 **Reversal symmetry and Max-SIFT**

We start from observing how SIFT, a typical handcrafted descriptor, changes with left–right image reversal. The structure of a SIFT descriptor is illustrated in Fig. 15.3. A patch is partitioned into 4×4 spatial grids, and in each grid a 8-dimensional gradient histogram is computed. Here we assume that spatial grids are traversed from top to bottom, then left to right, and gradient intensities in each grid is collected in a counter-clockwise order. When an image is left–right reversed, all the patches on it are reversed as well. In a reversed patch, both the order of traversing spatial grids and collecting gradient values are changed, although the absolute gradient values in the corresponding directions do not change. Taking the lower-right grid in the original SIFT descriptor (#15) as the example. When the image is reversed, this grid appears at the lower-left position (#12), and the order of collecting gradients in the grid changes from $(0, 1, 2, 3, 4, 5, 6, 7)$ to $(4, 3, 2, 1, 0, 7, 6, 5)$.

Denote the original SIFT as $\mathbf{d} = (d_0, d_1, \ldots, d_{127})$, in which $d_{i \times 8+j} = a_{i,j}$ for $i = 0, 1, \ldots, 15$ and $j = 0, 1, \ldots, 7$. As shown in Fig. 15.3, each index (0 to 127) of the original SIFT is mapped to another index of the reversed SIFT. For example, d_{117} ($a_{14,5}$, the bold arrow in Fig. 15.3) would appear at d_{111} ($a_{13,7}$) when the descriptor is reversed. Denote the *index mapping function* as $f(\cdot)$ (*e.g.*, $f(0) = 28$, $f(117) = 111$), so that the reversed SIFT can be computed as $\mathbf{d}^{\mathrm{R}} \doteq \mathbf{f}(\mathbf{d}) = \left(d_{f(0)}, d_{f(1)}, \ldots, d_{f(127)}\right)$.

Towards reversal invariance, we need to design a *descriptor transformation function* $\mathbf{r}(\mathbf{d})$, so that $\mathbf{r}(\mathbf{d}) = \mathbf{r}\left(\mathbf{d}^{\mathrm{R}}\right)$ for any descriptor $\mathbf{d}$. For this, we define $\mathbf{r}(\mathbf{d}) = \mathbf{s}(\mathbf{d}, \mathbf{d}^{\mathrm{R}})$, in which $\mathbf{s}(\cdot, \cdot)$ satisfies *symmetry*, *i.e.*, $\mathbf{s}(\mathbf{d}_1, \mathbf{d}_2) = \mathbf{s}(\mathbf{d}_2, \mathbf{d}_1)$ for any pair $(\mathbf{d}_1, \mathbf{d}_2)$. In this way reversal invariance is achieved: $\mathbf{r}(\mathbf{d}) = \mathbf{s}(\mathbf{d}, \mathbf{d}^{\mathrm{R}}) = \mathbf{s}(\mathbf{d}^{\mathrm{R}}, \mathbf{d}) = \mathbf{s}\left(\mathbf{d}^{\mathrm{R}}, (\mathbf{d}^{\mathrm{R}})^{\mathrm{R}}\right) = \mathbf{r}(\mathbf{d}^{\mathrm{R}})$. We use the fact that $(\mathbf{d}^{\mathrm{R}})^{\mathrm{R}} = \mathbf{d}$ holds for any descriptor $\mathbf{d}$.

There are a lot of symmetric function $\mathbf{s}(\cdot, \cdot)$, such as dimension-wise summation or maximization. Here we consider an extremely simple case named Max-SIFT, in which we choose the one in $\mathbf{d}$ and $\mathbf{d}^{\mathrm{R}}$ with the larger *sequential lexicographic order*. Here, by the sequential lexicographic order we mean to regard each SIFT descriptor as a sequence with length 128, and on each dimension (an element in the sequence), the larger value has a higher priority.

Therefore, to compute the Max-SIFT descriptor for $\mathbf{d}$, we only need to compare the dimensions of $\mathbf{d}$ and $\mathbf{d}^{\mathrm{R}}$ one by one and stop at the first difference. Let us denote the Max-SIFT descriptor of $\mathbf{d}$ by $\widehat{\mathbf{d}}$, and use the following notation:

$$\widehat{\mathbf{d}} = \mathbf{r}(\mathbf{d}) = \widehat{\max}\left\{\mathbf{d}, \mathbf{d}^{\mathrm{R}}\right\}, \tag{15.1}$$

where $\widehat{\max}\{\cdot, \cdot, \ldots, \cdot\}$ denotes the element with the maximal sequential lexicographic order. We point out that there are many other symmetric functions, but their performance is often inferior to Max-SIFT. For example, using Average-SIFT, *i.e.*, $\mathbf{r}(\mathbf{d}) = \frac{1}{2}\left(\mathbf{d} + \mathbf{d}^{\mathrm{R}}\right)$, leads to 1%–3% accuracy drop in image classification scenarios.

15.3.2 RIDE: generalized reversal invariance

Let us choose the descriptor from $\mathbf{d}$ and $\mathbf{d}^{\mathrm{R}}$ in a more generalized manner. In general, we define an *orientation quantization function* $q(\cdot)$, and choose the one in $\left\{\mathbf{d}, \mathbf{d}^{\mathrm{R}}\right\}$ with the larger function value. Ideally, $q(\cdot)$ can capture the orientation property of a descriptor, *e.g.*, $q(\mathbf{d})$ reflects the extent that $\mathbf{d}$ is oriented to the right. Recall that in the original version of SIFT [41], each descriptor is naturally assigned an orientation angle $\theta \in [0, 2\pi)$, so that we can simply take $q(\mathbf{d}) = \cos\theta$, but orientation is often ignored in the implementation of dense SIFT [6,63]. We aim at recovering the orientation with fast computations.

The major conclusion is that, the global orientation of a densely sampled SIFT descriptor can be estimated by accumulating clues from the local gradients. For each of the 128 dimensions, we take its gradient value and lookup for its (1 of 8) direction. The gradient value is then decomposed into two components along the x-axis and y-axis, respectively. The left/right orientation of the descriptor is then computed by collecting the x-axis components over all the 128 dimensions. Formally, we define 8 orientation vectors $\mathbf{u}_j$, $j = 0, 1, \ldots, 7$. According to the definition of SIFT in Fig. 15.3, we have $\mathbf{u}_j = (\cos(j\pi/4), \sin(j\pi/4))^\top$. The global gradient can be computed as $\mathbf{G}(\mathbf{d}) = (G_x, G_y)^\top = \sum_{i=0}^{15} \sum_{j=0}^{7} a_{i,j} \mathbf{u}_j$. The computation of $\mathbf{G}(\mathbf{d})$ is illustrated in Fig. 15.3.

We simply take G_x as the value of quantization function, *i.e.*, $q(\mathbf{d}) = G_x(\mathbf{d})$ for every $\mathbf{d}$. It is worth noting that $q(\mathbf{d}) = -q(\mathbf{d}^{\mathrm{R}})$ holds for any $\mathbf{d}$, therefore we can simply use the sign of $q(\mathbf{d})$ to compute the reversal-invariant descriptor transform $\widetilde{\mathbf{d}}$:

$$\widetilde{\mathbf{d}} = \mathbf{r}(\mathbf{d}) = \begin{cases} \mathbf{d} & q(\mathbf{d}) > 0, \\ \mathbf{d}^{\mathrm{R}} & q(\mathbf{d}) < 0, \\ \widehat{\max}\left\{\mathbf{d}, \mathbf{d}^{\mathrm{R}}\right\} & q(\mathbf{d}) = 0. \end{cases} \tag{15.2}$$

We name the algorithm RIDE (Reversal-Invariant Descriptor Enhancement). When $q(\mathbf{d}) = 0$, RIDE degenerates to Max-SIFT. Since Max-SIFT first compares d_0 and d_{28} ($f(0) = 28$), we can approximate it as a special case of RIDE, with $q(\mathbf{d}) = d_0 - d_{28}$.

We generalize RIDE to (a) other local descriptors and (b) more types of reversal invariance. When RIDE is applied on other dense descriptors, we can first extract SIFT descriptors on the same patches, then compute $\mathbf{G}$ to estimate the orientation of those patches, and perform reversal operation if necessary. The extra time overheads in this process mainly come from the computation of SIFT, which can be exempted in the case of using Color-SIFT descriptors. For example, RGB-SIFT is composed of three SIFT vectors $\mathbf{d}_{\mathrm{R}}$, $\mathbf{d}_{\mathrm{G}}$ and $\mathbf{d}_{\mathrm{B}}$, from the individual red, green and blue channels, therefore we can compute $\mathbf{G}_{\mathrm{R}}$, $\mathbf{G}_{\mathrm{G}}$ and $\mathbf{G}_{\mathrm{B}}$ individually, and combine them with $\mathbf{G} = 0.30\mathbf{G}_{\mathrm{R}} + 0.59\mathbf{G}_{\mathrm{G}} + 0.11\mathbf{G}_{\mathrm{B}}$. For other color SIFT descriptors, the only difference lies in the linear combination coefficients. By this trick we can perform RIDE on Color-SIFT descriptors very fast.

In the case that RIDE is applied to fast binary descriptors for image retrieval, we can obtain the orientation vector $\mathbf{G}$ without computing SIFT. Let us take the BRIEF descriptor [7] as an example. For a descriptor $\mathbf{d}$, $G_x(\mathbf{d})$ is obtained by accumulating the *binary tests*. For each tested pixel pair (p_1, p_2) with distinct x-coordinates, if the left pixel has a smaller intensity value, add 1 to $G_x(\mathbf{d})$, otherwise subtract 1 from $G_x(\mathbf{d})$. If the x-coordinates of p_1 and p_2 are the same, this pair is ignored. $G_y(\mathbf{d})$ is similarly computed. We still take $q(\mathbf{d}) = G_x(\mathbf{d})$ to quantize left–right orientation. This idea can also be generalized to other binary descriptors such as ORB [54], which is based on BRIEF.

RIDE is also capable of canceling out a larger family of reversal operations, including upside–down image reversal, and image rotation by 90°, 180° and 270°. For this we need to take more information from the global gradient vector $\mathbf{G} = \left(G_x, G_y\right)^{\top}$. Recall that limiting $G_x > 0$ selects 1 descriptor from 2 candidates, resulting in RIDE-2 (equivalent to RIDE mentioned previously) for left–right reversal invariance. Similarly, limiting $G_x > 0$ and $G_y > 0$ selects 1 from 4 descriptors, obtaining RIDE-4 for both left–right and upside–down reversal invariance, and limiting $G_x > G_y > 0$ obtains RIDE-8 for both reversal and rotation invariance. We do not use RIDE-4 and RIDE-8 in this chapter, since upside–down reversal and heavy rotations are not often observed, whereas the descriptive power of a descriptor is reduced by strong constraints.

15.3.3 Application to image classification

The benefit brought by the Max-SIFT and RIDE to image classification is straightforward. Consider an image $\mathbf{I}$, and a set of, say, SIFT descriptors extracted from the image: $\mathcal{D} = \{\mathbf{d}_1, \mathbf{d}_2, \ldots, \mathbf{d}_M\}$. When the image is left–right reversed, the set $\mathcal{D}$ becomes: $\mathcal{D}^{\mathrm{R}} = \left\{\mathbf{d}_1^{\mathrm{R}}, \mathbf{d}_2^{\mathrm{R}}, \ldots, \mathbf{d}_M^{\mathrm{R}}\right\}$. If the descriptors are not reversal-invariant, *i.e.*, $\mathcal{D} \neq \mathcal{D}^{\mathrm{R}}$, the feature representation produced by $\mathcal{D}$ and $\mathcal{D}^{\mathrm{R}}$ might be totally different. With Max-SIFT or RIDE, we have $\widehat{\mathbf{d}} = \widehat{\mathbf{d}^{\mathrm{R}}}$ or $\widetilde{\mathbf{d}} = \widetilde{\mathbf{d}^{\mathrm{R}}}$, for any $\mathbf{d}$; therefore $\widehat{\mathcal{D}} = \widehat{\mathcal{D}^{\mathrm{R}}}$ or $\widetilde{\mathcal{D}} = \widetilde{\mathcal{D}^{\mathrm{R}}}$. Consequently, we generate the same representation for an image and its reversed copy.

A simple trick applies when Max-SIFT or RIDE is adopted with Spatial Pyramid Matching [32]. Note that corresponding descriptors might have different x-coordinates on an image and its reversed copy, *e.g.*, a descriptor appearing at the upper-left corner of the original image can also be found at the upper-right corner of the reversed image, resulting in the difference in spatial pooling bin assignment. To cope with, we count the number of descriptors to be reversed, *i.e.*, those satisfying $\widehat{\mathbf{d}} \neq \mathbf{d}$ or $\widetilde{\mathbf{d}} \neq \mathbf{d}$. If the number is larger than half of the total number of descriptors, we left–right reverse the descriptor set by replacing the x-coordinate of each descriptor with $W - x$, where W is the image width. This is equivalent to predicting the orientation of an image using the orientation of SIFT descriptors. Despite this, a safer way is to use symmetric pooling bins. In our experiments (see Section 15.3.4), we use a spatial pyramid with four regions (the entire image and three horizontal stripes).

15.3.4 Experiments

We evaluate our algorithm on four publicly available fine-grained object recognition datasets, three scene classification datasets and one generic ob-

Table 15.1 Image classification datasets used in our experiments. Among them, four datasets are designed for fine-grained object recognition, three for scene classification and one for generic object recognition

Dataset	Abbreviation	# Classes	# Images	# Training samples / class
Pet-37 [48]	P-37	37	7390	100
Aircraft-100 [43]	A-100	100	10,000	67
Flower-102 [47]	F-102	102	8189	20
Bird-200 [64]	B-200	200	11,788	30
LandUse-21 [75]	L-37	21	2100	80
Indoor-67 [52]	I-67	67	15,620	80
SUN-397 [67]	S-397	397	108,754	50
Caltech256 [22]	C-256	257	30,607	60

ject classification dataset. The detailed information of the used datasets is summarized in Table 15.1.

Basic experimental settings follow the recent proposed BoVW model [56]. An image is scaled, with the aspect ratio preserved, so that there are 300 pixels on the larger axis. We only use the region within the bounding box if it is available. We use VLFeat [63] to extract dense RootSIFT [3] descriptors. The spatial stride and window size of dense sampling are 6 and 12, respectively. On the same set of patches, LCS, RGB-SIFT and Opponent-SIFT [62] descriptors are also extracted. Max-SIFT or RIDE is thereafter computed for each type of descriptors. In the former case, we can only apply Max-SIFT on SIFT-based descriptors, thus the LCS descriptors remain unchanged. When we apply RIDE to RootSIFT, we compute RIDE on the original SIFT, obtained by dimension-wise squaring RootSIFT.

The dimensions of SIFT, LCS and color SIFT descriptors are reduced by PCA to 64, 64 and 128, respectively. We cluster the descriptors with a GMM of 32 components, and use the improved Fisher vectors (IFV) for feature encoding. A spatial pyramid with four regions (the entire image and three horizontal stripes) is adopted. Features generated by SIFT and LCS descriptors are concatenated as the FUSED feature. The final vectors are square-root normalized followed by ℓ_2 normalized [31], and then fed into LibLINEAR [16], a scalable SVM implementation, with the slacking parameter $C = 10$. Averaged accuracy by category is reported on the fixed training/testing split provided by the authors.

Table 15.2 Classification accuracy (%) of different models. Evaluated features include SIFT (S), LCS (L), FUSED (F, where SIFT and LCS features are concatenated), RGB-SIFT (R) and OPP-SIFT (O) features, while models include using the original descriptors (**ORIG**), Max-SIFT (**MAX**), RIDE (**RIDE**) or data augmentation (**AUGM**). Max-SIFT does not work on LCS, thus the LCS part remains unchanged in the FUSED feature. **RIDE×2** denotes using RIDE with doubled codebook size

	ORIG	MAX	RIDE	AUGM	RIDE×2
S	37.92	41.78	42.28	42.24	**45.61**
L	43.25	–	44.27	45.12	**46.83**
F	52.06	53.92	54.69	54.67	**57.51**
R	44.90	46.73	47.35	46.98	**49.53**
O	46.53	48.39	49.01	48.72	**51.19**

(a) ***Pet-37 results***

	ORIG	MAX	RIDE	AUGM	RIDE×2
S	53.13	57.72	57.82	57.16	**60.14**
L	41.82	–	42.86	43.13	**44.81**
F	57.36	60.49	61.27	60.59	**63.62**
R	57.89	61.90	63.09	62.48	**65.11**
O	47.06	52.35	53.12	51.39	**55.79**

(b) ***Aircraft-100 results***

	ORIG	MAX	RIDE	AUGM	RIDE×2
S	53.68	58.12	59.12	58.01	**61.09**
L	73.47	–	75.30	75.88	**77.40**
F	76.96	79.59	80.51	79.49	**82.14**
R	71.52	74.00	74.97	74.18	**77.10**
O	76.12	78.40	79.68	78.83	**81.69**

(c) ***Flower-102 results***

	ORIG	MAX	RIDE	AUGM	RIDE×2
S	25.77	31.59	32.14	31.60	**34.07**
L	36.18	–	38.50	38.97	**40.16**
F	38.11	43.48	44.73	43.98	**46.38**
R	31.36	38.20	39.16	38.79	**41.73**
O	35.40	41.15	42.18	41.72	**44.30**

(d) ***Bird-200 results***

To compare our results with the state-of-the-art classification results, strong Fisher vectors are extracted by resizing the images to 600 pixels in the larger axis, using spatial stride 8, window size 16, and clustering 256 GMM components.

We first report fine-grained object recognition accuracy with different descriptors in Table 15.2. Beyond original descriptors, we implement Max-SIFT, RIDE and data augmentation. By data augmentation we mean to generate a reversed copy for each training/testing image, use the augmented set to train the model, test with both original and reversed samples, and predict the label with a soft-max function [49].

In Table 15.2, one can see that both Max-SIFT and RIDE produces consistent accuracy gain beyond original descriptors (**ORIG**). Moreover, when we use SIFT or Color-SIFT descriptors, **RIDE** also produces higher accuracy than using data augmentation (**AUGM**). When the LCS descriptors are used, **RIDE** works a little worse than **AUGM**, which is probably because the orientation of LCS (not a gradient-based descriptor) is not very well estimated with SIFT gradients.

We shall emphasize that data augmentation requires almost doubled computational costs compared to RIDE, since the time/memory complexity of many classification models is proportional to the number of training/testing images. To make fair comparison, we double the codebook size used in RIDE to obtain longer features, since it is a common knowledge that larger codebooks often lead to better classification results. Such a model, denoted by **RIDE×2**, works better than **AUGM** in every single case.

We also use strong features and compare Max-SIFT and RIDE with other reversal-invariant descriptors, namely MI-SIFT [42], FIND [23] and F-SIFT [82]. We compute these competitors for each SIFT component in RGB-SIFT, and leave LCS unchanged in the FUSED feature. Besides descriptor computation, all the other stages are exactly the same. Results are shown in Table 15.3. The consistent 3%–4% gain verifies that RIDE makes stable contribution to visual recognition.

It is worth noting that Bird-200 is a little bit different from other datasets, since part detection and description often play important roles in this fine-grained recognition task. Recently, researchers design complicated part-based recognition algorithms, including [8,18,69,78,81,77,37,80]. We also evaluate RIDE with SIFT on the detected parts provided by some of these approaches. RIDE boosts the recognition accuracy of [8] and [18] from 56.6% to 60.7% and from 65.3% to 67.4%, respectively. In comparison, [18] applies data augmentation to boost the accuracy from 65.3% to 67.0%. RIDE produces better results with only half time/memory consumption. With the parts learned by deep CNNs [77], the baseline performance is 71.5% with SIFT and LCS descriptors. With RIDE, we obtain 73.1%, which is close to the accuracy using deep features (73.9% reported in the paper).

To reveal that Max-SIFT and RIDE can be applied to generalized classification, we perform experiments on the scene classification and generic object recognition tasks. The FUSED (SIFT with LCS) features are used, and the results are summarized in Table 15.3. It is interesting to see that

Table 15.3 Classification accuracy (%) comparison with some recent work. We use RGB-SIFT on the Aircraft-100 dataset, and the FUSED (SIFT with LCS) features on other datasets. We implement MI-SIFT [42], FIND [23] and F-SIFT [82] by ourselves

	P-37	A-100	F-102	B-200	L-21	I-67	S-397	C256
ORIG	60.24	74.61	83.53	47.61	93.64	63.17	48.35	58.77
Max-SIFT	62.80	77.54	85.82	49.93	94.13	64.12	49.39	59.21
RIDE	**63.49**	**78.92**	**86.45**	**50.81**	**94.71**	**64.93**	**50.12**	**60.25**
MI-SIFT [42]	58.91	72.26	81.06	45.59	92.86	61.49	46.51	55.39
FIND [23]	59.63	74.06	82.91	47.49	93.14	62.91	47.87	56.72
F-SIFT [82]	61.06	75.95	84.72	48.21	93.64	63.36	48.61	58.57
[2]	54.30	–	80.66	–	–	–	–	–
[43]	–	48.69	–	–	–	–	–	–
[45]	56.8	–	84.6	33.3	–	–	–	–
[49]	–	–	–	45.2	–	–	–	–
[51]	–	–	–	44.2	–	–	–	–
[66]	59.29	–	75.26	–	–	–	–	–
[27]	–	–	–	–	–	63.10	–	–
[28]	–	–	–	–	92.8	63.4	46.1	57.4
[72]	–	–	–	–	–	63.48	45.91	–
[31]	–	–	–	–	–	–	49.5	–

Max-SIFT and RIDE also work well to outperform the recent competitors. Thus, although Max-SIFT and RIDE are motivated by the observation on the fine-grained case, it enjoys good recognition performance on a wide range of image datasets.

15.3.5 Summary

In this section, we explore reversal invariance in the context of the BoVW model. We propose the **Max-SIFT** descriptor and the **RIDE** (Reversal-Invariant Descriptor Enhancement) algorithm which bring reversal invariance to local descriptors. Our idea is inspired by the observation that most handcrafted descriptors are not reversal-invariant, whereas many fine-grained datasets contain objects with different left/right orientations. Max-SIFT and RIDE cancel out the impact of image/object reversal by estimating the orientation of each descriptor, and then forcing all the descriptors to have the same orientation. Experiments reveal that both of them significantly improve the accuracy of fine-grained object recognition and scene classification with very few computational costs. Both Max-SIFT and RIDE are robust to small image noise. Compared with data augmentation, RIDE produces better results with lower time/memory consumptions.

15.4 REVERSAL INVARIANCE IN CNN

In this section, we generalize the above ideas from BoVW to CNN by proposing a new convolution operation so that every neural network is naturally equipped with reversal-invariance.

15.4.1 Reversal-invariant convolution (RI-Conv)

To extract reversal-invariant deep features for each input image, the most straightforward idea is to average the feature vectors of the original image and its reversed copy. Despite its simplicity and effectiveness [73], this approach requires almost doubled computational costs in the feature extraction stage.

As an alternative solution, we show that directly training a reversal-invariant deep CNN is possible and more efficient. Here, we say a CNN model is *reversal-invariant* if it produces symmetric neural responses on each pair of symmetric images, *i.e.*, for an arbitrary image $\mathbf{I}$, if we take $\mathbf{I}$ and $\mathbf{I}^{\mathrm{R}}$ as the input, the neuron responses on each layer of the pre-trained network $\mathcal{M}$, *i.e.*, $\mathbf{f}_l(\mathbf{I}; \mathcal{M})$ and $\mathbf{f}_l\left(\mathbf{I}^{\mathrm{R}}; \mathcal{M}\right)$, are symmetric to each other: $\mathbf{f}_l^{\mathrm{R}}(\mathbf{I}; \mathcal{M}) = \mathbf{f}_l\left(\mathbf{I}^{\mathrm{R}}; \mathcal{M}\right)$. In a reversal-invariant network, when we extract features on a fully connected layer (*e.g.*, *fc-6* in the AlexNet), the original and reversed outputs are exactly the same since the spatial resolution is 1×1. If the features are extracted on an earlier layer (*e.g.*, *conv-5* in the AlexNet), we can also achieve reversal invariance by performing average-pooling or max-pooling over the responses at all spatial locations, similar to the strategy used in some previous publications [24].

The key to constructing a reversal-invariant CNN model is to guarantee that all the network layers are performing *symmetric* operations. Among the frequently used network operations (*e.g.*, convolution, pooling, normalization, non-linear activation, *etc.*), only convolution is non-symmetric, *i.e.*, a local patch and its reversed copy may produce different convolution outputs. We aim at designing a new *reversal-invariant* convolution operation to replace the original one.

Mathematically, let l be the index of a convolution layer with K_l convolution kernels, and $\mathbf{f}_{l-1} \doteq \mathbf{f}_{l-1}(\mathbf{I}; \mathcal{M})$ is the input of the lth layer. $\boldsymbol{\theta}_l \in \mathbb{R}^{W_l \times H_l \times K_l}$ and $\mathbf{b}_l \in \mathbb{R}^{K_l}$ are the weighting and bias parameters, respectively. The convolution operation takes a patch with the same spatial scale as the kernels, computes its inner-product with each kernel, and adds the bias to the result. For the kth kernel, $k = 1, 2, \ldots, K_l$, we have

$$f_l^{(a,b,k)}(\mathbf{I}; \mathcal{M}) = \left\langle \mathbf{p}_{l-1}^{(a,b)}, \boldsymbol{\theta}_l^{(k)} \right\rangle + b_l^{(k)}, \tag{15.3}$$

Here, $f_l^{(a,b,k)}$ denotes the unit at the spatial position (a, b), and convolved by the kth kernel, on the lth layer, and $\mathbf{p}_{l-1}^{(a,b)}$ denotes the related data patch on the previous layer. Note that $\mathbf{p}_{l-1}^{(a,b)}$ and $\boldsymbol{\theta}_l^{(k)}$ are of the same dimension. $\langle\cdot,\cdot\rangle$ denotes the inner-product operation.

Inspired by the reversal-invariant deep features, reversal invariance is achieved if we perform a symmetric operation on the neuron responses from a patch and its reversed copy. Again, we take the element-wise average and maximal responses, leading to the Average-Conv and the Max-Conv formulations:

$$\begin{aligned} r_{l,\text{AVG}}^{(a,b,k)}&(\mathbf{I};\mathcal{M}) \\ &= \frac{1}{2}\left[\left\langle \mathbf{p}_{l-1}^{(a,b)}, \boldsymbol{\theta}_l^{(k)} \right\rangle + \left\langle \mathbf{p}_{l-1}^{(a,b),\mathrm{R}}, \boldsymbol{\theta}_l^{(k)} \right\rangle\right] + b_l^{(k)} \\ &= \left\langle \frac{1}{2}\left[\mathbf{p}_{l-1}^{(a,b)} + \mathbf{p}_{l-1}^{(a,b),\mathrm{R}}\right], \boldsymbol{\theta}_l^{(k)} \right\rangle + b_l^{(k)}, \end{aligned} \tag{15.4}$$

$$\begin{aligned} r_{l,\text{MAX}}^{(a,b,k)}&(\mathbf{I};\mathcal{M}) \\ &= \max\left\{\left\langle \mathbf{p}_{l-1}^{(a,b)}, \boldsymbol{\theta}_l^{(k)} \right\rangle, \left\langle \mathbf{p}_{l-1}^{(a,b),\mathrm{R}}, \boldsymbol{\theta}_l^{(k)} \right\rangle\right\} + b_l^{(k)}. \end{aligned} \tag{15.5}$$

Since Average-Conv and Max-Conv simply perform the average/max pooling operation on the convolved data, it is straightforward to derive the formula of back-propagation. In the case of Average-Conv, we can accelerate both forward-propagation and back-propagation by modifying the input data (the original input $\mathbf{p}_{l-1}^{(a,b)}$ is replaced by the average of $\mathbf{p}_{l-1}^{(a,b)}$ and $\mathbf{p}_{l-1}^{(a,b),\mathrm{R}}$), thus the time-consuming convolution process is performed only once. In the case of Max-Conv, we need to create a mask blob to store the index of forward-propagated units, as in max-pooling layers.

In what follows, we will plug the reversal-invariant convolution modules into the conventional CNN models. We name a CNN model **RI-CNN** if all the convolution layers in it, *including the fully connected layers*, are made reversal-invariant. We start with discussing its property of reversal invariance, and the cooperation with data augmentation strategies.

15.4.2 Relationship to data augmentation

It is obvious that both Average-Conv and Max-Conv are symmetric operations. We prove that an RI-CNN is reversal-invariant, *i.e.*, the feature vectors extracted from an image and its reversed copy are identical.

We use mathematical induction, with the starting point that an image and its reversed copy are symmetric to each other, *i.e.*, $\mathbf{f}_0^{\mathrm{R}}(\mathbf{I};\mathcal{M}) = \mathbf{f}_0(\mathbf{I}^{\mathrm{R}};\mathcal{M})$.

Now, given that $\mathbf{f}_{l-1}^{\mathrm{R}}(\mathbf{I};\mathcal{M}) = \mathbf{f}_{l-1}\left(\mathbf{I}^{\mathrm{R}};\mathcal{M}\right)$, we derive that $\mathbf{f}_{l}^{\mathrm{R}}(\mathbf{I};\mathcal{M}) = \mathbf{f}_{l}\left(\mathbf{I}^{\mathrm{R}};\mathcal{M}\right)$, if both of them are computed with a reversal-invariant convolution operation on the lth layer. For this, we assume that when padding (increasing data width/height with 0-valued stripes) is used, the left and right padding width must be the same, *i.e.*, the geometric symmetry is guaranteed.

Consider a patch $\mathbf{p}_{l-1}^{(a,b)}(\mathbf{I};\mathcal{M})$. According to symmetry, we have $\mathbf{p}_{l-1}^{(a,b),\mathrm{R}}(\mathbf{I};\mathcal{M}) = \mathbf{p}_{l-1}^{(W_{l-1}-a-1,b)}\left(\mathbf{I}^{\mathrm{R}};\mathcal{M}\right)$, where $(W_{l-1}-a-1,b)$ is the left–right symmetric position to (a,b). These two patches are fed into the kth convolution kernel $\boldsymbol{\theta}_k$, and the outputs are $f_l^{(a,b,k)}(\mathbf{I};\mathcal{M})$ and $f_l^{(W_{l-1}-a-1,b,k)}\left(\mathbf{I}^{\mathrm{R}};\mathcal{M}\right)$. These two scalars equal to each other since both Average-Conv and Max-Conv are symmetric, thus the neuron responses on the lth layer are also symmetric: $\mathbf{f}_{l}^{\mathrm{R}}(\mathbf{I};\mathcal{M}) = \mathbf{f}_{l}\left(\mathbf{I}^{\mathrm{R}};\mathcal{M}\right)$. This finishes the induction, *i.e.*, the neuron responses of a pair of symmetric inputs are symmetric.

We point out that such a good property in feature extraction can be a significant drawback in the network training process, since an RI-CNN model suffers from the difficulty to cooperate with "reversal data augmentation". Here, by reversal data augmentation we mean to reverse each training sample with the probability of 50%. As an RI-CNN model generates exactly the same (symmetric) neuron responses for an image and its reversed copy, these two training samples actually produce the same gradients with respect to the network parameters on each layer. Consequently, reversing a training image cannot provide any "new" information to the network training process. Since using reversal-invariant convolution operations increases the capacity of the CNN model, the decrease of training data may cause overfitting, which harms the generalization ability of the model.

To deal with, we intentionally damage the reversal-invariant property of the network in the training process. For this, we crop the training image into a smaller size, so that the *geometric symmetry* does not hold any more. Taking the AlexNet as an example. The original input image size is 227×227, in which geometric symmetry holds on each convolutional/pooling layer. If the size becomes $S' \times S'$ where S' is a little smaller than 227, then in some layers, the padding margin on the left side is not the same as that on the right side. By the way, S' shall be at least 199, so that the input of the *fc-6* layer still has a spatial resolution of 6×6. In practice, we simply use $S' = 199$, so that we can generate as many training images as possible. As we shall see in Section 15.4.4, this strategy also improves the baseline accuracy slightly.

15.4.3 CIFAR experiments

The CIFAR10 and CIFAR100 datasets [29] are subsets of the 80 million tiny images database [60]. Both of them have 50,000 training samples and 10,000 testing samples, each of which is a 32×32 color image, uniformly distributed among all the categories (they have 10 and 100 categories, respectively). It is a popular dataset for training relatively small-scale neural networks for simple recognition tasks.

We use a modified version of the LeNet [33]. A $32 \times 32 \times 3$ image is passed through three units consisting of convolution, ReLU and max-pooling operations. Using abbreviation, the network configuration for CIFAR10 can be written as:

```
[C5(S1P2)@32-MP3(S2)]-[C5(S1P2)@32-MP3(S2)]-
[C5(S1P2)@64-MP3(S2)]-FC10.
```

Here, `C5(S1P2)` means a convolutional layer with a kernel size of 5, a spatial stride of 1 and a padding width of 2; `MP3(S2)` refers to a max-pooling layer with a kernel size of 5 and a spatial stride of 2, and `FC10` indicates a fully connected layer with 10 outputs. In CIFAR100, we replace `FC10` as `FC100` in order to categorize 100 classes. A 2-pixel wide padding is added to each convolution operation so that the width and height of the data remain unchanged. We do not produce multiple sizes of input images, since the LeNet is not symmetric itself: on each pooling layer, the left padding margin is 0 while the right margin is 1. We apply 120 training epochs with a learning rate of 10^{-3}, followed by 20 epochs with a learning rate of 10^{-4}, and another 10 epochs with a learning rate of 10^{-5}.

We train six different models individually, *i.e.*, training a network with the original version of convolution, Average-Conv or Max-Conv (three choices), and using data augmentation (probabilistic training image reversal) or not (two choices). We name these models LeNet, LeNet-AUGM ("AUGM" for augmentation), LeNet-AVG, LeNet-AVG-AUGM, LeNet-MAX and LeNet-MAX-AUGM, respectively. For instance, LeNet-MAX indicates the network with Max-Conv but without data augmentation. Note that reversal-invariant convolution also applies to the fully connected layer. To reveal the statistics significance, we train five independent models in each case, and report the average accuracy.

Results are summarized in Table 15.4. One can observe similar phenomena on both datasets. First, Average-Conv causes dramatic accuracy drop due to the lack of reasonability, as it assumes that the input pattern to be similar to both the original and reversed versions of the template, *i.e.*, this implies that

Table 15.4 CIFAR classification error rate (%) with respect to different versions of LeNet (left two columns) and BigNet (right two columns)

CIFAR10	LeNet w/o AUGM	LeNet w/ AUGM	BigNet w/o AUGM	BigNet w/ AUGM
Original	18.11 ± 0.20	16.99 ± 0.22	10.29 ± 0.29	7.80 ± 0.18
Avg-Conv	21.01 ± 0.35	20.99 ± 0.26	12.13 ± 0.19	9.53 ± 0.19
Max-Conv	16.93 ± 0.18	$\mathbf{16.64} \pm 0.17$	7.63 ± 0.12	$\mathbf{7.49} \pm 0.23$
CIFAR100	**LeNet w/o AUGM**	**LeNet w/ AUGM**	**BigNet w/o AUGM**	**BigNet w/ AUGM**
Original	46.08 ± 0.26	44.55 ± 0.10	34.51 ± 0.17	31.03 ± 0.24
Avg-Conv	47.79 ± 0.41	47.55 ± 0.31	36.74 ± 0.19	32.61 ± 0.30
Max-Conv	43.90 ± 0.19	$\mathbf{43.65} \pm 0.16$	30.92 ± 0.32	$\mathbf{30.76} \pm 0.24$

all convolutional kernels are symmetric and is not reasonable. On the other side, Max-Conv improves the recognition accuracy consistently, as it allows a convolutional kernel to be matched with a patch which is close to either itself or its reversed copy, *i.e.*, the descriptive ability of the same-sized kernel set is almost doubled. In the CIFAR10 dataset, both data augmentation and Max-Conv boost the accuracy by about 1%, and these two strategies cooperate with each other to outperform the baseline by 1.5%. In the CIFAR100 dataset, Max-Conv alone contributes a more-than-2% accuracy gain, which is higher than the 1.5% gain by data augmentation, and the combination gives a nearly 2.5% gain. As a last note, the improvement on CIFAR100 is much larger than that on **CIFAR10**, which indicates that CIFAR100 is a more challenging dataset (with more categories), and that Max-Conv increases the network capacity to benefit the recognition task.

We also evaluate our algorithm on a deeper network named BigNet. It is a 10-layer network [46] which resembles the VGGNet in using small convolutional kernels. We inherit all the settings from the original author, and report the accuracy produced by six different versions as above. Results summarized in Table 15.4 deliver the same conclusion as using the small network (LeNet).

We also put our result in the CIFAR datasets in the context of some recent publications in this dataset. Results are summarized in Table 15.5.

15.4.4 ILSVRC2012 classification experiments

We also evaluate our model on the ILSVRC2012 classification dataset [55], a subset of the ImageNet database [14] which contains 1000 object categories. The training set, validation set and testing set contain 1.3M, 50K and 150K images, respectively. We use the AlexNet (provided by the CAFFE library [26], sometimes referred to as the CaffeNet). The input image is of

Table 15.5 Comparison with some chain-shaped deep networks in CIFAR classification error rates (%)

	CIFAR10	CIFAR100
[20]	9.38	38.57
[39]	8.81	35.68
[35]	7.97	34.57
[38]	7.09	31.75
[34]	**6.05**	32.37
LeNet-MAX	16.64	43.65
BigNet-MAX	**7.49**	**30.76**

size 199 × 199, randomly cropped from the original 256 × 256 image. The AlexNet structure is abbreviated as:

```
[C11(S4)@96-MP3(S2)-LRN]-[C5(S1P2)@256-MP3(S2)-LRN]-
[C3(S1P1)@384]-[C3(S1P1)@384]-[C3(S1P1)@256-MP3(S2)]-
FC4096-D0.5-FC4096-D0.5-FC1000.
```

Here, `LRN` means local response normalization [30] and `D0.5` means Dropout with a drop ratio 0.5. Following the setting of CAFFE, a total of 450,000 mini-batches (approximately 90 epochs) are used for training, each of which has 256 image samples, with the initial learning rate 0.01, momentum 0.9 and weight decay 0.0005. The learning rate is decreased to 1/10 after every 100,000 mini-batches.

We individually train four models, *i.e.*, using original convolution or Max-Conv, using data augmentation or not. Similarly, we name these variations AlexNet, AlexNet-AUGM, AlexNet-MAX and AlexNet-MAX-AUGM, respectively. Considering the large computational costs, we only train two individual networks for each setting. We do not train models based on Average-Conv according to the dramatic accuracy drop in CIFAR experiments. Note that Max-Conv also applies to the fully connected layers.

Result are summarized in Table 15.6 and Fig. 15.4. As we have slightly modified the data augmentation strategy, the baseline performance (80.48% top-5 accuracy) is slightly better than that reported using the standard setting (approximately 80.1% top-5 accuracy[1]). With Max-Conv, the top-5 accuracy is boosted to 80.88%, which show that Max-Conv and data augmentation cooperate to improve the recognition performance. We emphasize that the 0.40% accuracy gain is not small, given that the network structure is unchanged. Meanwhile, the conclusions drawn in CIFAR experiments also hold in this large-scale image recognition task.

[1] https://github.com/BVLC/caffe/wiki/Models-accuracy-on-ImageNet-2012-val.

Table 15.6 ILSVRC2012 classification error rate (%) with respect to different versions of AlexNet

ILSVRC2012, top-1	w/o AUGM	w/ AUGM
AlexNet	43.05 ± 0.19	42.52 ± 0.17
AlexNet-MAX	42.16 ± 0.05	$\mathbf{42.10} \pm 0.07$
ILSVRC2012, top-5	**w/o AUGM**	**w/ AUGM**
AlexNet	20.62 ± 0.08	19.52 ± 0.05
AlexNet-MAX	19.42 ± 0.03	$\mathbf{19.12} \pm 0.07$

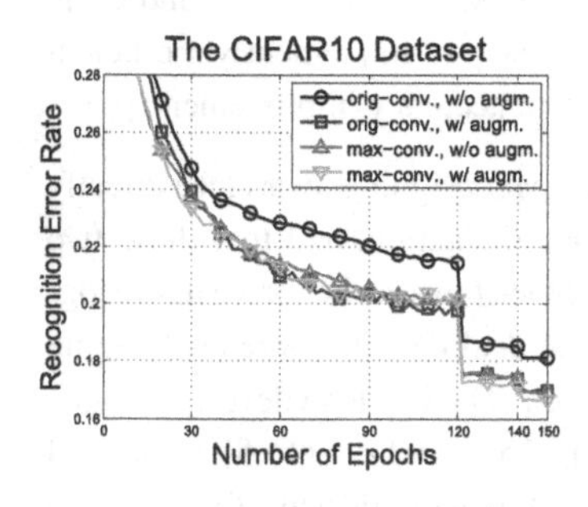

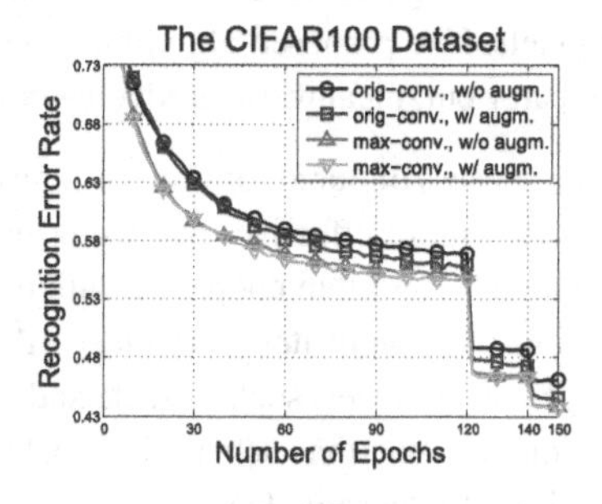

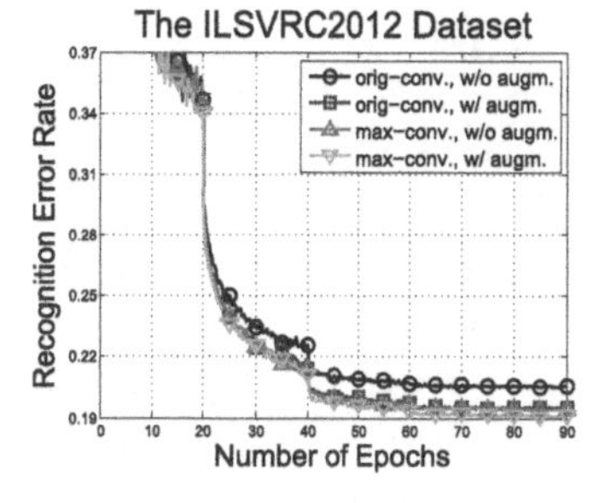

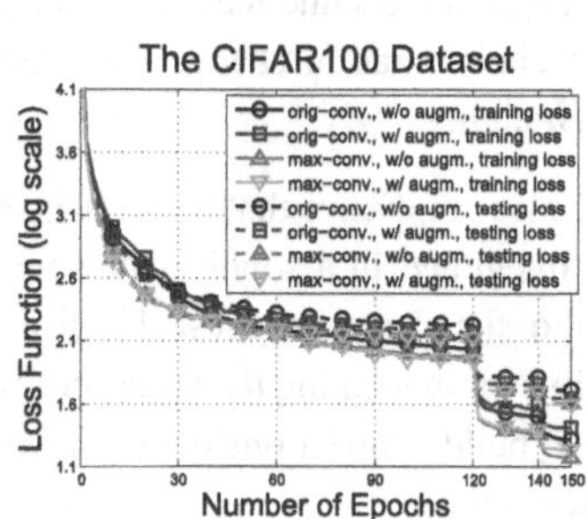

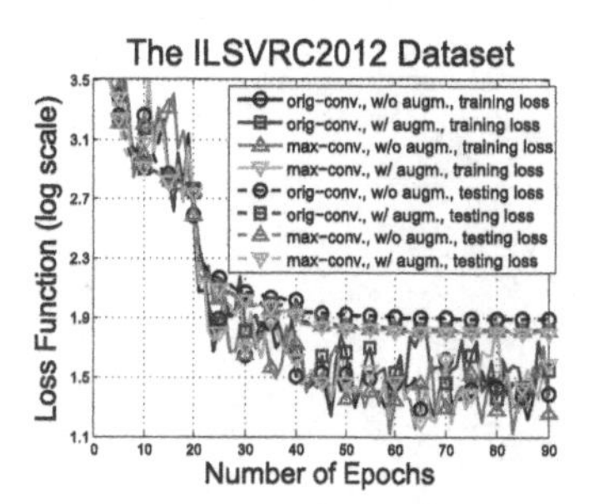

■ **FIGURE 15.4** Error rate curves and training/testing loss curves on the CIFAR datasets and the ILSVRC2012 dataset. We report top-1 and top-5 error rates in CIFAR and ILSVRC2012, respectively.

15.4.5 Summary

In this part, we generalize the idea of reversal-invariant representation from the BoVW model to deep CNNs, and verify that reversal invariance is also important in both deep feature extraction and deep network training. We propose **RI-Conv** as a reversal-invariant version of convolution. This is a small modification in convolution that leads to a deep network which is intrinsically reversal-invariant, which has larger capacity yet unchanged complexity, meanwhile makes the feature extraction more effective. Reversal-invariant convolution also cooperates well with data augmentation, creating the possibility of applying deep neural networks to even larger databases. Last but not least, the Max-Conv operator is easy to implement yet fast to carry out (less than 20% extra time is required).

15.5 CONCLUSIONS

As the main source of real world perception, visual understanding plays a crucial role in multi-modal analysis. In which, it is important to consider reversal invariance in order to achieve more robust image representation, but the conventional BoVW and CNN models often lack of an explicit implementation of reversal invariance. This chapter presents a basic idea that designs reversal-invariant local patterns, such as **Max-SIFT** and **RIDE** (local descriptors), as well as **RI-Conv** (convolution), so that reversal invariance is guaranteed in the representation based on the BoVW and CNN models. The proposed algorithms are very easy to implement yet efficient to carry out, meanwhile producing consistent accuracy improvement.

The success of our algorithms also reveals that designing invariance directly is often more effective than using data augmentation, and that these two strategies can often cooperate with each other towards better visual recognition. This motivates us to make efforts in dealing with more challenging types of variances, such as scale and rotation. The key is to develop a transformation function such as Eq. (15.1), Eq. (15.2) and Eq. (15.5), to map all variants to the same feature vector. This task is more difficult for scale and especially rotation changes, in which the dimension of variance becomes even higher.

We also emphasize that the invariance in audio and video understanding can be integrated in a sophisticated manner. For example, the intensity of the audio signal may suggest the distance to the target, and its direction can help a lot in locating the target in a panoramic image or video. This requires us to build connections between two or more submodules, which is left for future research.

REFERENCES

[1] A. Alahi, R. Ortiz, P. Vandergheynst, FREAK: fast retina keypoint, in: Computer Vision and Pattern Recognition, 2012.

[2] A. Angelova, S. Zhu, Efficient object detection and segmentation for fine-grained recognition, in: Computer Vision and Pattern Recognition, 2013.

[3] R. Arandjelovic, A. Zisserman, Three things everyone should know to improve object retrieval, in: Computer Vision and Pattern Recognition, 2012.

[4] H. Bay, A. Ess, T. Tuytelaars, L. Van Gool, Speeded-up robust features (SURF), Comput. Vis. Image Underst. 110 (3) (2008) 346–359.

[5] S. Belongie, J. Malik, J. Puzicha, Shape matching and object recognition using shape contexts, IEEE Trans. Pattern Anal. Mach. Intell. 24 (4) (2002) 509–522.

[6] A. Bosch, A. Zisserman, X. Munoz, Scene classification via pLSA, in: International Conference on Computer Vision, 2006.

[7] M. Calonder, V. Lepetit, C. Strecha, P. Fua, BRIEF: binary robust independent elementary features, in: European Conference on Computer Vision, 2010.

[8] Y. Chai, V. Lempitsky, A. Zisserman, Symbiotic segmentation and part localization for fine-grained categorization, in: International Conference on Computer Vision, 2013.

[9] K. Chatfield, V. Lempitsky, A. Vedaldi, A. Zisserman, The devil is in the details: an evaluation of recent feature encoding methods, in: British Machine Vision Conference, 2011.

[10] K. Chatfield, K. Simonyan, A. Vedaldi, A. Zisserman, Return of the devil in the details: delving deep into convolutional nets, in: British Machine Vision Conference, 2014.

[11] S. Clinchant, G. Csurka, F. Perronnin, J.M. Renders, XRCE's participation to ImagEval, in: ImagEval Workshop at CVIR, 2007.

[12] G. Csurka, C. Dance, L. Fan, J. Willamowski, C. Bray, Visual categorization with bags of keypoints, in: Workshop on Statistical Learning in Computer Vision, European Conference on Computer Vision, 2004.

[13] N. Dalal, B. Triggs, Histograms of oriented gradients for human detection, in: Computer Vision and Pattern Recognition, 2005.

[14] J. Deng, W. Dong, R. Socher, L.J. Li, K. Li, L. Fei-Fei, ImageNet: a large-scale hierarchical image database, in: Computer Vision and Pattern Recognition, 2009.

[15] J. Donahue, Y. Jia, O. Vinyals, J. Hoffman, N. Zhang, E. Tzeng, T. Darrell, DeCAF: a deep convolutional activation feature for generic visual recognition, in: International Conference on Machine Learning, 2014.

[16] R.E. Fan, K.W. Chang, C.J. Hsieh, X.R. Wang, C.J. Lin, LIBLINEAR: a library for large linear classification, J. Mach. Learn. Res. 9 (2008) 1871–1874.

[17] J. Feng, B. Ni, Q. Tian, S. Yan, Geometric Lp-norm feature pooling for image classification, in: Computer Vision and Pattern Recognition, 2011.

[18] E. Gavves, B. Fernando, C.G.M. Snoek, A.W.M. Smeulders, T. Tuytelaars, Local alignments for fine-grained categorization, Int. J. Comput. Vis. 111 (2) (2014) 191–212.

[19] R. Girshick, J. Donahue, T. Darrell, J. Malik, Rich feature hierarchies for accurate object detection and semantic segmentation, in: Computer Vision and Pattern Recognition, 2014.

[20] I.J. Goodfellow, D. Warde-Farley, M. Mirza, A. Courville, Y. Bengio, Maxout networks, in: International Conference on Machine Learning, 2013.

[21] K. Grauman, T. Darrell, The pyramid match kernel: discriminative classification with sets of image features, in: International Conference on Computer Vision, 2005.

[22] G. Griffin, Caltech-256 Object Category Dataset, Technical Report: CNS-TR-2007-001, 2007.

[23] X. Guo, X. Cao, FIND: a neat flip invariant descriptor, in: International Conference on Patter Recognition, 2010.

[24] K. He, X. Zhang, S. Ren, J. Sun, Spatial pyramid pooling in deep convolutional networks for visual recognition, IEEE Trans. Pattern Anal. Mach. Intell. 37 (9) (2015) 1904–1916.

[25] S. Ioffe, C. Szegedy, Batch normalization: accelerating deep network training by reducing internal covariate shift, in: International Conference on Machine Learning, 2015.

[26] Y. Jia, E. Shelhamer, J. Donahue, S. Karayev, J. Long, R. Girshick, S. Guadarrama, T. Darrell, CAFFE: convolutional architecture for fast feature embedding, in: ACM International Conference on Multimedia, 2014.

[27] M. Juneja, A. Vedaldi, C.V. Jawahar, A. Zisserman, Blocks that shout: distinctive parts for scene classification, in: Computer Vision and Pattern Recognition, 2013.

[28] T. Kobayashi, Dirichlet-based histogram feature transform for image classification, in: Computer Vision and Pattern Recognition, 2014.

[29] A. Krizhevsky, G. Hinton, Learning Multiple Layers of Features from Tiny Images, Technical report, University of Toronto, 2009.
[30] A. Krizhevsky, I. Sutskever, G.E. Hinton, ImageNet classification with deep convolutional neural networks, in: Advances in Neural Information Processing Systems, 2012, pp. 1097–1105.
[31] M. Lapin, B. Schiele, M. Hein, Scalable multitask representation learning for scene classification, in: Computer Vision and Pattern Recognition, 2014.
[32] S. Lazebnik, C. Schmid, J. Ponce, Beyond bags of features: spatial pyramid matching for recognizing natural scene categories, in: Computer Vision and Pattern Recognition, 2006.
[33] Y. LeCun, J.S. Denker, D. Henderson, R.E. Howard, W. Hubbard, L.D. Jackel, Handwritten digit recognition with a back-propagation network, in: Advances in Neural Information Processing Systems, 1990.
[34] C.Y. Lee, P.W. Gallagher, Z. Tu, Generalizing pooling functions in convolutional neural networks: mixed, gated, and tree, in: International Conference on Artificial Intelligence and Statistics, 2016.
[35] C.Y. Lee, S. Xie, P. Gallagher, Z. Zhang, Z. Tu, Deeply-supervised nets, in: International Conference on Artificial Intelligence and Statistics, 2015.
[36] S. Leutenegger, M. Chli, R.Y. Siegwart, BRISK: binary robust invariant scalable keypoints, in: International Conference on Computer Vision, 2011.
[37] L. Li, Y. Guo, L. Xie, X. Kong, Q. Tian, Fine-grained visual categorization with fine-tuned segmentation, in: International Conference on Image Processing, 2015.
[38] M. Liang, X. Hu, Recurrent convolutional neural network for object recognition, in: Computer Vision and Pattern Recognition, 2015.
[39] M. Lin, Q. Chen, S. Yan, Network in network, in: International Conference on Learning Representations, 2014.
[40] K. Liu, H. Skibbe, T. Schmidt, T. Blein, K. Palme, T. Brox, O. Ronneberger, Rotation-invariant hog descriptors using Fourier analysis in polar and spherical coordinates, Int. J. Comput. Vis. 106 (3) (2014) 342–364.
[41] D.G. Lowe, Distinctive image features from scale-invariant keypoints, Int. J. Comput. Vis. 60 (2) (2004) 91–110.
[42] R. Ma, J. Chen, Z. Su, MI-SIFT: mirror and inversion invariant generalization for SIFT descriptor, in: Conference on Image and Video Retrieval, 2010.
[43] S. Maji, E. Rahtu, J. Kannala, M. Blaschko, A. Vedaldi, Fine-Grained Visual Classification of Aircraft, Technical Report, 2013.
[44] J. Matas, O. Chum, M. Urban, T. Pajdla, Robust wide-baseline stereo from maximally stable extremal regions, Image Vis. Comput. 22 (10) (2004) 761–767.
[45] N. Murray, F. Perronnin, Generalized max pooling, in: Computer Vision and Pattern Recognition, 2014.
[46] Nagadomi, The Kaggle CIFAR10 network, https://github.com/nagadomi/kaggle-cifar10-torch7/, 2014.
[47] M.E. Nilsback, A. Zisserman, Automated flower classification over a large number of classes, in: International Conference on Computer Vision, Graphics & Image Processing, 2008.
[48] O.M. Parkhi, A. Vedaldi, A. Zisserman, C.V. Jawahar, Cats and dogs, in: Computer Vision and Pattern Recognition, 2012.
[49] M. Paulin, J. Revaud, Z. Harchaoui, F. Perronnin, C. Schmid, Transformation pursuit for image classification, in: Computer Vision and Pattern Recognition, 2014.
[50] F. Perronnin, J. Sánchez, T. Mensink, Improving the Fisher kernel for large-scale image classification, in: European Conference on Computer Vision, 2010.
[51] J. Pu, Y.G. Jiang, J. Wang, X. Xue, Which looks like which: exploring inter-class relationships in fine-grained visual categorization, in: European Conference on Computer Vision, 2014.

[52] A. Quattoni, A. Torralba, Recognizing indoor scenes, in: Computer Vision and Pattern Recognition, 2009.
[53] A.S. Razavian, H. Azizpour, J. Sullivan, S. Carlsson, CNN features off-the-shelf: an astounding baseline for recognition, in: Computer Vision and Pattern Recognition, 2014.
[54] E. Rublee, V. Rabaud, K. Konolige, G. Bradski, ORB: an efficient alternative to SIFT or SURF, in: International Conference on Computer Vision, 2011.
[55] O. Russakovsky, J. Deng, H. Su, J. Krause, S. Satheesh, S. Ma, Z. Huang, A. Karpathy, A. Khosla, M. Bernstein, A.C. Berg, L. Fei-Fei, ImageNet large scale visual recognition challenge, Int. J. Comput. Vis. (1) (2015) 1–42.
[56] J. Sanchez, F. Perronnin, T. Mensink, J. Verbeek, Image classification with the Fisher vector: theory and practice, Int. J. Comput. Vis. 105 (3) (2013) 222–245.
[57] K. Simonyan, A. Zisserman, Very deep convolutional networks for large-scale image recognition, in: International Conference on Learning Representations, 2015.
[58] C. Szegedy, W. Liu, Y. Jia, P. Sermanet, S. Reed, D. Anguelov, D. Erhan, V. Vanhoucke, A. Rabinovich, Going deeper with convolutions, in: Computer Vision and Pattern Recognition, 2015.
[59] G. Takacs, V. Chandrasekhar, S.S. Tsai, D. Chen, R. Grzeszczuk, B. Girod, Fast computation of rotation-invariant image features by an approximate radial gradient transform, IEEE Trans. Image Process. 22 (8) (2013) 2970–2982.
[60] A. Torralba, R. Fergus, W.T. Freeman, 80 million tiny images: a large data set for nonparametric object and scene recognition, IEEE Trans. Pattern Anal. Mach. Intell. 30 (11) (2008) 1958–1970.
[61] T. Tuytelaars, Dense interest points, Computer Vision and Pattern Recognition 32 (9) (2010) 1582–1596.
[62] K.E.A. van de Sande, T. Gevers, C.G.M. Snoek, Evaluating color descriptors for object and scene recognition, IEEE Trans. Pattern Anal. Mach. Intell. 32 (9) (2010) 1582–1596.
[63] A. Vedaldi, B. Fulkerson, VLFeat: an open and portable library of computer vision algorithms, in: ACM International Conference on Multimedia, 2010.
[64] C. Wah, S. Branson, P. Welinder, P. Perona, S. Belongie, The Caltech-UCSD Birds-200-2011 Dataset, Technical Report: CNS-TR-2011-001, 2011.
[65] J. Wang, J. Yang, K. Yu, F. Lv, T. Huang, Y. Gong, Locality-constrained linear coding for image classification, in: Computer Vision and Pattern Recognition, 2010.
[66] Z. Wang, J. Feng, S. Yan, Collaborative linear coding for robust image classification, Int. J. Comput. Vis. (1) (2014) 1–12.
[67] J. Xiao, J. Hays, K.A. Ehinger, A. Oliva, A. Torralba, SUN database: large-scale scene recognition from Abbey to zoo, in: Computer Vision and Pattern Recognition, 2010.
[68] L. Xie, R. Hong, B. Zhang, Q. Tian, Image classification and retrieval are ONE, in: International Conference on Multimedia Retrieval, 2015.
[69] L. Xie, Q. Tian, R. Hong, S. Yan, B. Zhang, Hierarchical part matching for fine-grained visual categorization, in: International Conference on Computer Vision, 2013.
[70] L. Xie, Q. Tian, M. Wang, B. Zhang, Spatial pooling of heterogeneous features for image classification, IEEE Trans. Image Process. 23 (5) (2014) 1994–2008.
[71] L. Xie, Q. Tian, B. Zhang, Simple techniques make sense: feature pooling and normalization for image classification, in: IEEE Trans. Circuits Syst. Video Technol., 2015.
[72] L. Xie, J. Wang, B. Guo, B. Zhang, Q. Tian, Orientational pyramid matching for recognizing indoor scenes, in: Computer Vision and Pattern Recognition, 2014.
[73] L. Xie, J. Wang, W. Lin, B. Zhang, Q. Tian, Towards reversal-invariant image representation, Int. J. Comput. Vis. 123 (2) (2017) 226–250.

[74] J. Yang, K. Yu, Y. Gong, T. Huang, Linear spatial pyramid matching using sparse coding for image classification, in: Computer Vision and Pattern Recognition, 2009.
[75] Y. Yang, S. Newsam, Bag-of-visual-words and spatial extensions for land-use classification, in: International Conference on Advances in Geographic Information Systems, 2010.
[76] M.D. Zeiler, R. Fergus, Visualizing and understanding convolutional networks, in: European Conference on Computer Vision, 2014.
[77] N. Zhang, J. Donahue, R. Girshick, T. Darrell, Part-based R-CNNs for fine-grained category detection, in: European Conference on Computer Vision, 2014.
[78] N. Zhang, R. Farrell, F. Iandola, T. Darrell, Deformable part descriptors for fine-grained recognition and attribute prediction, in: International Conference on Computer Vision, 2013.
[79] S. Zhang, Q. Tian, G. Hua, Q. Huang, S. Li, Descriptive visual words and visual phrases for image applications, in: ACM International Conference on Multimedia, 2009.
[80] X. Zhang, H. Xiong, W. Zhou, W. Lin, Q. Tian, Picking deep filter responses for fine-grained image recognition, in: Computer Vision and Pattern Recognition, 2016.
[81] X. Zhang, H. Xiong, W. Zhou, Q. Tian, Fused one-vs-all mid-level features for fine-grained visual categorization, in: ACM International Conference on Multimedia, 2014.
[82] W.L. Zhao, C.W. Ngo, Flip-invariant SIFT for copy and object detection, IEEE Trans. Image Process. 22 (3) (2013) 980–991.
[83] X. Zhou, K. Yu, T. Zhang, T.S. Huang, Image classification using super-vector coding of local image descriptors, in: European Conference on Computer Vision, 2010.

Chapter 16

Sentiment concept embedding for visual affect recognition

Victor Campos*, Xavier Giro-i-Nieto*,†, Brendan Jou‡, Jordi Torres*, Shih-Fu Chang‡

*Barcelona Supercomputing Center, Barcelona, Catalonia, Spain †Universitat Politecnica de Catalunya, Barcelona, Catalonia, Spain ‡Columbia University, New York City, NY, USA

CONTENTS

16.1 INTRODUCTION

Computer vision systems continue to see an increase in machine understanding of the semantics of objects, actions and more in images and videos, in large part thanks to advances in publicly available high-computing hardware and large-scale datasets [9,15]. However, the underlying assumption in many of such systems is that the class assignment for each sample is binary, i.e. each example does or does not belong to every category in the problem. However, in classification tasks with hundreds or even thousands of categories, two main challenges arise. First, the cost of collecting enough training samples for each category can become prohibitive. Second, the

Multimodal Behavior Analysis in the Wild. https://doi.org/10.1016/B978-0-12-814601-9.00018-3
Copyright © 2019 Elsevier Ltd. All rights reserved.

boundaries between some classes are fuzzy and the hard assignment assumption does not hold, as there are label correlations.

Affective labels especially suffer from the challenges of soft label boundaries given the abstract and complex concepts associated with sentiment and emotions. We propose to address this challenge by learning visual representations that map images into an output embedding space encoding relations between affective concepts. Our approach is inspired by the work in [10], where a Deep Visual-Semantic Embedding Model (DeViSE) was proposed to alleviate the problems of data scarcity for object classification problems. Although similar in spirit, our work differs from DeViSE in two important directions: (1) we use a concept pool of compound concepts, which is constructed using an additional combination step to build the output embedding space, and (2) inter-class similarities are explicitly integrated in the cost function used for training the visual models.

We formulate and test our approach with ANPs [4]. These mid-level affective representations try to bridge the *affective gap* between low-level image features and high-level affective semantics, a task that goes beyond overcoming the *semantic gap*, i.e. recognizing the objects in an image. We adopt ANPs to define an embedded representation of high-level affective concepts in an output space where similar compound concepts such as "*pretty lady*" and "*beautiful girl*" lie close to each other. Fig. 16.1 provides an overview of our approach.

Our contributions are three-fold: (1) we combine existing adjective and noun embeddings to define an output space for visual affect recognition, (2) we explore three loss functions for learning visual representations which are aware of class correlations, and (3) we demonstrate the clear gain that the proposed embedding model provides in visual emotion recognition and sentiment analysis tasks as compared to a classical training where classes are represented by means of one-hot vector encodings.

16.1.1 Embeddings for image classification

Methods for mapping images into spaces have been studied for zero-shot learning tasks [20,24,33] and attribute-based learning [1]. There the image is first encoded into a fixed input space defined by traditional visual feature extraction methods such as SIFT-based Bag of Words, and then a mapping between such fixed input space and the defined output space is learned. Frome et al. proposed DeViSE [10], where an output embedding is trained in an unsupervised manner using a large text corpus, and later a Convolutional Neural Network (CNN) is optimized to learn the mapping from the raw pixel values to the output space, without using any predefined

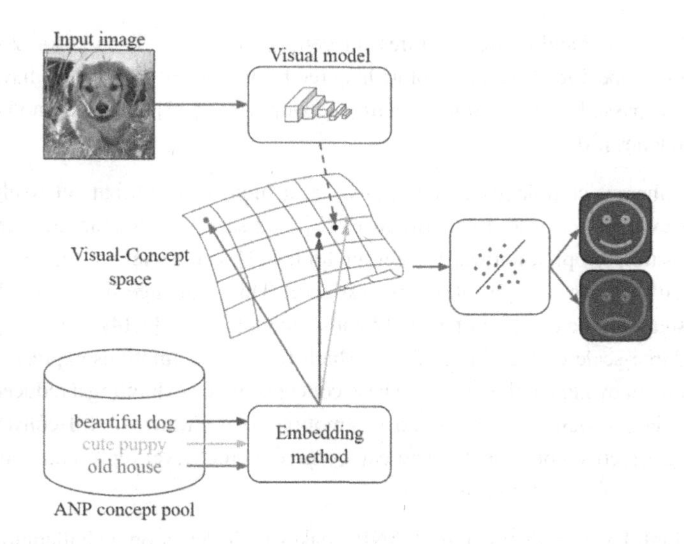

■ **FIGURE 16.1** Overview of the proposed system. First, an output embedding is generated by passing all adjectives and nouns in the ANP pool through a language model and combining their outputs. Then, a visual model learns to map the image to the output space while leveraging the fine-grained inter-class relationships through the embedded representations. Finally, the learned visual model is used as a mid-level representation for affect recognition.

input space for the image. DeViSE approaches state-of-the-art results in the 1000-class ILSVRC object recognition task [9], while making mistakes that are more semantically reasonable than those in other methods.

Our work differs from DeViSE in the fact that we explore how to build output spaces for categories that are combinations of two individual concepts instead of individual entities. In particular, we address the task of ANP detection, where each class is the combination of an adjective and a noun. With these embeddings, we explore how inter-class similarities may be introduced in the learning process without making the assumption that the classes are independent. This is a hypothesis that is reasonable for the ImageNet dataset [9] used for training and evaluating DeViSE [10], but not for a dataset such as the one considered in our experiments, where no mutual exclusion between the defined categories is imposed. Finally, the trained models are evaluated on visual sentiment and emotion prediction tasks using human-annotated datasets.

16.1.2 **Affective computing**

The challenge of predicting visual sentiments and emotions has received increasing interest from the research community during the past years. Methods based on hand-crafted features for generic computer vision tasks [23] and more specific visual descriptors inspired by artistic disciplines [16] have been surpassed by convolutional neural networks [6,28,31] and multi-modal approaches [30].

The inherent complexity to emotions, i.e. a high intensity, but relatively brief experience, onset by a stimuli [5,22], and sentiment, i.e. an attitude, disposition or opinion towards a certain topic [21], is reflected in categories that suffer from a large intra-class variance. This challenge has been addressed with the creation of Visual Sentiment Ontologies [4,14], consisting of a large-scale collection of ANPs, which focuses on emotions expressed by content owners in the images. These concepts, while exhibiting a reduced intra-class variance as compared to emotions or sentiments, still convey strong affective connotations and can be used as mid-level representations for visual affect related tasks.

The high level of abstraction of ANPs makes their detection a challenging task. It has been addressed with hand-crafted features and SVM classifiers [3] and state-of-the-art CNNs [7,13,14]. The connection between ANPs and multi-task learning has been exploited with factorized CNNs [19], coupled nets with mutual supervision [27] and deep cross-residual learning [12]. Despite the former works taking advantage of the relationship between adjectives and nouns, none of them exploits the similarities between the ANP categories in the concept pool, which exhibit fine-grained relationships that can be leveraged to train stronger classifiers. In this work, we propose to model the subtle dependencies between adjectives, nouns and ANPs as a whole through the creation of semantically meaningful output embeddings.

16.2 VISUAL SENTIMENT ONTOLOGY

A Multilingual Visual Sentiment Ontology (MVSO) with over 156,000 ANPs from 12 different languages was built by Jou et al. [14], extending the Visual Sentiment Ontology (VSO) by Borth et al. [4]. In order to guarantee a link between emotions and the concepts in the ontology, emotion keywords from a well-known emotion model derived from psychology studies, the *Plutchik Wheel of Emotions* model [22], were used to query the Flickr API[1] and retrieve a large corpus of images with related tags and metadata. After a data-driven filtering of ANP candidates, visual examples for these concepts were retrieved by querying the Flickr API for images containing them either in their tags or metadata.

[1] https://www.flickr.com/services/api.

The resulting MVSO dataset contains over 15M images, but also presents two major challenges when training visual concept detectors: (1) there is no human-level supervision for the ground truth, so that the annotations have to be considered as *weak labels*, and (2) there is no guarantee that the classes are mutually exclusive, meaning that some ANPs may be very similar or even synonymous. In this work, we mitigate the former issue by using a restricted set of the English partition of an MVSO, namely the tag-restricted subset [13] that contains over 1.2M images belonging to 1200 classes where the associated ANP was found in the image tags and is more likely to have reliable annotations. The second challenge is addressed by means of an output embedding and will be further developed in the following sections of this manuscript.

16.3 BUILDING OUTPUT EMBEDDINGS FOR ANPS

The discrete nature of ontologies imposes an inherent limitation in the finite number of concepts that they can contain, thus bounding their representational capacity. A natural solution to counteract such a limitation consists in extending the number of elements in the ontology so that a wider set of concepts is covered. However, the complexity of building ontologies does not scale well with respect to the number of items they contain, because of the increasing ambiguity in the relationship between concepts and the cost of collecting samples for every class, which may become prohibitive. Besides, typical classification schemes make the underlying assumption that all classes are independent and penalize all errors equally. This is a hypothesis that not necessarily holds for all ontologies. For these reasons, we propose to project the tag-restricted subset in the English partition of MVSO into a continuous output space defined by a data-driven word embedding model. By doing that, we can model the fine-grained class relationships between the ANPs.

Throughout the manuscript we will consider the skip-gram text modeling architecture [17,18]. This architecture is able to learn a fixed length representation for each word in large text collections in an unsupervised manner by predicting adjacent terms in the training corpus. The resulting representations, or *word embeddings*, show desirable properties such as synonyms falling in similar regions of the embedded space, or even linear relationships between concepts such that $vector(queen)$ is the closest representation to $vector(king) - vector(man) + vector(woman)$. In the following subsections we will study how embeddings learned with *word2vec* can be used to build an output space for the recognition of ANPs in images using CNNs.

16.3.1 Combining adjectives and nouns

A CNN capable of learning a mapping from raw pixel values into a position in the space defined by a word embedding was proposed in [10] to solve the 1000-class ImageNet object recognition challenge [9]. In that work, each class had a unique mapping defined by the concepts in a set of synonyms, i.e. *synset*. In our work, we address the challenge of combining the individual representations of adjectives and nouns into a single representation of fixed length for each ANP.

We study three combination methods based on different interpretations of each concept in an ANP:

Addition. The noun portion can be understood to ground the visual appearance of the entity, whereas the adjective polarizes the concept towards a positive or negative sentiment, or emotion [14]. For example, *happy face* and *sad face* may have a similar appearance but represent opposite emotional states. Such an interpretation leads to a representation of the ANP in the output space coming from the addition of the individual vectors for its adjective and noun:

$$\vec{y}_i = \vec{x}_{adj} + \vec{x}_{noun} \tag{16.1}$$

where $\vec{y}_i$, $\vec{x}_{adj}$ and $\vec{x}_{noun}$ represent the embedded vectors for the *i*th ANP and its corresponding adjective and noun, respectively.

Product. ANPs may be seen as the concurrence of two concepts, i.e. an adjective and a noun. This *logical AND* relationship can be modeled by the element-wise product of the individual parts:

$$\vec{y}_i = \vec{x}_{adj} \odot \vec{x}_{noun}. \tag{16.2}$$

Concatenation. The clear connection between ANPs and multi-task learning [12,19] suggests that each bi-concept may be modeled as the concatenation of its elements:

$$\vec{y}_i = \vec{x}_{adj} | \vec{x}_{noun}. \tag{16.3}$$

Others. More complex representations, such as adjectives being represented by matrices and nouns by vectors [2], have been proposed in the literature. However, since we aim to leverage the semantic information encoded in word embeddings, we will only consider representations learned using state-of-the-art methods such as *word2vec* and leave other types of representations as potential future research lines.

Table 16.1 Qualitative evaluation for different ANP embeddings. Vectors for adjectives and nouns are obtained using the word2vec model trained on the Google News corpora. Cosine similarity is used to determine the relationships between ANP concepts

ANP	Method	Ranked examples
Beautiful landscape	Addition	Most similar: *beautiful scenery, beautiful flowers, beautiful nature, snowy landscape*
		Most different: *senior adult, grilled chicken, broken arm, low income, alarm clock*
	Product	Most similar: *beautiful scenery, natural landscape, snowy landscape*
		Most different: *good cause, cute animal, old door, senior project, super trees*
	Concatenation	Most similar: *beautiful scenery, beautiful nature, snowy landscape*
		Most different: *criminal justice, senior adult, grilled chicken, domestic cat*
Funny dog	Addition	Most similar: *cute dog, funny cats, cute puppy, cute cat, mad dog*
		Most different: *traditional architecture, interior architecture, senior project*
	Product	Most similar: *funny cats, cute dog, cute puppy, cute cat, mad dog*
		Most different: *golden circle, common crane, formal dress, open days*
	Concatenation	Most similar: *funny cats, cute dog, cute puppy, mad dog, cute cat*
		Most different: *electrical engineering, traditional architecture, global gathering*

After projecting ANPs in the embedded space using any of the proposed combinations methods, the closeness between concepts can be measured by any vector similarity metric. In particular, we will consider the cosine similarity:

$$s_{ij} = \frac{\vec{y}_i \cdot \vec{y}_j}{\|\vec{y}_i\| \cdot \|\vec{y}_j\|}. \tag{16.4}$$

Table 16.1 illustrates with examples the behavior of the proposed combination methods. Although the three embeddings yield similar qualitative results, it is important to notice that concatenating the representations for adjectives and nouns leads to a multi-task learning scenario. In the next sections we explore how to train a CNN using the single-task embeddings, i.e. *addition* and *product*, while the more complex multi-task configuration remains an open challenge, to be studied in future work.

16.3.2 Loss functions for the embeddings

The top layer in CNN architectures for image classification is traditionally a softmax linear classifier, which produces outputs with a probabilistic meaning. These outputs can then be used to compute the cross-entropy loss with respect to the ground truth and backpropagate the gradients through the CNN. However, the former approach assumes a probabilistic nature for the ground truth as well, i.e. each ground truth vector represents the probability distribution of a sample over all the classes in the dataset. Such an assumption does not hold for the proposed approach, where the CNN maps every input image into a position vector in the output space. For this reason, two modifications to the traditional classification pipeline need to be done: (1) the softmax classifier is replaced by a projection matrix which maps the visual features extracted by the CNN into the output embedding, and (2) a loss function different from the cross-entropy is used to train the network. We have proposed three extensions to two different loss functions previously used to learn these mappings from the input to the output space:

L_2 Regression loss [24]. The regression loss aims to make the predicted and target vectors close to each other while remaining agnostic to the rest of the classes in the problem.

Similarity-L_2 regression loss. We modify the L_2 loss to take into account all classes in the dataset, as defined in Eq. (16.5),

$$loss(image, label) = \left\| M \cdot \vec{f}(image) - M \cdot \vec{y}_{label} \right\|^2 \tag{16.5}$$

where $\vec{f}(image)$ is a column vector obtained at the output of the CNN for the given image, $\vec{y}_{label}$ is the column vector embedding of class *label* in the output space, $M = [\vec{y}_0, ..., \vec{y}_{N-1}]^T$ and N is the number of classes in the training set. All vectors were constrained to have unit norm, so that the inner product between M and a vector in the output space results in an N-dimensional vector containing the similarity of the latter with respect to each class in the ground truth.

Hinge rank loss [10]. This ranking loss aims to minimize the distance between the output of the CNN and the target vector while isolating the former from *all* the other vectors, thus penalizing equally all errors. It is defined in Eq. (16.6),

$$loss(image, label) = \sum_{j \neq label} \max\left[0, m - \vec{y}_{label}^{T} \cdot \vec{f}(image) + \vec{y}_{j}^{T} \cdot \vec{f}(image)\right] \quad (16.6)$$

where $0 \leq m \leq 1$ is a margin constant and the sum is truncated after the first margin-violating term.

Weighted hinge rank loss. We argue that the hinge rank loss does not completely suit our problem, as some mistakes should have a larger penalization than others, e.g. mistaking a *happy boy* for a *happy child* is an acceptable error, but mistaking it for a *tropical house* should have a large associated cost. The *weighted hinge rank loss* scales the loss associated to each pair depending on the prior information given by the embedding, as defined in Eq. (16.7),

$$loss(image, label) = \sum_{j \neq label} d_{ij} \cdot \max\left[0, m - \vec{y}_{label}^{T} \cdot \vec{f}(image) + \vec{y}_{j}^{T} \cdot \vec{f}(image)\right] \quad (16.7)$$

where $d_{ij} = 1 - \vec{y}_{label}^{T} \cdot \vec{y}_j$, so that each pair has a different contribution to the overall loss, depending on how similar or dissimilar the classes in the pair are.

Variable margin hinge rank loss. We extend the weighted hinge rank loss by imposing more strict conditions for dissimilar classes, while relaxing the margin for similar ones, as described in Eq. (16.8),

$$loss(image, label) = \sum_{j \neq label} \max\left[0, d_{ij} - \vec{y}_{label}^{T} \cdot \vec{f}(image) + \vec{y}_{j}^{T} \cdot \vec{f}(image)\right] \quad (16.8)$$

where not only prior information is leveraged, but the margin hyperparameter, m, is removed as well.

16.4 EXPERIMENTAL RESULTS

The proposed approaches are evaluated using the English tag-restricted subset of MVSO [14] with a stratified split of 80–20 for training and validation, respectively, so that the two splits have the same class distribution. Output embeddings for the *addition* and *product* combination methods are built using a *word2vec* model trained on the Google News corpus, which contains 100 billion tokens and 3,000,000 unique words [18]. This model, which was trained to generate 300-dimensional outputs using a context window of 5 words, is publicly available and provided by [18]. Further experiments with a word2vec model trained on the Flickr 100M corpus are reported in the supplementary material.

The visual model is implemented by means of a CNN following the ResNet50 architecture [11] except for the last classification layer, which is replaced by a projection matrix as described in Section 16.3.2. The CNN weights are initialized using a model pre-trained on ILSVRC [9], a practice that has been proven beneficial even when training on large-scale datasets [29]. Optimization is performed using RMSProp [25] with a learning rate of 0.1, decay of 0.9 and $\epsilon = 1.0$ on batches of 128 samples. Data augmentation, consisting in random crops and/or horizontal flips on the input images, together with L2-regularization with a weight decay rate of 10^{-4} is used in order to reduce overfitting. The gradients are allowed to backpropagate through the whole CNN, unlike [10] where the model is trained in three steps: (1) the CNN is optimized for the traditional classification task using a softmax classification layer, (2) the parameters in the projection matrix are trained while freezing the rest of the model, and (3) gradients are allowed to backpropagate through the whole CNN only during the latest training stage. Model evaluations are performed using a running average of the parameters computed over time.

The whole system is implemented with TensorFlow[2] and trained on a machine with two NVIDIA K80 GPUs, using four model replicas with synchronous parameter updates. Our best models will be made publicly available upon publication of this work.

16.4.1 Adjective noun pair detection

Once an image has been projected in the output space by the CNN, it is possible to do a hard class assignment by finding the closest vector in the ground truth:

$$prediction = \max_{label} \left[\vec{y}_{label}^{T} \cdot \vec{f}(image) \right] \tag{16.9}$$

where all vectors are L_2-normalized, so that the inner product is equivalent to the cosine similarity.

The tag-restricted subset has almost 1.2M images, a figure too high to evaluate all the proposed loss functions and combination methods by grid search. For this reason, assuming that the different loss functions will have a similar behavior regardless of the combination method, we first evaluate all of them

[2]https://www.tensorflow.org.

Table 16.2 ANP classification accuracy comparison for the proposed loss functions. The output embedding is built by adding the *word2vec* representations for adjectives and nouns learned using the Google News corpus. For the ranking losses, a margin constant $m = 0.1$ was used

Loss function	top-1	top-5	top-10
Softmax baseline	0.228	0.398	0.452
L_2	0.186	0.276	0.314
Similarity-L_2 (Eq. (16.5))	0.140	0.234	0.279
Hinge rank (Eq. (16.6))	0.188	0.356	0.427
Weighted hinge rank (Eq. (16.7))	0.175	0.343	0.422
Variable margin hinge rank (Eq. (16.8))	0.189	0.300	0.347

for the *addition* embedding to get some insight on their performance. Table 16.2 summarizes the accuracy of the evaluated loss functions, together with the traditional classification baseline, which outperforms the embedding approach regardless of the cost objective. This behavior is somewhat expected, as the baseline's training objective matches the evaluation metric better than the rest of the approaches [10].

When comparing regression and ranking objectives, it is clear that the latter show a consistent superiority across all variants. Moreover, the proposed *similarity-*L_2 yields the least competitive results among all compared training objectives. We hypothesize that simultaneously taking into account all classes, though it theoretically would properly model the fine-grained relationship between all ANPs, in practice becomes a noisy objective when the number of classes is as large as 1200. This behavior can be experimentally observed by plotting the cost value for any possible mistake with respect to the similarity between the involved classes. Fig. 16.2 effectively demonstrates how the *similarity-*L_2 *loss* function is not injective with respect to the similarity between classes, i.e. two mistakes that should be penalized differently may get the same cost value, thus translating in poorer ANP detection results. The main source of this noise comes from the unbalanced number of similar and dissimilar classes: for each ANP, there are few synonyms and close concepts, but hundreds of dissimilar categories. Such an unbalance biases the similarity-L_2, as the contribution of potential positive categories is diluted. By reducing the number of considered classes, the range of cost values for each (i, j) pair is reduced, the L_2 loss function being the extreme case where only one category is considered (Fig. 16.2, bottom). However, when reducing the number of considered classes, the system becomes agnostic to vectors in the output space other than y_{label}. A good trade-off is achieved by the pairwise ranking losses, which provide a smoother objective for any (i, j) pair, while still considering a broader set of classes than the L_2 cost objective.

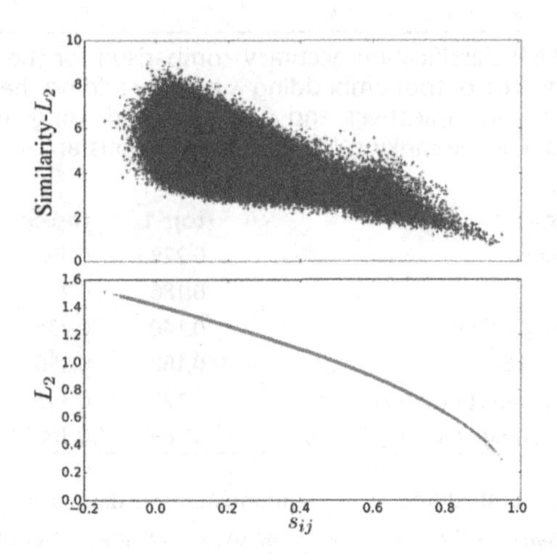

FIGURE 16.2 Comparison of regression loss functions. The graph shows the cost value for predicting class j when the real label is i as a function of the cosine similarity between classes, s_{ij}, for all possible combinations of i and j when using similarity-L_2 loss function (top) and L_2 (bottom).

Among the ranking losses, the *variable margin hinge rank loss* achieves the best top-1 accuracy, but for the rest of top-k metrics it is clearly under the other rankings alternatives. Since the margin for this cost function is, for most pairs, greater than the one used in the other losses, its behavior is similar to the L_2 loss because most pairs generate a non-zero loss. For the displayed metrics, the regular *hinge rank loss* is the one getting closest to the performance of the softmax baseline, even outperforming it in top-k accuracy for larger k values than those in Table 16.2. A similar behavior is observed for the *weighted hinge rank loss*, which becomes the best performing ranking model for larger k values, suggesting that it is able to better capture the fine-grained relationship between classes, a feature that is not observable by means of top-k accuracy for low values of k.

Table 16.3 compares the influence of the combination method for the best performing cost functions, i.e. the ranking training objectives. The results indicate that the *product* method consistently outperforms *addition* in terms of classification accuracy. We argue that the element-wise multiplication of adjectives and nouns results in a sparser output embedding because it imposes a stricter condition in each direction of the space, i.e. an ANP will have a large component in a particular direction of the space if, and only if, both adjective and noun do. On the contrary, this is achieved in the *addition* method even if only one of the concepts has a large component in a specific direction. A more detailed experimental evaluation of the resulting output embeddings is performed in the following section through zero-shot concept detection.

Table 16.3 ANP classification accuracy comparison for the proposed combination methods using ranking objectives. The way in which the *word2vec* representations for adjectives and nouns are combined to build the output embedding has influence on the classification accuracy of the resulting models

Loss function	Method	top-1	top-5	top-10
Hinge rank (Eq. (16.6))	Addition	0.188	0.356	0.427
	Product	0.196	0.368	0.448
Weighted hinge rank (Eq. (16.7))	Addition	0.175	0.343	0.422
	Product	0.192	0.362	0.444
Variable margin hinge rank (Eq. (16.8))	Addition	0.189	0.300	0.347
	Product	0.203	0.324	0.367

16.4.2 Zero-shot concept detection

One of the advantages of projecting the input image into a structured output embedding that leverages semantic knowledge from text sources is the system's potential for predicting classes for which no visual examples have been used during training, thanks to their relationship with other classes in the manifold [24,33]. This property is interesting for object recognition tasks with a large number of classes [10] to overcome the cost of collecting enough training samples for each type of object, but also when working with compound concepts, as new combinations of already seen concepts are allowed. In particular, for the task of ANP classification, there are adjectives and nouns that appear separately in the ontology, but never together, e.g. *ugly* and *lady* appear in the MVSO tag-restricted subset in ANPs such as *ugly doll* or *beautiful lady*, but never together as in *ugly lady*.

We build a large and challenging set of 200,000 images containing 1000 unseen ANPs with 200 images per class. These unseen labels are randomly sampled from the complete pool of 4342 ANPs in the English partition of MVSO, after filtering out those ANPs in the tag-restricted subset used for training and those with less than 200 samples. The resulting zero-shot set is heterogeneous with respect to the seen concepts: there are ANPs with either one or both concepts, i.e. adjective and noun, in the training set, whereas some others are completely new.

Class assignment for the zero-shot concept detection task is performed as described in Eq. (16.9), but evaluating the max (.) expression over the set of both seen and unseen classes, resulting in a 2200-way classification task.

Accuracy results for the different models in the zero-shot concept detection task are summarized in Table 16.4. Although the models using the *product* combination method exhibit a consistent improvement over those

Table 16.4 ANP classification accuracy results on the zero-shot concept detection task. The models predict one label among the set of 2200 seen and unseen ANPs

Model		top-1	top-5	top-10	top-20
Chance		0.04%	0.21%	0.42%	0.90%
Hinge rank (Eq. (16.6))	Addition	0.46%	3.84%	7.19%	11.90%
	Product	0.07%	1.43%	3.24%	5.92%
Weighted hinge rank (Eq. (16.7))	Addition	**0.53%**	**3.89%**	**7.29%**	**12.11%**
	Product	0.08%	1.44%	3.21%	5.95%
Variable margin hinge rank (Eq. (16.8))	Addition	0.13%	2.99%	5.82%	10.15%
	Product	0.003%	2.03%	4.11%	7.19%

using *addition* for the classification of seen ANPs, they show poorer generalization capabilities when extending the task to unseen classes. This result follows the intuition that the element-wise multiplication produces sparser embeddings where ANPs sharing one of the concepts are further away from each other. We hypothesize that adding more training classes, thus having more examples in the manifold, would help to better model the relationships between concepts and improve the zero-shot generalization capabilities of CNNs trained using these embeddings.

Another important difference with respect to the results of seen ANPs classification is the improvement of the proposed weighted hinge rank loss (Eq. (16.7)) over the regular hinge rank loss (Eq. (16.6)). This result, together with the improvement on the classification of seen ANPs for top-k metrics with large k values, suggests that the weighted hinge rank loss function may help to learn a more generic mapping that can be more easily extended to unseen classes.

16.5 VISUAL AFFECT RECOGNITION

Adjective Noun Pairs were originally conceived as a mid-level representation bridging the affective gap between raw pixel values and high level affective semantics such as emotions and sentiments [4,14]. In the following subsections we assess the performance of the compact and dense representations obtained by the proposed models for the tasks of visual emotion and sentiment prediction with human-annotated datasets, unlike the *weak labels* in MVSO. We compare three ResNet50 CNNs: (1) a classification model pre-trained on ILSVRC [9], (2) the ANP softmax baseline from Section 16.4.1, and (3) the *addition* ANP embedding model trained with hinge rank loss, as it presents the best trade-off between the accuracy on seen

ANPs and the generalization capabilities when dealing with unseen classes. The ILSVRC model is fine-tuned for the target task, and we consider both backpropagating gradients through the whole CNN and training only the final linear layer. Linear classifiers are trained on top of the ANP models; for the sake of comparison fairness with the ILSVRC model, two different setups are evaluated: (a) training the classifier on the output of the CNNs, and (b) removing the last layer and placing the classifier on top of the Global Average Pooling (GAP) layer, i.e. the second-to-last layer of the CNN that outputs 2048-d feature vectors. All approaches are trained using the setup described in Section 16.4.

16.5.1 Visual emotion prediction

The performance of the proposed methods for visual emotion prediction is evaluated on the Rochester Emotion dataset [32], the largest publicly available dataset for such a task. It contains over 23,000 images collected from social networks and annotated with eight emotion categories using the Amazon Mechanical Turk. The data is split into training and validation following a 80–20 proportion per class, so that the two sets have the same class distribution.

An additional setup with a 2-layer neural network with 300 hidden units on top of the ANP-embedding model was evaluated in order to compensate for the difference in dimensionality between the embedding (300-d) and the rest of approaches (1000-d for ILSVRC, 1200-d for ANP-softmax). As shown in Table 16.5, the three models yield a comparable performance in terms of absolute accuracy; however, when training only a simpler linear classifier on top of the penultimate layer of the CNNs, the ANP models reach a higher performance. This fact suggests that both ANP models have learned more affect-related features from the weakly labeled MVSO dataset, which can be used to train reliable emotion classifiers when the training data is scarce.

It is somewhat surprising that the model pre-trained on ILSVRC for the task of object recognition reaches the same accuracy as those pre-trained on an affect-related task. We argue that this behavior is due to a dataset bias: similarly to traditional computer vision datasets where some objects appear under a very restricted range of orientations or colors [26], the concept of dataset bias at affect level is related to some objects or patterns being always associated to the same emotion. This is likely to happen with images downloaded from social networks, where it is unlikely that an user will upload images of *pets* or *babies* associated to negative emotions. If such a bias is present, a simple algorithm mapping each object class to a single emotion would reach a high accuracy. On the contrary, the affective bias introduced

Table 16.5 Emotion classification accuracy on the Rochester Emotion dataset. In the central column, the word between brackets makes reference to the CNN layers that is used as input to the classifier. For the fusion results, the best performing method for each model was selected

Model	Method	Accuracy
ILSVRC	Full fine-tuning	0.636
	Linear (GAP)	0.605
ANP-softmax	Linear (output)	0.606
	Linear (GAP)	0.630
ANP-embedding	Linear (output)	0.379
	2-layer NN (output)	0.618
	Linear (GAP)	0.634
Fusion	ILSVRC + ANP-softmax	0.646
	ILSVRC + ANP-embedding	0.654
	ANP-softmax + ANP-embedding	0.641
	ILSVRC + ANP-softmax + ANP-embedding	**0.655**

by the adjective concept in ANPs models the range of affective states of each noun or object. The fact that emotion recognition can be improved by fusing the output scores of the fine-tuned ILSVRC CNN with that of the ANP models shows that they learn complementary representations. In particular, these results suggest that the ANP-embedding model is the best complement for the fine-tuned ILSVRC CNN, reaching roughly the same accuracy as the fusion of the three models.

16.5.2 **Visual sentiment prediction**

With the goal of overcoming the dataset bias, workers in the Amazon Mechanical Turk were asked to annotate the sentiment that they evoked in a scale between -2 and 2 for images sampled from ANPs in MVSO [8]. This sampling guarantees a wider range of affective connotations for different instances of the same concept, e.g. the concept *cat* appears in *beautiful cat* or *cute cat*, but also in *evil cat*, so that mapping the noun *cat* to a positive sentiment is not necessarily the correct decision for the classifier.

Over 8000 images were annotated by three workers each. After filtering out those for which the annotators determined that the ANP label did not match the content of the image, we average the three scores per sample and assign them to three classes: negative if $score < 0.5$, positive if $score > 0.5$, or neutral otherwise. A stratified 80–20 split is then applied to generate training and validation sets.

Table 16.6 Sentiment classification accuracy results. In the central column, the word between brackets makes reference to the CNN layers that is used as input to the classifier

Model	Method	Accuracy
ILSVRC	Full fine-tuning	0.571
	Linear (GAP)	0.583
ANP-softmax	Linear (output)	0.581
	Linear (GAP)	0.586
ANP-embedding	Linear (output)	0.581
	Linear (GAP)	**0.593**

Table 16.6 compares the accuracy of the different methods for visual sentiment prediction. When compared to those for visual emotion prediction in Table 16.5, the first noticeable difference is the drop in performance of the ILSVRC fine-tuning as compared to training only the linear classifier. During training, a stronger overfitting was observed in the former, where the training accuracy keeps improving up to very large figures after the validation accuracy has saturated, suggesting that the pre-trained object recognition model does not completely suit this challenging dataset. Despite the ANP-embedding approach reaching a higher accuracy, the performance of the ANP-softmax and ILSVRC models are comparable. We attribute this gap between ANP models to the fact that the dataset was built by sampling images for all ANPs and not only those in the tag pool subset, meaning that the capabilities of the embedding approach for generalizing to unseen classes is conferring it an advantage when dealing with concepts that were not considered in the ontology used for training.

16.6 CONCLUSIONS AND FUTURE WORK

In this work, we explored how to generate output embeddings for a pool of concepts so that the range of detectable elements is widened in an unsupervised manner. In particular, the challenge of building embeddings for ANPs, affect-related representations which are composed by two concepts, has been addressed by studying different combination methods. We explored several loss functions that allow the system to learn the mapping from the input image to the output embedding in an end-to-end fashion. Also we showed how the resulting models are able to perform zero-shot concept detection on a challenging dataset of unseen classes, thus expanding the domain where these mid-level representations can be applied. Finally, the proposed system is included in visual emotion and sentiment prediction pipelines, showing how they complement or even outperform state-of-the-art domain transferred models for such tasks.

In the future, we will address the task of jointly learning the visual and textual models in order to obtain output embeddings that are optimized for the target task, as well as training the combination method that outputs a fixed length representation for an arbitrary number of attributes. Besides, improving the visual detectability of ANPs remains an open challenge that we believe should be addressed by designing algorithms that are able to reach a higher level of abstraction, together with a refinement of the weak labels used for training those models.

REFERENCES

[1] Zeynep Akata, Florent Perronnin, Zaid Harchaoui, Cordelia Schmid, Label-embedding for image classification, in: IEEE Transactions on Pattern Analysis and Machine Intelligence, 2016.
[2] Marco Baroni, Roberto Zamparelli, Nouns are vectors, adjectives are matrices: representing adjective-noun constructions in semantic space, in: EMNLP, 2010.
[3] Damian Borth, Tao Chen, Rongrong Ji, Shih-Fu Chang, Sentibank: large-scale ontology and classifiers for detecting sentiment and emotions in visual content, in: ACM MM, 2013.
[4] Damian Borth, Rongrong Ji, Tao Chen, Thomas Breuel, Shih-Fu Chang, Large-scale visual sentiment ontology and detectors using adjective noun pairs, in: ACM MM, 2013.
[5] Michel Cabanac, What is emotion?, Behav. Process. 60 (2) (2002).
[6] Victor Campos, Amaia Salvador, Brendan Jou, Xavier Giro-i-Nieto, Diving deep into sentiment: understanding fine-tuned CNNs for visual sentiment prediction, in: ASM, ACM MM Workshops, 2015.
[7] Tao Chen, Damian Borth, Trevor Darrell, Shih-Fu Chang, DeepSentiBank: visual sentiment concept classification with deep convolutional neural networks, arXiv: 1410.8586, 2014.
[8] Vaidehi Dalmia, Hongyi Liu, Shih-Fu Chang, Columbia MVSO image sentiment dataset, arXiv preprint, arXiv:1611.04455, 2016.
[9] Jia Deng, Wei Dong, Richard Socher, Li-Jia Li, Kai Li, Li Fei-Fei, ImageNet: a large-scale hierarchical image database, in: CVPR, 2009.
[10] Andrea Frome, Greg S. Corrado, Jon Shlens, Samy Bengio, Jeff Dean, Tomas Mikolov, et al., Devise: a deep visual-semantic embedding model, in: NIPS, 2013.
[11] Kaiming He, Xiangyu Zhang, Shaoqing Ren, Jian Sun, Deep residual learning for image recognition, in: CVPR, 2016.
[12] Brendan Jou, Shih-Fu Chang, Deep cross residual learning for multitask visual recognition, in: ACM MM, 2016.
[13] Brendan Jou, Shih-Fu Chang, Going deeper for multilingual visual sentiment detection, arXiv preprint, arXiv:1605.09211, 2016.
[14] Brendan Jou, Tao Chen, Nikolaos Pappas, Miriam Redi, Mercan Topkara, Shih-Fu Chang, Visual affect around the world: a large-scale multilingual visual sentiment ontology, in: ACM MM, 2015.
[15] Andrej Karpathy, George Toderici, Sanketh Shetty, Thomas Leung, Rahul Sukthankar, Li Fei-Fei, Large-scale video classification with convolutional neural networks, in: CVPR, 2014.
[16] Jana Machajdik, Allan Hanbury, Affective image classification using features inspired by psychology and art theory, in: ACM MM, 2010.
[17] Tomas Mikolov, Kai Chen, Greg Corrado, Jeffrey Dean, Efficient estimation of word representations in vector space, in: ICLR, 2013.

[18] Tomas Mikolov, Ilya Sutskever, Kai Chen, Greg S. Corrado, Jeff Dean, Distributed representations of words and phrases and their compositionality, in: NIPS, 2013.

[19] Takuya Narihira, Damian Borth, Stella X. Yu, Karl Ni, Trevor Darrell, Mapping images to sentiment adjective noun pairs with factorized neural nets, arXiv preprint, arXiv:1511.06838, 2015.

[20] Mohammad Norouzi, Tomas Mikolov, Samy Bengio, Yoram Singer, Jonathon Shlens, Andrea Frome, Greg S. Corrado, Jeffrey Dean, Zero-shot learning by convex combination of semantic embeddings, arXiv preprint, arXiv:1312.5650, 2013.

[21] Bo Pang, Lillian Lee, Opinion mining and sentiment analysis, Inf. Retr. 2 (1–2) (2008).

[22] Robert Plutchik, Emotion: A Psychoevolutionary Synthesis, Harper & Row, 1980.

[23] Stefan Siersdorfer, Enrico Minack, Fan Deng, Jonathon Hare, Analyzing and predicting sentiment of images on the social web, in: ACM MM, 2010.

[24] Richard Socher, Milind Ganjoo, Christopher D. Manning, Andrew Ng, Zero-shot learning through cross-modal transfer, in: NIPS, 2013.

[25] Tijmen Tieleman, Geoffrey Hinton, Lecture 6.5-rmsprop: divide the gradient by a running average of its recent magnitude, COURSERA: Neural Netw. Mach. Learn. 4 (2012).

[26] Antonio Torralba, Alexei A. Efros, Unbiased look at dataset bias, in: CVPR, 2011.

[27] Jingwen Wang, Jianlong Fu, Yong Xu, Tao Mei, Beyond object recognition: visual sentiment analysis with deep coupled adjective and noun neural networks, in: IJCAI, 2016.

[28] Can Xu, Suleyman Cetintas, Kuang-Chih Lee, Li-Jia Li, Visual sentiment prediction with deep convolutional neural networks, arXiv preprint, arXiv:1411.5731, 2014.

[29] Jason Yosinski, Jeff Clune, Yoshua Bengio, Hod Lipson, How transferable are features in deep neural networks?, in: NIPS, 2014.

[30] Quanzeng You, Liangliang Cao, Hailin Jin, Jiebo Luo, Robust visual-textual sentiment analysis: when attention meets tree-structured recursive neural networks, in: ACM MM, 2016.

[31] Quanzeng You, Jiebo Luo, Hailin Jin, Jianchao Yang, Robust image sentiment analysis using progressively trained and domain transferred deep networks, in: AAAI, 2015.

[32] Quanzeng You, Jiebo Luo, Hailin Jin, Jianchao Yang, Building a large scale dataset for image emotion recognition: the fine print and the benchmark, in: AAAI, 2016.

[33] Ziming Zhang, Venkatesh Saligrama, Zero-shot learning via semantic similarity embedding, in: CVPR, 2015.

Chapter 17

Video-based emotion recognition in the wild

Albert Ali Salah*,†, Heysem Kaya‡, Furkan Gürpınar*

*Boğaziçi University, Department of Computer Engineering, Bebek, Istanbul, Turkey †Nagoya University, Future Value Creation Research Center (FV-CRC), Nagoya, Japan ‡Namık Kemal University, Department of Computer Engineering, Çorlu, Tekirdağ, Turkey

CONTENTS

17.1 INTRODUCTION

Audiovisual emotion recognition is not a new problem. There has been a lot of work in visual pattern recognition for facial emotional expression recognition, as well as in signal processing for audio-based detection of emotions, and many multimodal approaches combining these cues [85]. However, improvements in hardware, availability of datasets and wide-scale annotation infrastructure made it possible to create real affective systems a reality, and we now see applications across many domains, including robotics, HCI, healthcare, and multimedia.

In this chapter, we use the term "emotion" to represent a subjective state displayed via social signals. This is markedly different than, say, recognition of the emotional content of a multimedia clip, which deals with emotions that the clip may evoke in its viewers. In this work, we assume that emotions are short-term phenomena with manifest (and possibly idiosyncratic) expressions [68], which we analyze via pattern recognition techniques to

Multimodal Behavior Analysis in the Wild. https://doi.org/10.1016/B978-0-12-814601-9.00031-6
Copyright © 2019 Elsevier Ltd. All rights reserved.

determine their possible nature. Depending on the application, these expressions can be defined in a continuous dimensional space (the most frequently used representation being the Valence-Arousal space), or in terms of discrete categories, basic or otherwise. We briefly discuss the issue of representation at the end of the chapter.

Audio- and video-based emotion recognition "in the wild" is still a challenging task. In the wild (as opposed to controlled) refers to real-life and uncontrolled conditions of data acquisition, including changing and challenging lighting conditions, indoor and outdoor scenarios, sensor and environmental noise and resolution issues, motion blur, occlusions and pose changes, as well as uncontrolled scenarios from which the emotional expressions are produced. In this chapter, we briefly describe recent advances in audiovisual emotion recognition in the wild, and propose a general framework for tackling this problem.

In Section 17.2 we describe related recent work in multimodal emotion recognition from video. Section 17.3 sketches our general framework for processing. Section 17.4 details the framework on three problems with in-the-wild conditions, illustrating the flexibility of the approach. Our technical conclusions and a broadly conceived discussion are given in Section 17.5.

17.2 RELATED WORK

In-the-wild conditions may refer to a number of challenges in emotion estimation. The acquisition of the video may be under non-controlled illumination conditions, either indoors or outdoors. The elicited emotions may be spontaneous, thus the manifestation of the expression could be uncontrolled. In the literature, an example of such data are acquired from talk shows, and other naturalistic interactions. The recording medium may be uncontrolled, for instance data acquired with webcams by individuals can be gathered and evaluated. Not all these challenges will be present at the same time. For instance, a dataset collected from movie clips will feature actors, posing for emotional expressions (in a way that can be identified by human viewers), but it will have uncontrolled illumination conditions. Good summaries of available video-based affect databases and their acquisition conditions can be found in [51,85]. Table 17.1 summarizes the most frequently used video-, audio- and image-based databases. Purely audio-based affective databases collected in the wild are rarely found in the literature, datasets such as TESS [21] and SAVEE [32] are acquired under controlled conditions.

One of the important drivers of research in this area has been the Emotion Recognition in the Wild (EmotiW) Challenges, which introduced and de-

Table 17.1 Popular and recent affect databases with "in-the-wild" conditions

Database	Type	# of items	Affect modeling
AFEW [17]	Video	330	7 emotion categories
AffectNet [58]	Image	~450.000	8 emotion categories; Valence and arousal
Aff-Wild [84]	Video	500	Valence and arousal
AVEC 2014 [77]	Video	300	Continuous valence, activation and dominance
EmoChildRU [48]	Audio	1116	3 affect categories
EmotioNet [5]	Image	~100.000	12 AUs annotated
FAU-AIBO [74]	Audio	18,216	11 emotion categories
FER-2013 [28]	Image	35.887	7 emotion categories
FER-Wild [59]	Image	~24.000	7 emotion categories

veloped an out of laboratory dataset—Acted Facial Expressions in the Wild (AFEW)—, collected from videos that mimic real life, and posing difficult and realistic conditions of analysis [14–16].

The EmotiW Challenge, which started in 2013, aims to overcome the challenges of data collection, annotation, and estimation for multimodal emotion recognition in the wild. The challenge uses the AFEW corpus, which mainly consists of movie excerpts with uncontrolled conditions [16]. The top performing system of the first challenge employed visual bag of words features, gist features, paralinguistic audio features and Local Phase Quantization from Three Orthogonal Planes (LPQ-TOP) features, which were separately processed by RBF kernels and combined with a multi kernel support vector machine [72].

In its subsequent years, the EmotiW Challenge saw a marked increase in neural network-based solutions. For instance, recurrent neural networks (RNNs) were used to model the dynamics of facial expressions [22]. While temporal models improved the performance, they also typically have more parameters and run the risk of overfitting. In the 2015 EmotiW Challenge, the best performing system used a decision fusion approach, with linear SVMs on multiple visual scales and audio [83]. The approach we describe in this work also uses deep neural networks in a transfer learning setting.

Most entries to the EmotiW Challenge describe combinations of weaker learners, such as the pipeline we describe in this chapter. We observe that such weak learners are redundant to some degree, and process similar representations. Nonetheless, their combination is empirically shown to be useful. Training an end-to-end deep neural network (so far) did not improve over such ensemble approaches. [67] reports experiments with

VGG-16 [73], VGG-19 [73], ResNet [33], and Xception [11] architectures. [75] reports results with VGG-19, ResNet and BN-Inception [38]. Individually, these models do not reach the accuracy of the ensemble and combined approaches.

Visual processing for emotions mostly focuses on faces. Several off the shelf face methods are used to detect and align faces in the literature [4,56,79, 82,86]. [65] experimented with the open-source TinyFace detector [35], and reported improved face detection results, but emotion estimation from small faces did not work. However, it is still a good idea to combine multiple face detectors to make this stage more robust [43]. It was shown that accurate registration and alignment of faces was crucial for image-based processing pipelines [53].

When implementing face analysis, a commonly used pre-processing stage is detecting facial landmarks, and providing a supervised learning pipeline to process landmark positions and movements [49,67]. Such geometric representations are considered to be complementary to entirely image-based representations, and are also amenable to temporal modeling.

Audio emotion recognition gained momentum towards maturity in the last two decades, driving other affect related paralinguistic research including, but not limited to, laughter [76], personality [31,42] and depression [46,77]. Speech emotion recognition has manifold applications such as evaluating customer satisfaction in call centers and some early research is dedicated to emotion recognition in dialogues [52].

The main problem in the past decade for audio-based affect recognition was the scarcity of public data: the available corpora were poor both in terms of quantity (i.e. the total duration of annotated utterances) and quality (not reflecting in-the-wild conditions). The majority of corpora were portrayed and recorded in lab environment [9,23,47,55]. One of the first publicly available speech emotion corpora exhibiting richness in both aforementioned dimensions was the FAU AIBO corpus [74], where 10–13 years old German children were expected to navigate an AIBO robot with speech commands. The corpus was delivered and used in the 2009 Computational Paralinguistics (ComParE) Challenge [70], which was the first of the series held at INTERSPEECH conferences since then. The ComParE series boosted the paralinguistic studies by introducing new corpora and by providing a common benchmark for comparability and reproducibility of results.

The INTERSPEECH challenges highlight the advancement of the state-of-the-art in audio-based emotion recognition in the last decade. In ComParE 2009, the first audio-based in-the-wild emotion recognition challenge, the majority of participants employed the popular Mel Frequency Cepstral Coefficients (MFCC) [0–12], augmented with their first and second order temporal derivatives together with Gaussian Mixture Models (GMM) [8,20,50,80]. Kockmann et al., who achieved the best results in Open-Performance Subchallenge for 5-class emotion recognition task, further improved their GMM systems based on discriminative training and Joint Factor Analysis (JFA) [50]. The contribution of Bozkurt et al. additionally investigated Line Spectral Frequency (LFS) representation of speech, which is related to formants [8]. Based on the motivating results, LSF and formants are further investigated on the same in-the-wild acoustic emotion recognition task [6, 7]. We should note that the challenge test set scores of the participants were quite similar, even though they were free to choose their own set of features and classifiers [69]. However surprising, the simple Naive Bayes classifier rendered the best results in the "Classifier Subchallenge", while none of the feature sets proposed by participants was able to outperform the baseline 384-dimensional acoustic feature set in the "Feature Subchallenge" [69].

The number of acoustic features that were extracted (using the openSMILE tool [26]) and presented as baseline in ComParE series has increased over the years, yielding more competitive scores. In the 2013–2017 challenges, the baseline set for utterance-level challenges covered 6373 suprasegmental features [71]. Comparing the winning systems' performance with that of the baselines, we can see that in most cases beating the baseline system (openSMILE features and SVM classifier) was hard. Thus, brute-forced functional encoding of Low Level Descriptors (LLDs; such as F0, energy, MFCCs) can be considered as a state-of-the-art acoustic feature representation for a range of tasks.

In-the-wild acoustic data are naturally noisy. Consequently, more elaborate feature representation methods, as well as robust learners need to be investigated. In the 2015 ComParE Challenge, Kaya et al. [45] proposed combining MFCCs and RASTA-Style Linear Prediction Cepstral Coefficients (LPCC) using the Fisher Vector (FV) representation, a popular LLD encoding method used in computer vision [64]. The FV encoding is also employed in subsequent challenges of the ComParE series, where the fusion systems yielded outstanding results in affective tasks [44].

We should note that the popular set of LLDs extracted from a speech frame has dramatically lower dimensionality compared to those extracted from an image. This makes functional-based brute-forcing feasible. While it is still possible to delve into extracting more emotionally informative LLDs, investigating LLD representation methods (such as FV encoding or Vector of Locally Aggregated Descriptors-VLAD [3]) have a higher potential for advancing the state-of-the-art in in-the-wild speech emotion recognition.

Multimodal approaches to emotional expression recognition leverage paralinguistic audio cues, but also synchronization between modalities to improve robustness. Early multimodal approaches focused on coarse affective states (e.g. positive and negative states), because data collection and annotation for natural or spontaneous scenarios was difficult (see [85] for a comprehensive survey of earlier approaches, and [81] for available databases). In the EmotiW Challenges, a general observation was that the contribution of audio was small, but consistent. Weighted approaches that automatically selected the relative importance of a modality allowed the inspection of how much audio contributed over the visual information. The reported difference was sometimes as high as 30 times in favor of visual information [83]. The weight for features extracted over the entire scene was even smaller than the audio. However, when the analyzed dataset includes movie scenes, audio processing can learn to classify emotions like "fear" from the soundtrack.

Starting with the image based group happiness intensity estimation task in 2016, the newer editions of EmotiW include group emotion estimation tasks [1,13,36,49,65,67,75,78]. Predicting the prevalent emotions in a group of people can simply mean the average emotion value of the individuals in the group. There is not much work yet on the complex interactions of the group, and on group-level features. Cerekovic used spatial distribution of the faces as a group-level descriptor to estimate group happiness [10]. Other work integrated scene context by extracting deep neural network features from the scene by using pre-trained models for object recognition [1].

In the next section, we describe an approach for video-based emotion estimation, where a single individual is featured, as opposed to groups of people.

17.3 PROPOSED APPROACH

We give an overview flow for the proposed approach in Fig. 17.1, adapted from [43]. Since the face contains the most prominent visual cues, the first step in the pipeline is the detection and alignment (or registration) of the face. Even for deep neural network models, this step cannot be neglected, and will have a significant impact on the results. To process the face with transfer learning approaches, we propose to use a deep neural network ini-

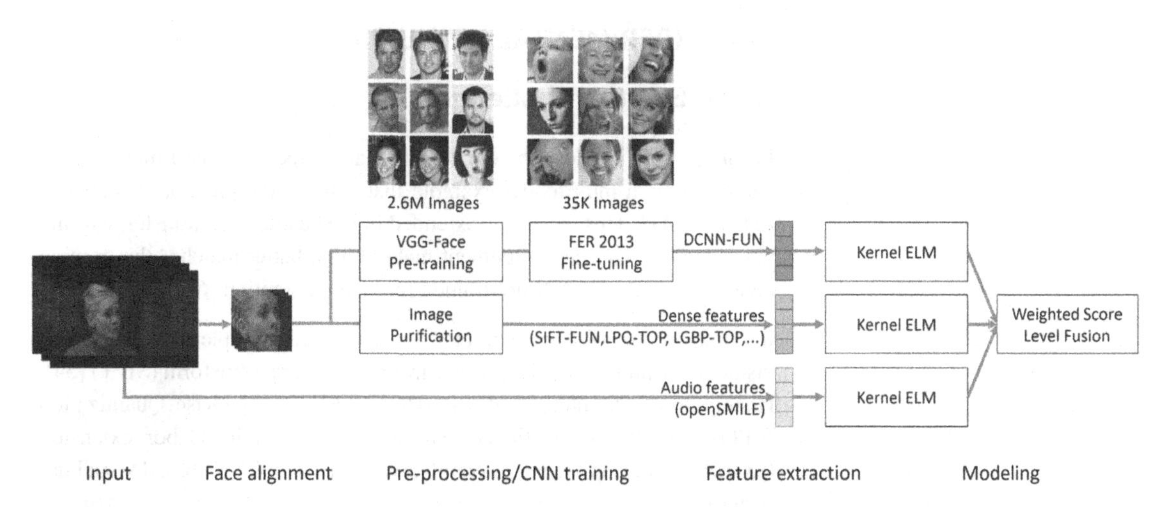

FIGURE 17.1 Overview of the proposed approach (adapted from [43]).

tially trained for face recognition, but fine-tuned for emotion estimation. A reason for this choice is the availability of larger amounts of data for training recognition models. Both problems have the same underlying logic of augmenting details in faces for fine-level classification.

In this model, we have multiple visual classifiers, combined to make use of the strengths of each. The audio modality is processed separately and fused at the decision level. If other modalities are available (for instance the scene features can be added as a context), they will also be processed in separate pathways. The first pipeline in the figure shows a deep neural network for emotion estimation, used as a feature extractor. The second pipeline consists of features extracted via popular image processing algorithms. Such features are observed to be complementary to deep neural network features. Finally, the third pipeline is for processing the audio information, and we use the OpenSMILE library for this purpose. More details will be given in the next section.

Submissions to most recent EmotiW Challenge (2017) use similar combinations of multiple channels for increased robustness. For instance, [49] uses four image-based and three audio-based approaches in parallel, and performs adaptive fusion for classification. Temporal information is integrated in both audio and vision modalities via Long Short-Term Memory (LSTM) networks.

We use kernel ELM classifiers for fast training in this work [37]. Since these classifiers have relatively fewer tunable parameters, they are also shown to be resistant against overlearning. A weighted score-level fusion is proposed at the end of the pipeline. Here, several fusion strategies can be tested.

17.4 EXPERIMENTAL RESULTS

17.4.1 EmotiW Challenge

In this section, we describe our system that competed in the EmotiW 2015 Challenge as a multimodal example that follows the pipeline depicted in Fig. 17.1 [41]. This work was extended in [43] using deep transfer learning via an appropriate face alignment method, that better matches the training conditions of the DCNN pre-trained for face recognition [63].

In our approach, we have extracted and compared multiple commonly used visual descriptors, including Scale Invariant Feature Transform (SIFT) [54], Histogram of Oriented Gradients (HOG) [12], Local Phase Quantization (LPQ) [34,40], Local Binary Patterns (LBP) [62] its Gabor extension (LGBP), and Deep Convolutional Neural Network (CNN) based visual descriptors.

For temporal modeling of video clips, the Three Orthogonal Planes (TOP) feature extraction method is commonly used with LPQ, LBP and LGBP [40, 87]. This description method extracts the visual features from the XY, XT and YT planes (where T represents time) independently, and concatenates the resulting feature vectors. The LGBP-TOP descriptor was used as a baseline feature in the AVEC 2014 challenge [77]. In our implementation, we enhance temporal modeling by dividing the videos into two equal parts and extract the spatio-temporal TOP features from each part separately.

The restricted amount of data for training extensive systems is the first major challenge one has to overcome for systems in the wild. Transfer learning is the most popular answer to this hurdle. In our approach, we started from a pre-trained CNN model (VGG-Face) [63] and fine-tuned this model with the FER 2013 dataset for emotion estimation [28]. A series of preliminary experiments confirmed the assumption that a model initially developed for face recognition would serve better in feature extraction compared to the more popular ImageNet, trained for generic object recognition. However, there are examples where such networks are successfully employed, as their feature set is quite rich. An example is [60], which uses pre-trained models trained on ImageNet dataset, followed by two stages of fine-tuning, on FER 2013 and Static Facial Expression Sub-Challenge (SFEW) Dataset from the EmotiW 2015 Challenge, respectively [17].

Our preliminary experiments on DCNN based feature extraction revealed that employing a relatively loose alignment with dynamic parts model (DPM) based face detection [56] results in a dramatically higher performance (about 30% relative improvement) compared to Mixture of Parts (MoP)-based alignment [27].

Table 17.2 Validation and test set accuracies of the systems submitted to EmotiW 2015. First part: top two systems submitted and evaluated in the official challenge [41] without CNN features. Second part: Systems using CNN features. WF: weighted fusion, FF: feature-level fusion, *MoP*: MoP-based alignment, *DPM*: DPM-based alignment

System	Val	Test
WF (Audio, LBP-TOP$_{MoP}$, LGBP-TOP$_{MoP}$)	50.14%	50.28%
WF (Audio, LGBP-TOP$_{MoP}$, HOG-FUN$_{MoP}$, SIFT-FV$_{MoP}$, LBP-TOP$_{MoP}$, LPQ-TOP$_{MoP}$)	**52.30%**	**53.62%**
CNN-FUN$_{MoP}$	44.47%	42.86%
CNN-FUN$_{DPM}$	51.60%	51.39%
WF (Audio, CNN-FUN$_{MoP}$, SIFT-FV$_{MoP}$, LBP-TOP$_{MoP}$, LGBP-TOP$_{MoP}$, HOG-FUN$_{MoP}$)	53.70%	51.76%
WF (Audio, CNN-FUN$_{DPM}$, LGBP-TOP$_{MoP}$, HOG-FUN$_{MoP}$)	**57.02%**	**54.55%**

To enhance the diversity of learners, we have used functionals on frame-level features and Fisher vector encoding of low-level descriptors. State-of-the-art acoustic feature extraction pipelines commonly use a large set of summarizing functionals over low-level descriptor contours. In our system, we used the mean and the standard deviation, and three functionals based on polynomials, fit to each descriptor contour.

To learn a classification model, we have employed kernel extreme learning machine (ELM) [37] and Partial Least Squares (PLS) regression. These approaches do not require long training time and extensive parameter optimization, and consequently are more adequate than CNN or SVM models for fusion classifiers. We should however mention that under suitable training conditions, end-to-end multimodal deep neural networks have achieved impressive results [61].

The performance of systems submitted to the EmotiW 2015 Challenge and subsequently developed systems combining fine-tuned DCNN features are presented in Table 17.2. We observe that DCNN features obtained by presenting DPM aligned videos to fine-tuned VGG-Face achieve the highest single-modality, single-feature type performance for this corpus. Moreover, when DCNN features are fused with audio and other visual features at score level, the system achieves a test set accuracy of 54.55%, which outperforms the top submission to EmotiW 2015 [83].

The EmotiW 2016 Challenge extended the corpus of 2015 with videos of people with different nationality. The newly added training/test data were found to worsen generalization in this corpus, as the test data contained videos with different recording conditions. We observe that training with the 2015 Challenge corpus renders a test set accuracy of 52.11% (over a baseline score of 40.47%), while the same fusion scheme with training with the 2016 Challenge corpus gives a test set performance of 48.40%.

17.4.2 ChaLearn Challenges

In this subsection, we report results with the same general pipeline on two related challenges. While the aim in these challenges is not emotion estimation, the processing pipeline is very similar, and it is instructive to study them, as they feature in-the-wild conditions.

The first of these, ChaLearn-LAP First Impressions Challenge, contains 10,000 video clips collected from YouTube, annotated via Amazon Mechanical Turk for the apparent personality traits of the people featured in them. The Big-Five scale was used in these annotations (Openness, Conscientiousness, Extroversion, Agreeableness and Neuroticism). Each clip takes 15 s and features a single person, generally with a close to frontal pose. However, clips present a large variety of backgrounds (e.g. shot in car, while walking, in bed, or taken from a TV show) featuring people from different ethnic origins. For more details, the reader is referred to the challenge overview paper [25].

The pipeline of the system that won the challenge is similar to the one shown in Fig. 17.1, with an additional channel for the scene features, extracted from a VGG-VD-19 network pre-trained for object detection [73]. The scene channel uses the first frame (including the face) of each video and is shown to have: (i) a high accuracy of predicting the first impressions and (ii) complementary information to face and audio based features. It serves as a crude approximation of the context, as discussed before. The overall system fuses two face features (LGBP-TOP and DCNN—via our fine-tuned version of VGG-Face), a deep scene feature set from VGG-VD-19, and a standard openSMILE acoustic feature set presented in the INTERSPEECH 2013 ComParE Challenge [71].

The proposed system rendered an average accuracy of 0.913 on the test set, and was the top system in the ChaLearn LAP Challenge organized at ICPR 2016. We observe that the distribution of estimations of the automatic algorithm were much more conservative than the actual annotations (clustered tightly around the mean).

The personality trait estimation challenge was extended for an automated job interview candidate screening challenge, organized at CVPR/IJCNN in 2017 [24]. The same dataset was used, together with the addition of a new target variable that represents whether the subject would be invited to a job interview or not. The quantitative part of the challenge aimed to estimate this new variable along with the personality traits. The interesting part of this challenge was that the algorithm that predicted the target variable was expected to produce an explanation on the decision as well. This is well aligned with recent concerns about algorithmic accountability, and against black-box AI systems with millions of parameters, and hard to explain decisions.

For the quantitative (prediction) stage, we employed the same audio and visual feature sets that were employed in our winning entry to ChaLearn LAP-FI Challenge, however with a different fusion scheme. We first combined face features (CNN and LGBP-TOP) in one channel, while scene and audio features were combined in a parallel channel, and both channels were fed into Kernel ELM regressors. The personality trait and job interview variable outputs of the ELM regressors are later combined ($6 \times 2 = 12$ high-level features) and stacked with a Random Forest regressor, which gives the final scores. The automatically generated explanation (i.e. the qualitative part of the challenge) was implemented with a binary decision tree that treated each estimated personality trait as a high or low value. The proposed system ranked first in both qualitative and quantitative parts of the challenge [42].

17.5 CONCLUSIONS AND DISCUSSION

The success of the approach on the various challenges illustrates the flexibility of the proposed pipeline. Treating the individual modalities separately early on makes it easier to automatically generate explanations for the given decisions.

Our findings on the EmotiW Challenge confirmed that multimodality brings in diminishing returns for natural (or seminatural) data, but it is nonetheless essential to achieve higher performance [18].

Audio and visual subsystems have different performances on individual affect classes. For instance, it is easier to recognize "fear" from the audio modality, but for "disgust," visual cues seem to be more informative. Affective displays that rely on subtle cues are affected much more by the difficult conditions posed by "in-the-wild" data [78].

In our experiments on different problems, we have repeatedly observed the impact of flexible alignment (and fine-tuning) on transfer learning for CNN models. We proposed using summarizing functionals on CNN features, which gave us good single-modality classifiers.

Recent work on emotion recognition focuses on dimensional and continuous approaches [29,85]. There is relevant work for both image based and video based estimation in the wild [5,59,84]. However, categorical and discrete approaches that go beyond the six (or seven, if we include "contempt") basic expressions are still very relevant, as their interpretation is more natural for humans. Also, it is difficult to reduce a complex emotion to a point or region in the valence-arousal space. For instance, we can claim that "love" is a positive emotion, and has a positive valence, but this is not always the case. Consider the loving, concerned expression of a mother, looking at her sick child, and this becomes obvious. Subsequently, a given image or video can be annotated in the continuous space, but there is still need for mapping its points to a semantic space, where it can be properly interpreted.

The protagonist of "The Face of Another," written by the Japanese novelist Kobo Abe, is a scientist who forges an expressive mask for his badly burned face [2]. While preparing the mask, he draws a list of tentative emotions, with corresponding proportions of their exhibit: Concentration of interest (16 percent), curiosity (7 percent), assent (10 percent), satisfaction (12 percent), laughter (13 percent), denial (6 percent), dissatisfaction (7 percent), abhorrence (6 percent), doubt (5 percent), perplexity (6 percent), concern (3 percent), anger (9 percent). The author adds that "[it] cannot be considered satisfactory to analyze such a complicated and delicate thing as expression into these few components. However, by combining just this many elements on my palette, I should be able to get almost any shade." (p. 102). While this list is not based on a scientific classification, it is clear that when we seek for emotions in the wild, we get many more colors than basic expressions. Work on compound facial emotions [19] is a step in this direction. We note here the difficulties associated with such finer grained categorization. As an example, while the FER-2013 database was collected using 184 emotion-related keywords on the Google image search API, the provided annotations were limited to 7 categories [28].

The annotation and ground truth acquisition of "in-the-wild" data are clearly problematic. For instance research on empathy demonstrated very clearly that the historical and social context need to be evaluated carefully in interpreting the emotional state of a subject [66]. But how should this context be codified into computational models? The necessity of grounding of such information in complex semantic relations creates a significant problem for today's dominant machine learning paradigms, which are only able to learn certain spatial and temporal relationships, and only when the programmer takes the necessary precautions and provides appropriate mechanisms for their representation. Thus, our study on first impressions took in and evaluated a rudimentary form of context by encoding the background of the subject in a deep neural network trained to recognize objects in the scene [30]. While a marginal improvement in recognition accuracy is obtained by simple contextual features, it is clear that this representation is not an adequate proxy for the rich cultural backdrop that will influence the exhibit of emotions in a myriad of ways.

The second problem that aggravates the situation for the computational models is that we rely on "expert" annotations for ground truth, which however—in most studies—completely ignore the proper context of the subject. The ground truth under such conditions is barely an apparent emotional display according to the socio-cultural context of the annotator. Also, any cultural biases that are in the annotator will be transferred to the automatic algorithm due to machine learning.

As an example, consider the algorithm we have developed for predicting whether someone will be invited to a job interview, or not [42]. There are two major problems in the training of this system. When we annotate the database for gender and ethnicity, and look at the distribution of labels, we can see a small bias for preferring white subjects over African Americans, and a small bias for females over males. This comes directly from the annotations, and if the ultimate aim is to predict the human judgment, the algorithm indeed should learn this bias. There are applications where this is not desirable. For emotion estimation, we have ample empirical evidence that different cultures interpret facial expressions differently [39]. According to these findings, establishing ground truth for a facial-expression database requires the annotation of both the subject and the annotator's cultural background. If the same database is annotated by, say, Japanese and American subjects, we may get different ratings, unless a very clear, discrete categorization is used. The typical scenario of letting the annotators choose from a closed set of annotation labels may mask the differences in perception, and even when a closed set is used, we see empirical evidence for these differences. In an early study, Matsumoto indeed experimented with American and Japanese subjects and obtained as low as 54.5% agreement for "fear" annotations and 64.2% agreement for "anger" annotations in Japanese subjects, even with a closed set of discrete labels for annotation [57]. In any case, we may or may not want the algorithm to learn the biases of the annotators, depending on the application. If an algorithm is pre-screening applicants for a job interview selection decision (a highly undesired situation, but maybe conceivable for job posts with tens of thousands of applicants), we do not want any biases there. If the algorithm is used to understand the actual decisions of human screeners, however, we may want to model this.

The third problem in the annotations is the lack of ecological validity. The act of annotation does not involve the proper context of an emotional expression, and this results in a systematic bias. Recording physiological signals could be an alternative, but this also damages the ecological validity to a large extent, and also does not completely escape the reductionism of the previous approach.

We believe that the next decade of affective computing will face each of these challenges, including proper internal models that will give affective systems emotional palettes similar to or different from humans, and we will have rich toolboxes to help us design more natural interactions with intelligent systems in the wild.

ACKNOWLEDGMENTS

This research is supported by Boğaziçi University project BAP 12060 and by the BAGEP Award of the Science Academy.

REFERENCES

[1] Asad Abbas, Stephan K. Chalup, Group emotion recognition in the wild by combining deep neural networks for facial-expression classification and scene-context analysis, in: Proc. ICMI, 2017, pp. 561–568.

[2] Kobo Abe, The Face of Another, Vintage, 2011.

[3] Relja Arandjelovic, Andrew Zisserman, All about VLAD, in: Proc. CVPR, 2013, pp. 1578–1585.

[4] Tadas Baltrušaitis, Peter Robinson, Louis-Philippe Morency, Openface: an open source facial behavior analysis toolkit, in: Proc. WACV, IEEE, 2016, pp. 1–10.

[5] C.F. Benitez-Quiroz, R. Srinivasan, A.M. Martinez, EmotioNet: an accurate, real-time algorithm for the automatic annotation of a million facial expressions in the wild, in: Proc. CVPR, 2016.

[6] Elif Bozkurt, Engin Erzin, Çiğdem Eroğlu Erdem, A. Tanju Erdem, Use of line spectral frequencies for emotion recognition from speech, in: 2010 20th Intl. Conf. on Pattern Recognition, August 2010, pp. 3708–3711.

[7] Elif Bozkurt, Engin Erzin, Çiğdem Eroğlu Erdem, A. Tanju Erdem, Formant position based weighted spectral features for emotion recognition, Speech Commun. 53 (9–10) (November 2011) 1186–1197.

[8] Elif Bozkurt, Engin Erzin, Ç. Eroğlu Erdem, Tanju Erdem, Improving automatic emotion recognition from speech signals, in: Proc. INTERSPEECH, Brighton, 2009, pp. 324–327.

[9] Felix Burkhardt, Astrid Paeschke, Miriam Rolfes, Walter F. Sendlmeier, Benjamin Weiss, A database of German emotional speech, in: Proc. of Interspeech, 2005, pp. 1517–1520.

[10] Aleksandra Cerekovic, A deep look into group happiness prediction from images, in: Proc. ICMI, 2016, pp. 437–444.

[11] François Chollet, Xception: deep learning with depthwise separable convolutions, arXiv preprint, arXiv:1610.02357, 2016.

[12] Navneet Dalal, Bill Triggs, Histograms of oriented gradients for human detection, in: Proc. CVPR, 2005, pp. 886–893.

[13] Abhinav Dhall, Roland Goecke, Shreya Ghosh, Jyoti Joshi, Jesse Hoey, Tom Gedeon, Group-level emotion recognition using transfer learning from face identification, in: Proc. ICMI, 2017, pp. 524–528.

[14] Abhinav Dhall, Roland Goecke, Jyoti Joshi, Karan Sikka, Tom Gedeon, Emotion recognition in the wild challenge 2014: baseline, data and protocol, in: Proceedings of the 16th ACM International Conference on Multimodal Interaction, 2014, pp. 461–466.

[15] Abhinav Dhall, Roland Goecke, Jyoti Joshi, Michael Wagner, Tom Gedeon, Emotion recognition in the wild challenge 2013, in: Proceedings of the 15th ACM International Conference on Multimodal Interaction, 2013, pp. 509–516.

[16] Abhinav Dhall, Roland Goecke, Simon Lucey, Tom Gedeon, Collecting large, richly annotated facial-expression databases from movies, IEEE Multimed. 19 (3) (July 2012) 34–41.

[17] Abhinav Dhall, O.V. Ramana Murthy, Roland Goecke, Jyoti Joshi, Tom Gedeon, Video and image based emotion recognition challenges in the wild: EmotiW 2015, in: Proc. ACM ICMI, ICMI '15, ACM, New York, NY, USA, 2015, pp. 423–426.

[18] Sidney D'Mello, Jacqueline Kory, Consistent but modest: a meta-analysis on unimodal and multimodal affect detection accuracies from 30 studies, in: Proc. ACM ICMI, ACM, 2012, pp. 31–38.

[19] Shichuan Du, Yong Tao, Aleix M. Martinez, Compound facial expressions of emotion, Proc. Natl. Acad. Sci. 111 (15) (2014) E1454–E1462.

[20] Pierre Dumouchel, Najim Dehak, Yazid Attabi, Reda Dehak, Narjes Boufaden, Cepstral and long-term features for emotion recognition, in: Proc. INTERSPEECH, Brighton, 2009, pp. 344–347.

[21] Kate Dupuis, M. Kathleen Pichora-Fuller, Toronto Emotional Speech Set (TESS), University of Toronto, Psychology Department, 2010.

[22] Samira Ebrahimi Kahou, Vincent Michalski, Kishore Konda, Roland Memisevic, Christopher Pal, Recurrent neural networks for emotion recognition in video, in: Proc. ACM ICMI, ACM, 2015, pp. 467–474.

[23] Inger Engberg, Anya Hansen, Documentation of the Danish Emotional Speech Database (DES), Center for Person Kommunikation, Denmark, 1996.

[24] Hugo Jair Escalante, Isabelle Guyon, Sergio Escalera, Julio Jacques, Meysam Madadi, Xavier Baró, Stephane Ayache, Evelyne Viegas, Yağmur Güçlütürk, Umut Güçlü, et al., Design of an explainable machine learning challenge for video interviews, in: Proc. IJCNN, 2017, pp. 3688–3695.

[25] Hugo Jair Escalante, Víctor Ponce-López, Jun Wan, Michael Riegler, Baiyu Chen, Albert Clapes, Sergio Escalera, Isabelle Guyon, Xavier Baró, Paal Halvorsen, Henning Müller, Martha Larson, ChaLearn joint contest on multimedia challenges beyond visual analysis: an overview, in: Proc. ICPR, 2016, pp. 67–73.

[26] Florian Eyben, Martin Wöllmer, Björn Schuller, Opensmile: the Munich versatile and fast open-source audio feature extractor, in: Proc. of the Intl. Conf. on Multimedia, ACM, 2010, pp. 1459–1462.

[27] Pedro F. Felzenszwalb, Daniel P. Huttenlocher, Pictorial structures for object recognition, Int. J. Comput. Vis. 61 (1) (2005) 55–79.

[28] Ian J. Goodfellow, Dumitru Erhan, Pierre Luc Carrier, Aaron Courville, Mehdi Mirza, Ben Hamner, Will Cukierski, Yichuan Tang, David Thaler, Dong-Hyun Lee, et al., Challenges in representation learning: a report on three machine learning contests, Neural Netw. 64 (2015) 59–63.

[29] Hatice Gunes, Björn Schuller, Categorical and dimensional affect analysis in continuous input: current trends and future directions, Image Vis. Comput. 31 (2) (February 2013) 120–136.

[30] Furkan Gürpınar, Heysem Kaya, Albert Ali Salah, Combining deep facial and ambient features for first impression estimation, in: Computer Vision–ECCV 2016 Workshops, Springer, 2016, pp. 372–385.

[31] Furkan Gürpınar, Heysem Kaya, Albert Ali Salah, Multimodal fusion of audio, scene, and face features for first impression estimation, in: Proc. ICPR, 2016, pp. 43–48.

[32] Sanaul Haq, Philip J.B. Jackson, Multimodal emotion recognition, in: Machine Audition: Principles, Algorithms and Systems, 2010, pp. 398–423.
[33] Kaiming He, Xiangyu Zhang, Shaoqing Ren, Jian Sun, Deep residual learning for image recognition, in: Proc. CVPR, 2016, pp. 770–778.
[34] Janne Heikkilä, Ville Ojansivu, Esa Rahtu, Improved blur insensitivity for decorrelated local phase quantization, in: Proc. ICPR, 2010, pp. 818–821.
[35] Peiyun Hu, Deva Ramanan, Finding tiny faces, arXiv preprint, arXiv:1612.04402, 2016.
[36] Ping Hu, Dongqi Cai, Shandong Wang, Anbang Yao, Yurong Chen, Learning supervised scoring ensemble for emotion recognition in the wild, in: Proc. ICMI, 2017, pp. 553–560.
[37] Guang-Bin Huang, Hongming Zhou, Xiaojian Ding, Rui Zhang, Extreme learning machine for regression and multiclass classification, IEEE Trans. Syst. Man Cybern., Part B, Cybern. 42 (2) (2012) 513–529.
[38] Sergey Ioffe, Christian Szegedy, Batch normalization: accelerating deep network training by reducing internal covariate shift, in: Proc. ICML, 2015, pp. 448–456.
[39] Rachael E. Jack, Oliver G.B. Garrod, Hui Yu, Roberto Caldara, Philippe G. Schyns, Facial expressions of emotion are not culturally universal, Proc. Natl. Acad. Sci. 109 (19) (2012) 7241–7244.
[40] Bihan Jiang, Michel Valstar, Brais Martinez, Maja Pantic, A dynamic appearance descriptor approach to facial actions temporal modeling, IEEE Trans. Cybern. 44 (2) (2014) 161–174.
[41] Heysem Kaya, Furkan Gürpınar, Sadaf Afshar, Albert Ali Salah, Contrasting and combining least squares based learners for emotion recognition in the wild, in: Proc. ACM ICMI, ICMI '15, ACM, New York, NY, USA, 2015, pp. 459–466.
[42] Heysem Kaya, Furkan Gürpınar, Albert Ali Salah, Multi-modal score fusion and decision trees for explainable automatic job candidate screening from video CVs, in: CVPR 2017 Workshops, 2017.
[43] Heysem Kaya, Furkan Gürpınar, Albert Ali Salah, Video-based emotion recognition in the wild using deep transfer learning and score fusion, Image Vis. Comput. 65 (2017) 66–75.
[44] Heysem Kaya, Alexey A. Karpov, Fusing acoustic feature representations for computational paralinguistics tasks, in: INTERSPEECH, 2016, pp. 2046–2050.
[45] Heysem Kaya, Alexey A. Karpov, Albert Ali Salah, Fisher vectors with cascaded normalization for paralinguistic analysis, in: INTERSPEECH, Dresden, Germany, 2015, pp. 909–913.
[46] Heysem Kaya, Albert Ali Salah, Eyes whisper depression: a CCA based multimodal approach, in: Proc. of the 22nd Intl. Conf. on Multimedia, Proc. ACM MM, Orlando, Florida, USA, ACM, 2014.
[47] Heysem Kaya, Albert Ali Salah, Sadik Fikret Gurgen, Hazim Ekenel, Protocol and baseline for experiments on Boğaziçi University Turkish emotional speech corpus, in: IEEE Signal Processing and Communications Applications Conf., April 2014, pp. 1698–1701.
[48] Heysem Kaya, Albert Ali Salah, Alexey Karpov, Olga Frolova, Aleksey Grigorev, Elena Lyakso, Emotion, age, and gender classification in children's speech by humans and machines, Comput. Speech Lang. 46 (2017) 268–283.
[49] Dae Ha Kim, Min Kyu Lee, Dong Yoon Choi, Byung Cheol Song, Multi-modal emotion recognition using semi-supervised learning and multiple neural networks in the wild, in: Proc. ICMI, 2017, pp. 529–535.
[50] Marcel Kockmann, Lukáš Burget, Jan Černocký, Brno University of Technology system for interspeech 2009 emotion challenge, in: Proc. INTERSPEECH, Brighton, 2009, pp. 348–351.

[51] Jean Kossaifi, Georgios Tzimiropoulos, Sinisa Todorovic, Maja Pantic, AFEW-VA database for valence and arousal estimation in-the-wild, Image Vis. Comput. 65 (2017) 23–36.

[52] Chul Min Lee, Shrikanth S. Narayanan, Toward detecting emotions in spoken dialogs, IEEE Trans. Speech Audio Process. 13 (2) (2005) 293–303.

[53] Mengyi Liu, Ruiping Wang, Zhiwu Huang, Shiguang Shan, Xilin Chen, Partial least squares regression on Grassmannian manifold for emotion recognition, in: Proc. ICMI, ACM, 2013, pp. 525–530.

[54] David G. Lowe, Distinctive image features from scale-invariant keypoints, Int. J. Comput. Vis. 60 (2) (2004) 91–110.

[55] Veronika Makarova, Valery A. Petrushin, RUSLANA: a database of Russian emotional utterances, in: INTERSPEECH, Denver, Colorado, USA, 2002, pp. 2041–2044.

[56] Markus Mathias, Rodrigo Benenson, Marco Pedersoli, Luc Van Gool, Face detection without bells and whistles, in: Proc. ECCV, Springer International Publishing, 2014, pp. 720–735.

[57] David Matsumoto, American–Japanese cultural differences in the recognition of universal facial expressions, J. Cross-Cult. Psychol. 23 (1) (1992) 72–84.

[58] Ali Mollahosseini, Behzad Hasani, Mohammad H. Mahoor, AffectNet: a database for facial expression, valence, and arousal computing in the wild, in: IEEE Trans. Affect. Comput., 2017.

[59] Ali Mollahosseini, Behzad Hasani, Michelle J. Salvador, Hojjat Abdollahi, David Chan, Mohammad H. Mahoor, Facial expression recognition from world wild web, in: Proc. CVPRW, 2016, pp. 58–65.

[60] Hong-Wei Ng, Viet Dung Nguyen, Vassilios Vonikakis, Stefan Winkler, Deep learning for emotion recognition on small datasets using transfer learning, in: Proc. ACM ICMI, ACM, 2015, pp. 443–449.

[61] Jiquan Ngiam, Aditya Khosla, Mingyu Kim, Juhan Nam, Honglak Lee, Andrew Y. Ng, Multimodal deep learning, in: Proc. ICML, 2011, pp. 689–696.

[62] Timo Ojala, Matti Pietikainen, Topi Maenpaa, Multiresolution gray-scale and rotation invariant texture classification with local binary patterns, IEEE Trans. Pattern Anal. Mach. Intell. 24 (7) (2002) 971–987.

[63] O.M. Parkhi, A. Vedaldi, A. Zisserman, Deep face recognition, in: British Machine Vision Conference, 2015.

[64] Florent Perronnin, Christopher Dance, Fisher kernels on visual vocabularies for image categorization, in: Proc. CVPR, 2007.

[65] Stefano Pini, Olfa Ben Ahmed, Marcella Cornia, Lorenzo Baraldi, Rita Cucchiara, Benoit Huet, Modeling multimodal cues in a deep learning-based framework for emotion recognition in the wild, in: Proc. ICMI, 2017, pp. 536–543.

[66] Stephanie D. Preston, Alicia J. Hofelich, The many faces of empathy: parsing empathic phenomena through a proximate, dynamic-systems view of representing the other in the self, Emot. Rev. 4 (1) (2012) 24–33.

[67] Alexandr Rassadin, Alexey Gruzdev, Andrey Savchenko, From individual to group-level emotion recognition: EmotiW 5.0, in: Proc. ICMI, 2017, pp. 544–548.

[68] Klaus R. Scherer, What are emotions? and how can they be measured?, Soc. Sci. Inf. 44 (4) (2005) 695–729.

[69] Björn Schuller, Anton Batliner, Stefan Steidl, Dino Seppi, Recognising realistic emotions and affect in speech: state of the art and lessons learnt from the first challenge, Speech Commun. 53 (9) (2011) 1062–1087.

[70] Björn Schuller, Stefan Steidl, Anton Batliner, The INTERSPEECH 2009 emotion challenge, in: Proc. INTERSPEECH, Brighton, UK, ISCA, September 2009, pp. 312–315.

[71] Björn Schuller, Stefan Steidl, Anton Batliner, Alessandro Vinciarelli, Klaus Scherer, Fabien Ringeval, Mohamed Chetouani, Felix Weninger, Florian Eyben, Erik Marchi, Marcello Mortillaro, Hugues Salamin, Anna Polychroniou, Fabio Valente, Samuel Kim, The INTERSPEECH 2013 computational paralinguistics challenge: social signals, conflict, emotion, autism, in: Proc. INTERSPEECH, Lyon, France, ISCA, ISCA, August 2013, pp. 148–152.

[72] Karan Sikka, Karmen Dykstra, Suchitra Sathyanarayana, Gwen Littlewort, Marian Bartlett, Multiple kernel learning for emotion recognition in the wild, in: Proc. ICMI, 2013, pp. 517–524.

[73] Karen Simonyan, Andrew Zisserman, Very deep convolutional networks for large-scale image recognition, CoRR arXiv:1409.1556, 2014.

[74] Stefan Steidl, Automatic Classification of Emotion Related User States in Spontaneous Children's Speech, PhD thesis, Logos Verlag, Berlin, FAU Erlangen-Nuremberg, 2009.

[75] Lianzhi Tan, Kaipeng Zhang, Kai Wang, Xiaoxing Zeng, Xiaojiang Peng, Yu Qiao, Group emotion recognition with individual facial emotion CNNs and global image based CNNs, in: Proc. ICMI, 2017, pp. 549–552.

[76] Khiet P. Truong, David A. van Leeuwen, Automatic detection of laughter, in: Proc. INTERSPEECH, Lisbon, Portugal, 2005, pp. 485–488.

[77] Michel Valstar, Björn Schuller, Kirsty Smith, Timur Almaev, Florian Eyben, Jarek Krajewski, Roddy Cowie, Maja Pantic, AVEC 2014: 3D dimensional affect and depression recognition challenge, in: Proc. of the 4th ACM Intl. Workshop on Audio/Visual Emotion Challenge, 2014.

[78] Valentin Vielzeuf, Stéphane Pateux, Frédéric Jurie, Temporal multimodal fusion for video emotion classification in the wild, in: Proc. ICMI, 2017, pp. 569–576.

[79] Paul Viola, Michael Jones, Rapid object detection using a boosted cascade of simple features, in: Proc. CVPR, 2001.

[80] Vlasenko Bogdan, Andreas Wendemuth, Processing affected speech within human machine interaction, in: Proc. INTERSPEECH, Brighton, 2009, pp. 2039–2042.

[81] Chung-Hsien Wu, Jen-Chun Lin, Wen-Li Wei, Survey on audiovisual emotion recognition: databases, features, and data fusion strategies, APSIPA Trans. Signal Inf. Process. 3 (2014) e12.

[82] Xuehan Xiong, Fernando De la Torre, Supervised descent method and its applications to face alignment, in: Proc. CVPR, 2013, pp. 532–539.

[83] Anbang Yao, Junchao Shao, Ningning Ma, Yurong Chen, Capturing AU-aware facial features and their latent relations for emotion recognition in the wild, in: Proc. ACM ICMI, ACM, 2015, pp. 451–458.

[84] Stefanos Zafeiriou, Athanasios Papaioannou, Irene Kotsia, Mihalis Nicolaou, Guoying Zhao, Facial affect "in-the-wild", in: Proc. CVPRW, 2016, pp. 36–47.

[85] Zhihong Zeng, Maja Pantic, Glenn I. Roisman, Thomas S. Huang, A survey of affect recognition methods: audio, visual, and spontaneous expressions, IEEE Trans. Pattern Anal. Mach. Intell. 31 (1) (2009) 39–58.

[86] Kaipeng Zhang, Zhanpeng Zhang, Zhifeng Li, Yu Qiao, Joint face detection and alignment using multitask cascaded convolutional networks, IEEE Signal Process. Lett. 23 (10) (2016) 1499–1503.

[87] Guoying Zhao, Matti Pietikainen, Dynamic texture recognition using local binary patterns with an application to facial expressions, IEEE Trans. Pattern Anal. Mach. Intell. 29 (6) (2007) 915–928.

Chapter 18

Real-world automatic continuous affect recognition from audiovisual signals

Panagiotis Tzirakis*, Stefanos Zafeiriou*,†, Björn Schuller*,‡

*Imperial College London, Department of Computing, London, UK †University of Oulu, Center for Machine Vision and Signal Analysis, Oulu, Finland ‡University of Augsburg, Chair of Embedded Intelligence for Health Care and Wellbeing, Augsburg, Germany

CONTENTS

18.1 INTRODUCTION

Affect recognition is vital to human-to-human communication, and to accomplish a complete interaction between humans and machines, emotion recognition needs to be considered. Nowadays, more and more intelligent

Multimodal Behavior Analysis in the Wild. https://doi.org/10.1016/B978-0-12-814601-9.00028-6
Copyright © 2019 Elsevier Ltd. All rights reserved.

systems are using emotion recognition models to improve their interaction with humans. This is important, as the systems can adapt their responses and behavioral patterns according to the emotions of the humans and make the interaction more *natural*. One application can be found in an automatic tutoring system, where the system adjusts the level of the tutorial depending on the user's affective state, such as excitement or boredom. In another application, a fatigue state can be predicted by utilizing affect states [28]. A further example can be found in the speech recognition domain, and more particularly in call centers. The goal is to successfully predict the affective state of the caller or the agent and provide feedback for the quality of the service [7]. But these are just a few of the manifold applications where emotion recognition can and very likely very soon will be used.

However, recognizing emotions is challenging, as human emotions lack of temporal boundaries. In addition, emotions can be expressed in various ways among different individuals [2] and cultures, as different persons can react differently in the same situation. Also, the perception of emotion can be highly subjective which makes the task even more challenging. Moreover, emotions can be expressed and sensed with and via various modalities such as facial expressions, speech, or other biosignals. Considering different modalities is vital when designing intelligent systems, as some emotions can be identified more easily by a specific modality. For example, arousal can be detected easily through speech acoustics whilst valence is mostly better assessed by speech linguistics or through visual information.

In this chapter, we aim at presenting the audiovisual cues that convey emotions and how emotions can be modeled. Our second purpose is to describe the advances of the affective computing community towards continuous emotion recognition facing real-world conditions. To this end, we present a multimodal emotion recognition approach for continuous affect recognition.

The chapter is organized as follows. Section 18.2 provides the main differences between real-world and laboratory settings. Section 18.3 describes how emotions can be perceived from audio and visual information, and the main categories emotions can be quantified in. Section 18.4 presents multimodal fusion strategies, and describes the related work for continuous emotion recognition using audio, visual and audiovisual information. In addition, it mentions well-known databases and the competitions that have emerged for continuous affect recognition or for real-world settings. Section 18.5 describes a representative system for continuous affect recognition from audiovisual signals, and, finally, Section 18.6 concludes this chapter.

18.2 REAL WORLD VS LABORATORY SETTINGS

Before describing the background on emotions, it is important to clarify the differences between real-world/in-the-wild and laboratory/posed settings. These include the following:

- **Environmental conditions.** Laboratory settings consider controlled environmental conditions, i.e., without noise or reverberation. However, in real world settings the environmental conditions are not controlled. In addition, there may be different kinds of occlusions, which do not occur in a controlled environment.
- **Subject movements and linguistics.** Restrictions may be applied to the subject in terms of bodily movements, either moving head/hands or changing its position, in the laboratory settings. However, no such constrains apply in real-world situations. The subject can freely make any movement he/she desires.
- **Subject's expressions.** Posed settings may limit the expressions the participants can perform. On the other hand, in a realistic environment humans do not restraint themselves and can perform any kind of expression. Similarly, the verbal content may be prompted or predefined in lab settings and thus limited in some way. Obviously, this would be different in real-world conditions.

Due to the fact that real-world settings pose such difficult challenges, the affective computing community has mostly dealt with posed settings. However, in the last years a shift towards real-world conditions is observed. This is more apparent with the increased number of databases and challenges (competitions) that have been proposed.

18.3 AUDIO AND VIDEO AFFECT CUES AND THEORIES OF EMOTION

Humans can automatically recognize emotions from different modalities such as visual or audio cues. In this section, we present the signals that convey emotion in both the visual and audio modalities. Then, we will give a brief on how to quantize emotion for usage in technical system.

18.3.1 Audio signals

Speech is a vital modality for affect recognition. The information that can be conveyed from speech can be linguistic and acoustic. Considering only the linguistic information in speech for emotion recognition can be misleading, as the words can be misinterpreted. For example, it is very difficult to understand sarcasm only from the spoken words. In addition, the choice of the words is human dependent, meaning that different humans use different words for exactly the same thing.

On the other hand, acoustically encoded information is often very significant for affect recognition. For example, the intonation, intensity, and speaking rate of the voice convey vital information for the arousal of the speaker. More particularly, common findings in the literature suggest that the mean of the fundamental frequency (F0), the speech rate, and the mean energy of the speech signal are positively correlated with the arousal of an individual [8, 27]. However, the reality of today's engines is much more complex than that with several tens to even thousands of characteristics being used to model speech acoustics for emotion recognition.

18.3.2 Visual signals

Similar to speech, visual information provides a range of useful insights for automatic affect recognition. Facial expressions, such as a smile or frown, are one of the most essential visual signals humans rely on for emotion interpretation. Ekman and Fiesen [18] created the well-known facial action coding system (FACS) that describes the muscle movements of the face in terms of predefined action units (AUs). However, their work focused on a discrete number of emotions. A major problem with this work is the existence of a high inter-individual variability. In another study, Russell [47] created a two dimensional plane (arousal and valence) to express facial expressions.

Contextual information is an important factor for recognizing emotions [1]. Facial movement can be split in four possible temporal phases: neutral, onset, apex, and offset [17]. In the neutral phase, the face is relaxed, i.e., facial muscles are inactive. In the onset phase, the muscles start to contract and increase in intensity. In the apex phase, the muscle activity reaches its peak and lastly, in the offset phase, the muscles start to relax and move back to the neutral state.

Beyond facial expressions, one finds a rich body of literature on body posture [6,10], and even gait analysis [37] to identify human emotion. For example, an interesting finding by Bull [6] was that agreement (disagreement) and interest (boredom) are associated with specific body movements. In a more recent study, Kleinsmith et al. [32] found that sad/depressed categories can be recognized from postures among the cultures Japanese, Sri Lankan, and American.

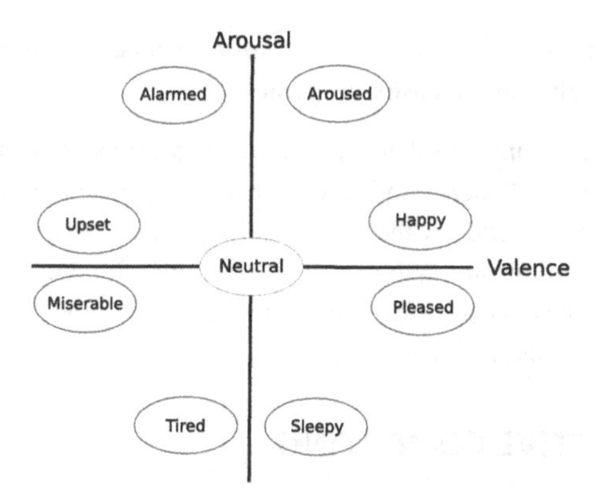

■ **FIGURE 18.1** Circumplex of Affect [47]. The horizontal axis represents the valence and the vertical the arousal.

18.3.3 Quantifying affect

Quantifying emotions has long been in the interest of various research areas like psychology and neuroscience. According to Grandjean et al. [22], there are three major approaches to characterize emotional models: categorical, dimensional, and appraisal-based. And indeed, these are the ones mainly used in technical approaches these days, whereas in particular categorical and dimensional approaches are pre-dominating.

The categorical approach indicates that there are a few emotions that are basic, and can be recognized universally independent of the race. This theory has been extensively studied in the literature for many years. The most famous model is the model of Ekman [16], which suggests that there are six basic emotions that can be recognized universally, namely, surprise, fear, disgust, anger, joy, and sadness. However, many emotion researchers believe that the affective states are a lot more complex, and that the emotions are not independent from one another, but related in a systematic manner. Hence, a small number of states cannot capture the complexity of the emotions.

The dimensional approach suggests that emotions are not independent, but related to one another. Russell [47] proposed to characterize emotions with a small number of latent dimensions. To this end, he describes emotions with a circular configuration called Circumplex of Affect. This model represents the affective states in a two dimensional bipolar space. These dimensions are the arousal, which indicates how relaxed (i.e., bored) or aroused or active (i.e., excited) the subject is, and valence, which indicates the degree of pleasantness or unpleasantness of feelings. Fig. 18.1 depicts this model. Another well-known dimensional model is the PAD emotion space [29]. This model uses three dimensions to represent the affective state by adding dominance to the pack. However, further additions can be encountered such as surprise or novelty, and intensity dimensions.

A relatively new approach that models affect in terms of technical usage was introduced by Scherer and is called the Component Process Model (CPM) [48]. It is based on the appraisal-based approach [20,22]. In this method, the emotions are assumed to be generated through the evaluation of events from the outside world and the internal state of the individual (e.g., motivation, physiological reactions).

18.4 AFFECTIVE COMPUTING

This part of the chapter focuses on presenting the progress achieved to day in the real-world continuous affect recognition community using audio and visual information. First, we describe the main techniques to fuse different modalities and then, we present recent studies for emotion recognition. Finally, after we show databases that can be used for continuous or in-the-wild emotion recognition, we present current competitions that have been run for this task.

18.4.1 Multimodal fusion techniques

Models for affect recognition tasks can be benefited by fusing the information that is extracted from different modalities (i.e., audio, visual, physiology, or alike). For this purpose, there are three common fusion strategies that exist in the literature, namely, feature-level fusion, decision- or score-level fusion, and model-level fusion.

The feature-level fusion combines the features extracted from different modalities by concatenating them in a single feature vector before passing it to a single prediction algorithm. The process of this method is similar to the human 'mechanism'. That is, humans first extract features from modalities such as audio or visual cues, and then map these features to the same space before their prediction. This method requires large training datasets as more information is captured than using a single modality. An additional requirement for this method is that the modalities should be in correspondence. The main advantage of this strategy is that a prediction will be available even if one modality's data is missing.

In the decision-level fusion, each modality is utilized independently for prediction and then, the scores of each modality are combined. The main disadvantage of this method is that if the data of one modality are missing, then its full capabilities cannot be exploited. For classification tasks, the fusion can be performed with a simple majority vote (albeit in more sophisticated versions, weights can be introduced and even learnt), while for regression tasks, a linear regressor can be trained using the prediction for each modality and its weights can be used for the fusion.

The model- or hybrid-level fusion combines the advantages of both the feature- and the decision-level fusion strategies. An example of a model-level fusion is to perform a feature-level fusion of 'some' modalities and combine its prediction with scores from different modalities that were processed independently. A recent example of such a method was proposed by Wang et al. [58].

18.4.2 **Related work**

There are a number of studies that deal with continuous affect prediction from speech (e.g., [24,54]). In fact, the body of literature is too rich, to give a complete overview here. Just to name one more recent approach, Schmitt et al. [49] used a bag-of-audio-words (BoAW) created from mel-frequency cepstral coefficients (MFCCs) and energy low-level descriptors (LLDs), as feature vector. Then, a Support Vector Regression (SVR) is used to predict the arousal and valence dimensions. In much other work, brute-force larger feature spaces are employed rather than going directly from LLDs or BoAWs. In the last few years, also deep learning techniques have increasingly dominated the field of emotion recognition from speech. The first study that utilized a deep learning architecture, by Wöllmer et al. [60], used functionals of acoustic LLDs as input to a three layer long-short term memory (LSTM) recurrent neural network (RNN) for continuous emotion recognition. Next Stuhlsatz et al. [53], also used deep neural networks (DNNs) and more specifically restricted Boltzmann Machines (RBMs) for discriminative feature extraction from raw data and proposed a Generalized Discriminant Analysis (GerDA). In another study, Han et al. [24] combined the models of SVR and bi-directional LSTM RNNs for continuous affect recognition. More particularly, they first trained one of the models using 13 LLDs (12 MFCCs and 1 log energy coefficient) and 26 statistical features. The output of this model is concatenated with the hand-crafted features and input to the second one. The total number of combinations would be four, from which the authors considered three (SVR-SVR is excluded). In more recent studies, Trigeorgis et al. [54] proposed the first end-to-end model for this task in the literature that uses a convolutional neural network (CNN) to extract features from the raw signal. The network is comprised of a two layer convolution max-pooling layout. The extracted features are then passed to a two layers LSTM to capture the contextual information in the data.

The visual modality has also been explored for continuous emotion recognition. For instance, Nicolaou et al. [38] proposed an Output-Associative

Relevance Vector Machine (OA-RVM) regression framework. Their work expanded the RVM regression so as to be able to learn non-linear dependencies from the input and the output. In another work, Kipp and Martin [31] investigated whether association between hand gestures and the PAD emotional dimensions exists. In a more recent study, Kollias et al. [33] explored training end-to-end architectures based on CNNs with a RNN. In their paper, the authors experimented with networks that were based on a ResNet-50 and on VGG-Face/VGG-16 architectures. Their results indicated that VGG-Face produced the best results. Experimenting also with a CNN-RNN architecture and utilizing as RNN either a LSTM or gated recurrent unit (GRU), with 2 layers, they concluded that GRUs produce best results. Another interesting study was proposed by Chen et al. [9], where the authors first restore 3D facial models from 2D images and then use a variant of the random-forest algorithm. The interested reader is referred to [63] for more studies that can be found in the literature exploiting visual information for affect recognition.

More recent work focuses on emotion recognition using multiple modalities. For instance, Kim et al. [30] utilized dynamic Bayesian networks (DBNs) and proposed four architectures. One of them was considered as the basis which is comprised of a two-layer DBN where the features are extracted for each modality (audio and video) separately, before being concatenated in a single vector and passed on to a support vector machine (SVM) classifier. The other three DBNs were variations of the basic one. A more recent work was proposed by Ringeval et al. [40], where the authors used a bi-directional LSTM-RNN to capture the contextual information using multimodal features (video, but also audio, and even physiological data) extracted from the data. In another work, Han et al. [23] proposed a strength modeling framework, which can be used in a feature-level or decision-level fusion strategy. This framework comprises two regression models, where the first model's predictions are concatenated with the original feature vector and fed to the second regression model for the final prediction.

18.4.3 Databases

Databases are vital to train and evaluate models for affect recognition. The methods of collecting affect data can be categorized as follows [61]: (i) posed, (ii) induced, and (iii) spontaneous. The posed method indicates that an actor was used to express different emotions. In the induced class, a subject's emotions are triggered by, for example, she/he is watching TV or listening to music or imagining a story, etc. Finally, the spontaneous databases are collected when a spontaneous emotion is expressed by a subject. Due to the fact that the discussion in this chapter is focused on real-world continuous emotion recognition, we present well-known databases focusing on in-the-wild conditions and/or continuous affect prediction for audiovisual purposes. For further study the interested reader is forwarded to [61].

Numerous databases have been proposed in the literature for continuous emotion recognition. One of the first databases for continuous affect recognition was the Humaine database [15], which includes 50 recordings with multiple modalities, namely speech, language, gesture, face and physiological. Another database is the Sustained Emotionally colored Machine–human Interaction using Nonverbal Expression Dataset (SEMAINE) [36], which contains 959 conversations from 150 participants where each participant is interacting with an automatic agent. The modalities it supports are audiovisual. The Belfast induced natural emotion database [51] also includes audiovisual clips for various tasks. However, few of them are for continuous emotion recognition. Another database is the MAHNOB mimicry dataset [4], which was the first natural human-to-human interaction and consists of 54 audiovisual recordings (11 h) from 12 confederates and 48 counterparts.

A more recent broadly used dataset is the REmote COLlaborative and Affective (RECOLA) database introduced by Ringeval et al. [45] for continuous affect recognition. The number of modalities that are included are four, namely, audio, video, electro-cardiogram (ECG), and electro-dermal activity (EDA). The emotions investigated in this database are spontaneous and natural. For the audio and visual modalities, continuous emotions (arousal and valence) are provided. In total, there were 46 French-speaking participants, from which 17 were French, three German, and three Italian. Overall, 9.5 h were recorded and the recordings were annotated for 5 min each by six French-speaking annotators (three male, three female). The dataset is split into three partitions—train (16 subjects), validation (15 subjects) and test (15 subjects)—by stratifying (i.e., balancing) the gender and the age of the speakers. This database was used in the Audio/Visual Emotion Challenges (AVEC) in the years 2015 [42] and 2016 [56].

A very recent 'in-the-wild' database was further introduced by Zafeiriou et al. [63]. This database consists of more than 500 YouTube videos which show subjects conveying spontaneous facial behavior affect while they are watching video clips, performing an activity or being surprised.

As an example that focuses on multiple cultures and dyadic conversations, the Sentiment Analysis in the Wild (SEWA)[1] database consists of audiovi-

[1] https://sewaproject.eu/.

sual recordings that were collected from web-cameras and microphones and captured spontaneous and naturalistic behavior, i.e., under real-world conditions. It contains 32 subject pairs with age ranging from 18 to 60 years. The participants watched in pairs a 90 s commercial video and the subsequent task was to discuss the content with their partner for maximum 3 min. Six cultures/backgrounds across Europe and Chinese are contained. Conversation was run over IP. Part of the data was used in the 2017 AVEC competition.

As a final example, different versions of the Acted Facial Expressions in the Wild (AFEW) [14] dataset have been used in all the Emotion Recognition in the Wild (EmotiW) challenges. The dataset is a facial-expression data corpus. In a late edition, it is comprised of 1426 video clips extracted from movies and annotated into six emotions plus a neutral: anger, disgust, fear, sadness, happiness, surprise, and neutral. Static frames were extracted from the video clips of AFEW, and the Static Facial Expressions in the Wild (SFEW) was created.

Many further sets could be mentioned as, thankfully, the amount of available data has started to grow recently.

18.4.4 Affect recognition competitions

The importance of recognizing emotions motivated the creation of a number of affect recognition competitions, such as the first ever Interspeech 2009 Emotion Challenge which was run for speech only. It featured classes, and was followed by the first ever dimensional Interspeech 2010 Paralinguistics Challenge, which had a continuous interest recognition task.[2] These were followed by another first of its kind Audio/Visual Emotion Challenge and Workshop (AVEC), which was held in 2011 [50]. Its goal is to compare the audio, visual, and later also physiological modalities to recognize continuous emotions, as well as to estimate the severity of depression. To give just but a bit of insight into participants' contributions, in one of the submitted models for the 2016 challenge [56], Huang et al. [26] proposed to use variants of a Relevance Vector Machine (RVM) for modeling audio, video, and audiovisual data. In another work, a model by Weber et al. [59] used high-level geometry features for predicting dimensional features. Brady et al. [5] also used low- and high-level features for modeling emotions. In a different study, Povolny et al. [39] complemented original baseline features that had been given by the organizers for both audio and video to perform emotion recognition. Somandepalli et al. [52] also used additional features, but only

[2] Cf. www.compare.openaudio.eu.

for the audio modality. As stated, the latest AVEC 2017 challenge [43] used the SEWA database in order to compare models in real-life conditions for dimensional affect recognition.

Another challenge that followed the example is the Emotion Recognition in the Wild (EmotiW) [12] challenge. The first challenge was organized in 2013 [13] and its main purpose is to challenge the community on 'in-the-wild' data for emotion recognition. In 2016, two sub-challenges were proposed [12]. The first used audiovisual data for emotion recognition (VReco), and the second consisted of images with groups of people and the task was to infer the emotional state of the whole group (GReco). The winner of the GReco challenge used a CNN to extract useful information from the faces in the image and then a RNN to combine the information from the different extracted features for group emotion prediction. The winner of the VReco subchallenge used: (i) a CNN with a RNN on top, and (ii) a 3D CNN to extract spatiotemporal features from the visual information of the video. These networks were trained separately with a linear classifier. In addition, one used the openSMILE [19] toolkit which has served as baseline toolkit across AVEC and EmotiW (and further challenges) to extract features from the audio and train a SVM. In the final step, one combined their predictions in a score-level fusion to obtain the final prediction.

A very recent competition for facial affect/behavior analysis is the Affect-in-the-Wild (Aff-W) [62]. The database [63] was obtained mostly from the YouTube video web-site and consists of 298 videos. Overall there are 200 subjects (130 males and 70 females), who show affect when performing an activity or when they watch a movie or trailer. Another challenge that uses facial expressions for emotion recognition is the Facial-Expression Recognition (FERA) challenge [57]. Two subchallenges were defined: the detection of AU occurrence, and the estimation of AU intensity.

Finally, the Multimodal Emotion Challenge (MEC) [35] series is held for the second time at the time of writing. The organizers chose the Chinese Natural Audio-Visual Emotion Database (CHEAVD) [3], which was obtained from Chinese movies and TV programs.

18.5 AUDIOVISUAL AFFECT RECOGNITION: A REPRESENTATIVE END-TO-END LEARNING SYSTEM

In this section, we present a system for continuous affect recognition using audiovisual raw input. This is a very new approach, and allows to learn directly from labeled data without taking care of specific features such as pitch or energy, or alike. Rather, all is learned directly from the data in terms of representation and decision. Obviously, this is an extremely powerful tool, as it allows to add further modalities without the need or only little need of expert knowledge on these—only the design of a suited neural networks layout remains, which is mainly a matter of estimation based on the amounts of data available, complexity of the task, and experience. This is usually a choice of CNN layers, pooling blocks, and final RNN layers with LSTM leaving their number, size, etc. to be decided based on the named. The method introduced in the following extracts features from each modality independently and then combines these in a single feature vector (feature-level fusion). In order to consider the contextual information in the data, LSTM neural networks are utilized.

18.5.1 Proposed model

Our proposed exemplary model to build up an audiovisual emotion recognizer is comprised of (i) a visual network, which is used to extract useful features from the face of the subject, (ii) a speech network, that extracts information from the speech, and a (iii) RNN which captures the temporal dynamics in the signal.

18.5.1.1 Visual network

For the visual modality, we utilize a deep residual network (ResNet) of 50 layers [25]. The input to the network is the pixel intensities from the cropped faces of the subject's video. Deep residual networks adopt residual learning by stacking building blocks of the form

$$\mathbf{y}_k = \mathcal{F}(\mathbf{x}_k, \{W_k\}) + \mathbf{h}(x_k), \tag{18.1}$$

where $\mathbf{x}$ and $\mathbf{y}$ are the input and output of the layer k, $\mathcal{F}(\mathbf{x}_k, \{W_k\})$ is the residual function to be learned, and $h(\mathbf{x}_k)$ can be either an identity mapping or a linear projection to match the dimensions of the function $\mathcal{F}$ and the input $\mathbf{x}$.

The ResNet-50 is comprised of: (i) 1 convolution layer of 64 feature maps with filter size 7×7, (ii) a max-pooling layer of size 3×3, and (iii) 4 bottleneck layers. Each bottleneck layer contains 3 convolutional layers of sizes 1×1, 3×3, and 1×1. Table 18.1 shows the number of replications and the number of feature maps for each convolution layer of the bottleneck layers.

18.5.1.2 Speech network

The input to the speech network is a standardized (i.e., zero mean, unit variance) raw signal of 6 s long, which corresponds to a 96 000-dimensional

Table 18.1 The replication of each bottleneck architecture of the ResNet-50 along with the size of the features maps of the convolutions [55]

Bottleneck layer	Replication	Number of feature maps (1 x 1, 3 x 3, and 1 x 1)
1st	3	64, 64, 256
2nd	4	128, 128, 512
3rd	6	256, 256, 1024
4th	3	512, 512, 2048

vector for a signal with frequency of 16 kHz. The layers of the network are the following:

- **(i) Convolution**: the signal is convolved with a filter of size 80, which corresponds to a window of 5 ms.
- **(ii) Pooling**: the signal is passed to a max-pooling layer of size 2, to reduce its sampling rate to 8 kHz.
- **(iii) Convolution**: a convolution of the signal is performed with a filter of size 4000, which corresponds to a 500 ms window.
- **(iv) Pooling**: a final max-pooling layer that is performed across the channel domain with size of 10.

Dropout regularization is adopted with a probability of 0.5 because of the large number of parameters of the network (≈1.1M).

18.5.1.3 Objective function

Most of the studies in the literature minimize the Mean Squared Error (MSE) for training, and the evaluation is performed on the concordance correlation coefficient (ρ_c) [41,44]. However, we propose to include to the objective function the ρ_c [34] for training of the networks. This comes as in value and time continuous emotion recognition—a highly subjective task as noted; the raters tend to make different use of the scale. Likewise, a compromise between optimal correlation and minimal offset in terms of mean error is desirable. This is different from tasks such as age recognition where an objective ground truth exists, and one merely optimizes for minimal error. Since the objective function is a cost function, we define the loss $\mathcal{L}_c$ to be minimized as follows:

$$\begin{aligned} \mathcal{L}_c = 1 - \rho_c &= 1 - \frac{2\sigma_{xy}^2}{\sigma_x^2 + \sigma_y^2 + (\mu_x - \mu_y)^2} \\ &= 1 - 2\sigma_{xy}^2 \psi^{-1}, \end{aligned} \tag{18.2}$$

where $\psi = \sigma_x^2 + \sigma_y^2 + (\mu_x - \mu_y)^2$ and $\mu_x = \mathbb{E}(\mathbf{x})$, $\mu_y = \mathbb{E}(\mathbf{y})$, $\sigma_x^2 = \text{var}(\mathbf{x})$, $\sigma_y^2 = \text{var}(\mathbf{y})$ and $\sigma_{xy}^2 = \text{cov}(\mathbf{x}, \mathbf{y})$.

The gradient of the weights of the last layer with respect to $\mathcal{L}_c$ that is back-propagated is defined as follows:

$$\frac{\partial \mathcal{L}_c}{\partial \mathbf{x}} \propto 2\frac{\sigma_{xy}^2(\mathbf{x}-\mu_y)}{\psi^2} + \frac{\mu_y - \mathbf{y}}{\psi}, \tag{18.3}$$

where all vector operations are performed element-wise.

18.5.1.4 Network training

To have a good initialization point for both the visual and speech network when training them together, we trained each modality-specific network separately. Then, we used the weights learned to initialize the weights in the multimodal network.

In addition, to train the visual network in unimodal settings (i.e., visual), the initialization of the parameters can be performed using a pre-trained network to cater for optimal initialization such as the ResNet-50 learnt on the ImageNet 2012 [46] classification dataset that is comprised of 1000 classes as used in our example here. In our example, the training of both, the visual and speech networks, is performed with a 2-layer LSTM stack on top of them, which is comprised of 256 cells.

Finally, the multimodal network is initialized with the trained visual and speech networks. In order to capture the contextual information in the data, a 2-layer LSTM RNN is used with 256 cells, which is initialized following initialization based on Glorot [21]. The network is trained in an end-to-end manner. Fig. 18.2 shows the multimodal network.

The unimodal networks and the multimodal one minimize the following objective

$$\mathcal{L}_c = \frac{\mathcal{L}_c^a + \mathcal{L}_c^v}{2}, \tag{18.4}$$

where $\mathcal{L}_c^a$ and $\mathcal{L}_c^v$ are the concordance losses of the arousal and valence, respectively.

For the recurrent layers of the speech, visual, and audiovisual networks, we segment the 6 s sequences to 150 smaller sub-sequences to match the granularity of the annotation frequency of 40 ms.

18.5.2 **Experiments & results**

Let us now see how such an end-to-end learning approach may perform. For our experiments, the RECOLA database is utilized. As mentioned earlier, RECOLA provides both audio and visual modalities for continuous affect

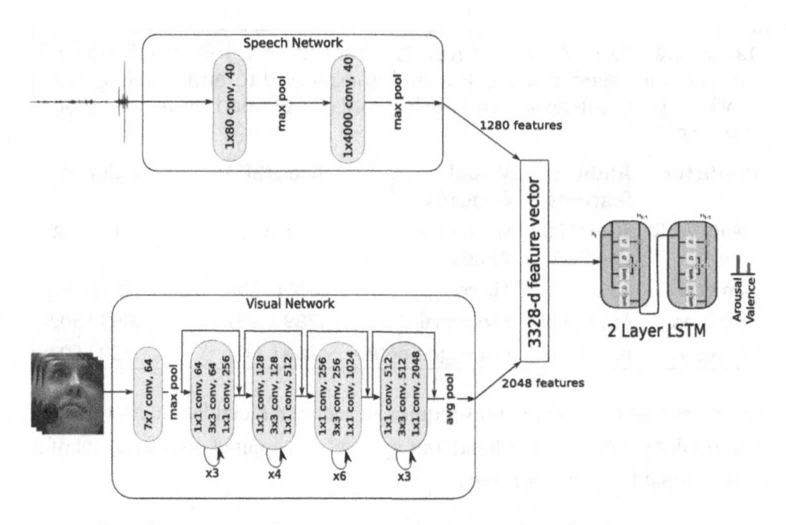

■ **FIGURE 18.2** The network comprises two parts: the multimodal feature extraction part and the RNN part. The multimodal part extracts features from raw speech and visual signals. The extracted features are concatenated and used to feed two recurrent LSTM layers. These are used to capture the contextual information in the data [55].

recognition (arousal and valence) and part of it was used in the AVEC challenge series in 2015 and 2016. For our purposes, we used the full dataset which comprised 46 subjects.

We compare our model with other models in the literature that use the entire RECOLA database. Only two of them were found, namely, the Output-Associative Relevance Vector Machine Staircase Regression (OA RVM-SR) [26], and the strength modeling system proposed by Han et al. [23]. For fair comparison with the other models, that use both training and validation sets in the training process, we utilize the same training data. A major problem occurs with this approach, as for deep learning models, the validation set is used to stop the training of the model, and in our case we use it in the training data. To tackle this, we first train the model using only the training set, and we stop the training according to the results obtained on the validation set. Utilizing also the validation set in the training process, we stop the training of our model when we reach twice the number of epochs required to train the model when only the training set is used. We perform this, as the number of instances in the validation set are almost equal to the number of the training examples, hence, our training data are doubled.

Results are depicted in Table 18.2. Our model outperforms the other two models in both, the arousal and valence dimensions; especially for the valence dimension indeed with a high magnitude. It is important to note that the input to our system are the raw pixels, while the other two systems made use of a number of geometric features (e.g., 2D/3D facial landmarks).

Table 18.2 RECOLA dataset results (in terms of ρ_c) for prediction of arousal and valence using the multimodal end-to-end learning network. In parentheses are the performances obtained on the development set [55]

Predictor	Audio features	Visual features	Arousal	Valence
OA RVM-SR	eGeMAPS ComParE	Geometric Appearance	.770 (.855)	.545 (.642)
Han et al.	13 LLDs	Mixed	.610 (.728)	.463 (.544)
Proposed	raw signal	raw signal	**.789** (.731)	**.691** (.502)
Proposed	raw signal	raw + geometric	**.788** (.731)	**.732** (.502)

The extraction of these features require an accurate facial landmark tracking methodology. On the other hand, our system was applied on the results of a conventional face detector only.

Our model can easily incorporate these features by concatenating them to the input feature vector of the recurrent model. For the extraction of facial landmarks, we use the face alignment method by Deng et al. [11] and apply Procrustes alignment (i.e., removing scale, rotation, and translation). Then, we train only the recurrent model, i.e., fixing the parameters of unimodal networks. Results are depicted in Table 18.2. As expected, our model is benefited with these features in the prediction of the valence dimension while the arousal is unchanged.

18.6 CONCLUSIONS

In this chapter, we first presented audiovisual cues that convey emotions, and we described how emotions can be quantified. For decades, categorical models were mostly used. However, in the recent years, there is a shift in the affective computing community towards continuous emotion recognition so that more complex emotions can be identified and predicted. The most widely used continuous affect model is the two dimensional (arousal and valence) circumplex model of affect.

The second part of the chapter focused on the advances in the affective computing community. A number of databases have been proposed in the literature that focus on real-world conditions for either categorical or continuous affect recognition. However, the in-the-wild conditions pose many challenges for this task. For example, the audiovisual signals can be noisy because of occlusions or recording devices. To further advance the methods for emotion recognition, a number of emotion recognition challenges have been organized, where numerous researches are submitting their models. As a consequence, numerous models have been suggested that use both audio and visual information as input to improve their predictions. We presented such a multimodal model for continuous emotion recognition that can learn end-to-end thus omitting the need for high expert knowledge on the nitty gritty details of a modality. In fact, in our example, it outperformed alternative approaches showing the huge potential of such an approach.

As this chapter shows, we have entered an exciting phase in automatic emotion recognition, as by end-to-end learning more researchers can easily enter the multimodal approach that synergistically combines strengths of individual information sources. This renders more efficient approaches in terms of exploitation of limited training data—an ever present major bottleneck in the field—as a primary concern for such novel architectures besides elegant ways of choosing suited network topologies—ideally in an automatic machine learning fashion. Solutions are, however, already well under way, such as generative adversarial networks for the first problems of efficient data exploitation or evolutionary learning in the case of self-optimizing networks.

REFERENCES

[1] N. Ambady, R. Rosenthal, Half a minute: predicting teacher evaluations from thin slices of nonverbal behavior and physical attractiveness, J. Pers. Soc. Psychol. 64 (1993) 431.
[2] C.N. Anagnostopoulos, T. Iliou, I. Giannoukos, Features and classifiers for emotion recognition from speech: a survey from 2000 to 2011, Artif. Intell. Rev. 43 (2015) 155–177.
[3] W. Bao, Y. Li, M. Gu, M. Yang, H. Li, L. Chao, J. Tao, Building a Chinese natural emotional audio-visual database, in: Proceedings of the International Conference on Signal Processing, 2014, pp. 583–587.
[4] S. Bilakhia, S. Petridis, A. Nijholt, M. Pantic, The MAHNOB mimicry database: a database of naturalistic human interactions, Pattern Recognit. Lett. 66 (2015) 52–61.
[5] K. Brady, Y. Gwon, P. Khorrami, E. Godoy, W. Campbell, C. Dagli, T.S. Huang, Multi-modal audio, video and physiological sensor learning for continuous emotion prediction, in: Proceedings of the International Workshop on Audio/Visual Emotion Challenge, 2016, pp. 97–104.
[6] P. Bull, Posture and Gesture, vol. 16, Pergamon Books, 1987.
[7] F. Burkhardt, J. Ajmera, R. Englert, J. Stegmann, W. Burleson, Detecting anger in automated voice portal dialogs, in: Annual Conference of the International Speech Communication Association, 2006, pp. 1053–1056.
[8] R.A. Calvo, S. D'Mello, Affect detection: an interdisciplinary review of models, methods, and their applications, Trans. Affect. Comput. 1 (2010) 18–37.
[9] H. Chen, J. Li, F. Zhang, Y. Li, H. Wang, 3d model-based continuous emotion recognition, in: Proceedings of the Conference on Computer Vision and Pattern Recognition, 2015, pp. 1836–1845.
[10] Mark Coulson, Attributing emotion to static body postures: recognition accuracy, confusions, and viewpoint dependence, J. Nonverbal Behav. 28 (2004) 117–139.
[11] J. Deng, G. Trigeorgis, Y. Zhou, S. Zafeiriou, Joint multi-view face alignment in the wild, arXiv preprint, arXiv:1708.06023, 2017.
[12] A. Dhall, R. Goecke, J. Joshi, J. Hoey, T. Gedeon, EmotiW 2016: video and group-level emotion recognition challenges, in: Proceedings of the International Conference on Multimodal Interaction, 2016, pp. 427–432.

[13] A. Dhall, R. Goecke, J. Joshi, M. Wagner, T. Gedeon, Emotion recognition in the wild challenge (EmotiW) challenge and workshop summary, in: International Conference on Multimodal Interaction, 2013, pp. 371–372.
[14] A. Dhall, R. Goecke, S. Lucey, T. Gedeon, Collecting large, richly annotated facial-expression databases from movies, J. Multimed. 19 (2012) 34–41.
[15] E. Douglas-Cowie, R. Cowie, I. Sneddon, C. Cox, O. Lowry, M. Mcrorie, J.C. Martin, L. Devillers, S. Abrilian, A. Batliner, N. Amir, K. Karpouzis, The Humaine database: addressing the collection and annotation of naturalistic and induced emotional data, in: Proceedings of the International Conference on Affective Computing and Intelligent Interaction, 2007, pp. 488–500.
[16] P. Ekman, An argument for basic emotions, Cogn. Emot. 6 (1992) 169–200.
[17] P. Ekman, Emotional and conversational nonverbal signals, in: Proceedings of the Language, Knowledge, and Representation, 2004, pp. 39–50.
[18] P. Ekman, W.V. Friesen, Facial Action Coding System: a Technique for the Measurement of Facial Movement, Consulting Psychologists Press, 1978.
[19] F. Eyben, M. Wöllmer, B. Schuller, openSMILE: the Munich versatile and fast open-source audio feature extractor, in: Proceedings of the International Conference on Multimedia, 2010, pp. 1459–1462.
[20] J. Fontaine, K. Scherer, E. Roesch, P. Ellsworth, The world of emotions is not two-dimensional, Psychol. Sci. 18 (2007) 1050–1057.
[21] X. Glorot, Y. Bengio, Understanding the difficulty of training deep feedforward neural networks, in: Proceedings of the International Conference on Artificial Intelligence and Statistics, 2010, pp. 249–256.
[22] D. Grandjean, D. Sander, K.R. Scherer, Conscious emotional experience emerges as a function of multilevel, appraisal-driven response synchronization, Conscious. Cogn. 17 (2008) 484–495.
[23] J. Han, Z. Zhang, N. Cummins, F. Ringeval, B. Schuller, Strength modelling for real-world automatic continuous affect recognition from audiovisual signals, Image Vis. Comput. 65 (2017) 76–86.
[24] J. Han, Z. Zhang, F. Ringeval, B. Schuller, Prediction-based learning for continuous emotion recognition in speech, in: Proceedings of the International Conference on Acoustics, Speech, and Signal Processing, 2017, pp. 5005–5009.
[25] K. He, X. Zhang, S. Ren, J. Sun, Deep residual learning for image recognition, in: Proceedings of the Conference on Computer Vision and Pattern Recognition, 2016, pp. 770–778.
[26] Z. Huang, B. Stasak, T. Dang, K. Wataraka Gamage, P. Le, V. Sethu, J. Epps, Staircase regression in OA RVM, data selection and gender dependency in AVEC 2016, in: Proceedings of the International Workshop on Audio/Visual Emotion Challenge, 2016, pp. 19–26.
[27] G.L. Huttar, Relations between prosodic variables and emotions in normal American English utterances, J. Speech Lang. Hear. Res. 11 (1968) 481–487.
[28] Q. Ji, Z. Zhu, P. Lan, Real-time nonintrusive monitoring and prediction of driver fatigue, Trans. Veh. Technol. 53 (2004) 1052–1068.
[29] J. Jia, S. Zhang, F. Meng, Y. Wang, L. Cai, Emotional audio-visual speech synthesis based on pad, Trans. Audio Speech Lang. Process. 19 (2011) 570–582.
[30] Y. Kim, H. Lee, E.M. Provost, Deep learning for robust feature generation in audiovisual emotion recognition, in: Proceedings of the International Conference on Acoustics, Speech and Signal Processing, 2013, pp. 3687–3691.
[31] M. Kipp, J.C. Martin, Gesture and emotion: can basic gestural form features discriminate emotions?, in: Proceedings of the International Conference on Affective Computing and Intelligent Interaction and Workshops, 2009, pp. 1–8.
[32] A. Kleinsmith, P.R. De Silva, N. Bianchi-Berthouze, Recognizing Emotion from Postures: Cross-Cultural Differences in User Modeling, Springer, 2005, pp. 50–59.

[33] D. Kollias, M.A. Nicolaou, I. Kotsia, G. Zhao, S. Zafeiriou, Recognition of affect in the wild using deep neural networks, in: Proceedings of the Conference on Computer Vision and Pattern Recognition Workshops, 2017, pp. 1972–1979.
[34] I. Lawrence, K. Lin, A concordance correlation coefficient to evaluate reproducibility, Biometrics 45 (1989) 255–268.
[35] Y. Li, J. Tao, B. Schuller, S. Shan, D. Jiang, J. Jia, MEC 2016: the multimodal emotion recognition challenge of CCPR 2016, in: Proceedings of the Chinese Conference on Pattern Recognition, 2016, pp. 667–678.
[36] G. McKeown, M. Valstar, R. Cowie, M. Pantic, M. Schroder, The Semaine database: annotated multimodal records of emotionally colored conversations between a person and a limited agent, Trans. Affect. Comput. 3 (2012) 5–17.
[37] M.J. Montepare, S.B. Goldstein, A. Clausen, The identification of emotions from gait information, J. Nonverbal Behav. 11 (1987) 33–42.
[38] M.A. Nicolaou, H. Gunes, M. Pantic, Output-associative RVM regression for dimensional and continuous emotion prediction, Image Vis. Comput. 30 (2012) 186–196.
[39] F. Povolny, P. Matejka, M. Hradis, A. Popková, L. Otrusina, P. Smrz, I. Wood, C. Robin, L. Lamel, Multimodal emotion recognition for AVEC 2016 challenge, in: Proceedings of the International Workshop on Audio/Visual Emotion Challenge, 2016, pp. 75–82.
[40] F. Ringeval, F. Eyben, E. Kroupi, A. Yuce, J.-P. Thiran, T. Ebrahimi, D. Lalanne, B. Schuller, Prediction of asynchronous dimensional emotion ratings from audiovisual and physiological data, Pattern Recognit. Lett. 66 (2015) 22–30.
[41] F. Ringeval, F. Eyben, E. Kroupi, A. Yuce, J.P. Thiran, T. Ebrahimi, D. Lalanne, B. Schuller, Prediction of asynchronous dimensional emotion ratings from audiovisual and physiological data, Pattern Recognit. Lett. 66 (2015) 22–30.
[42] F. Ringeval, B. Schuller, M. Valstar, R. Cowie, M. Pantic, AVEC 2015: the 5th international audio/visual emotion challenge and workshop, in: Proceedings of the International Workshop on Audio/Visual Emotion Challenge, ACM Multimedia, 2015, pp. 1335–1336.
[43] F. Ringeval, B. Schuller, M. Valstar, J. Gratch, R. Cowie, S. Scherer, S. Mozgai, N. Cummins, M. Schmi, M. Pantic, AVEC 2017—real-life depression, and affect recognition workshop and challenge, in: Proceedings of the International Workshop on Audio/Visual Emotion Challenge, ACM Multimedia, 2017, pp. 3–9.
[44] F. Ringeval, B. Schuller, M. Valstar, S. Jaiswal, E. Marchi, D. Lalanne, R. Cowie, M. Pantic, AV+EC 2015—the first affect recognition challenge bridging across audio, video, and physiological data, in: Proceedings of the International Workshop on Audio/Visual Emotion Challenge, ACM Multimedia, 2015, pp. 3–8.
[45] F. Ringeval, A. Sonderegger, J. Sauer, D. Lalanne, Introducing the RECOLA multimodal corpus of remote collaborative and affective interactions, in: Proceedings of the International Conference and Workshops on Automatic Face and Gesture Recognition, 2013, pp. 1–8.
[46] O. Russakovsky, J. Deng, H. Su, J. Krause, S. Satheesh, S. Ma, Z. Huang, A. Karpathy, A. Khosla, M. Bernstein, A.C. Berg, L. Fei-Fei, ImageNet large scale visual recognition challenge, Int. J. Comput. Vis. 115 (2015) 211–252.
[47] J. Russell, A circular model of affect, J. Pers. Soc. Psychol. 39 (1980) 1161–1178.
[48] Klaus R. Scherer, On the nature and function of emotion: a component process approach, in: Approaches to Emotion, vol. 2293, 1984, p. 317.
[49] M. Schmitt, F. Ringeval, B. Schuller, At the border of acoustics and linguistics: bag-of-audio-words for the recognition of emotions in speech, in: Proceedings of the Annual Conference of the International Speech Communication Association, 2016, pp. 495–499.
[50] B. Schuller, M. Valstar, F. Eyben, G. McKeown, R. Cowie, M. Pantic, AVEC 2011—the first international audio/visual emotion challenge, in: Affective Computing and Intelligent Interaction, vol. 6975, 2011, pp. 415–424.

[51] I. Sneddon, M. McRorie, G. McKeown, J. Hanratty, The Belfast induced natural emotion database, Trans. Affect. Comput. 3 (2012) 32–41.
[52] K. Somandepalli, E. Gupta, M. Nasir, B.M. Booth, S. Lee, S. Narayanan, Online affect tracking with multimodal Kalman filters, in: Proceedings of the International Workshop on Audio/Visual Emotion Challenge, 2016, pp. 59–66.
[53] A. Stuhlsatz, C. Meyer, F. Eyben, T. Zielke, G. Meier, B. Schuller, Deep neural networks for acoustic emotion recognition: raising the benchmarks, in: Proceedings of the International Conference on Acoustics, Speech and Signal Processing, 2011, pp. 5688–5691.
[54] G. Trigeorgis, F. Ringeval, R. Brueckner, E. Marchi, M.A. Nicolaou, B. Schuller, S. Zafeiriou, Adieu features? End-to-end speech emotion recognition using a deep convolutional recurrent network, in: Proceedings of the International Conference on Acoustics, Speech and Signal Processing, 2016, pp. 5200–5204.
[55] P. Tzirakis, G. Trigeorgis, M.A. Nicolaou, B. Schuller, S. Zafeiriou, End-to-end multimodal emotion recognition using deep neural networks, IEEE J. Sel. Top. Signal Process. 11 (8) (2017) 1301–1309.
[56] M. Valstar, J. Gratch, B. Schuller, F. Ringeval, D. Lalanne, M. Torres Torres, S. Scherer, G. Stratou, R. Cowie, M. Pantic, AVEC 2016: depression, mood, and emotion recognition workshop and challenge, in: Proceedings of the International Workshop on Audio/Visual Emotion Challenge, ACM Multimedia, 2016, pp. 3–10.
[57] M. Valstar, E. Sánchez-Lozano, J. Cohn, L. Jeni, J. Girard, Z. Zhang, L. Yin, M. Pantic, FERA 2017-addressing head pose in the third facial expression recognition and analysis challenge, in: Proceedings of the International Conference on Automatic Face Gesture Recognition, 2017, pp. 839–847.
[58] Y. Wang, C.v.d. Weth, Y. Zhang, K.H. Low, V.K. Singh, M. Kankanhalli, Concept based hybrid fusion of multimodal event signals, in: International Symposium on Multimedia, Proceedings, 2016, pp. 14–19.
[59] R. Weber, V. Barrielle, C. Soladié, R. Séguier, High-level geometry-based features of video modality for emotion prediction, in: Proceedings of the International Workshop on Audio/Visual Emotion Challenge, 2016, pp. 51–58.
[60] M. Wöllmer, F. Eyben, S. Reiter, B. Schuller, C. Cox, E. Douglas-Cowie, R. Cowie, Abandoning emotion classes-towards continuous emotion recognition with modelling of long-range dependencies, in: Proceedings of the Annual Conference of the International Speech Communication Association, 2008, pp. 597–600.
[61] C.H. Wu, J.C. Lin, W.L. Wei, Survey on audiovisual emotion recognition: databases, features, and data fusion strategies, Trans. Signal Inform. Process. 3 (2014) 572–587.
[62] S. Zafeiriou, D. Kollias, M.A. Nicolaou, A. Papaioannou, G. Zhao, I. Kotsia, Aff-wild: valence and arousal in-the-wild challenge, in: Proceedings of the Conference on Computer Vision and Pattern Recognition Workshops, 2017, pp. 1980–1987.
[63] S. Zafeiriou, A. Papaioannou, I. Kotsia, M. Nicolaou, G. Zhao, Facial affect "in-the-wild": a survey and a new database, in: Proceedings of the Conference on Computer Vision and Pattern Recognition Workshops, 2016, pp. 36–47.

Chapter 19

Affective facial computing: Generalizability across domains

Jeffrey F. Cohn*, Itir Onal Ertugrul[†], Wen-Sheng Chu[‡], Jeffrey M. Girard[§], László A. Jeni[†], Zakia Hammal[†]

*Department of Psychology, University of Pittsburgh, Pittsburgh, PA, USA [†]Robotics Institute, Carnegie Mellon University, Pittsburgh, PA, USA [‡]Carnegie Mellon University, Pittsburgh, PA, USA [§]Language Technology Institute, Carnegie Mellon University, Pittsburgh, PA, USA

CONTENTS

19.1 INTRODUCTION

Affective facial computing (AFC) refers to the use of computer-based techniques to automatically analyze and synthesize facial behavior that may be related to emotion and other cognitive-affective states. Fields where AFC can be fruitfully applied include (but are not limited to) psychology, medicine, security, marketing, and consumer electronics. In this chapter, we focus on applications of AFC that analyze (rather than synthesize) facial behavior.

Multimodal Behavior Analysis in the Wild. https://doi.org/10.1016/B978-0-12-814601-9.00026-2
Copyright © 2019 Elsevier Ltd. All rights reserved.

AFC originated in part through a seminal convergence of computer vision, machine learning, and behavioral science in the early 1990s [1]. The field remains highly interdisciplinary and is central to affective computing and computational behavioral science. Initial work in the field focused on distinguishing between posed (i.e., deliberate) facial behaviors. With time, AFC progressed to the challenge of analyzing spontaneous facial behavior in more real-world settings. It has been applied to further our understanding of emotion, physical pain, aging, gender, culture, neuromuscular and craniofacial impairment, deception, depression, and psychological distress [2–11].

The utility of an AFC system designed to analyze facial behavior depends on its ability to robustly and reliably detect specific behaviors and/or their underlying cognitive-affective states. This ability is typically evaluated by comparing the system's predictions to trusted labels collected from human annotators. As will be discussed below, various methods exist for quantifying an AFC system's predictive performance. They all attempt to capture the degree to which the predictions agree with (i.e., they lead to the same inferences as) the trusted labels.

To the extent that a system demonstrates high performance when evaluated, there is evidence that it is capable of correctly detecting the behaviors and/or states of interest. However, such systems may struggle to *generalize*, i.e., to maintain the same performance when applied in different settings. Thus, a system that is capable of nearly perfect performance in ideal conditions may drop to near chance performance in more challenging ones. Put another way, a system may have "operating parameters" or "domains of application" within which it performs well but beyond which it struggles. For anyone interested in applying such a system, the important question is not "How well can this system perform under ideal conditions?" but rather "How well can this system perform in the domain we want to apply it in?"

There are several potential approaches to answering this important question. One powerful, but rarely used, approach is to run experiments and estimate statistical models to explore the influence of changes in operating parameters on a system's predictive performance. For example, Girard et al. examined how performance was influenced by changes in illumination and head pose, as well as by participants' gender and skin color [12]. Another approach, which is more commonly used, is to evaluate a system's predictive performance in one or more databases that are different from the one it was trained on. This approach yields estimates of cross-database performance, which can be compared to estimates of within-database performance. High cross-database performance suggests that a system can successfully generalize to the unseen databases and inspires confidence that it may also generalize to novel domains that are similar to these databases.

This chapter evaluates the state-of-the-art in the *generalizability of AFC systems across domains* and also reviews approaches that are designed to improve such generalizability. We address two questions that we regard as critical to the field: (1) To what extent do AFC systems generalize to new data sources, or domains? (2) What approaches show the most promise in improving generalizability?

We begin with a brief description of approaches to AFC and annotation schemes used for training and testing AFC. We next consider psychometric issues critical to training, testing, and comparing results from different studies. We then review 39 studies that have reported results for cross-database performance and consider promising approaches for improving the generalizability.

19.2 OVERVIEW OF AFC

AFC systems almost always employ the supervised learning approach, wherein a predictive algorithm is developed using a subset of data called the "training set" for which both features (i.e., quantitative descriptions of the data) and annotations (i.e., trusted labels of the behaviors or states to be predicted) are available. The algorithm uses mathematical techniques to learn a high-dimensional mapping between the features and annotations; this mapping can then be used to generate predicted annotations for novel data with the same type of features. The algorithm's performance is often optimized using an independent set of data called the "validation set" and its ultimate performance is evaluated in another independent dataset called the "testing set."

Most approaches to supervised learning in AFC conform to one of two types. *Shallow learning* extracts "hand-crafted" features which have been designed (in a top-down or theory-driven sense); these features are registered to a canonical orientation to account for differences in head pose/orientation [13–15] and input to a separate predictive algorithm. An alternative approach is *deep learning* [16], in which features are learned (in a bottom-up or data-driven sense) and classified in a hierarchical series of steps. Because deep learning scales to larger amounts of data and is data driven, it has been hypothesized to afford greater generalizability than shallow learning [17]. This hypothesis is evaluated below.

A prototypical example of a shallow-learning pipeline is depicted in Fig. 19.1 (originally printed in [18]). The left side of this figure shows the training process. With deep learning, feature extraction and classification would be combined in a hierarchical series of steps. The right side of the figure depicts the process of generalizing the trained AFC algorithm to novel

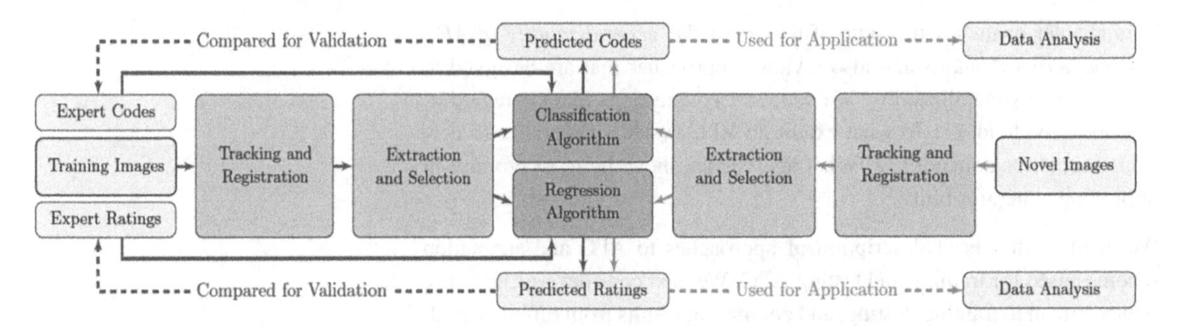

FIGURE 19.1 Standard shallow-learning pipeline for training, validating, and applying AFC. From [18]. In deep learning, feature selection, or representation, and classification are optimized in a hierarchical series of steps, which is one reason they are referred as "deep learning."

data. Note that this generalization process is fully automatic and does not require annotations to be collected (although such annotations are necessary for evaluating the algorithm's cross-domain performance). Trained algorithms can be purchased or downloaded from a variety of commercial and research sources (e.g., Emotient https://imotions.com/emotient/ and OpenFace https://github.com/TadasBaltrusaitis/OpenFace). Given the high cost and difficulty of developing such an algorithm, many users are interested in applying these algorithms to analyze their own data. However, as mentioned above, there is no guarantee that algorithms will successfully generalize and yield valid predictions in such settings. Generalizability must be evaluated empirically.

19.3 APPROACHES TO ANNOTATION

As noted above, a major use of AFC is to analyze visual data (i.e., images and videos) for behaviors and cognitive-affective states. This analysis takes the form of generating predicted labels for each image or video frame; these labels are also called annotations, codes, or ratings. Although these terms have different connotations in some fields [19], we use them in this chapter interchangeably. Two primary types of annotations have become popular in AFC: message-based and sign-based.

Message-based annotation. In *message-based* annotation [20], the goal is inferential. That is, the label assigned to an image or video is meant to capture the holistic meaning of the behavior (i.e., the message being expressed or communicated by the behavior). Thus, message-based annotation assumes that behaviors have a particular meaning and that this meaning is inferable. An important distinction can be made between message-based approaches that make inferences in terms of discrete categories and those that do so in terms of continuous dimensions.

The categorical approach emphasizes differences between affective states and makes inferences about behaviors using one or more discrete categories, which typically correspond to "basic" emotions (e.g., anger, disgust, fear, joy, sadness, or surprise) or other cognitive-affective states (e.g., concentration, depression, or physical pain). Categorical approaches may focus on the occurrence (i.e., presence versus absence) of a category or on its intensity (e.g., slight, moderate, intense), but they always use specific categories that are separate from one another. Among the various category sets used in AFC, Ekman's [21] basic emotion set has been used most often.

The dimensional approach emphasizes similarities between affective states and makes inferences about behaviors using one or more continuous dimensions. Different dimensions can be used to capture the message of a behavior, but they typically derive from factor analytic or multidimensional scaling models of emotion [22–24]. The most commonly used dimensions are valence (i.e., pleasant-versus-unpleasant) and arousal (i.e., active-versus-inactive), which come from Russell's [23] circumplex model of emotion. Dimensional measurement in AFC is reviewed by Pantic (this volume).

Sign-based annotation. In *sign-based* annotation, the goal is purely descriptive. That is, the label assigned to an image or video is meant to precisely describe the behaviors present in it without attempting to infer what they might mean. This approach is meant to be more objective than message-based annotation and tends to emphasize small units of behavior (i.e., signs) such as individual motions or actions. The Facial Action Coding System (FACS) [25] is the most comprehensive sign-based approach for facial behavior. By combining its more than 30 anatomically based action units (AUs), FACS can describe nearly all possible facial behaviors [25,26]. Each AU corresponds to the contraction of specific facial muscle groups (e.g., the tightening of the lower eye lids) and, like message-based approaches, may be annotated for occurrence (presence versus absence) or intensity (on an ordinal scale). For an interactive introduction to FACS, see http://www.pitt.edu/~jeffcohn/FACSmodule.html.

19.4 RELIABILITY AND PERFORMANCE

Reliability of measurement. Reliability in the context of AFC refers to the extent to which labels from different sources (but of the same images or videos) are consistent. A useful distinction can be made between inter-observer reliability and inter-system reliability.

Inter-observer reliability refers to the extent to which labels assigned by different human annotators are consistent with one another. For example, inter-observer reliability is high if the annotators tended to assign images or videos the same labels (e.g., AUs). Note that, for inter-observer reliability, the "true" label of the image or video is often not knowable, so we are primarily interested in how much the annotators agreed with one another. Because such labels are used to train and evaluate supervised learning systems, inter-observer reliability matters.

Inter-system reliability refers to the extent to which labels assigned by AFC systems are consistent with labels assigned by human annotators. Inter-system reliability is also called "criterion validity" as the human labels are taken to be the gold standard or criterion measure. Inter-system reliability is the primary measure for the performance of an AFC system.

Inter-observer reliability of training data likely serves as an upper bound for what inter-system reliability is possible, and inter-observer reliability often exceeds inter-system reliability by a considerable margin [27–30]. To fulfill the goal of creating an AFC system that is interchangeable with (or perhaps even more accurate and consistent than) a trained human annotator, both forms of reliability must be maximized. Furthermore, the generalizability of the system (i.e., its inter-system reliability in novel domains) must be maximized.

Level of measurement. AFC systems typically analyze behaviors in single images or video frames, and reliability is calculated on this level of measurement. In the studies reviewed below, frame-level performance is almost always the focus. However, other levels of measurement are also possible and evaluating reliability on these levels may be appropriate for certain tasks or applications. For instance, behavioral *events*, which span multiple contiguous frames, may be the focus when the dynamic unfolding of behavior is of interest [31–34]. Alternatively, when behavioral tendencies over longer periods of time are of interest, a more molar approach that aggregates many behaviors (e.g., into counts, means, or proportions) may be appropriate [35]. Note that reliability may differ between levels of measurement. Aggregated annotations are often more reliable than frame-level annotations [27], but they are also less detailed. It is important to match the analyzed level of measurement to the particular use-case of the system.

Metrics for quantifying reliability. Many metrics have been proposed for estimating reliability. When categorical labels are used, percentage agreement or accuracy (i.e., the proportion of objects that were assigned the same label) is an intuitive and popular option. However, accuracy is a poor choice when the categories are highly imbalanced, such as when a facial behavior has a very high (or very low) occurrence rate and the algorithm is trying to predict when the behavior did and did not occur. In such cases, a naïve algorithm that simply guessed that every image or video contained (or did not contain) the behavior would have a high accuracy.

To attempt to resolve this issue, a number of alternative metrics have been developed including the F-score, receiver operator characteristic (ROC) curve analyses, and various chance-adjusted agreement measures. In this paper, we focus on the three most popular metrics: accuracy, the $F1$ score, and 2AFC. Accuracy, as stated earlier, is the percentage of agreement. The $F1$ score or balanced F-score is the harmonic mean of precision and recall. Finally, 2AFC is resampling-based estimate of the area under the receiver operating characteristic (ROC) curve.

When dimensional labels are used, correlation coefficients (i.e., standardized covariances) are popular options [36]. The Pearson correlation coefficient (PCC) is a linearity index that quantifies how well two vectors can be equated using a linear transformation (i.e., with the addition of a constant and scalar multiplication). The consistency intra-class correlation coefficients (also known as ICC-C) are additivity indices that quantify how well two vectors can be equated with only the addition of a constant. Finally, the agreement intra-class correlation coefficients (also known as ICC-A) are identity indices that quantify how well two vectors can be equated without transformation. The choice of correlation type should depend on how measurements are obtained and how they will be used.

The existence and use of so many different metrics makes comparison between studies and approaches quite difficult. Different metrics are not similarly interpretable and may behave differently in response to imbalanced categories (Fig. 19.2) [37]. As such, we compare performance scores within metrics but never across them, and we acknowledge that differences in occurrence rates between studies may unavoidably confound some comparisons.

19.5 FACTORS INFLUENCING PERFORMANCE

Differences in illumination, cameras, and orientation of the face are well-known sources of variability in computer vision but rarely are considered as sources of error in AFC. They may affect both within- and between-domain performance. The face looks different under different illuminations; resolution and motion vary with camera characteristics; and the face looks different from different angles and when posed in different ways. 2D alignment, which is typical in AFC systems, treats the face as a 2D object, flat like a sheet of paper. That works well provided images are frontal or nearly so and pitch and yaw remain modest [12]. In real-world conditions, these constraints often are violated. To increase the robustness of AFC systems to pose variation, 3D alignment and data augmentation have been proposed

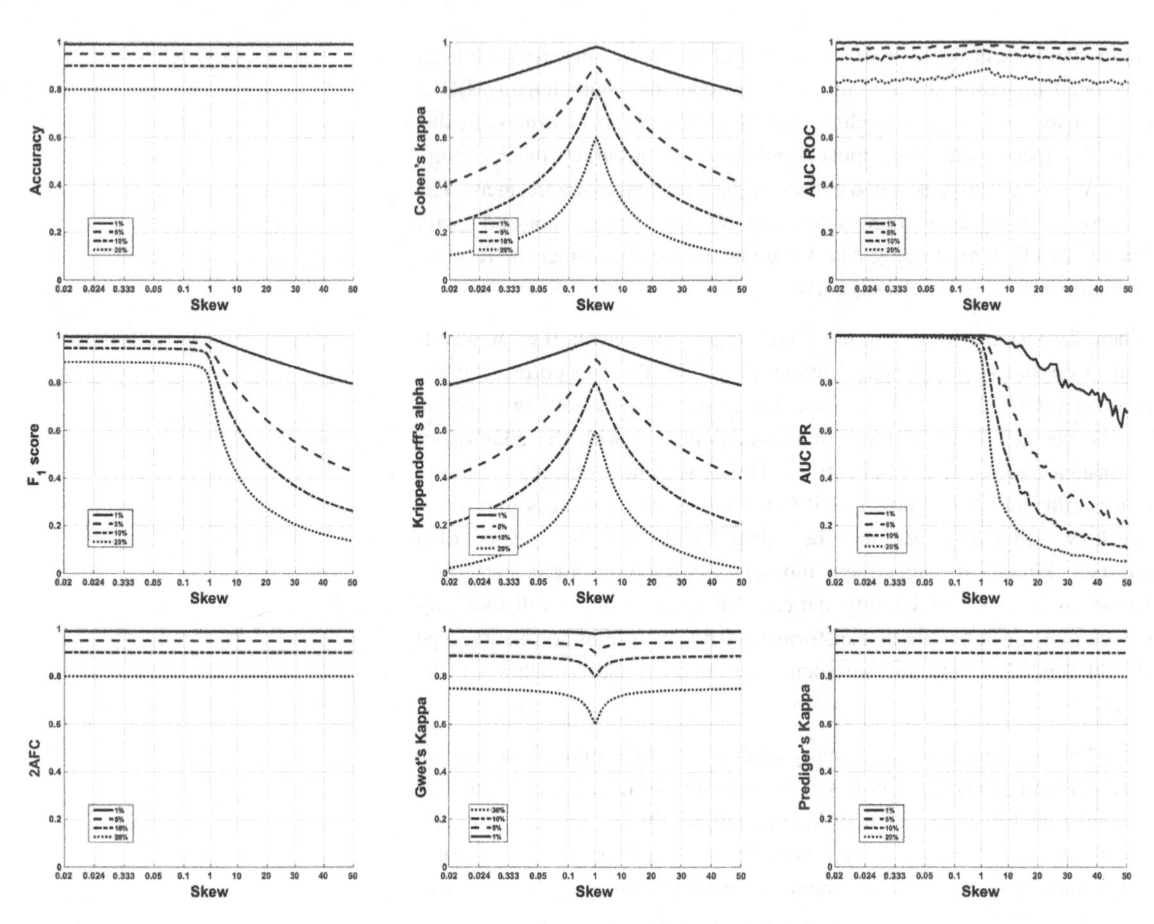

■ **FIGURE 19.2** The behavior of different metrics using simulated classifiers. The horizontal axis depicts the skew ratio while the vertical axis shows the given metric score. The different lines show the relative misclassification rates of the simulated classifiers. Adapted from [37].

(Fig. 19.3). These approaches estimate the depth structure of the face from a variety of sensor data (e.g., 2D images, motion capture, or structured light). The estimated 3D information makes it possible to extend the tracking range and to enforce consistency of landmark locations across viewpoints [38]. The combination of the 3D shape and the 2D appearance of the face is well suited to synthesize an almost arbitrary set of orientations and illumination conditions [13,39], which can enrich training data to cover previously unseen conditions [13].

The quality and diversity of the training data influence predictive performance. As stated earlier, predictive performance is likely limited by the inter-observer reliability of the training data. Girard et al. [40] found that performance was related to the diversity of the training data, as quantified by the number of unique participants included in the training set. The amount of examples per participant, however, mattered less. In their experiment, as

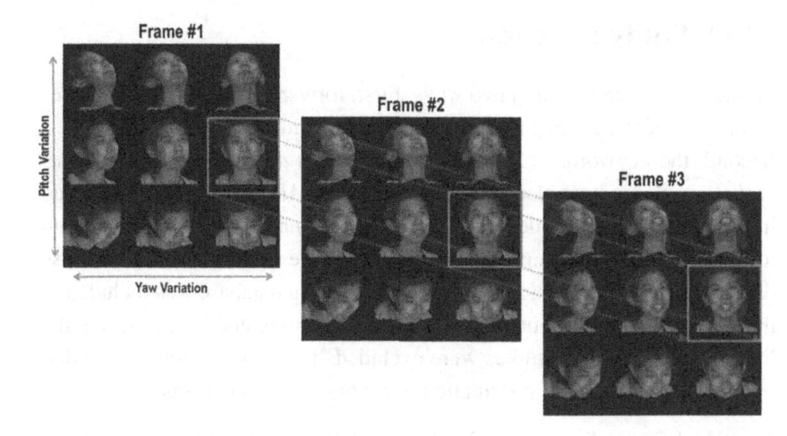

■ **FIGURE 19.3** Using 3D data, for each of three video frames, images of three different pitch and yaw angles are synthesized, resulting in 9 different poses, including frontal. The pitch angles are 40, 0, and −40 degrees. The yaw angles are 40, 20, and 0 degrees (as faces are vertically symmetric). From [13].

little as 15 s of annotated video per participant could suffice (as long as many participant were included).

Individual differences in people (e.g., gender, age, and skin color) can also influence performance [12,41]. There are structural differences between male and female faces and differences in texture. Cosmetic use is also more common in women. In infants, relative to adults, faces are rounder and smoother, their eyebrows are fainter or absent altogether, they have fat pads in their cheeks, and they show movements that would be unusual for adults (e.g., "munchies") [42]. Similarly, as adults age, their faces become narrower, more wrinkled, and lose muscle tone. Recent work has begun to examine the robustness of AFC to the effects of aging [3]. Differences in skin color can be another challenge for predictive performance. Recent work in face recognition suggests that the accuracy for dark-skinned faces is much less than that for lighter-skinned faces [43].

19.6 SYSTEMATIC REVIEW OF STUDIES OF CROSS-DOMAIN GENERALIZABILITY

To evaluate cross-domain generalizability, we identified published studies in which AFC systems were trained and tested across different domains (i.e., databases). Because most studies focused on either message-based annotation of holistic expressions or sign-based (FACS AU) annotation of facial actions, we limited our scope to the detection of holistic expressions, the detection of AU occurrence, and the detection of AU intensity.

19.6.1 Study selection

Studies were ascertained in two ways. First, forward and backward searches from state-of-the-art papers were performed to identify candidate studies. Second, the keywords *cross-database* and *cross-domain* together with *facial expression recognition, AU occurrence*, and *AU intensity* were searched in Google Scholar. Studies were included for analysis if they trained and tested algorithms on independent databases. Some studies were found to use data augmentation in which data from the training database was included in the test database. To ensure independence of training and testing sets in the literature review, such studies were excluded. They are included in the discussion that follows the systematic review of cross-domain studies.

A total of 39 studies using 22 different databases met inclusion criteria. The databases included both posed and spontaneous facial behavior. Among the 39 qualifying studies, 16 performed holistic facial expression recognition, 19 performed AU occurrence estimation, three performed AU intensity estimation, and one performed both AU occurrence and AU intensity estimation.

19.6.2 Databases

Three types of databases were used in studies of cross-domain generalizability: *Posed*, in which facial behavior (e.g., holistic expressions or AUs) were deliberately performed; *spontaneous*, in which facial behavior occurred in response to emotion inductions or social interaction; and *mixed*, which included both posed and spontaneous facial behavior. Derived databases in which face images have been synthesized from existing sources were not included. Although synthesized data are beginning to appear in the AFC literature [13], they have not been used yet in cross-domain studies and may differ from that found in the wild.

Table 19.1 describes the databases in terms of subject characteristics, annotation characteristics, and the number of available images or video frames. Note that not all relevant information was reported for all databases. When a subset of a database was analyzed, a range of frames is noted.

Posed. Because posed expressions tend to be more extreme than ones that occur spontaneously, we anticipated that recognition of holistic expressions would be less challenging for posed than those for spontaneous facial behavior.

Spontaneous. These studies sample a variety of emotion inductions. With the exception of DISFA, all occur in an interpersonal context and thus involve some degree of social interaction. While interpersonal context is

Table 19.1 Databases used in cross-domain studies

	Year	N	Age	% not European–American	% Female	Sequences	Frames (used)	Holistic expressions	AU occur	AU intensity	Reliability
Posed											
Bosphorus	2008	105	25 to 35	NA	43%	–	4666	Y	Y	Y	NA
BU-3DFE	2006	100	18 to 70	NA	44%	NA	(2500)	Y	N	N	NA
CK	2000	97	18 to 50	19%	69%	486	(280 to 8000)	Y	Y	N	$k = 0.82$
G-FERA	2010	10	NA	NA	50%	158	(7000)	Y	Y	N	$PCC = 0.87$
ISL	NA	10	NA	NA	NA	42	NA	N	Y	N	NA
Jaffe	1997	10	NA	100%	100%	NA	213	Y	N	N	NA
KDEF	1998	70	NA	0%	50%	NA	490	Y	N	N	$PCC = 0.72$
MPIE	2010	337	$\mu = 27.9$	40%	30%	NA	750,000 (200,000)	Y	N	N	NA
PICS	2008	23	NA	NA	NA	NA	552	Y	N	N	NA
RaFD	2010	67	NA	0%	37%	NA	8040	Y	N	N	$k = 0.82$
SFEW	2011	95	NA	NA	NA	NA	700 (663)	Y	Y	N	NA
Mixed											
CK+	2010	123	18 to 50	81%	69%	593	(309 to 1236)	Y	Y	N	$k = 0.75$
FFD07	2007	NA	NA	NA	NA	NA	NA	N	Y	N	NA
MMI	2002	75	19 to 62	NA	50%	2894	(384 to 11,500)	Y	Y	N	NA

(continued on next page)

Table 19.1 *(continued)*

	Year	N	Age	% not European–American	% Female	Sequences	Frames (used)	Holistic expressions	AU occur	AU intensity	Reliability
Spontaneous											
BP4D	2013	41	18 to 29	52%	44%	328	368,036 (98,400 to 150,000)	N	Y	Y	$k = 0.68$ to 0.98
DISFA	2013	27	18 to 50	28%	56%	NA	130,000 (4845 to 130,000)	N	Y	Y	$ICC = 0.80$ to 0.94
FEED	2006	18	23 to 38	NA	NA	54	NA	Y	N	N	NA
GFT	2017	96	21 to 28	85%	42%	NA	172,800	N	Y	Y	$S = 0.65$ to 1.00
RU-FACS	2006	100	18 to 30	NA	38%	NA	(5000 to 8000)	Y	Y	N	NA
SAL	2008	10	NA	NA	NA	77	NA	N	Y	N	NA
SEMAINE	2012	150	22 to 60	NA	38%	NA	(90,000)	Y	Y	N	NA
FER2013	2013	NA	NA	NA	NA	NA	35,887 (28,907 to 35,887)	Y	N	N	NA
UNBC Pain Archive	2011	25	NA	NA	51%	200	49,398 (9032 to 49,398)	N	Y	Y	$PCC = 0.80$

Note. Posed: Bosphorus [96,97], BU-3DFE [98], Cohn–Kanade (CK) [99], GEMEP [100], ISL [101], JAFFE [102], KDEF [103], Multi-PIE [104], PICS [105], RaFD [106], and SFEW [107]. Spontaneous: BP4D [108], DISFA [109], FEED [110], RU-FACS [111,112], SAL [113], Sayette GFT [27], SEMAINE [114], and UNBC Pain Archive [9]. Mixed: Extended Cohn–Kanade (CK+) [46], MMI [47, 48], and FFD07 [49]. ***k****: Cohen's kappa [115,116].* ***S*** *is free-marginal kappa [117,118].* ***PCC*** *is Pearson correlation.* ***ICC*** *is intraclass correlation. NA: Not reported in the source publication.*

known to increase the intensity of facial actions [44], the intensity of facial actions that was observed tended to be low to moderate [9,27,45]. Spontaneous actions or expressions tend to be more difficult to detect than posed ones. Spontaneous facial actions also are more likely to have more variable face orientation than those that are posed. Multi-PIE and BU-3DFE are exceptions, as these databases systematically sampled variations of orientation.

Mixed. Mixed databases included in this review were Extended Cohn–Kanade (CK+) [46], MMI [47,48], and FFD07 [49]. Although CK+ includes spontaneous facial behavior, its principal advantage may be that the emotion labels are better validated. MMI includes an increased number of spontaneous actions, although it is not always reported to what extent they are included in studies. FFD07 is a combination of Ekman–Hager, CK, MMI, RU-FACS and two non-public datasets collected by the United States government that are similar in nature to RU-FACS.

The databases have various strengths and limitations. With exception of the posed databases Bosphorus, BU-3DFE, and Multi-PIE, face orientations are frontal or within about 15 degrees of frontal. Many of the databases have relatively few subjects, and ethnic and racial variability often are limited or unreported. The relative absence of African–American subjects is a concern in that skin color is a potential challenge for AFC. To date, only one study has examined robustness to skin color in a small number of participants [12]. Another concern is the paucity of information as regards the reliability of the annotation used for training and testing AFC. Without such information, it is impossible to know to what extent performance is impaired by error in the annotation. Performance metrics also vary, which complicates comparisons between studies.

19.6.3 **Cross-domain generalizability**

We evaluated cross-domain generalizability for each of three tasks: Detection of holistic expressions, detection of AU occurrence, and estimation of AU intensity. Within each task, we could only report average performance across behaviors (e.g., expressions or AUs). Reporting only average performance was unavoidable as most studies failed to report behavior-specific results. A further problem was that different studies often analyzed different sets of behaviors, thus rendering comparison across studies problematic. One study might achieve better results simply because it chose "easier" rather than "harder" behaviors to detect. For example, the holistic expressions of fear and surprise are often difficult to distinguish, which leads some researchers to remove or collapse certain classes.

Metrics. Performance metrics varied among types of tasks. For holistic expressions, 16 studies used accuracy (also called hit rate or recognition rate), which as noted above is biased by skew and chance agreement. Because no more rigorous metric was used by all studies, only the accuracy could be reported.

For studies of AU occurrence,13 studies used $F1$ and five used 2AFC. Because these metrics lack comparability, we report the results separately for each. We omit results from three studies of AU occurrence that used accuracy [17,50,51] and one that used AUC [52]. These metrics lack direct comparability with the $F1$ and 2AFC scores used by the majority of studies.

For AU intensity, studies used either PCC or one of the ICC formulations. Because these metrics are closely related, we report results for all of them and note differences in interpretation.

Generalizability for holistic expression. For holistic expressions, each of the studies used posed expressions or emotion portrayals. With few exceptions, all achieved strong within-database accuracy. Six of seven studies of CK+ achieved accuracy greater than 90% (Table 19.2).

Cross-domain performance, however, was lower and more variable. Generalizability between CK+ and MMI ranged from about 49 to 71%. The generalizability from FER2013 to CK+ was 76%. When multiple datasets were used for training (Table 19.3), the results remained similar. Increasing the number of databases used in training resulted in disappointing increases in cross-domain performance. Cross-domain generalizability for holistic expressions was modest at best.

Generalizability for AU occurrence. Databases included both posed and spontaneous behavior. Table 19.4 shows $F1$ scores for AU detection within- and between-databases. Within-databases, $F1$ scores were variable, ranging from a low of 0.27 to a high of 0.83. For both CK/CK+ and BP4D, one or more studies reported $F1$ greater 0.80, which represents strong inter-system reliability.

In all but one database, $F1$ scores for cross-domain generalizability were substantially lower than those for reliability within database. Only in the ISL database was generalizability comparable to that for within-domain reliability. These findings suggest caution when applying AU detectors trained in one database to video from a new database.

For 2AFC, within-database performance was lower in posed than in spontaneous databases and variable within both categories. Cross-domain generalizability was lower and more variable still (Table 19.5). Cross-database

Table 19.2 Holistic facial expression recognition results (accuracy)

	Study	CK+	MMI	BU-3DFE	MPIE	JAFFE	FEED	G-FERA	PICS	SFEW	FER2013	DISFA	KDEF	RaFD
CK+	[53]	0.93	–	–	–	–	–	–	–	–	–	–	–	–
	[54]	0.93	–	–	–	–	–	–	–	–	–	–	–	–
	[55]	0.95	0.55	–	–	–	–	–	–	–	–	–	–	–
	[57]	0.93	–	–	–	–	–	–	–	–	–	–	–	–
	[58]	0.76	0.53	–	–	–	0.32	–	–	–	–	–	–	–
	[59]	0.94	0.67	–	–	–	–	–	–	–	–	–	–	–
	[67]	0.91	0.51	–	–	0.41	–	–	–	–	–	–	–	–
	[68]	N.R.	–	–	–	0.72	0.44	–	0.58	–	–	–	–	–
	[69]	0.92	–	–	–	0.56	–	–	–	–	–	–	–	–
	[119]	N.R.	0.56	–	–	0.59	–	–	–	–	–	–	–	–
	[120]	0.97	–	–	–	0.46	–	–	–	–	–	–	–	–
	[121]	0.62	–	–	–	–	–	–	–	–	–	–	–	–
MMI	[53]	–	0.78	–	–	–	–	–	–	–	–	–	–	–
	[54]	–	0.79	–	–	–	–	–	–	–	–	–	–	–
	[55]	0.71	0.70	–	–	–	–	–	–	–	–	–	–	–
	[57]	–	0.78	–	–	–	–	–	–	–	–	–	–	–
	[58]	0.49	0.62	–	–	–	0.43	–	–	–	–	–	–	–
	[59]	0.61	0.93	–	–	–	–	–	–	–	–	–	–	–
	[67]	–	0.87	–	–	–	–	–	–	–	–	–	–	–

(continued on next page)

Table 19.2 (*continued*)

	Study	CK+	MMI	BU-3DFE	MPIE	JAFFE	FEED	G-FERA	PICS	SFEW	FER2013	DISFA	KDEF	RaFD
BU-3DFE	[70]	–	–	N.R.	–	0.42	–	–	0.36	0.21	–	–	0.46	0.55
	[71]	–	–	N.R.	0.60	–	–	–	–	–	–	–	–	–
	[72]	–	–	N.R.	0.64	–	–	–	–	–	–	–	–	–
MPIE	[53]	–	–	–	0.95	–	–	–	–	–	–	–	–	–
	[71]	–	–	0.79	N.R.	–	–	–	–	–	–	–	–	–
	[72]	–	–	0.67	N.R.	–	–	–	–	–	–	–	–	–
JAFFE	[67]	–	–	–	–	0.81	–	–	–	–	–	–	–	–
	[68]	–	–	–	–	0.61	–	–	–	–	–	–	–	–
	[69]	0.54	–	–	–	0.90	–	–	–	–	–	–	–	–
	[120]	–	–	–	–	0.92	–	–	–	–	–	–	–	–
	[121]	–	–	–	–	0.36	–	–	–	–	–	–	–	–
FEED	[58]	0.42	0.49	–	–	–	0.55	–	–	–	–	–	–	–
	[121]	0.69	–	–	–	0.62	N.R.	–	–	–	–	–	–	–
G-FERA	[53]	–	–	–	–	–	–	0.77	–	–	–	–	–	–
	[54]	–	–	–	–	–	–	0.67	–	–	–	–	–	–
	[57]	–	–	–	–	–	–	0.77	–	–	–	–	–	–
PICS	[68]	–	–	–	–	–	–	–	0.55	–	–	–	–	–
SFEW	[53]	–	–	–	–	–	–	–	–	0.48	–	–	–	–
	[68]	–	–	–	–	–	–	–	–	0.46	–	–	–	–
FER2013	[53]	–	–	–	–	–	–	–	–	–	0.66	–	–	–
	[56]	0.76	–	–	–	0.51	–	–	–	–	0.70	–	–	–
DISFA	[53]	–	–	–	–	–	–	–	–	–	–	0.55	–	–
	[57]	–	–	–	–	–	–	–	–	–	–	0.58	–	–

Note. *Train set reliability was unavailable for KDEF and RaFD. N.R.: Not reported. For G-FERA and DISFA cross-database results were obtained for multiple training databases and are reported in Table 19.4.*

Table 19.3 Holistic facial expression recognition results (accuracy) – multiple training datasets

	Study	CK+	MMI	MPIE	JAFFE	FER2013	G-FERA	DISFA	SFEW
CK+ + MMI	[54]	–	–	–	–	–	0.53	–	–
	[119]	–			0.58				
CK+ + DISFA	[54]	–	0.69	–	–	–	–	–	–
MMI + DISFA	[54]	0.74	–	–	–	–	–	–	–
MMI + G-FERA + DISFA	[57]	0.68	–	–	–	–	–	–	–
CK+ + G-FERA + DISFA	[57]	–	0.55	–	–	–	–	–	–
CK+ + MMI + DISFA	[57]	–	–	–	–	–	0.42	–	–
CK+ + MMI + G-FERA	[57]	–	–	–	–	–	–	0.41	–
MPIE + MMI + DISFA + G-FERA + SFEW + FER2013	[53]	0.64	–	–	–	–	–	–	–
CK+ + MMI + DISFA + G-FERA + SFEW + FER2013	[53]	–	–	0.46	–	–	–	–	–
MPIE + CK+ + DISFA + G-FERA + SFEW + FER2013	[53]	–	0.56	–	–	–	–	–	–
MPIE + MMI + DISFA + CK+ + SFEW + FER	[53]	–	–	–	–	–	0.39	–	–
MPIE + MMI + CK+ + G-FERA + SFEW + FER2013	[53]	–	–	–	–	–	–	0.38	–
MPIE + MMI + DISFA + CK+ + G-FERA + FER2013	[53]	–	–	–	–	–	–	–	0.40
MPIE + MMI + DISFA + CK+ + G-FERA + SFEW	[53]	–	–	–	–	0.34	–	–	–

Note. Results of within-database training where reported can be found in Table 19.3.

Table 19.4 AU occurrence estimation results (average $F1$-score values)

	Study	CK/CK+	MMI	DISFA	UNBC Pain Archive	RU-FACS	BP4D-FERA	GFT	SEM.	SAL	G-FERA	ISL
CK/CK+	[61]	0.81	–	–	–	–	–	–	0.52	–	–	–
	[75]	0.78	–	–	–	–	–	–	–	–	0.47	–
	[76]	0.69	–	0.39	0.16	–	–	–	–	–	0.45	–
	[77]	0.80	–	–	–	–	–	–	–	–	–	0.85
	[78]	0.80	–	–	–	–	–	–	–	–	–	0.85
	[122]	0.72	–	–	–	–	–	–	–	–	–	–
	[123]	0.60	0.37	–	–	–	–	–	–	–	–	–
	[124]	0.77	–	–	–	–	–	–	–	–	–	–
MMI	[73]	0.51	0.68	–	–	–	–	–	–	–	–	–
	[122]	0.56	0.65	–	–	–	–	–	–	–	–	–
	[123]	0.47	0.53	–	–	–	–	–	–	–	–	–
DISFA	[76]	0.60	–	0.43	0.18	–	–	–	–	–	0.45	–
	[64]	–	–	0.47	–	–	0.47	–	0.38	–	–	–
	[65]	–	–	0.48	–	–	0.54	–	–	–	–	–
UNBC Pain Archive	[76]	0.49	–	0.44	0.27	–	–	–	–	–	0.43	–
RU-FACS	[124]	–	–	–	–	0.54	–	–	–	–	0.54	–
BP4D-FERA	[62]	–	–		–	–	0.83	0.33	–	–	–	–
	[63]	–	–	0.27	–	–	0.48			–	–	–
	[64]	–	–	0.40	–	–	0.58		0.36	–	–	–
	[65]	–	–	0.50	–	–	0.63		–	–	–	–

(continued on next page)

Table 19.4 *(continued)*

	Study	CK/CK+	MMI	DISFA	UNBC Pain Archive	RU-FACS	BP4D-FERA	GFT	SEM.	SAL	G-FERA	ISL
GFT	[62]	–	–	–	–		0.43	0.66	–	–	–	–
	[124]					0.42		N.R.				
SEM.	[61]	–	–		–	–			0.57	–	–	–
	[64]			0.35			0.35		0.49			
SAL	[73]	–	–	–	–	–	–	–	0.64	0.69	–	–
G-FERA	[75]	0.74	–	–	–	–	–	–	–	–	0.53	–
	[76]	0.55		0.39	0.16						0.52	
FFD07	[49]	–	–	–	–	–	–	–	–	–	0.57	–

Note. Reliability of training set is presented along the main diagonal and highlighted.

Table 19.5 AU occurrence estimation results (average 2AFC scores)

		CK/CK+	MMI	SAL	G-FERA	BP4D	DIS-FA	SE-MAINE
CK+	[125]	0.91	–	–	0.66	–	–	–
MMI	[73]	0.80	0.87	–	–	–	–	–
SAL	[73]	–	–	0.81	–	–	–	0.73
G-FERA	[50]	–	–	–	0.71	–	–	–
	[125]	0.70	–	–	0.63	–	–	–
BP4D	[60]	0.76	–	–	–	0.75	0.73	–
DISFA	[60]	0.78	–	–	–	0.69	0.76	–
FFD07	[49]	–	–	–	0.75	–	–	–
Bosphorus + CK	[50]	–	–	–	0.66	–	–	–

Table 19.6 AU intensity estimation results

	Study	DISFA	UNBC Pain Archive	BP4D	FERA2015
DISFA	[64]	N.R.	–	0.65^{PCC}	–
	[126]	0.69^{ICC}	0.49^{ICC}	–	–
	[127]	0.84^{ICC}	0.49^{ICC}	0.81^{ICC}	–
	[128]	0.63^{PCC}			
UNBC Pain Archive	[126]	0.51^{ICC}	0.62^{ICC}	–	–
	[127]	–	0.59^{ICC}	–	–
BP4D	[64]	–	–	0.70^{PCC}	–
	[127]	0.74^{ICC}	0.50^{ICC}	0.85^{ICC}	–
	[128]	0.64^{PCC}	–	0.64^{PCC}	–
BP4D + DISFA	[64]	–	–	–	0.62^{ICC}
	[127]	–	0.49^{ICC}	–	–

Note. Results are averages of one or more action units. Because different studies used different action units, correlations even for the same metric are not necessarily comparable. AUs used were AU 6, 12, and 17 [128]; AU 4, 6, 9, 12, 20, 25, and 26 [126]; AU 6, 10, 12, 14, and 17 [64]; and AU 12 [127]. PCC is the Pearson correlation coefficient. ICC is the intra-class correlation coefficient. Type of ICC (for agreement or for consistency) is not reported.

2AFCs ranged from 0.66 to 0.80, which is relatively low, especially when considering that a 2AFC of 0.5 represents chance performance.

Generalizability for AU intensity. Within-database performance ranged from 0.59 to 0.85 on the various correlation coefficients and generalizability between datasets approached these values as well (Table 19.6). Between DISFA and BP4D, within- and between-database correlations were similar. Generalizability between BP4D and Pain Archive was about 0.10 lower than within-database reliability. Training on BP4D and DISFA yielded moderate generalizability (0.62). While one must be cautious given the difference in metrics and in distributions, these findings suggest that generalizability for intensity may be closer to adequate than was observed for AU occurrence. Appearance features vary significantly with intensity (e.g., (the depth and shape of the nasolabial furrow varies markedly with increasing intensity of several action units), which could contribute to error when only occurrence/non-occurrence is considered.

19.6.4 **Studies using deep- vs. shallow learning**

Deep learning approaches [16] seek common representations across domains, hierarchically select features and weight classifiers, and can better avoid the need to subsample video to avoid the curse of dimensionality. These and other advantages of deep learning lead to the hypothesis that deep learning can achieve higher generalizability between source and target domains. We examined this hypothesis by comparing studies that used deep learning and shallow learning.

For holistic expression, five of 16 studies [53–57] used deep architectures. Two of these studies [53,54] compared their results with those of studies that used shallow learning. The results from these studies were consistent with the hypothesis that deep learning has higher generalizability than shallow learning. Several factors, however, qualify this conclusion. One, deep- and shallow-learning approaches were trained on different databases. Because the deep- and shallow-learning models were trained on different datasets, differences between datasets confounded differences between learning models. From Table 19.3 we can observe that one [55] performed better when trained on CK+ and tested on MMI compared to non-deep studies, but was not among the best when trained on MMI and tested on CK+. Comparisons in any case are fraught by differences in databases and source and target sets. Even when studies used CK+ and MMI, they used different number of subjects from MMI (31 subjects [55], 19 subjects [58], 30 subjects [59]) and CK+ (118 subjects [55], 182 subjects [58], 106 subjects [59]).

Four studies [60–63] used deep approaches for AU occurrence estimation. In one [63], authors trained their model on BP4D and tested it on DISFA. Studies proposing shallow models [64,65] also reported results after training their models on BP4D and testing them on DISFA. Comparisons are confounded by differences in the AUs that were sampled. One study [64] computed the average $F1$ score over AU6, AU12 and AU17 while another [63] used AU1, AU2, AU4, AU6, AU9, AU12, AU25 and AU26 and yet another [65] used AU1, AU2, AU4, AU6, AU12, AU15 and AU17. All three studies, however, included AU 12. For AU 12, the study that used deep learning [63] performed worse ($F1 = 37.7$) than the ones that used shallow approaches [64] ($F1 = 59.0$) and [65] ($F1 = 63.9$). Still, the number of frames employed by these studies are not the same. Therefore, we cannot make certain conclusions about whether deep approaches are better or not.

So far, none of the AU intensity estimation studies have employed deep methods for cross-domain experiments. The hypothesis that deep learning affords greater generalizability in comparison with shallow learning remains inconclusive.

19.6.5 **Discussion**

We reviewed within- and between-domain comparisons from 39 studies. Within-domain reliability was variable. Within-domain performance appears to be yet unsolved. Across domains, few studies obtained good results when generalizing methods to new domains. Based on these findings, we recommend caution in applying AFC to domains in which it has not been trained. Until commercial systems and shareware systems publish peer-reviewed findings to the contrary, our caution also extends to use of commercial and shareware systems as well.

The literature review is limited by multiple factors. Because the dominant metric for each of the tasks differed, performance for any one task (e.g., detection of AU occurrence) could not be directly compared to that for the others (e.g., AU intensity estimation). Different metrics were also used within tasks. For AU occurrence, $F1$ was the most used metric and is reported in the table. As consequence, results of 7 studies could not be shared in Table 19.5 since they used a measure other than $F1$. For intensity, some studies used PCC while others used ICCs. Because these metrics quantify covariance very differently, we urge caution in making comparisons between them. Even ICC coefficients may not be fully comparable with each other (e.g., ICC-A vs. ICC-C).

Another limiting factor was the lack of statistical comparisons. Because all results are subject to sampling error, differences between methods or domains may be due to measurement error or other stochastic processes. Methods are needed to quantify the precision of performance estimates. More generally, statistical models enable evaluation of operational parameters that may affect outcomes.

Common confounds were differences in the selection of AUs among databases and differences in occurrence rates even when same AUs were used. For example, some studies [58,66] reported average results for six basic expressions (anger, disgust, fear, happy, sadness, surprise), while others [53–55,67–70] also included a neutral expression in their experiments.

Some studies [71,72] focused on estimation of different subsets of expressions. The number of AUs considered in studies also varied greatly. Considering that some AUs (such as AU12) are easier to recognize compared to the others (such as AU 1), average test statistics were not comparable. An option might have been to report results for AUs or expressions common across studies, but most studies reported only aggregate results [61,73–78]. For this reason, it was not possible to compare studies directly for the same AUs or expressions.

Direct comparisons between within- and between-domain experiments conducted by the same studies often were confounded by differences in between-domain occurrence rates. Unless occurrence rates are accounted for in some way, test statistics may vary for this reason alone. Future studies could consider constraining occurrence rates to achieve comparability between train and test sets, bearing in mind that occurrence rate differences may be an important difference between domains. For instance, smiling is likely to have a lower occurrence rate in a clinical setting (e.g., hospital) than in a non-clinical setting (e.g., school). Use of metrics that are robust to variation in occurrence rates is another option, although there may be other statistical limitations to these. In any case, it is important to explicitly report occurrence rates in all domains, and to take occurrence rate into consideration when interpreting performance.

To provide an environment in which different studies can be compared for within- and between-domain performance, it is essential to standardize selection of subjects in train and test sets, the sequences selected for analysis, and the labels to be included in the analyses. Further, for comparability of findings, common metrics are essential. Because no single metric is likely to be widely adopted, we recommend investigators to report multiple metrics.

19.7 NEW DIRECTIONS

Most efforts to improve generalizability have focused on features and classifiers within the realm of shallow learning. Their success has been modest at best. Within- and cross-domain performance using shallow learning remain far apart. Proponents of deep learning have hypothesized that its approach to features and classifiers can narrow the disparity between within- and between-domain performance. Deep neural networks (DNNs) have notable advantages that might improve generalizability. DNNs can encode the highly non-linear nature of AFC data, their architecture jointly learns feature representations and classifiers in a series of iterations, and their stochastic optimization algorithms enable model training with larger datasets. Despite

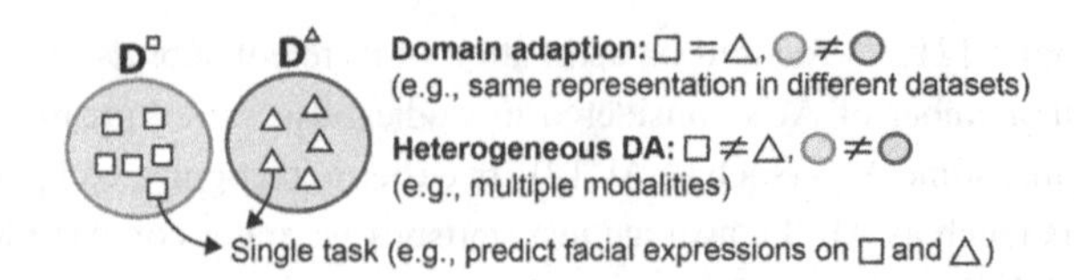

■ **FIGURE 19.4** An illustration of domain adaptation (DA). Denote two domains as $\mathbf{D}^{\square}$ and $\mathbf{D}^{\triangle}$, where □ and △ indicate feature representations in each domain, and the red/green colors indicate the distributions from which samples in each domain are drawn. As in most AFC scenarios (e.g., predict facial expressions), DA assumes the tasks on two domains to be the same, yet the data distributions are different due to selection bias or distribution mismatch (e.g., different datasets). Heterogeneous DA is a special case when two domains have different feature representations (e.g., multiple modalities such as viewpoints or recording variations). From [79].

the advantages, initial results for deep learning are not substantially better than those for shallow learning.

While further work may yet discover the improved generalizability of DNNs, we would suggest that DNNs themselves are not sufficient. All machine learning methods, whether shallow or deep, implicitly assume that representations and classifiers are drawn from the same domains [79]. When that assumption is violated, additional learning is required. To generalize from one domain to another likely requires a new approach.

To learn adaptation to new domains, several types of transfer learning have been proposed [79]. Transductive transfer applies when tasks (e.g., AU detection) are the same but domains are different. Inductive transfer applies when one wishes to adapt a new task to same or different domain. An example would be learning to detect AU 6 (cheek raising) when only AU 12 notation (lip-corner pull) is available in the source domain. Unsupervised transfer applies when labels are available in neither source nor target domains. Examples would be clustering, dimension reduction, and density estimation. Of these three, transductive transfer learning is the most relevant to AFC systems that must *learn to adapt* to new domains.

DNNs appear to be well suited for this purpose (Fig. 19.4). Several approaches may be considered.

Multi-layer neural networks minimize reconstruction error between input image pairs from different domains and their corrupted versions (i.e., randomly zero out partial inputs). One approach is denoising autoencoders (DAE). They reduce reconstruction error by avoiding the over-complete representations that are common in conventional auto-encoders. Given input image pairs, DAE learns to reconstruct inputs by finding a latent representation shared between different domains. Several variants of DAE may be noted. Stacked DAE [80], which stacks the ordinary DAEs, establishes common representations that improve performance even when combined with shallow classifiers. Marginalized DAE [81], which yields a unique optimal solution with closed-form updates between layers, marginalizes random corruptions. Multi-task DAE [82], which substitutes artificial corruptions with inter-domain variability in natural images, learns representations that generalize across domains.

Siamese architecture [83] is another promising approach to learning cross-domain representations. A Siamese architecture consists of two identical networks that take inputs from two streams (representing the source and the target domains) and a contrastive loss to measure the similarity between the outputs of the two streams. Inspired by domain adaptation procedures from shallow-learning methods, the two-stream design allows for learning domain-invariant representations.

Most studies on DA leverage the basic Siamese network by imposing a classification loss and a discrepancy loss (i.e., domain loss). The classification loss ensures low classification errors on the labeled source data (assuming target labels are unavailable or too few); the discrepancy loss aims to improve domain invariance by reducing the distance between the learned representations from different streams.

Studies following the design of separable losses can accommodate both supervised (with target labels) and unsupervised (without target labels) adaptation. The discrepancy is typically implemented by variants of Maximum Mean Discrepancy (MMD), which are applied to constrain the distribution of activations from one (e.g., Deep Domain Confusion [84]) or more top layers (e.g., Deep Adaptation Networks [85]). Instead of directly measuring discrepancy on activation maps using MMD, Deep CORAL [86] uses a correlation loss to reduce the discrepancy between activation statistics of the source and target distributions.

Generative adversarial networks (GANs) [87] are another promising approach. Standard GANs comprise two networks: A generator network that tries to generate an image from a noise vector drawn from a pre-determined distribution, and a discriminator network to tell real images from the generated images. The "adversary" competition between the two networks drives each network to improve itself until the counterfeits are indistinguishable from the genuine cases. A generator learns to adapt images in the source domain to images in a target domain; a discriminator learns to identify whether an image is from the source or target domain. Cross-domain GANs provide mechanisms for transforming images to different domains and alleviate the need for annotated target data.

PixelGAN [88] directly learns to modify a source image to appear as a target image while maintaining the original content. PixelGAN accomplishes this task without knowing target annotations. Similarly, the Domain Transfer Network [89] exploits a compound loss that contains a multi-class GAN loss; an f-consistency loss enforces the output of a given function f to remain unchanged, no matter whether the input is from the target or source, and a total variation loss to encourage smooth results. Without assuming correspondence between the input source and target images, Coupled GANs (CoGANs) [90] learn a joint distribution of images from multiple domains by enforcing a weight-sharing constraint. The learned joint distribution can then be used to generate novel tuples of images (e.g., faces with different attributes). More recent variants include adversarial discriminative domain adaptation [91], DiscoGAN [92], DC-GAN [93], and others.

A final approach is *heterogeneous domain adaptation*, in which the feature space differs between the source and target domains. Examples include multiple modalities (e.g., textual-audio-visual) and multiple views (e.g., recording scenarios under different resolutions, illumination, poses, expressions, etc.). Heterogeneous domain adaptation offers an opportunity to learn complementary representations that improve the robustness of existing models trained with only one representation. Most existing approaches to heterogeneous domain adaptation aim to learn a feature mapping across feature spaces, where the learned mapping enables interchangeable representations between the source and target domains.

A variant of heterogeneous domain adaptation, transfer neural trees [94], jointly learn the cross-domain feature mapping and classification, following a two-stream design (one for each domain); weights of the top layers are shared. Similar to feature transfer, weakly-shared deep transfer networks [95] learns to transfer label information across heterogeneous domains with intermediate "weakly-shared" layers of features. The sharable features and labels mitigate the problem of insufficient data and allow joint modeling of complex data representation from different domains.

More than one approach may be needed to optimize generalizability across domains. The key, however, is to pose the problem as one of *learning* to adapt AFC between different domains. Simply training in one domain and hoping for generalizability to another domain has not worked in the past and is unlikely to succeed in the future. Generalizability is a central problem for AFC and an open research question. Deep learning appears to be well suited to this challenge and numerous approaches appear promising for use in AFC. The time is right to pursue the problem of minimizing error between domains. The approaches reviewed have potential to train models in source domains and optimize their transfer to target domains.

19.8 SUMMARY

AFC systems have made major advances in detection of the occurrence and intensity of facial action units and holistic emotion, depression and other distress disorders, pain, and user engagement. In almost all of these applications, AFC systems have been trained and applied within the same domains. Strong interest is emerging among researchers and practitioners in various fields for commercial systems that they may use in their own domains. To evaluate the readiness of AFC systems for applications in domains other than those in which they have been trained, we evaluated 39 studies of cross-domain generalizability.

With the exception of posed or deliberate facial actions, cross-domain generalizability significantly lagged behind within-domain performance. Few studies of action unit occurrence achieved adequate cross-domain results outside of posed data. For action unit intensity, the results were somewhat better. Two studies achieved intraclass correlations above 0.70 for cross-domain generalizability. Even then, however, lack of detail qualifies this finding. None of the studies available for review reported results for individual labels (holistic emotions or action units); all reported only aggregate findings. The generalizability of AFC systems remains an open question.

Research to date has focused on transferring trained classifiers directly to new domains. A more fruitful approach is suggested by recent advances in domain adaption. Domain adaption explicitly addresses differences in feature representations, probability distributions, and labels across domains. Given the success of domain adaptation in object recognition and related areas, the next generation of research in domain generalization for AFC would do well to pursue one or more of the approaches we discussed.

In designing further research, a number of lingering issues would be well to address, as well. These include the need to report results for individual labels rather than only aggregates, use of standard train and test protocols where possible, and statistical modeling of operating parameters (such as participant characteristics) in relation to within and between domain generalizability. A final issue is that of performance metrics. Because different metrics quantify different aspects of performance and are differentially affected by class imbalance, investigators are strongly encouraged to report multiple measures of performance. Cross-domain generalizability is critical to impact of AFC systems for research, clinical, educational, and commercial use. With the introduction of domain adaptation approaches, research on domain generalizability may yield high impact across wide range of fields.

ACKNOWLEDGMENTS

This research was sponsored in part by NSF award 1418026. Preparation of this article was supported in part by the National Institute of Mental Health of the National Institutes of Health under Award Number MH096951 and the National Science Foundation under award CNS 1629716. The content is solely the responsibility of the authors and does not necessarily represent the views of the National Institutes of Health or the National Science Foundation.

REFERENCES

[1] P. Ekman, T.S. Huang, T.J. Sejnowski, Final Report to NSF of the Planning Workshop on Facial Expression Understanding, National Science Foundation, Washington, D.C., 1992, July 30 to August 1, 1992.

[2] H. Dibeklioglu, Z. Hammal, J.F. Cohn, Dynamic multimodal measurement of depression severity using deep autoencoding, IEEE J. Biomed. Health Inform. 22 (2) (2018) 525–536.

[3] D. McDuff, Smiling from adolescence to old age: a large observational study, in: Proceedings of the Affective Computing and Intelligent Interaction, San Antonio, TX, 2017.

[4] D. McDuff, J.M. Girard, R. el Kalioubi, Large-scale observational evidence of cross-cultural differences in facial behavior, J. Nonverbal Behav. 41 (1) (2016) 1–19.

[5] M.F. Valstar, J. Gratch, B. Schuller, F. Ringeval, D. Lalanne, M.T. Torres, et al., AVEC 2016: depression, mood, and emotion recognition workshop and challenge, in: Proceedings of the 6th International Workshop on Audio/Visual Emotion Challenge, 2016, pp. 3–10.

[6] B. Martinez, M.F. Valstar, B. Jiang, M. Pantic, Automatic analysis of facial actions: a survey, IEEE Trans. Affect. Comput. (2018), https://doi.org/10.1109/TAFFC.2017.2731763 (early access).

[7] Z. Hammal, J.F. Cohn, Automatic, objective, and efficient measurement of pain using automated face analysis, in: K. Prkachin, K.D. Craig (Eds.), Social and Interpersonal Processes in Pain: We Don't Suffer Alone, Springer, 2018, https://doi.org/10.1007/978-3-319-78340-6_7, in press.

[8] G.S. Wachtman, J.F. Cohn, J.M.V. Swearingen, E.K. Manders, Automated tracking of facial features in patients with facial neuromuscular dysfunction, Plast. Reconstr. Surg. 107 (April 15 2001) 1124–1133.

[9] P. Lucey, J.F. Cohn, K.M. Prkachin, P. Solomon, I. Matthews, Painful data: the UNBC-McMaster shoulder pain expression archive database, Image Vis. Comput. 30 (2012) 197–205.

[10] M. Georgopoulos, Y. Panagakis, M. Pantic, Modelling of facial aging and kinship: a survey, http://arxiv.org/pdf/1802.04636.pdf, 2018.

[11] M.S. Bartlett, G.C. Littlewort, M.G. Frank, K. Lee, Automatic decoding of facial movements reveals deceptive pain expressions, Curr. Biol. 24 (7) (2014) 738–743, https://doi.org/10.1016/j.cub.2014.02.009.

[12] J.M. Girard, J.F. Cohn, L.A. Jeni, M.A. Sayette, F. De la Torre, Spontaneous facial expression in unscripted social interactions can be measured automatically, Behav. Res. Methods 47 (2015) 1136–1147.

[13] M.F. Valstar, E. Sanchez-Lozano, J.F. Cohn, L.A. Jeni, J.M. Girard, Z. Zhang, et al., FERA 2017 – addressing head pose in the third facial expression recognition and analysis challenge, in: Proceedings of the International Conference on Automatic Face and Gesture Recognition, Washington, DC, 2017.

[14] S. Tulyakov, L.A. Jeni, N. Sebe, J.F. Cohn, Viewpoint-consistent 3D face alignment, IEEE Trans. Pattern Anal. Mach. Intell. 40 (9) (2017) 2250–2264, https://doi.org/10.1109/TPAMI.2017.2750687.
[15] L.A. Jeni, J.F. Cohn, T. Kanade, Dense 3d face alignment from 2d videos for real time use, Image Vis. Comput. 58 (2017) 13–24.
[16] Y. LeCun, Y. Bengio, G. Hinton, Deep learning, Nature 521 (2015) 436–444.
[17] S. Ghosh, E. Laksana, S. Scherer, L.-P. Morency, A multi-label convolutional neural network approach to cross-domain action unit detection, in: Proceedings of the Affective Computing and Intelligent Interaction, Xi'an, China, 2015.
[18] J.M. Girard, J.F. Cohn, Automated audiovisual depression analysis, Curr. Opin. Psychol. 4 (2015) 75–79.
[19] J.M. Girard, J.F. Cohn, A primer on observational measurement, Assessment 23 (2016) 404–413.
[20] J.F. Cohn, P. Ekman, Measuring facial action by manual coding, facial EMG, and automatic facial image analysis, in: J.A. Harrigan, R. Rosenthal, K.R. Scherer (Eds.), Handbook of Nonverbal Behavior Research Methods in the Affective Sciences, Oxford, New York, 2005, pp. 9–64.
[21] P. Ekman, Basic emotions, in: T. Dalagleish, M. Power (Eds.), Handbook of Cognition and Emotion, John Wiley & Sons, UK, 1999, pp. 45–60.
[22] D. Watson, A. Tellegen, Toward a consensual structure of mood, Psychol. Bull. 98 (1985) 219–235.
[23] J.A. Russell, A circumplex model of affect, J. Pers. Soc. Psychol. 39 (1980) 1161–1178.
[24] H. Schlosberg, Three dimensions of emotion, Psychol. Rev. 61 (1954) 81–88.
[25] P. Ekman, W.V. Friesen, J.C. Hager, Facial Action Coding System, Research Nexus, Network Research Information, Salt Lake City, UT, 2002.
[26] J.F. Cohn, Z. Ambadar, P. Ekman, Observer-based measurement of facial expression with the facial action coding system, in: J.A. Coan, J.J.B. Allen (Eds.), The Handbook of Emotion Elicitation and Assessment, in: Oxford University Press Series in Affective Science, Oxford University, New York, NY, 2007, pp. 203–221.
[27] J.M. Girard, W.-S. Chu, L.A. Jeni, J.F. Coh, F. De la Torre, M.A. Sayette, Sayette Group Formation Task (GFT) spontaneous facial expression database, in: Proceedings of the IEEE International Conference on Automatic Face and Gesture Recognition, 2017.
[28] X. Zhang, L. Yin, J.F. Cohn, S. Canavan, M. Reale, A. Horowitz, et al., BP4D-spontaneous: a high-resolution spontaneous 3D dynamic facial expression database, Image Vis. Comput. 32 (2014) 692–706.
[29] X. Zhang, L. Yin, J.F. Cohn, J.M. Girard, Multimodal spontaneous human emotion corpus for human behavior analysis, in: Proceedings of the IEEE International Conference on Computer Vision and Pattern Recognition, Las Vegas, NV, 2016.
[30] Z. Hammal, E.R. Wallace, C.L. Heike, M.L. Speltz, Automatic action unit detection in infants using convolutional neural network, in: Proceedings of the International Conference on Affective Computing and Intelligent Interaction, San Antonio, TX, 2017.
[31] Z. Ambadar, J.F. Cohn, L.I. Reed, All smiles are not created equal: morphology and timing of smiles perceived as amused, polite, and embarrassed/nervous, J. Nonverbal Behav. 33 (2009) 17–34.
[32] K.L. Schmidt, Z. Ambadar, J.F. Cohn, L.I. Reed, Movement differences between deliberate and spontaneous facial expressions: zygomaticus major action in smiling, J. Nonverbal Behav. 30 (2006) 37–52.
[33] J.F. Cohn, L.I. Reed, T. Moriyama, J. Xiao, K.L. Schmidt, Z. Ambadar, Multimodal coordination of facial action, head rotation, and eye motion, in: Proceedings of the Sixth IEEE International Conference on Automatic Face and Gesture Recognition, Seoul, Korea, 2004.

[34] M.F. Valstar, M. Pantic, Z. Ambadar, J.F. Cohn, Spontaneous vs. posed facial behavior: automatic analysis of brow actions, in: Proceedings of the ACM International Conference on Multimodal Interfaces, Banff, Canada, 2006.

[35] J.M. Girard, J.F. Cohn, M.H. Mahoor, S.M. Mavadati, Z. Hammal, D. Rosenwald, Nonverbal social withdrawal in depression: evidence from manual and automatic analyses, Image Vis. Comput. 32 (2014) 641–647.

[36] K. McGraw, S.P. Wong, Forming inferences about some intraclass correlation coefficients, Psychol. Methods 1 (1996) 30–46.

[37] L.A. Jeni, J.F. Cohn, F. De la Torre, Facing imbalanced data recommendations for the use of performance metrics, in: Proceedings of the Humaine Association Conference on Affective Computing and Intelligent Interaction, Geneva, Switzerland, 2013.

[38] L.A. Jeni, J.F. Cohn, T. Kanade, Dense 3d face alignment from 2d videos in real-time, in: IEEE International Conference and Workshops on Automatic Face and Gesture Recognition, 2015, pp. 1–8.

[39] C. Tang, W. Zheng, J. Yan, Q. Li, Y. Li, T. Zhang, et al., View-independent facial action unit detection, in: Proceedings of the IEEE International Conference on Automatic Face and Gesture Recognition, 2017.

[40] J.M. Girard, J.F. Cohn, L.A. Jeni, S. Lucey, F. De la Torre, How much training data for facial action unit detection?, in: Proceedings of the IEEE International Conference on Automatic Face and Gesture Recognition, Ljubljana, Slovenia, 2015.

[41] W.-S. Chu, F. De la Torre, J.F. Cohn, Selective transfer machine for personalized facial action unit detection, IEEE Trans. Pattern Anal. Mach. Intell. 39 (2017) 529–545.

[42] H. Oster, Baby FACS: Facial Action Coding System for Infants and Young Children, New York University, New York, 2001.

[43] J. Buolamwini, T. Gebru, Gender shades: intersectional accuracy disparities in commercial gender classification, Proc. Mach. Learn. Res. 81 (2018) 1–15.

[44] A.J. Fridlund, Sociality of solitary smiling: potentiation by an implicit audience, J. Pers. Soc. Psychol. 60 (1991) 229–240.

[45] M.F. Valstar, T. Almaev, J.M. Girard, G. McKeown, M. Mehu, L. Yin, et al., FERA 2015 – second facial expression recognition and analysis challenge, in: Proceedings of the IEEE International Conference on Automatic Face and Gesture Recognition, Ljubljana, Slovenia, 2015.

[46] P. Lucey, J.F. Cohn, T. Kanade, J.M. Saragih, Z. Ambadar, I. Matthews, The extended Cohn–Kanade dataset (CK+): a complete facial expression dataset for action unit and emotion-specified expression, in: Third IEEE Workshop on CVPR for Human Communicative Behavior Analysis (CVPR4HB 2010), June 18, 2010, pp. 1–8.

[47] M. Pantic, M. Valstar, R. Rademaker, L. Maat, Web-based database for facial expression analysis, in: Proceedings of the IEEE International Conference on Multimedia and Expo, 2005.

[48] M.F. Valstar, M. Pantic, Induced disgust, happiness and surprise: an addition to the MMI Facial Expression Database, in: Proceedings of the Third International Workshop on Emotion (Satellite of LREC): Corpora for Research on Emotion and Affect, Valletta, Malta, 2010.

[49] T. Wu, N.J. Butko, P. Ruvolo, J. Whitehill, M.S. Bartlett, J.R. Movellan, Action unit recognition transfer across datasets, in: 2011 IEEE International Conference on Automatic Face and Gesture Recognition and Workshops, FG 2011, 2011, pp. 889–896.

[50] T. Senechal, V. Rapp, H. Salam, R. Seguier, K. Bailly, L. Prevost, Facial action recognition combining heterogeneous features via multikernel learning, IEEE Trans. Syst. Man Cybern., Part B, Cybern. 42 (2012) 993–1005.

[51] R. Walecki, O. Rudovic, V. Pavlovic, M. Pantic, Variable-state Latent Conditional Random Field models for facial expression analysis, Image Vis. Comput. 58 (2017) 25–37.

[52] A. Mohammadian, H. Aghaeinia, F. Towhidkhah, S.Z. Seyyedsalehi, Subject adaptation using selective style transfer mapping for detection of facial action units, Expert Syst. Appl. 56 (2016) 282–290.

[53] A. Mollahosseini, D. Chan, M.H. Mahoor, Going deeper in facial expression recognition using deep neural networks, in: 2016 IEEE Winter Conference on Applications of Computer Vision (WACV), 2016, pp. 1–10.

[54] B. Hasani, M.H. Mahoor, Spatio-temporal facial expression recognition using convolutional neural networks and conditional random fields, in: Automatic Face & Gesture Recognition (FG 2017), 2017 12th IEEE International Conference, May 2017, pp. 790–795.

[55] Z. Meng, P. Liu, J. Cai, S. Han, Y. Tong, Identity-aware convolutional neural network for facial expression recognition, in: 2017 12th IEEE International Conference on Automatic Face & Gesture Recognition (FG 2017), 2017, pp. 558–565.

[56] G. Wen, Z. Hou, H. Li, D. Li, L. Jiang, E. Xun, Ensemble of deep neural networks with probability-based fusion for facial expression recognition, Cogn. Comput. 9 (2017) 597–610.

[57] B. Hasani, M.H. Mahoor, Facial expression recognition using enhanced deep 3D convolutional neural networks, in: IEEE Conference on Computer Vision and Pattern Recognition Workshops (CVPRW), July 2017, pp. 2278–2288.

[58] C. Mayer, M. Eggers, B. Radig, Cross-database evaluation for facial expression recognition, Pattern Recognit. Image Anal. 24 (2014) 124–132.

[59] X. Zhang, M.H. Mahoor, S.M. Mavadati, Facial expression recognition using lp-norm MKL multiclass SVM, Mach. Vis. Appl. 26 (2015) 467–483.

[60] S. Ghosh, E. Laksana, S. Scherer, L.P. Morency, A multi-label convolutional neural network approach to cross-domain action unit detection, in: International Conference on Affective Computing and Intelligent Interaction (ACII), September 2015, pp. 609–615.

[61] Y. Wu, Q. Ji, Constrained joint cascade regression framework for simultaneous facial action unit recognition and facial landmark detection, in: 2016 IEEE Conference on Computer Vision and Pattern Recognition (CVPR), 2016, pp. 3400–3408.

[62] W.-S. Chu, F. De la Torre, J.F. Cohn, Learning spatial and temporal cues for multi-label facial action unit detection, in: 2017 12th IEEE International Conference on Automatic Face & Gesture Recognition (FG 2017), 2017, pp. 25–32.

[63] K. Zhao, W.-S. Chu, H. Zhang, Deep region and multi-label learning for facial action unit detection, in: 2016 IEEE Conference on Computer Vision and Pattern Recognition (CVPR), 2016, pp. 3391–3399.

[64] T. Baltrusaitis, M. Mahmoud, P. Robinson, Cross-dataset learning and person-specific normalisation for automatic action unit detection, in: 2015 11th IEEE International Conference and Workshops on Automatic Face and Gesture Recognition (FG), 2015, pp. 1–6.

[65] S. Eleftheriadis, O. Rudovic, M.P. Deisenroth, M. Pantic, Gaussian process domain experts for modeling of facial affect, IEEE Trans. Image Process. 26 (2017) 4697–4711.

[66] X. Zhang, M.H. Mahoor, S.M. Mavadati, Facial expression recognition using l_p-norm MKL multiclass-SVM, Mach. Vis. Appl. 26 (2015) 467–483.

[67] C. Shan, S. Gong, P.W. McOwan, Facial expression recognition based on Local binary patterns: a comprehensive study, Image Vis. Comput. 27 (2009) 803–816.

[68] R. Zhu, G. Sang, Q. Zhao, Discriminative feature adaptation for cross-domain facial expression recognition, in: International Conference on Biometrics (ICB 2016), June 2016, pp. 1–7.

[69] W. Gu, C. Xiang, Y.V. Venkatesh, D. Huang, H. Lin, Facial expression recognition using radial encoding of local Gabor features and classifier synthesis, Pattern Recognit. 45 (2012) 80–91.
[70] M.K. Abd El Meguid, M.D. Levine, Fully automated recognition of spontaneous facial expressions in videos using random forest classifiers, IEEE Trans. Affect. Comput. 5 (2014) 141–154.
[71] W. Zheng, Y. Zong, X. Zhou, M. Xin, Cross-domain color facial expression recognition using transductive transfer subspace learning, IEEE Trans. Affect. Comput. 3045 (2016) 1.
[72] K. Yan, W. Zheng, Z. Cui, Y. Zong, Cross-database facial expression recognition via unsupervised domain adaptive dictionary learning, in: International Conference on Neural Information Processing, pp. 427–434.
[73] B. Jiang, M. Valstar, B. Martinez, M. Pantic, A dynamic appearance descriptor approach to facial actions temporal modeling, IEEE Trans. Cybern. 44 (2014) 161–174.
[74] S. Koelstra, M. Pantic, I. Patras, A dynamic texture-based approach to recognition of facial actions and their temporal models, IEEE Trans. Pattern Anal. Mach. Intell. 32 (2010) 1940–1954.
[75] Y. Li, J. Chen, Y. Zhao, Q. Ji, Data-free prior model for facial action unit recognition, IEEE Trans. Affect. Comput. 4 (2013) 127–141.
[76] A. Ruiz, X. Binefa, From emotions to action units with hidden and semi-hidden-task learning, in: Proceedings of the International Conference on Computer Vision, Araucano Park, Las Condes, Chile, 2015.
[77] S. Wang, Q. Gan, Q. Ji, Expression-assisted facial action unit recognition under incomplete AU annotation, Pattern Recognit. 61 (2017) 78–91.
[78] J. Wang, S. Wang, Q. Ji, Facial action unit classification with hidden knowledge under incomplete annotation, in: Proceedings of the 5th ACM on International Conference on Multimedia Retrieval, 2015, pp. 75–82.
[79] S.J. Pan, Q. Yang, A survey on transfer learning, IEEE Trans. Knowl. Data Eng. 22 (2009) 1345–1359.
[80] P. Vincent, H. Larochelle, I. Lajoie, Y. Bengio, P.-A. Manzagol, Stacked denoising autoencoders: learning useful representations in a deep network with a local denoising criterion, J. Mach. Learn. Res. 11 (2010) 3371–3408.
[81] M. Chen, Z.E. Xu, K.Q. Weinberger, F. Sha, Marginalized denoising autoencoders for domain adaptation, in: Proceedings of the 29th International Conference on Machine Learning, Edinburgh, Scotland, UK.
[82] M. Ghifary, W.B. Kleijn, M. Zhang, D. Balduzzi, Domain generalization for object recognition with multi-task autoencoders, in: Proceedings of the International Conference on Computer Vision, 2015.
[83] J. Bromley, I. Guyon, Y. LeCun, E. Sickinger, R. Shah, Signature verification using a Siamese time delay neural network, in: Proceedings of the 6th International Conference on Neural Information Processing Systems, Denver, Colorado, 1993.
[84] E. Tzeng, J. Hoffman, N. Zhang, K. Saenko, T. Darrell, Deep domain confusion: maximizing for domain invariance, arXiv:1412.3474, 2014, pp. 1–9.
[85] M. Long, Y. Cao, J. Wang, M.I. Jordan, Learning transferable features with deep adaptation networks, in: International Conference on Machine Learning, 2015.
[86] B. Sun, K. Saenko, Deep coral: correlation alignment for deep domain adaptation, in: Proceedings of the European Conference on Computer Vision, 2016.
[87] I.J. Goodfellow, J. Pouget-Abadie, B.X. Mehdi Mirza, David Warde-Farley, S. Ozair, A. Courville, Y. Bengio, Generative adversarial nets, in: Proc. Adv. Neural Inf. Process., 2014.
[88] B. Konstantinos, S. Nathan, D. David, E. Dumitru, K. Dilip, Unsupervised pixel-level domain adaptation with generative adversarial networks, in: Proceedings of the IEEE International Conference on Computer Vision and Machine Learning, 2017.

[89] Y. Taigman, A. Polyak, L. Wolf, Unsupervised cross-domain image generation, arXiv:1611.02200, 2017.
[90] M.-Y. Liu, O. Tuzel, Coupled generative adversarial networks, in: Advances in Neural Information Processing Systems, 2017, pp. 1–9.
[91] E. Tzeng, J. Hoffman, K. Saenko, T. Darrel, Adversarial discriminative domain adaptation, in: Proceedings of the IEEE International Conference on Computer Vision and Pattern Recogntion, 2017.
[92] T. Kim, M. Cha, H. Kim, J.K. Lee, J. Kim, Learning to discover cross-domain relations with generative adversarial networks, in: Proceedings of the IEEE International Conference on Computer Vision and Pattern Recognition, 2017.
[93] A. Radford, L. Metz, S. Chintala, Unsupervised representation learning with deep convolutional generative adversarial networks, arXiv:1511.06434, 2015.
[94] W.-Y. Chen, T.-M.H. Hsu, Y.-H.H. Tsai, Y.-C.F. Wang, M.-S. Chen, Transfer neural s trees for heterogeneous domain adaptation, in: Proceedings of the European Conference on Computer Vision, 2016.
[95] X. Shu, G.-J. Qi, J. Tang, J. Wang, Weakly-shared deep transfer networks for heterogeneous-domain knowledge propagation, in: Proceedings of the ACM International Conference on Multimedia, Brisbane, Australia, 2015.
[96] A. Savran, N. Alyuz, H. Dibeklioglu, O. Celiktutan, B. Gokberk, B. Sankur, et al., Bosphorus database for 3D face analysis, in: Proceedings of the First COST 2101 Workshop on Biometrics and Identity Management, Boskilde University, Denmark, 2008.
[97] N. Alyüz, B. Gökberk, H. Dibeklioğlu, A. Savran, A.A. Salah, L. Akarun, et al., 3D face recognition benchmarks on the Bosphorus database with focus on facial expressions, in: Proceedings of the First COST 2101 Workshop on Biometrics and Identity Management, Roskilde University, Denmark, 2008.
[98] L. Yin, X. Wei, Y. Sun, J. Wang, M.J. Rosato, A 3D facial expression database for facial behavior research: BU-3DFE (Binghamton University 3D facial expression) database, in: 7th IEEE International Conference on Automatic Face and Gesture Recognition (FG 2006), Southampton, UK, 2006, pp. 211–216.
[99] T. Kanade, J.F. Cohn, Y. Tian, Comprehensive database for facial expression analysis, in: Proceedings of the Fourth International Conference on Automatic Face and Gesture Recognition, Grenoble, 2000.
[100] T. Bänziger, K. Scherer, Using actor portrayals to systematically study multimodal emotion expression: the GEMEP corpus, in: Proceedings of the Affective Computing and Intelligent Interaction (ACII 2007), 2007.
[101] Q. Ji, ISL facial expression databases, Rensslear Polytechnic Institute, https://www.ecse.rpi.edu/~cvrl/database/database.html.
[102] M. Lyons, S. Akamasku, M. Kamachi, J. Gyoba, Coding facial expressions with Gabor wavelets, in: Proceedings of the Third International Conference on Automatic Face and Gesture Recognition (FG 1998), Nara, Japan, 1998.
[103] E. Goeleven, R. De Raedt, L. Leyman, B. Verschuere, The Karolinska directed emotional faces: a validation study, Cogn. Emot. 22 (2008) 1094–1118.
[104] R. Gross, I. Matthews, J.F. Cohn, T. Kanade, S. Baker, Multi-PIE, Image Vis. Comput. 28 (2010) 807–813.
[105] P. Hancock, Psychological Image Collection at Stirling (PICS), University of Stirling, 2008.
[106] O. Langner, R. Dotsch, G. Bijlstra, D.H.J. Wigboldus, S.T. Hawk, A. van Knippenberg, Presentation and validation of the Radboud Faces Database, Cogn. Emot. 24 (2009) 1377–1388.
[107] A. Dhall, R. Goecke, S. Lucey, T. Gedeon, Static facial expression analysis in tough conditions: data, evaluation protocol and benchmark, in: IEEE International Conference on Computer Vision Workshops, Barcelona, Spain, 2011.

[108] X. Zhang, L. Yin, J.F. Cohn, S. Canavan, M. Reale, A. Horowitz, et al., BP4D-spontaneous: a high-resolution spontaneous 3d dynamic facial expression database, Image Vis. Comput. 32 (2014) 692–706.

[109] S.M. Mavadati, M.H. Mahoor, K. Bartlett, P. Trinh, J.F. Cohn, DISFA: a spontaneous facial action intensity database, IEEE Trans. Affect. Comput. 4 (2013) 151-160.

[110] F. Wallhoff, Facial Expression and Emotion Database, 2006.

[111] M.G. Frank, J. Movellan, M.S. Bartlett, G.C. Littlewort, RU-FACS-1 database, Machine Perception Laboratory, U.C., San Diego.

[112] M. Bartlett, G. Littlewort, M. Frank, C. Lainscsek, B. Fasel, J. Movellan, Automatic recognition of spontaneous expressions, J. Multimed. 1 (2006) 22–35.

[113] E. Douglas-Cowie, R. Cowie, C. Cox, N. Amier, D.K.J. Heylen, The sensitive artificial listener: an induction technique for generating emotionally coloured conversation, in: LREC Workshop on Corpora for Research on Emotion and Affect, Elra, Paris, 2008.

[114] G. McKeown, M. Valstar, R. Cowie, M. Schroder, The SEMAINE database: annotated multimodal records of emotionally colored conversations between a person and a limited agent, IEEE Trans. Affect. Comput. 3 (2012) 5–17.

[115] J. Cohen, Nominal scale agreement with provision for scaled disagreement or partial credit, Psychol. Bull. 70 (1968) 213–220.

[116] J. Cohen, A coefficient of agreement for nominal scales, Educ. Psychol. Meas. 20 (1960) 37–46.

[117] R.L. Brennan, D.J. Prediger, Coefficient kappa: some issues, misuses, and alternatives, Educ. Psychol. Meas. 41 (1981) 687–699.

[118] E.M. Bennett, R. Alpert, A.C. Goldstein, Communication through limited response questioning, Public Opin. Q. 18 (1954) 303–308.

[119] Y.Q. Miao, R. Araujo, M.S. Kamel, Cross-domain facial expression recognition using supervised kernel mean matching, in: Proceedings – 2012 11th International Conference on Machine Learning and Applications, ICMLA 2012, vol. 2, 2012, pp. 326–332.

[120] J. Zhou, T. Xu, J. Gan, Facial expression recognition based on local directional pattern using SVM decision-level fusion, in: CCA 2013, Astl, vol. 17, 2013, pp. 126–132.

[121] R. Zhu, T. Zhang, Q. Zhao, Z. Wu, A transfer learning approach to cross-database facial expression recognition, in: Proceedings of the International Conference on Biometrics (ICB), 2015.

[122] S. Koelstra, M. Pantic, I. Patras, A dynamic texture based approach to recognition of facial actions and their temporal models, IEEE Trans. Pattern Anal. Mach. Intell. 32 (2010) 1940–1954.

[123] M.F. Valstar, M. Pantic, Fully automatic recognition of the temporal phases of facial actions, IEEE Trans. Syst. Man Cybern., Part B, Cybern. 42 (2012) 28–43.

[124] W.S. Chu, F. De La Torre, J.F. Cohn, Selective transfer machine for personalized facial expression analysis, IEEE Trans. Pattern Anal. Mach. Intell. 39 (2017) 529–545.

[125] T. Gehrig, H.K. Ekenel, Facial action unit detection using kernel partial least squares, in: IEEE International Conference on Automatic Face and Gesture Recognition and Workshops (ICCV Workshops), November 2011, pp. 2092–2099.

[126] O. Rudovic, V. Pavlovic, M. Pantic, Context-sensitive dynamic ordinal regression for intensity estimation of facial action units, IEEE Trans. Pattern Anal. Mach. Intell. 37 (2015) 944–958.

[127] I. Hupont, M. Chetouani, Region-based facial representation for real-time action units intensity detection across datasets, Pattern Anal. Appl. (2017) 1–13.
[128] T. Baltrusaitis, L. Liandong, L.-P. Morency, Local-global ranking for facial expression intensity estimation, in: International Conference on Affective Computing and Intelligent Interaction (ACII), October 2017, pp. 111–118.

Chapter 20

Automatic recognition of self-reported and perceived emotions

Biqiao Zhang, Emily Mower Provost
University of Michigan, Computer Science and Engineering, Ann Arbor, MI, USA

CONTENTS

20.1 INTRODUCTION

Emotion is an essential component in our interactions with others. It transmits information that helps us interpret the meaning behind an individual's behavior. The goal of *automatic emotion recognition* is to provide this information, distilling emotion from behavioral data. Yet, emotion may be defined in multiple manners: recognition of a person's true felt sense, how that person believes his/her behavior will be interpreted, or how others actually do interpret that person's behavior. The selection of a definition fundamentally impacts system design, behavior, and performance. The goal of this

Multimodal Behavior Analysis in the Wild. https://doi.org/10.1016/B978-0-12-814601-9.00027-4
Copyright © 2019 Elsevier Ltd. All rights reserved.

chapter is to provide an overview of the theories, resources, and ongoing research related to automatic emotion recognition that considers multiple definitions of emotion.

Emotion recognition systems are often considered to be "omniscient." These systems are charged with inferring an individual's experienced emotion. They have numerous uses; for example, in augmented driving, a car could detect a driver's emotion and could provide warnings or additional assistance given observations of anger, stress, or fatigue [49,52,132] or in deception detection, a system could detect whether or not an individual is lying [82,85,124]. However, this style of emotion recognition is extremely challenging because individuals may intentionally mask their state, resulting in large difference between measured behavior and label. Further, if systems attempt to detect emotions that users are intentionally masking, users may think that the systems are overly intrusive.

Emotion recognition systems can also be designed to recognize how outside observers perceive the behavior of an individual or how an individual believes that others would perceive his/her behavior. This style of recognition provides information about how communicated emotion is perceived and is useful for automated agents. For example, in augmented homes (e.g., Siri, Alexa, or Google Assistant), emotion awareness could provide enhanced understanding of a user's behavior and could help in facilitating more natural and human-like interactions. In call centers, systems that are emotion-aware could better understand a caller's state and ensure that frustrated individuals are routed to agents that can more directly meet their needs [24,73,80,81].

Fundamentally, it is the definition of the *goal* of emotion recognition that drives system design. Therefore, the definition of emotion, often referred to as the *ground truth*, must be at the forefront of the design of automated systems. In this chapter, we focus on links between an individual's *self-reported* emotion and the emotion perception of others (*perceived emotion*). To clarify, perceived emotion labels are obtained from a collection of evaluators perceiving and annotating the behavior of a different individual (not themselves), while *self-reports* are evaluations by the individual that produced the emotion. It should be noted that self-report subsumes both *felt sense* (i.e., the individuals reporting what they felt) and self-perception (i.e., the individual reporting what they perceived from observing their own behavior). Self-report is different from *target emotion* labels, which are generally defined as the emotion that an individual is trying to portray.

In this chapter, we discuss methods for obtaining emotion labels, the connections and differences between types of emotion labels, available datasets, and the efforts that have been made to recognize and compare multiple definitions of emotion (i.e., self-reported and perceived). The rest of this chapter is arranged as follows:

- Section 20.2 discusses methods to describe emotion and psychological theories of emotion production and perception.
- Section 20.3 discusses empirical observations about the differences between self-reported emotion and perceived emotion.
- Section 20.4 discusses the process of collecting emotionally expressive data and annotation tools for self-reported and perceived emotion.
- Section 20.5 discusses publicly available datasets that have self-report and perceived emotion labels.
- Section 20.6 discusses research in automatic emotion recognition that has compared the two types of emotion labels.
- Section 20.7 discusses some of the challenges in data collection and automatic recognition of self-reported and perceived emotions.

20.2 EMOTION PRODUCTION AND PERCEPTION

20.2.1 Descriptions of emotion

There are two prevailing frameworks for describing emotion: the categorical (or discrete) view [27,45–47,75,78,83,117,125] and the dimensional view [68,94,108,128]. There is an active debate regarding the appropriateness of each approach, which will not be directly addressed in this chapter (see [28,96] for a detailed discussion). Yet, both categorical and dimensional descriptions are widely used in the emotion recognition community. We briefly introduce both approaches.

Categorical descriptions of emotion posit that there exists a small number of "basic" emotions. In his seminal work in 1884, James identified fear, grief, love, and rage as basic emotions [47], work that has been extended in [45,46,75,78,83,117,125]. Ekman defined a basic emotion as one that is differentiable from other emotions across a set of properties, including automatic appraisal, distinctive physiology, distinctive universal signals, distinctive universals in antecedent events, coherence among emotional response, presence in other primates, quick onset, brief duration, and unbidden occurrence [27]. Basic emotions define the basis of the human emotional space, with other more complex emotions described as combinations of these bases [27]. However, one challenge associated with basic emotions as a construct is the variable nature of the sets of emotion identified. For example, Watson proposed fear, love and rage [125]. Panksepp proposed expectancy, fear, rage and panic [78]. Plutchik proposed acceptance, anger, anticipation, disgust, joy, fear, sadness, and surprise [83]. Izard proposed anger, contempt, disgust, distress, fear, guilt, interest, joy, shame, and surprise [45, 46]. Tomkins argued for nine out of the above ten (excepting guilt) [117]. Oatley and Johnson-Laird proposed anger, anxiety, disgust, happiness, and sadness [75]. Ekman proposed anger, disgust, fear, happiness, sadness, and surprise [27], which are the most widely used in the field of automatic emotion recognition.

Dimensional descriptions of emotion instead characterize emotion as points in a continuous space. These descriptions, originally introduced by Wundt in 1897, included three dimensions as the basis of emotion and feeling, which were pleasant–unpleasant, tension–relaxation, and excitement–calm [128]. Schlosberg proposed that emotion could instead be described on a circular surface with pleasantness–unpleasantness, attention–rejection and level of activation as axes [108]. The two dimensions proposed by Russell [94], valence and arousal (also called activation), are the most commonly used dimensions in emotion recognition. A third dimension, dominance (dominant–submissive), is often used to distinguish emotions that are fundamentally different, but cannot be distinguished by valence and activation [68]. For example, both anger and fear have negative valence and high activation, but anger is a dominant emotion, while fear is more passive.

Dimensional characterizations of emotion remove the contextualized component of emotion. Instead this description focuses on the "core affect" of a display, a term coined by Russell as "a simple and non-reflective feeling" that integrates two dimensions: valence (pleasure–displeasure) and arousal (sleepy–activated) to describe common emotional states [97,98]. This mitigates well-known problems, such as the lack of basic emotion universality. For example, some of the common basic emotions, such as fear and anger, do not have exact equivalents in all languages [95]. In [97,98], Russell further argued that it is problematic to use concepts such as anger and fear as psychological primitives because they are object-directed and imply a cognitive structure. He suggested the categorical emotions are more like constellations rather than stars, because they are categorized by their prototypical examples, and different cultures identified them differently. The prototypes of emotions such as anger, happiness, sadness, and fear are the occasional co-occurrence of a set of events that fit the pattern defined by the culture.

20.2.2 Brunswik's functional lens model

Brunswik's functional lens model [9] provides an explanation for how emotion is produced, transmitted, and perceived (Fig. 20.1). It contains two entities: the individual who produces the message ("encoder") and one or more

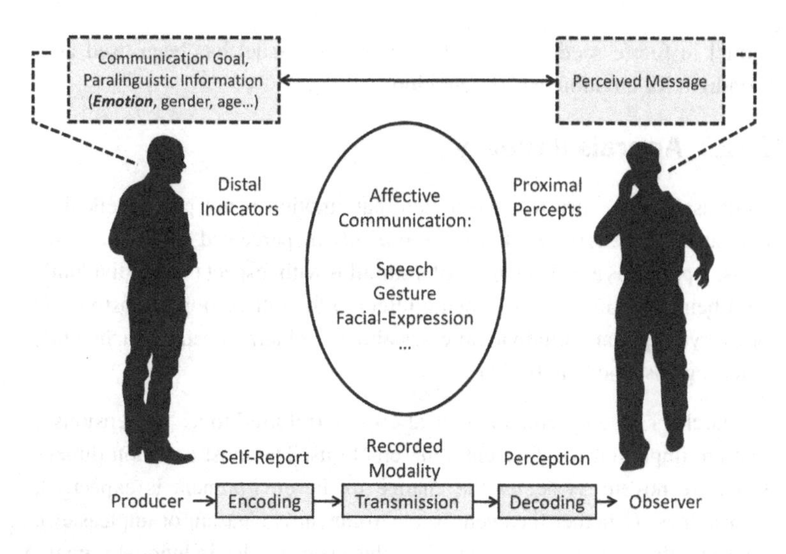

■ **FIGURE 20.1** The Brunswik's functional lens model.

individuals who perceive the message ("decoders"). The emotion communication process starts with the encoder producing a message that conveys his or her communicative goals, together with various paralinguistic properties, such as emotion, age, and gender. This message is encoded in distal indicators, which are a set of cues that are expressed over multiple channels, such as voice, facial expressions, or gestures. These cues are then transmitted to the decoders and referred to as proximal percepts. Proximal percepts serve as signals to the observers who then perceive and interpret the affective information. Finally, the decoders arrive at a higher-level judgment of the encoder's emotion by combining percepts from multiple modalities. Research has used adaptations of the Brunswik's functional lens model to explain the emotion communication process [4,51,103,106].

The Brunswik's functional lens model both confirms the link between distal indicators and proximal percepts and highlights the fact that distal indicators produced by the encoder are not necessarily the same as the proximal percepts perceived by the decoders [103]. It provides a critical tool to understand the differences that exist between self-report and perceived emotion labels. For example, self-report tasks often ask individuals to recall their distal indicators or to observe their proximal percepts, noting the feeling(s) that they experienced. On the other hand, perceived emotion tasks ask groups of annotators to observe the same proximal percepts and to infer how that individual felt. In both cases, the transmission process can lead to information loss because the available proximal percepts are only those modalities that were specifically recorded. We touch on the Brunswik's functional lens model in future sections, describing how this model has been used as a paradigm for emotion experimentation.

20.2.3 Appraisal theory

Appraisal theory forwards the notion that emotion is not purely reflexive, but rather responses result from *appraisals* of perceived events or situations. Appraisals are evaluations of a stimulus with respect to an individual's well-being. In this case, well-being refers to the satisfaction or obstruction of everything that an individual cares about, including needs, attachments, values, goals, and beliefs [71].

Researchers have specified appraisal criteria (referred to as "dimensions") that are important for differentiating emotions. The most common dimensions are novelty (whether the change in the environment is expected), pleasantness (whether the event or environment is pleasant or unpleasant), goal significance (the importance of the event to the individual's goal), agency (whether the individual is responsible for the event), legitimacy (whether the action of the individuals fits moral standards or social norms) [36,102,109,113].

Appraisal theory is commonly used in affective computing [63]. The Ortony–Clore–Collins (OCC) model [76] is one of the most commonly used Appraisal Theories in this domain. It posits that an individual's emotions are the result of his or her valenced appraisal of the current situation with respect to events, actors, and objects. The OCC model clearly defines 22 emotion categories as conjunctions of situation appraisals [63] and provides tools that enable the prediction of an individual's emotion given knowledge of his/her goals and perception of events [20].

20.3 OBSERVATIONS FROM PERCEPTION EXPERIMENTS

Observers are not omniscient decoders of expressed emotion. For example, when actors are asked to portray a set of emotions, groups of observers correctly interpret the emotion only 50% to 80% of the time, where accuracy is defined in terms of correctly identifying the actors' intentions [28,103]. This demonstrates that emotion annotation is imperfect, even given clear emotional expression goals.

Emotion perception is influenced by the clarity of emotional displays. For example, individuals may intentionally suppress their emotion. Gross and Levenson conducted a study investigating the influence of emotion suppression on expressive behavior [41]. The participants of the study were asked to watch a short disgust-inducing video in two conditions: (1) with no suppression, and (2) behaving "in such a way that a person watching you would not know you were feeling anything." They found that in (2), the expression behavior of the participants was reduced, compared to (1). Another example is the existence of micro-expressions. A micro-expression is a fleeting facial expression, which reveals a true emotion that a person is trying to suppress [29]. They are hard to detect by the naked eye because of their brevity (they last less than 0.5 s) [130].

Emotion perception is also influenced by differences between the individuals who are tasked with perceiving the emotion. The ability to decode emotion, often referred to as "emotion sensitivity" differs across individuals [10,92]. Past research has developed various methods for detecting the differences in emotion sensitivity, such as the Affective Sensitivity Test [18] and the Brief Affect Recognition Test [30], among other work [65,77].

Finally, emotion perception is influenced by additional factors, such as culture and gender differences [66,135]. For example, Ekman et al. conducted cross-cultural experiments on emotion perception from facial expressions [32]. They found that there were differences in the judgments of the absolute level of emotional intensity across cultures. The work of Elfenbein and Ambady found a within-group advantage for emotion perception [34,35]. More specifically, they found that individuals can recognize the emotion of people from the same ethnic, national, or regional group more accurately than the emotion of people from different backgrounds. In addition, people may depend on different signals during emotion perception. Yuki, Maddux, and Masud found that individuals from cultures that encourage emotional display rely more on the mouth region than the eyes region, while individuals from cultures that control for emotional display focus more heavily on the eyes, compared to the mouth [135]. Further, Hall found that females can convey emotion through facial expression better than males [43]. Rotter and Rotter found that women are better at recognizing emotion expression of both males and females in general [93].

The links and differences between emotion expression and perception can be studied within Brunswik's theoretical framework. For example, Laukka et al. found that there are many acoustic cues that are correlated with both the intended emotion of musicians (linked to distal indicators) and the perceptions of listeners (linked to proximal percepts) [56]. Biersack and Kempe investigated whether the self-reported happiness of speakers and the perception of happiness by outside evaluators were linked to the same vocal cues [7]. They identified differences between distal indicators and proximal percepts. For example, perceived happiness was associated with higher pitch and pitch range, higher F1, faster speech rate, and lower jitter. However, only the wider pitch range of the female speaker and higher F1 were found to be related to the self-report happiness of the speakers.

These observations provide evidence supporting the differences that exist between self-report and perceived emotion. It highlights the importance of carefully considering label type when designing automatic emotion recognition systems.

20.4 COLLECTION AND ANNOTATION OF LABELED EMOTION DATA

The collection of emotionally expressive data and the acquisition of emotion labels are essential for creating automatic systems for recognizing self-reported or perceived emotions. In this section, we briefly discuss the methods commonly used for emotion-elicitation and the annotation tools for collecting the two types of labels.

20.4.1 Emotion-elicitation methods

The difficulty of obtaining emotionally expressive data is a well-known problem in the field of automatic emotion recognition. Researchers have adopted various methods that allow for different level of control and naturalness for eliciting emotions, such as acting, emotion-induction using images, music, or videos, and interactions (human–machine interactions, human–human conversations).

One of the earliest emotion-elicitation method was to ask actors to perform using stimuli with fixed lexical content [12,37] or to create posed expressions [50,62,133,134]. One of the advantages of this approach is that researchers have more control over the microphone and camera positions, lexical content, and the distributions of emotions. A criticism of acted or posed emotion is that there are differences between spontaneous and acted emotional displays [19,122].

However, many of these criticisms can be mitigated by changing the manner in which acted data are collected. Supporters have argued that the main problem may not be the use of actors themselves, but the methodologies and materials used in the early acted corpora [15]. They argue that the quality of acted emotion corpora can be enhanced by changing the acting styles used to elicit the data and by making a connection to real-life scenarios, including human interaction. Recent acted emotion datasets have embraced these alternative collection paradigms and have moved towards increased naturalness and subtlety [14,17]. One example is the popular audiovisual emotion corpus, Interactive Emotion Motion Capture (IEMOCAP), which was collected by recruiting trained actors, creating contexts for the actors using a dyadic setting, and allowing for improvisation [14]. The same elicitation strategy has been successfully employed in the collection of the MSP-Improv corpus [17].

Various methods have been proposed for eliciting spontaneous emotional behavior. One popular approach is the passive emotion-induction method [54,58,100,111,121,129,131]. The goal is to cause a participant to feel a given emotion using selected images, music, short videos, or movie clips known a priori to induce certain emotions. This method is more suitable for analysis of modalities aside from audio, because the elicitation process often does not involve active verbal responses from the participants. A more active version is induction by image/music/video using emotion-induction events. For example, Merkx et al. used a multi-player first-person shooter computer game for eliciting emotional behaviors [69]. The multi-player setting of the game encouraged conversation between players, and the game was embedded with emotion-inducing events.

Another method often adopted by audiovisual emotion corpora is emotion-elicitation through interaction, including human–computer interaction and human–human interaction. For example, participants were asked to interact with a dialog system [57,59]. The FAU-Aibo dataset collected the interactions between children and a robot dog [6]. The Sensitive Artificial Listener (SAL) portion of the SEMAINE dataset used four agents with different characters played by human operators for eliciting different emotional responses from the participants [67]. Examples of human–human interaction includes interviews of topics related to personal experience [91] and collaborative tasks [88].

An emerging trend is that researchers are moving emotional data collections outside of laboratory settings, a paradigm referred to as *in the wild.* For example, the Acted Facial Expressions In The Wild (AFEW) dataset selected video clips from movies [26], the Vera am Mittag (VAM) dataset [40] and the Belfast Naturalistic dataset [110] used recordings from TV chat shows. User-generated content on social networks has become another source for creating text-emotion datasets. Researchers curate the contents published on social networks, such as blogs, Twitter, and Facebook for emotion detection [86,87,89].

Finally, researchers use combinations of known "best practices" for collecting behaviors associated with different emotions. For example, in the BP4D-Spontaneous dataset, the spontaneous emotion of the participants was using a series of tasks, including interview, video-clip viewing and discussion, startle probe, improvisation, threat, cold pressor, insult, and smell

Table 20.1 Common emotion annotations tools. Label type: type of label originally designed to collect (SR: self-report; PE: perceived emotion); Emotion descriptor: the type of emotion descriptor (i.e., categorical or dimensional); Verbal description: if the tool includes verbal description; Graphical illustration: if the tool uses graphical illustration; Continuous: if the tool is used for collect time-continuous labels

Tool	Label type	Emotion descriptor	Verbal descriptions	Graphical illustrations	Continuous
SAM [55]	SR	dim.	N	Y	N
GEW [104]	SR	both	Y	Y	N
MECAS [25]	PE	cat.	Y	N	N
Feeltrace [22]	PE	dim.	Y	Y	Y
EMuJoy [74]	PE	dim.	N	Y	Y

[139]. These tasks were intended to induce happiness or amusement, sadness, surprise or startle, embarrassment, fear or nervous, physical pain, anger or upset, and disgust, respectively.

20.4.2 Data annotation tools

Emotion annotation is a critical component of emotion recognition. Yet, it is time-consuming and potentially error-prone [136]. Researchers have developed several tools designed to improve the quality of emotion labels and to reduce the annotators' cognitive load. These tools are generally designed to annotate either self-report or perceived emotion (see Table 20.1), but they can be used in either context.

Self-Assessment Manikins (SAMs) [55] belong to the most widely used tools for collecting dimensional emotion descriptions. SAMs are a pictorial grounding of N-point Likert scales. The common set is across three dimensions: valence (positive vs. negative), activation/arousal (calm vs. excited), and dominance (dominant vs. submissive). SAMs reduce variability in emotion annotation [39] and make it easier to obtain evaluations across cultures and languages [72].

The Geneva Emotion Wheel (GEW), originally proposed in [104] and updated to the most recent version in [105], is available in several languages. GEW arranges twenty emotion families according to their levels of valence and activation, with an intensity scale for each emotion family. GEW effectively combines categorical and dimensional descriptions of emotion and enables annotation using a wide range of emotions, together with intensity.

The Multi-level Emotion and Context Annotation Scheme (MECAS) is a hierarchical framework for collecting categorical emotion labels [25]. In this scheme, a set of emotion labels is organized into different layers of granularity (e.g., coarse-grained and fine-grained). This hierarchy mitigates challenges due to the scarcity of the fine-grained emotions.

Feeltrace is a tool that permits continuous annotation of emotion on the valence-activation space [22]. It provides visual feedback to the annotators by displaying a trace of their ratings. A more general version of Feeltrace called "General trace" was proposed in [23]. It allows people to create their own dimensions. The EMuJoy [74] is a similar approach to annotation that has been used in the music community.

20.5 EMOTION DATASETS

Emotion is communicated through both verbal and non-verbal cues. Research in automatic emotion recognition has historically focused on subsets of these cues, including: text, audio, visual, and physiological data. In this section, we discuss some of the publicly available emotion datasets. We separate our discussion into two parts: (1) text datasets and (2) audio, visual, physiological, and multi-modal datasets, due to the differences between the data, the collection, and the curation processes. The datasets in (2) are not further separated due to the non-disjoint nature of the constituent modalities.

20.5.1 Text datasets

Sentiment analysis and textual emotion recognition are closely related. Sentiment analysis is target-oriented, aiming to identify opinions or attitudes towards topics or entities (e.g., product, movie). Emotion recognition, on the other hand, focuses on recognizing either the emotion expressed in text or evoked by the text, with no attachment to a specific target.

There are two common types of data for emotion analysis in text: word-emotion lexicons and text corpora. Word-emotion lexicons provide ratings of the affective content of individual words, generally in terms of either categorical or dimensional descriptors. The ratings are curated by asking groups of annotators to rate the lists of words in terms of how they felt while reading each word. Some of the most influential word-emotion lexicons include Affective Norms for English Words (ANEW) [8], WordNet-Affect [116], and Linguistic Inquiry and Word Count (LIWC) [79]. There are also data-driven curations of word-emotion lexicons [70,112].

Text-emotion corpora are relatively scarce, compared to word-emotion lexicons. The sparseness is due, in part, to the relative youth of the text-emotion recognition field [11]. The ISEAR databank is one of the earliest text-emotion corpora [101]. It was created to enable cross-cultural studies of

Table 20.2 An overview of public text datasets for emotion analysis and detection. Label type: type of label collected by the datasets (SR: self-report; Oth: perception of others)

Dataset	Label type	Source	Emotion descriptor
ISEAR [101]	SR	experience-telling	categorical
Hashtag Emotion Corpus [70]	SR	Twitter	categorical
Alm et al. [1]	Oth	fairy tales	categorical
SemEval'07 [115]	Oth	news headlines	both
Aman et al. [2]	Oth	blogs	categorical
FB [86]	Oth	Facebook	dimensional

emotion antecedents and reactions. Students were asked to report situations during which they had experienced seven different emotions (anger, disgust, fear, guilt, joy, sadness, and shame). The final dataset consists of reports from about 3000 participants in 37 countries. Large scale text-emotion dataset creation has expanded further given the growing popularity of social networks. The Hashtag Emotion Corpus is a curation of Twitter posts that were self-tagged with emotion hashtags, which were then used as the self-report emotion labels [70]. The emotion hashtags originally included the six basic emotions proposed by Ekman [27]: anger, disgust, fear, happy, sadness, and surprise, and have now been expanded to include hundreds of emotions. Other works that used a similar approach include [38,89]. Alm et al. created a emotion corpus consisting of emotionally annotated sentences from fairy tales [1]. They selected approximately 185 stories by the Brothers Grimm, H.C. Andersen, and B. Potter. The stories were labeled at sentence-level with eight emotions, including angry, disgusted, fearful, happy, sad, positively surprised, negatively surprised, and neutral. The sentences were annotated from the perspective of the "feeler" in the text. Therefore, these labels reflect how the annotators perceived the emotion of the characters in the stories, rather than the emotion of the writers. The corpus for the SemEval 2007 Task on Affective Text consists of newspaper headlines [115]. These headlines were labeled by six annotators using Ekman's six basic emotions. Aman et al. selected emotion-rich sentences from blogs [2]. These sentences were annotated by four evaluators using Ekman's six basic emotions and the emotion of neurality. The Valence and Arousal Facebook Posts (FB) consists of 3120 Facebook posts. Two annotators with a training in psychology rated the posts along valence and arousal [86]. We list some of the public text-emotion datasets in Table 20.2, with characteristics including the type of emotion label (i.e., self-report or perception of others), the source of the texts (e.g., whether the text was written by professional writers or from experience-telling, blog, Twitter, or Facebook).

Table 20.3 An overview of public visual-only emotion datasets. Label type: type of label collected by the dataset (SR: self-report; Tar: target emotion of acting; Oth: label provided by others)

Dataset	Label type	Emotion elicitation	Emotion descriptor
JAFFE [62]	Oth	acted	categorical
CK [50]	Tar	acted	categorical
BU-3DFE [134]	Tar, Oth	acted	categorical
BU-4DFE [133]	Tar	acted	categorical
Multi-PIE [42]	Tar	acted	categorical
CK+ [61]	Oth[a]	acted	categorical
MMI [121]	Oth	spontaneous	categorical
NVIE [123]	Tar, SR[b]	acted, spontaneous	both
BP4D [139]	SR, Oth	spontaneous	categorical
CASME [131]	Oth + SR	spontaneous	categorical
CASME II [129]	Oth + SR	spontaneous	categorical
SMIC [58]	Oth + SR	spontaneous	categorical

[a] *Emotion labels were deduced from AU coding and confirmed by emotion researchers by visual inspection.*
[b] *Target emotion labels were used for acted facial expressions; self-report using both categorical and dimensional labels were collected for spontaneous facial expressions.*

20.5.2 **Audio, visual, physiological, and multi-modal datasets**

Visual-only datasets are often labeled using either the actors' target emotion (CK [50], BU-3DFE [134], BU-4DFE [133], Multi-PIE [42]), based on structured coding manuals (JAFFE [62], CK+ [61], MMI [121]), or on a combination of structured manuals and self-report (NVIE [123], BP4D [139], CASME [131], CASME II [129], SMIC [58]). These labeling approaches are strongly influenced by Ekman's theory of universal basic emotions [28], and consider the Facial Action Coding System (FACS) [31] the "gold standard" for both the encoding (posing) and decoding (labeling) process. We list some of the public visual-only emotion datasets in Table 20.3.

A subset of visual-only datasets use self-reported labels [123,139]. In NVIE, the spontaneous emotional facial expressions were induced by film clips and the participants were asked to report their emotion experience using both dimensional descriptions and categorical emotions [123]. The BP4D-spontaneous database collected both the participants' self-report and observers' perception [139]. The participants were regularly asked to report their emotion. These labels were augmented with the perception of five observers who coded the segments with AUs and emotion. Self-report is also commonly used in micro-expression recognition (CASME [131], CASME II [129], and SMIC [58]). While all three datasets used annota-

Table 20.4 An overview of public audio-only and audiovisual emotion datasets. Label type: type of label collected by the dataset (SR: self-report; Oth: label provided by others); Modality: available modality (A: audio; V: visual; mocap: motion capture; gesture)

Dataset	Label type	Modality	Emotion elicitation	Emotion descriptor
SmartKom [107]	Oth	AV, gesture	spontaneous	categorical
EmoDB [12]	Oth	A	acted	categorical
eNTERFACE'05 [64]	Oth	AV	acted	categorical
IEMOCAP [14]	Oth, SR	AV, mocap	acted, improvised	both
VAM [40]	Oth	AV	spontaneous	dimensional
FAU Aibo [6]	Oth	A	spontaneous	categorical
SAVEE [44]	Oth	AV	acted	categorical
GEMEP [3]	Oth	AV	acted	categorical[a]
Belfast Induced [110]	Oth, SR[b]	AV	spontaneous	both
Belfast Naturalistic [110]	Oth	AV	spontaneous	both
AFEW [26]	Oth	AV	acted[c]	categorical
SEMAINE [67]	Oth	AV	spontaneous	both
RECOLA [88]	Oth, SR[d]	AV	spontaneous	dimensional
MSP-Improv [17]	Oth	AV	acted, improvised, spontaneous	both

[a] *Geneva Wheel like categorical labels, can be separated into binary valence and activation.*
[b] *Self-reports are categorical emotion with intensity, perceived labels are dimensional (valence and intensity).*
[c] *Taken from movie.*
[d] *Self-report and perception labels are on different time-scale.*

tors for assessing the duration, coding the AUs, and labeling the emotion of the micro-expression, the final ground-truth labels combined the opinions of the annotators and self-report of the participants.

The common practice for audio-only and audiovisual emotion datasets is to collect perceived emotion labels. All the datasets listed in Table 20.4 provide emotion perception labels from sets of observers. In a subset of the datasets, the evaluator population is fixed (see, e.g., [88]), while in others, the population varies over the data instances (see, e.g., [17]). Annotation may include multiple components of perceived emotion, for example, both dimensional and categorical labeling (see, e.g., [14,110]). Datasets may also be downsampled from their original collection, leaving only examples in which the annotation matched the original targets of the actors' (see, e.g., [64]). Finally, these datasets may be evaluated using all available modalities in addition to only subsets (see, e.g., [3,44]).

Table 20.5 An overview of public emotion datasets with physiological signals. Label type: type of label collected by the dataset (SR: self-report; Tar: target of emotion-induction; Modality: available modality; V: visual; Phys: physiological signals)

Dataset	Label type	Modality	Emotion elicitation	Emotion descriptor
eNTERFACE-savran [100]	SR	V, Phys	spontaneous	dimensional
MAHNOB-HCI [111]	SR	V, Phys[a]	spontaneous	both
DEAP [54]	SR	V, Phys	spontaneous	dimensional
MMSE [140]	Tar, SR	V, Phys	spontaneous	categorical

[a] *Audio was recorded, but did not contain enough speech information.*

A subset of the audio-only and audiovisual emotion datasets also provide self-reported emotion labels. In IEMOCAP, six of the actors that participated in the recording were asked to evaluate their own improvised sessions using both categorical and dimensional descriptors [16]. The Belfast Induced Natural Emotion Dataset consists of three sets of recordings, which contain emotion-inducing tasks (active/social tasks and film-viewing tasks) [110]. The participants were asked to rate their own behavior. In RECOLA, self-reports were completed by participants at three points: at the beginning, after mood induction (watching videos), and at the end [88].

Datasets that feature physiological signals generally collect emotion labels from the participants themselves. All the four datasets listed in Table 20.5, including eNTERFACE-savran [100], MAHNOB-HCI [111], DEAP [54], and MMSE [140] provide self-report labels.

20.6 RECOGNITION OF SELF-REPORTED AND PERCEIVED EMOTION

Most research in automatic emotion recognition focuses on using a single type of emotion label. Coverage of single-label methods is outside the scope of this chapter. We refer the reader to the following survey concentrated on automatic emotion recognition from speech [33], facial expression [21,99], audiovisual [127,136], and physiological signals [48] for detailed information.

Research in automatic recognition of both self-reported emotion and perceived emotion is relatively scarce, in part due to the lack of appropriate datasets. The few works that used both types of emotion labels with visual-only data and physiological data include [139] and [60]. Zhang et al. collected the visual-only BP4D-spontaneous dataset, including both self-report and perceived emotion labels [139]. These labels were each compared to the known target labels, the emotion that was induced. However, there was no comparison between the self-reported and perceived emotion labels directly. Liu et al. created affect recognition systems using physiological signals for children with Autism Spectrum Disorder (ASD) [60]. They collected data using computer-based cognitive tasks to elicit the affective states of liking, anxiety, and engagement from participants with ASD. They annotated the emotion using the participants themselves and two types of observers: the parents and therapists. They compared the pairwise agreement using the Kappa-coefficient between participants, parents, and therapists for each participant. They found that the agreement between parents and therapists was the highest, while the participant-parent and participant-therapist agreement were both relatively low. Their recognition system depended entirely on the therapist's perceptions.

The majority of the research focusing on a comparison between self-report and perceived emotion has used audiovisual data. Truong et al. conducted a series of experiments, comparing self-reported and perceived emotion using the TNO-GAMING corpus, which they collected. This corpus consists of audiovisual recordings of teams playing a multiplayer video game. The experimenters augmented the emotional experience by: (1) creating incentives, such as rewards for the best performing teams, (2) generating surprising events, such as sudden deaths, sudden appearances of threats, and (3) hampering the mouse and keyboard controls during the game. Afterwards, each participant observed his/her recorded behavior, contextualized by the content of the game, and annotated his/her experienced emotion using categorical labels and semi-continuous dimensional labels of valence and arousal. These labels were augmented by perceived emotion labels under six different conditions: audio-only, video-only, audio and game context, video and game context, audio-visual, and audiovisual + game context. In the audiovisual + game context setting, the annotators have access to the same amount of material as in the self-report setting [118–120].

One of the primary challenges of working with both self-report and perceived emotion labels is the low levels of agreement between multiple observers (referred to as "inter-observer agreement"), and the agreement between the outside observers and the individual him/herself (referred to as "self-observer agreement"). Truong et al. measured the inter-observer agreement and self-observer agreement on the TNO-GAMING corpus using Krippendorff's α statistics [118]. They found that the inter-observer agreement is higher in the multi-modal evaluation setting, compared to the uni-modal settings. They then added self-report to the ratings of the observers and found that the agreement across individuals decreased, compared to only using the annotations of the observers. As a result, they constructed emotion recognition systems for the self-report and perceived emotion labels separately. They trained two sets of recognizers to detect levels of valence and activation, one with self-report labels and the other with perceived emotion labels using Support Vector Regression [120]. Their experiments suggested that it was easier to predict the perceived emotion labels, compared to self-report, for both valence and activation.

Truong et al. extended this work, conducting a more detailed comparison between self-report and perceived emotion and a comprehensive analysis on the performance of the systems when detecting the two types of emotion labels [119]. They observed that the emotions reported by the participants themselves tended to be more extreme than the average ratings of the outside observers. They predicted the valence and arousal ratings using both acoustic and lexical features and again found that perceived emotion could be predicted more accurately, compared to self-report. They argued that these results suggested that the emotions felt by the participants were not always perceivable by the observers. In addition, they conducted cross-label experiments: training on one type of label and testing on the other type. They found that the systems trained on the average of the perceived labels performed well on both perceived label and self-reported label, while systems trained using self-report labels had lower performance in comparison [119].

The mismatch between self-report and perceived emotion has been supported by other research groups. Busso and Narayanan compared the self-report and perceived labels of the IEMOCAP database [14] across categorical and dimensional emotion descriptors [16]. They again found support for the mismatch between the self-reported and perceived emotion labels. Similar to [114,118], they measured the agreement of labels obtained from different individuals using an entropy-based metric and the Kappa-statistic for evaluation agreement. They found that the level of agreement was significantly lower when the perceived emotion labels were augmented with the self-report labels. They also supported the finding that self-reported labels tend to have more extreme values for the dimensional descriptors, compared to the perceived emotion labels [16].

Research has identified differences between self-report and perceived emotion labels. Yet, questions remained whether the information present in each type of label provided complementary information that could be used to improve the performance of systems designed to recognize each type of label. In our recent work, we investigated how the two types of labels could be jointly modeled to improve overall recognition ability. We proposed to jointly recognize both self-reported and perceived emotion using a multitask learning framework. Our hypothesis was that the perceived emotion labels could act as stabilizers to reduce fluctuations caused by individual speaker differences and that the self-report labels could function as stabilizers to control for inter-personal emotion expression differences. We experimented on the IEMOCAP database using both the audio and video modalities and showed that the multi-task learning approach, combined with unsupervised feature learning, could improve the performance of emotion recognition systems for both types of labels. In addition, our results demonstrated that the perceived emotion tasks had the largest performance gain for emotionally subtle examples (i.e., examples without complete agreement between observers) when jointly modeling the two types of emotion. The self-reported emotion task, on the other hand, saw the greatest performance improvement in emotionally clear examples (i.e., examples with a single self-reported emotion) [137]. In the context of the Brunswik's lens model, this may be because that information from self-report can help distinguish between expressions that are easily confused during decoding process, but the information from the outside observers does not reduce confusion that arises from the encoding process.

The field of automatic emotion recognition has increasingly embraced the notion of modeling self-reported and perceived emotion labels, understanding the differences and building models that can benefit from their distinct properties. This research provides insight into the link between emotion recognition system performance and emotion theory. However, there are challenges that must be addressed in order to advance the performance of emotion recognition systems. Additional work is needed to collect self-report and perceived labels that are more reliable and comparable (i.e., on the same data, using the same protocol, on the same time-scale) and investigate the influence of label type on automatic emotion recognition performance. We summarize the challenges and prospects for the field in the next section.

20.7 CHALLENGES AND PROSPECTS

Automatic recognition of self-reported and perceived emotion faces various challenges. One set of challenges relates to our very understanding of self-report labels: there is a fundamental mismatch between their definition and their nature. Self-reported emotion may be an incomplete description of an emotional experience if an individual is unable or unwilling to categorize his or her own emotion in an objective manner. Take, for example, this scenario presented by Russell and Barrett [98]: Ralph yells at his wife for chatting with an attractive stranger at a party. An objective observer may deduce that Ralph was jealous, but Ralph may be unwilling to admit to the emotion himself. Challenges stem from social desirability concerns [53] also relating to the fact that individuals may not be willing to report their true feelings for sensitive topics [84]. Besides willingness, another challenge is the presence of a cognitive bias: self-reports can capture only those emotions of which an individual is aware [84]. However, even emotions that are not consciously experienced can alter observed behaviors [126]. Further, there is a distinction between an individual's emotion and beliefs about emotion: the former is episodic, experiential, and contextual, while the latter is semantic, conceptual, and decontextualized [90]. As a result, there are discrepancies between self-reports on feelings that individuals are currently experiencing and experienced in the past [90].

Other challenges of self-reports of emotion may stem from issues related to certain labeling protocols. For example, when an open-ended questionnaire is used for collecting self-report, the labels are influenced by both feelings and understanding of words. Besides, the choice of words reflect personal preferences: some individuals' self-reports are valence-focused, while others are arousal-focused [5]. Further, different experiments may use different definitions of self-report, in one experiment indicating the reported feeling of the participants (see, e.g., [54]) and in another, the self-perception of the participants (see, e.g., [14]).

Perceived emotion labels are also associated with variability. One common method to decrease variability is to use of expert coders [61,121,131]. However, the use of expert coders may increase both the time to complete annotation and the overall cost. Further, their ratings may not reflect the perception of naïve observers. Alternatively, experimenters may solicit assessments from multiple annotators and then merge these annotations into a single label, for example, using majority vote or the mean operator [14,88]. In the collection of the MSP-Improv dataset, the authors designed a new annotation technique designed to mitigate the variability seen over multiple annotators. They evaluated the performance of the annotators in real-time to identify unreliable evaluators. Annotators were stopped when their agreement with the reference ground truth set fell below a certain level. The technique was designed to mitigate unwanted variability while allowing for variability due to emotional subtlety [13]. However, the final ground truth was still a single point estimate. This assumes that any disagreement in perceived emotion labels is unwanted "noise" (e.g., annotators not paying attention). In our recent work, we took an alternative view of evaluator disagreement, arguing that this variability reveals information about the subtlety or clarity of emotion. Both the majority vote and mean operator mask this information [138].

Once the challenges associated with both self-report and perceived emotion labels are embraced, there still remains another, possibly more fundamental challenge: while there are many datasets that have *either* self-report or perceived emotion labels, datasets with both are few. Even when a dataset collects both types of labels other problems may arise. Some datasets have dual evaluations of only a subset of the data [14,110]. In others, the time-scales of the annotation may be different. For example, in the RECOLA dataset, the self-reports were collected a three discrete times: at the beginning, after emotion-induction, and at the end of the experiments, while the perceived emotion labels were collected time-continuously [88]. Finally, the meanings of the labels themselves may vary over the experiment. For example, the Belfast Induced Natural Emotion Database consists of three sets of recordings collected using different sets of active/social and passive/social tasks [110]. In the first set, the female participants were asked to describe their emotions in an open-ended way, along with intensity rating. The male participants were given a list of emotion words to choose from, and were also required to rate the intensity of the selected emotions. In the second and third sets, all participants were given a list of emotion words and asked to rate the intensity of each word.

The majority of the research in self-report and perceived emotion recognition have been in the audiovisual emotion expression domain, influenced largely by the relative prevalence of audiovisual datasets. Therefore, it is difficult to assess whether the classification trends of the perceived emotions vs. self-report and single-task vs. multi-task will extend to other modality combinations. For example, would the inter-annotator agreement for self-report vs. perceived emotion from text, visual-only, and physiological data follow the same trends seen in the audiovisual domain? Further, would the perception of others, collected over observable signals, be easier to predict than self-report for other modalities? Research that answers this questions will give us a better understanding of the relationship between multi-modal cues and different types of emotion labels.

Another direction worth exploring is how the discrepancy between self-reported and perceived emotions will influence automatic emotion recognition systems. A number of works have demonstrated that the two types of labels are different. Future research must investigate if the two label types can both be used, subject to careful experimental design, when training systems. For example, it is important to know if the findings in Truong et al. [119] about the performance of cross-label training generalize across datasets and modalities. In addition, it is important to know if there is a way to effectively integrate the knowledge from both types of emotion labels in a single system. The answers to these questions are important in the creation of general purpose emotion recognition systems.

20.8 CONCLUDING REMARKS

In this chapter, we provided an overview of the emotion theories, empirical observation, resources, and research related to automatic emotion recognition that considers self-reported and perceived emotion. We discussed methods to describe emotion, psychological theories of emotion production and perception and empirical observations about the differences between self-reported and perceived emotion. We summarized the process of collecting emotionally expressive data, annotation tools for self-reported and perceived emotion, and discussed publicly available datasets that have self-report and perceived emotion labels. Finally, we highlighted research in automatic emotion recognition that has compared the two types of emotion labels. We noted that challenges in both resources and systems. The collection of reliable self-report and perceived emotion labels is still a challenging problem. Moreover, while there are a variety of datasets using either type of labels, datasets that provide both self-reported and perceived labels collected using the same protocol and time-scale are rare. Work in automatic emotion recognition, on the other hand, needs to further investigate the influence of label type on the performance of automatic emotion recognition systems. This work will be crucial for advancing emotion recognition systems.

ACKNOWLEDGMENTS

This work was supported by the National Science Foundation (CAREER-1651740).

REFERENCES

[1] Cecilia Ovesdotter Alm, Dan Roth, Richard Sproat, Emotions from text: machine learning for text-based emotion prediction, in: Conference on Human Language Technology and Empirical Methods in Natural Language Processing, 2005, pp. 579–586.

[2] Saima Aman, Stan Szpakowicz, Identifying expressions of emotion in text, in: Text, Speech and Dialogue, 2007, pp. 196–205.

[3] Tanja Bänziger, Marcello Mortillaro, Klaus R. Scherer, Introducing the Geneva Multimodal expression corpus for experimental research on emotion perception, Emotion 12 (5) (2012) 1161.

[4] Tanja Bänziger, Sona Patel, Klaus R. Scherer, The role of perceived voice and speech characteristics in vocal emotion communication, J. Nonverbal Behav. 38 (1) (2014) 31–52.

[5] Lisa Feldman Barrett, Feelings or words? Understanding the content in self-report ratings of experienced emotion, J. Pers. Soc. Psychol. 87 (2) (2004) 266.

[6] A. Batliner, S. Steidl, E. Nöth, Releasing a thoroughly annotated and processed spontaneous emotional database: the FAU Aibo Emotion Corpus, in: Satellite Workshop of LREC, 2008, pp. 28–31.

[7] Sonja Biersack, Vera Kempe, Tracing vocal expression of emotion along the speech chain: do listeners perceive what speakers feel?, in: ISCA Workshop on Plasticity in Speech Perception, 2005.
[8] Margaret M. Bradley, Peter J. Lang, Affective Norms for English Words (ANEW): Instruction Manual and Affective Ratings, Technical report, 1999.
[9] Egon Brunswik, Representative design and probabilistic theory in a functional psychology, Psychol. Rev. 62 (3) (1955) 193.
[10] Ross Buck, The Communication of Emotion, Guilford Press, 1984.
[11] Sven Buechel, Udo Hahn, Emotion analysis as a regression problem-dimensional models and their implications on emotion representation and metrical evaluation, in: ECAI, 2016, pp. 1114–1122.
[12] Felix Burkhardt, Astrid Paeschke, Miriam Rolfes, Walter F. Sendlmeier, Benjamin Weiss, A database of German emotional speech, in: Interspeech, vol. 5, 2005, pp. 1517–1520.
[13] Alec Burmania, Srinivas Parthasarathy, Carlos Busso, Increasing the reliability of crowdsourcing evaluations using online quality assessment, IEEE Trans. Affect. Comput. 7 (4) (2016) 374–388.
[14] Carlos Busso, Murtaza Bulut, Chi-Chun Lee, Abe Kazemzadeh, Emily Mower, Samuel Kim, Jeannette N. Chang, Sungbok Lee, Shrikanth S. Narayanan, IEMOCAP: interactive emotional dyadic motion capture database, Lang. Resour. Eval. 42 (4) (2008) 335.
[15] Carlos Busso, Shrikanth Narayanan, Recording audio-visual emotional databases from actors: a closer look, in: Workshop on Emotion: Corpora for Research on Emotion and Affect, International Conference on Language Resources and Evaluation, 2008, pp. 17–22.
[16] Carlos Busso, Shrikanth S. Narayanan, The expression and perception of emotions: comparing assessments of self versus others, in: Interspeech, 2008, pp. 257–260.
[17] Carlos Busso, Srinivas Parthasarathy, Alec Burmania, Mohammed AbdelWahab, Najmeh Sadoughi, Emily Mower Provost, MSP-Improv: an acted corpus of dyadic interactions to study emotion perception, IEEE Trans. Affect. Comput. 8 (1) (2017) 67–80.
[18] Robert J. Campbell, Norman Kagan, David R. Krathwohl, The development and validation of a scale to measure affective sensitivity (empathy), J. Couns. Psychol. 18 (5) (1971) 407.
[19] Jeffrey F. Cohn, Karen L. Schmidt, The timing of facial motion in posed and spontaneous smiles, Int. J. Wavelets Multiresolut. Inf. Process. 2 (02) (2004) 121–132.
[20] Cristina Conati, Probabilistic assessment of user's emotions in educational games, Appl. Artif. Intell. 16 (7–8) (2002) 555–575.
[21] Ciprian Adrian Corneanu, Marc Oliu Simon, Jeffrey F. Cohn, Sergio Escalera Guerrero, Survey on RGB, 3d, thermal, and multimodal approaches for facial expression recognition: history, trends, and affect-related applications, IEEE Trans. Pattern Anal. Mach. Intell. 38 (8) (2016) 1548–1568.
[22] Roddy Cowie, Ellen Douglas-Cowie, Susie Savvidou, Edelle McMahon, Martin Sawey, Marc Schröder, Feeltrace: an instrument for recording perceived emotion in real time, in: ISCA Tutorial and Research Workshop on Speech and Emotion, 2000.
[23] Roddy Cowie, Martin Sawey, GTrace-general trace program from Queen's, Belfast, 2011.
[24] Laurence Devillers, Laurence Vidrascu, Real-life emotions detection with lexical and paralinguistic cues on human–human call center dialogs, in: International Conference on Spoken Language Processing, 2006.
[25] Laurence Devillers, Laurence Vidrascu, Lori Lamel, Challenges in real-life emotion annotation and machine learning based detection, Neural Netw. 18 (4) (2005) 407–422.

[26] Abhinav Dhall, Roland Goecke, Simon Lucey, Tom Gedeon, et al., Collecting large, richly annotated facial-expression databases from movies, IEEE Multimed. 19 (3) (2012) 34–41.
[27] Paul Ekman, An argument for basic emotions, Cogn. Emot. 6 (3–4) (1992) 169–200.
[28] Paul Ekman, Strong evidence for universals in facial expressions: a reply to Russell's mistaken critique, Psychol. Bull. 115 (2) (1994) 268–287.
[29] Paul Ekman, Telling Lies: Clues to Deceit in the Marketplace, Politics, and Marriage, revised edition, WW Norton & Company, 2009.
[30] Paul Ekman, Wallace V. Friesen, Nonverbal behavior and psychopathology, in: The Psychology of Depression: Contemporary Theory and Research, 1974, pp. 3–31.
[31] Paul Ekman, Wallace V. Friesen, Manual for the Facial Action Coding System, Consulting Psychologists Press, 1978.
[32] Paul Ekman, Wallace V. Friesen, Maureen O'Sullivan, Anthony Chan, Irene Diacoyanni-Tarlatzis, Karl Heider, Rainer Krause, William Ayhan LeCompte, Tom Pitcairn, Pio E. Ricci-Bitti, et al., Universals and cultural differences in the judgments of facial expressions of emotion, J. Pers. Soc. Psychol. 53 (4) (1987) 712.
[33] Moataz El Ayadi, Mohamed S. Kamel, Fakhri Karray, Survey on speech emotion recognition: features, classification schemes, and databases, Pattern Recognit. 44 (3) (2011) 572–587.
[34] Hillary Anger Elfenbein, Nalini Ambady, On the universality and cultural specificity of emotion recognition: a meta-analysis, Psychol. Bull. 128 (2) (2002) 203.
[35] Hillary Anger Elfenbein, Nalini Ambady, When familiarity breeds accuracy: cultural exposure and facial emotion recognition, J. Pers. Soc. Psychol. 85 (2) (2003) 276.
[36] Phoebe C. Ellsworth, Klaus R. Scherer, Appraisal processes in emotion, in: Handbook of Affective Sciences, 2003, pp. 572–595.
[37] Inger S. Engberg, Anya Varnich Hansen, Ove Andersen, Paul Dalsgaard, Design, recording and verification of a danish emotional speech database, in: European Conference on Speech Communication and Technology, 1997.
[38] Gonzalo Blázquez Gil, Antonio Berlanga de Jesús, José M. Molina Lopéz, Combining machine learning techniques and natural language processing to infer emotions using Spanish twitter corpus, in: International Conference on Practical Applications of Agents and Multi-Agent Systems, 2013, pp. 149–157.
[39] Michael Grimm, Kristian Kroschel, Evaluation of natural emotions using self assessment manikins, in: IEEE Workshop on Automatic Speech Recognition and Understanding, 2005, pp. 381–385.
[40] Michael Grimm, Kristian Kroschel, Shrikanth Narayanan, The Vera am Mittag German audio-visual emotional speech database, in: International Conference on Multimedia and Expo, 2008, pp. 865–868.
[41] James J. Gross, Robert W. Levenson, Emotional suppression: physiology, self-report, and expressive behavior, J. Pers. Soc. Psychol. 64 (6) (1993) 970.
[42] Ralph Gross, Iain Matthews, Jeffrey Cohn, Takeo Kanade, Simon Baker, Multi-pie, Image Vis. Comput. 28 (5) (2010) 807–813.
[43] Judith A. Hall, Nonverbal Sex Differences: Accuracy of Communication and Expressive Style, Johns Hopkins University Press, 1990.
[44] Sanaul Haq, Philip J.B. Jackson, J. Edge, Speaker-dependent audio-visual emotion recognition, in: AVSP, 2009, pp. 53–58.
[45] Carroll E. Izard, The Face of Emotion, 1971.
[46] Carroll E. Izard, Human Emotions, Plenum, New York, 1977.
[47] William James, What is an emotion?, Mind (34) (1884) 188–205.
[48] S. Jerritta, M. Murugappan, R. Nagarajan, Khairunizam Wan, Physiological signals based human emotion recognition: a review, in: International Colloquium on Signal Processing and its Applications, 2011, pp. 410–415.

[49] Qiang Ji, Peilin Lan, Carl Looney, A probabilistic framework for modeling and real-time monitoring human fatigue, IEEE Trans. Syst. Man Cybern., Part A, Syst. Hum. 36 (5) (2006) 862–875.

[50] Takeo Kanade, Jeffrey F. Cohn, Yingli Tian, Comprehensive database for facial expression analysis, in: International Conference on Automatic Face and Gesture Recognition, 2000, pp. 46–53.

[51] Arvid Kappas, Ursula Hess, Klaus R. Scherer, Voice and emotion, in: Fundamentals of Nonverbal Behavior, 1991, p. 200, Ch. 6.

[52] Christos D. Katsis, Nikolaos Katertsidis, George Ganiatsas, Dimitrios I. Fotiadis, Toward emotion recognition in car-racing drivers: a biosignal processing approach, IEEE Trans. Syst. Man Cybern., Part A, Syst. Hum. 38 (3) (2008) 502–512.

[53] Maryon F. King, Gordon C. Bruner, Social desirability bias: a neglected aspect of validity testing, Psychol. Mark. 17 (2) (2000) 79–103.

[54] Sander Koelstra, Christian Muhl, Mohammad Soleymani, Jong.-Seok Lee, Ashkan Yazdani, Touradj Ebrahimi, Thierry Pun, Anton Nijholt, Ioannis Patras, DEAP: a database for emotion analysis, using physiological signals, IEEE Trans. Affect. Comput. 3 (1) (2012) 18–31.

[55] Peter J. Lang, Behavioral treatment and bio-behavioral assessment: computer applications, in: Technology in Mental Health Care Delivery Systems, Ablex, Norwood, NJ, 1980, pp. 119–137.

[56] Petri Laukka, Tuomas Eerola, Nutankumar S. Thingujam, Teruo Yamasaki, Grégory Beller, Universal and culture-specific factors in the recognition and performance of musical affect expressions, Emotion 13 (3) (2013) 434.

[57] Chul Min Lee, Shrikanth S. Narayanan, Toward detecting emotions in spoken dialogs, IEEE Trans. Speech Audio Process. 13 (2) (2005) 293–303.

[58] Xiaobai Li, Tomas Pfister, Xiaohua Huang, Guoying Zhao, Matti Pietikäinen, A spontaneous micro-expression database: inducement, collection and baseline, in: International Conference and Workshops on Automatic Face and Gesture Recognition, 2013, pp. 1–6.

[59] Diane J. Litman, Kate Forbes-Riley, Predicting student emotions in computer–human tutoring dialogues, in: Annual Meeting on Association for Computational Linguistics, 2004, p. 351.

[60] Changchun Liu, Karla Conn, Nilanjan Sarkar, Wendy Stone, Physiology-based affect recognition for computer-assisted intervention of children with autism spectrum disorder, Int. J. Hum.-Comput. Stud. 66 (9) (2008) 662–677.

[61] Patrick Lucey, Jeffrey F. Cohn, Takeo Kanade, Jason Saragih, Zara Ambadar, Iain Matthews, The extended Cohn–Kanade dataset (CK+): a complete dataset for action unit and emotion-specified expression, in: Computer Society Conference on Computer Vision and Pattern Recognition Workshops, 2010, pp. 94–101.

[62] Michael Lyons, Shigeru Akamatsu, Miyuki Kamachi, Jiro Gyoba, Coding facial expressions with Gabor wavelets, in: International Conference on Automatic Face and Gesture Recognition, 1998, pp. 200–205.

[63] Stacy Marsella, Jonathan Gratch, Paolo Petta, et al., Computational models of emotion, in: A Blueprint for Affective Computing: A Sourcebook and Manual, 2010, pp. 21–46.

[64] Olivier Martin, Irene Kotsia, Benoit Macq, Ioannis Pitas, The eNTERFACE'05 audio-visual emotion database, in: International Conference on Data Engineering Workshops, 2006.

[65] Rod A. Martin, Glen E. Berry, Tobi Dobranski, Marilyn Horne, Philip G. Dodgson, Emotion perception threshold: individual differences in emotional sensitivity, J. Res. Pers. 30 (2) (1996) 290–305.

[66] David Matsumoto, Sachiko Takeuchi, Sari Andayani, Natalia Kouznetsova, Deborah Krupp, The contribution of individualism vs. collectivism to cross-national differences in display rules, Asian J. Soc. Psychol. 1 (2) (1998) 147–165.

[67] Gary McKeown, Michel Valstar, Roddy Cowie, Maja Pantic, Marc Schroder, The SEMAINE database: annotated multimodal records of emotionally colored conversations between a person and a limited agent, IEEE Trans. Affect. Comput. 3 (1) (2012) 5–17.

[68] Albert Mehrabian, Basic dimensions for a general psychological theory implications for personality, social, environmental, and developmental studies, Oelgeschlager, Gunn & Hain, Cambridge, 1980.

[69] P.P.A.B. Merkx, Khiet P. Truong, Mark A. Neerincx, Inducing and measuring emotion through a multiplayer first-person shooter computer game, in: Computer Games Workshop, 2007.

[70] Saif M. Mohammad, Svetlana Kiritchenko, Using hashtags to capture fine emotion categories from tweets, Comput. Intell. 31 (2) (2015) 301–326.

[71] Agnes Moors, Phoebe C. Ellsworth, Klaus R. Scherer, Nico H. Frijda, Appraisal theories of emotion: state of the art and future development, Emot. Rev. 5 (2) (2013) 119–124.

[72] Jon D. Morris, Observations: SAM: the self-assessment manikin; an efficient cross-cultural measurement of emotional response, J. Advert. Res. 35 (6) (1995) 63–68.

[73] Donn Morrison, Ruili Wang, Liyanage C. De Silva, Ensemble methods for spoken emotion recognition in call-centres, Speech Commun. 49 (2) (2007) 98–112.

[74] Frederik Nagel, Reinhard Kopiez, Oliver Grewe, Eckart Altenmüller, Emujoy: software for continuous measurement of perceived emotions in music, Behav. Res. Methods 39 (2) (2007) 283–290.

[75] Keith Oatley, Philip N. Johnson-Laird, Towards a cognitive theory of emotions, Cogn. Emot. 1 (1) (1987) 29–50.

[76] Andrew Ortony, Gerald L. Clore, Allan Collins, The Cognitive Structure of Emotions, Cambridge University Press, 1990.

[77] Maureen O'Sullivan, Measuring the ability to recognize facial expressions of emotion, Emot. Hum. Face 2 (1982) 281–317.

[78] Jaak Panksepp, Toward a general psychobiological theory of emotions, Behav. Brain Sci. 5 (03) (1982) 407–422.

[79] James W. Pennebaker, Ryan L. Boyd, Kayla Jordan, Kate Blackburn, The Development and Psychometric Properties of LIWC2015, Technical report, 2015.

[80] Valery Petrushin, Emotion in speech: recognition and application to call centers, in: Proceedings of Artificial Neural Networks in Engineering, vol. 710, 1999.

[81] Valery A. Petrushin, Emotion recognition in speech signal: experimental study, development, and application, Studies 3 (4) (2000).

[82] Tomas Pfister, Xiaobai Li, Guoying Zhao, Matti Pietikäinen, Recognising spontaneous facial micro-expressions, in: IEEE International Conference on Computer Vision, 2011, pp. 1449–1456.

[83] Robert Plutchik, A general psychoevolutionary theory of emotion, Theor. Emot. 1 (1980) 3–31.

[84] Karolien Poels, Siegfried Dewitte, How to capture the heart? Reviewing 20 years of emotion measurement in advertising, J. Advert. Res. 46 (1) (2006) 18–37.

[85] Stephen Porter, Leanne Ten Brinke, Brendan Wallace, Secrets and lies: involuntary leakage in deceptive facial expressions as a function of emotional intensity, J. Nonverbal Behav. 36 (1) (2012) 23–37.

[86] Daniel Preoţiuc-Pietro, H. Andrew Schwartz, Gregory Park, Johannes Eichstaedt, Margaret Kern, Lyle Ungar, Elisabeth Shulman, Modelling valence and arousal in Facebook posts, in: Workshop on Computational Approaches to Subjectivity, Sentiment and Social Media Analysis, 2016, pp. 9–15.

[87] Changqin Quan, Fuji Ren, A blog emotion corpus for emotional expression analysis in Chinese, Comput. Speech Lang. 24 (4) (2010) 726–749.

[88] Fabien Ringeval, Andreas Sonderegger, Juergen Sauer, Denis Lalanne, Introducing the recola multimodal corpus of remote collaborative and affective interactions, in: International Conference and Workshops on Automatic Face and Gesture Recognition, 2013, pp. 1–8.
[89] Kirk Roberts, Michael A. Roach, Joseph Johnson, Josh Guthrie, Sanda M. Harabagiu, Empatweet: annotating and detecting emotions on twitter, in: LREC, vol. 12, 2012, pp. 3806–3813.
[90] Michael D. Robinson, Gerald L. Clore, Belief and feeling: evidence for an accessibility model of emotional self-report, Psychol. Bull. 128 (6) (2002) 934.
[91] Glenn I. Roisman, Jeanne L. Tsai, Kuan-Hiong Sylvia Chiang, The emotional integration of childhood experience: physiological, facial expressive, and self-reported emotional response during the adult attachment interview, Dev. Psychol. 40 (5) (2004) 776.
[92] Robert Rosenthal, Sensitivity to Nonverbal Communication: The PONS Test, Johns Hopkins University Press, 1979.
[93] Naomi G. Rotter, George S. Rotter, Sex differences in the encoding and decoding of negative facial emotions, J. Nonverbal Behav. 12 (2) (1988) 139–148.
[94] James A. Russell, A circumplex model of affect, J. Pers. Soc. Psychol. 39 (6) (1980) 1161–1178.
[95] James A. Russell, Culture and the categorization of emotion, Psychol. Bull. 110 (1991) 426–450.
[96] James A. Russell, Is there universal recognition of emotion from facial expression? A review of the cross-cultural studies, Psychol. Bull. 115 (1) (1994) 102.
[97] James A. Russell, Core affect and the psychological construction of emotion, Psychol. Rev. 110 (1) (2003) 145.
[98] James A. Russell, Lisa Feldman Barrett, Core affect, prototypical emotional episodes, and other things called emotion: dissecting the elephant, J. Pers. Soc. Psychol. 76 (5) (1999) 805.
[99] Evangelos Sariyanidi, Hatice Gunes, Andrea Cavallaro, Automatic analysis of facial affect: a survey of registration, representation, and recognition, IEEE Trans. Pattern Anal. Mach. Intell. 37 (6) (2015) 1113–1133.
[100] Arman Savran, Koray Ciftci, Guillaume Chanel, Javier Mota, Luong Hong Viet, Blent Sankur, Lale Akarun, Alice Caplier, Michele Rombaut, Emotion detection in the loop from brain signals and facial images, in: eNTERFACE, 2006.
[101] K. Scherer, H. Wallbott, The ISEAR Questionnaire and Codebook, Geneva Emotion Research Group, 1997.
[102] Klaus R. Scherer, On the nature and function of emotion: a component process approach, in: Approaches to Emotion, vol. 2293, 1984, p. 317.
[103] Klaus R. Scherer, Vocal communication of emotion: a review of research paradigms, Speech Commun. 40 (1) (2003) 227–256.
[104] Klaus R. Scherer, What are emotions? And how can they be measured?, Soc. Sci. Inform. 44 (4) (2005) 695–729.
[105] Klaus R. Scherer, Vera Shuman, Johnny R.J. Fontaine, Cristina Soriano, The grid meets the wheel: assessing emotional feeling via self-report, in: Components of Emotional Meaning: A Sourcebook, 2013, pp. 281–298.
[106] K.R. Scherer, Emotion in action, interaction, music, and speech, in: Language, Music, and the Brain: A Mysterious Relationship, 2013, pp. 107–139.
[107] Florian Schiel, Silke Steininger, Ulrich Türk, The SmartKom multimodal corpus at bas, in: LREC, 2002.
[108] Harold Schlosberg, Three dimensions of emotion, Psychol. Rev. 61 (2) (1954) 81.
[109] Craig A. Smith, Phoebe C. Ellsworth, Patterns of cognitive appraisal in emotion, J. Pers. Soc. Psychol. 48 (4) (1985) 813.

[110] Ian Sneddon, Margaret McRorie, Gary McKeown, Jennifer Hanratty, The Belfast induced natural emotion database, IEEE Trans. Affect. Comput. 3 (1) (2012) 32–41.
[111] Mohammad Soleymani, Jeroen Lichtenauer, Thierry Pun, Maja Pantic, A multimodal database for affect recognition and implicit tagging, IEEE Trans. Affect. Comput. 3 (1) (2012) 42–55.
[112] Jacopo Staiano, Marco Guerini, DepecheMood: a lexicon for emotion analysis from crowd-annotated news, arXiv preprint arXiv:1405.1605, 2014.
[113] Alexander Staller, Paolo Petta, Towards a tractable appraisal-based architecture for situated cognizers, in: C. Numaoka, D. Canamero, P. Petta (Eds.), Grounding Emotions in Adaptive Systems, SAB 98, 1998.
[114] Stefan Steidl, Michael Levit, Anton Batliner, Elmar Noth, Heinrich Niemann, "Of all things the measure is man" automatic classification of emotions and inter-labeler consistency, in: International Conference on Acoustics, Speech, and Signal Processing, vol. 1, 2005, p. I-317.
[115] Carlo Strapparava, Rada Mihalcea, SemEval-2007 task 14: affective text, in: International Workshop on Semantic Evaluations, 2007, pp. 70–74.
[116] Carlo Strapparava, Alessandro Valitutti, et al., WordNet affect: an affective extension of WordNet, in: LREC, vol. 4, 2004, pp. 1083–1086.
[117] Silvan S. Tomkins, Affect theory, in: Approaches to Emotion, vol. 163, 1984, p. 195.
[118] Khiet P. Truong, Mark A. Neerincx, David A. Van Leeuwen, Assessing agreement of observer- and self-annotations in spontaneous multimodal emotion data, in: Interspeech, 2008, pp. 318–321.
[119] Khiet P. Truong, David A. Van Leeuwen, Franciska M.G. De Jong, Speech-based recognition of self-reported and observed emotion in a dimensional space, Speech Commun. 54 (9) (2012) 1049–1063.
[120] K.P. Truong, D.A. van Leeuwen, M.A. Neerincx, F.M.G. de Jong, Arousal and valence prediction in spontaneous emotional speech: felt versus perceived emotion, in: Interspeech, 2009.
[121] Michel Valstar, Maja Pantic, Induced disgust, happiness and surprise: an addition to the MMI facial expression database, in: International Workshop on Emotion: Corpora for Research on Emotion and Affect, 2010, p. 65.
[122] Michel F. Valstar, Maja Pantic, Zara Ambadar, Jeffrey F. Cohn, Spontaneous vs. posed facial behavior: automatic analysis of brow actions, in: International Conference on Multimodal Interfaces, 2006, pp. 162–170.
[123] Shangfei Wang, Zhilei Liu, Siliang Lv, Yanpeng Lv, Guobing Wu, Peng Peng, Fei Chen, Xufa Wang, A natural visible and infrared facial expression database for expression recognition and emotion inference, IEEE Trans. Multimed. 12 (7) (2010) 682–691.
[124] Gemma Warren, Elizabeth Schertler, Peter Bull, Detecting deception from emotional and unemotional cues, J. Nonverbal Behav. 33 (1) (2009) 59–69.
[125] John Broadus Watson, Behaviorism, 1930.
[126] Piotr Winkielman, Kent C. Berridge, Julia L. Wilbarger, Unconscious affective reactions to masked happy versus angry faces influence consumption behavior and judgments of value, Pers. Soc. Psychol. Bull. 31 (1) (2005) 121–135.
[127] Chung-Hsien Wu, Jen-Chun Lin, Wen-Li Wei, Survey on audiovisual emotion recognition: databases, features, and data fusion strategies, APSIPA Trans. Signal Inform. Process. 3 (2014).
[128] Wilhelm Max Wundt, Outlines of Psychology, W. Engelmann, 1897.
[129] Wen-Jing Yan, Xiaobai Li, Su-Jing Wang, Guoying Zhao, Yong-Jin Liu, Yu-Hsin Chen, Xiaolan Fu, CASME II: an improved spontaneous micro-expression database and the baseline evaluation, PLoS ONE 9 (1) (2014).
[130] Wen-Jing Yan, Qi Wu, Jing Liang, Yu-Hsin Chen, Xiaolan Fu, How fast are the leaked facial expressions: the duration of micro-expressions, J. Nonverbal Behav. 37 (4) (2013) 217–230.

[131] Wen-Jing Yan, Qi Wu, Yong-Jin Liu, Su-Jing Wang, Xiaolan Fu, CASME database: a dataset of spontaneous micro-expressions collected from neutralized faces, in: International Conference and Workshops on Automatic Face and Gesture Recognition, 2013, pp. 1–7.

[132] Guosheng Yang, Yingzi Lin, Prabir Bhattacharya, A driver fatigue recognition model based on information fusion and dynamic bayesian network, Inf. Sci. 180 (10) (2010) 1942–1954.

[133] Lijun Yin, Xiaochen Chen, Yi Sun, Tony Worm, Michael Reale, A high-resolution 3d dynamic facial expression database, in: International Conference on Automatic Face and Gesture Recognition, 2008, pp. 1–6.

[134] Lijun Yin, Xiaozhou Wei, Yi Sun, Jun Wang, Matthew J. Rosato, A 3d facial expression database for facial behavior research, in: International Conference on Automatic Face and Gesture Recognition, 2006, pp. 211–216.

[135] Masaki Yuki, William W. Maddux, Takahiko Masuda, Are the windows to the soul the same in the east and west? Cultural differences in using the eyes and mouth as cues to recognize emotions in Japan and the United States, J. Exp. Soc. Psychol. 43 (2) (2007) 303–311.

[136] Zhihong Zeng, Maja Pantic, Glenn I. Roisman, Thomas S. Huang, A survey of affect recognition methods: audio, visual, and spontaneous expressions, IEEE Trans. Pattern Anal. Mach. Intell. 31 (1) (2009) 39–58.

[137] Biqiao Zhang, Georg Essl, Emily Mower Provost, Automatic recognition of self-reported and perceived emotion: does joint modeling help?, in: International Conference on Multimodal Interaction, 2016, pp. 217–224.

[138] Biqiao Zhang, Georg Essl, Emily Mower Provost, Predicting the distribution of emotion perception: capturing inter-rater variability, in: International Conference on Multimodal Interaction, 2017.

[139] Xing Zhang, Lijun Yin, Jeffrey F. Cohn, Shaun Canavan, Michael Reale, Andy Horowitz, Peng Liu, Jeffrey M. Girard, BP4D-spontaneous: a high-resolution spontaneous 3d dynamic facial expression database, Image Vis. Comput. 32 (10) (2014) 692–706.

[140] Zheng Zhang, Jeff M. Girard, Yue Wu, Xing Zhang, Peng Liu, Umur Ciftci, Shaun Canavan, Michael Reale, Andy Horowitz, Huiyuan Yang, et al., Multimodal spontaneous emotion corpus for human behavior analysis, in: Computer Vision and Pattern Recognition, 2016, pp. 3438–3446.

Index

A

B

Printed in the United States
By Bookmasters